UNIVERSITY OF ROCHESTER LIBRARIES
3 9087 01016469 9

D1553439

INTERFERONS AND OTHER REGULATORY CYTOKINES

INTERFERONS AND OTHER REGULATORY CYTOKINES

Edward De Maeyer

Jaqueline De Maeyer-Guignard

Institut Curie, Orsay, France

A Wiley-Interscience Publication

JOHN WILEY & SONS

New York Chichester Brisbane Toronto Singapore

Copyright © 1988 by John Wiley & Sons, Inc.

All rights reserved. Published simultaneously in Canada.

Reproduction or translation of any part of this work
beyond that permitted by Section 107 or 108 of the
1976 United States Copyright Act without the permission
of the copyright owner is unlawful. Requests for
permission or further information should be addressed to
the Permissions Department, John Wiley & Sons, Inc.

Library of Congress Cataloging in Publication Data:

DeMaeyer, E. M.

Interferons and other regulatory cytokines/Edward De Maeyer and Jaqueline De Maeyer-Guignard.

p. cm.
"A Wiley-Interscience publication."
Includes index.
ISBN 0-471-82887-4

1. Interferon. 2. Cytokines. I. De Maeyer-Guignard, Jaqueline.
II. Title.
QR187.5.D46 1988
616'.0194—dc19

Printed in the United States of America

10 9 8 7 6 5 4 3 2 1

QW
800
D369i
1988

To the memory of Pieter De Somer and John F. Enders, and to John F. Crigler Jr., who put us on the path of investigation

PREFACE

This book has been written to provide the reader with a comprehensive up-to-date view of a specific cytokine family—the interferons. It assumes no prior knowledge of interferons and can be easily read by someone not familiar with the field. We have considered interferons in their general context in the organism, and, when possible, have tried to provide an integrated view with sufficient background information.

When appropriate, particular emphasis has been placed on interactions with other polypeptide effector molecules, for example, interleukins and growth factors, which make up the cytokine network in the organism. Important insights have been gained in this rapidly moving and very exciting area; some aspects are understood at the molecular level, many others are still at the descriptive stage. We have devoted to them as much space as was compatible with keeping the book a reasonable size.

The chapters are not written with the specialists of each particular area in mind, but address a much larger audience. We think there is a need for a book that is intelligible to readers with a general interest in interferons as models of multifunctional cytokines. We are indeed dealing with a group of substances that have an amazingly wide spectrum of activities and are of concern to scientists and clinicians involved in such different areas as virology, immunology, oncology, pathology, hematology, parasitology, molecular biology, and cell biology in general.

One of the pleasures of writing this book has been the collaboration of many friends and colleagues who have commented on different chapters in their area of expertise, resulting in many improvements. We are particularly indebted to Jean-François Bach (Paris), Georges Calothy (Orsay), Michael Clemens (London), Jean Content (Brussels), Hervé Fridman (Paris), Robert Friedman (Bethesda, Maryland), Ion Gresser (Villejuif), Jacques Hatzfeld (Villejuif), Ara Hovanessian (Paris), Ernest Knight, Jr. (Wilmington, Delaware), Peter Krammer (Heidelberg), François Lemonnier (Marseille), Philip Marcus (Storrs, Connecticut), Gilles Marchal (Paris), Geneviève Milon (Paris), Erik Mogensen (Villejuif), Luc Montagnier (Paris), Catherine Sautes (Paris), Isabelle Seif (Orsay), Markus Simon (Freiburg), Gerald Sonnenfeld (Louisville, Kentucky), Kendall Smith (Hanover, New Hampshire), Joyce

Taylor-Papadimitriou (London), Michael Tovey (Villejuif), Philippe Vigier (Orsay), Jan Vilcek (New York), Jean-Louis Virelizier (Paris), and Katherine Zoon (Bethesda, Maryland). Without their collaboration, we would not have been able to cast such a wide net.

EDWARD DE MAEYER
JAQUELINE DE MAEYER-GUIGNARD

CONTENTS

INTERFERONS AND OTHER REGULATORY CYTOKINES

1 FROM AN ANTIVIRAL FACTOR TO A FAMILY OF MULTIFUNCTIONAL CYTOKINES

Interferon was discovered during the study of viral interference, in which previous infection with an avirulent or inactivated virus protects animals or cells against subsequent infection with a more virulent virus. It was found that the interfering virus induces a substance that confers protection against subsequent challenge with another virus (Isaacs and Lindenmann, 1957; Lindenmann et al., 1957). The association of the putative substance with the phenomenon of viral interference gave rise to the name "interferon," which was coined more as a convenient laboratory jargon than as the result of a deliberate exercise in nomenclature (Lindenmann, 1982). For many years the study of interferons was the reserved domain of a few virologists who were working well away from the mainstream. It was customary to refer to "interferon," rather than "interferons," because for a long time it was not realized that there was more than one substance.

Thirty years later, the concept of "interferon" has taken on a molecular identity and corresponds to a family of many structurally and functionally related molecules, the interferons. The awareness that there is more than one interferon first developed gradually, through laborious and not always clear-cut characterization of different preparations with antisera, and then all of a sudden matured practically overnight, as a result of the molecular cloning of the different interferon species.

The idea, proposed soon after the discovery of interferon that interferons are major contributors to the first line of defense against virus infections, has withstood the test of time (Isaacs, 1961). This property by itself would be sufficient to make interferons of great interest, but it has also gradually become clear that interferons, in addition to inhibiting virus replication, exert many other important effects on cells, and that some of these effects have potential clinical applications. It has, furthermore, become evident that inducing interferon synthesis in cells is not an exclusive property of viruses, but rather, is a property of many other infectious agents as well as of a variety of different substances, ranging from simple chemical compounds to growth factors, cytokines and nucleic acids. Consequently, the study of interferons has not remained in the hands of the virologists; it has become part of a host of other disciplines, including genetics, immunology, oncology, and cell biology in general. Interferons are not exclusively antiviral agents; they are a family of substances, involved in many different functions that are currently being investigated. It is no exaggeration to say that an understanding of the molecular mechanisms regulating the induction of interferons, and the characterization of the many genes activated in interferon-treated cells, has bearing on the whole of cell biology.

Interferons belong to a network of regulatory cytokines that are all involved in the homeostatic control of cellular function and replication under normal conditions, and that become active participants in host defense when an emergency arises because of infection. The cytokines, which make up this network of intercellular messengers, are also called lymphokines when derived from lymphocytes and monokines when made by monocytes and macrophages. However, because the production of some lymphokines and monokines is not restricted to lymphocytes or macrophages, it seems simpler to refer to all these polypeptides as cytokines, with the understanding that this designation also covers lymphokines and monokines. A few cytokines are also called interleukins, a name originally coined to designate substances acting as signals between leukocytes. Since this definition has rapidly proved too restrictive, a substance can now be called an interleukin as long as it is produced by leukocytes, either exclusively or in addition to other cells, and also provided it functions during the immune or inflammatory response. According to this definition, interferons are also interleukins.

In several respects the difference between interferons and many other cytokines is becoming blurred, as it is realized that important properties, first described for interferons, are shared with a number of polypeptide effector molecules. In any case categories are artificial constructions, resulting from the necessity to classify in order to study and understand, but nature usually works with a continuum in which differences are only obvious at the extreme ends. The criteria for calling a substance an interferon are therefore admittedly somewhat arbitrary. Nevertheless, they are useful, because ideally it should be possible to deduce the most important properties of a substance from its name, and furthermore, these properties should be shared by other substances bearing the same name. A substance should not be called an interferon if, in addition to its many other biological activities, it does not have significant broad spectrum antiviral action. The antiviral effect should result from the interaction of the polypeptide with specific high affinity plasma membrane receptors—resulting in the activation of cellular genes, encoding proteins directly responsible for the antiviral state—and not via another plasma membrane receptor. This is the basic distinction between interferons and some polypeptide interferon inducers, for example, growth factors or interleukins, which can also act as antiviral proteins, but do so indirectly via the induction of interferon (IFN).

The interaction of IFNs and the immune system has received more attention than any other aspect of IFN activity. It is hoped that a better understanding of immunomodulation by IFNs will lead to a more efficient clinical use of IFNs, or, at the least, provide guidelines for research on other lymphokines and cytokines with clinical potential. There is also the fascination and pleasure of unraveling the complexity of the interaction between these two systems.

By some criteria the IFN system is distinct from the immune system, and by other criteria, it is an integral part of the immune system. As far as we know, any nucleated cell in the body is able to produce IFN-α/β, and thus is capable of directly influencing the immune system. Although cells of the immune system itself can make IFN-α and β, some of them can, in addition, produce their "own" IFN, IFN-γ, which has important effects not only on immune cells, but on many other cells in the organism. Making allowance for a few important exceptions, it is rather remarkable that the biological activities of IFN-γ are on the whole quite similar to those of IFNs α and β. This lymphokine, which was baptized an interferon primarily because of its antiviral activity, has subsequently fully stood up to its name, since whenever a new and interesting activity was described for IFN-α or IFN-β, the same activity was frequently found with IFN-γ.

By virtue of its mode of production, antigen-specific induction of sensitized T cells, IFN-γ acts from within the immune system, whereas IFNs α and β, made by macrophages and lymphocytes as well as by any other nucleated cell in the organism, act from within and from without. All three IFNs exert a multitude of effects on the many different cells that make up the immune system.

It is customary to refer to the "IFN system," and indeed, experimental conditions can be set up for studying different aspects of IFN production and action, with IFNs as the only variables as much as possible. Analytical reductionism is a scientific and methodological necessity, but it presents the danger of creating the impression that there really is an independent IFN system, whereas in reality the organism is not comprised of independent systems, but instead, consists of intertwined and continuously interacting components.

Host response to infection is an integration of humoral and cellular immunity, inflammatory mediators, and other cytokines. Cytokines influence the production and action of IFNs, and IFNs exert effects on the production and activity of other cytokines. The interaction can be either synergistic or antagonistic, and the outlines of a highly complex regulatory network are emerging from these studies. Since most of this work has been performed in vitro, we should exert due caution in the interpretation of data concerning the activity of a highly complex system based on analysis of only a small part of its components, frequently in concentrations considerably above physiological levels and usually removed from its natural context. René Dubos has well stated the reason for such caution: "In the most common and probably the most important phenomena of life, the constituent parts are so interdependent that they lose their character, their meaning, and indeed their very existence when dissected from the functional whole'' (Dubos, 1965).

REFERENCES

Dubos, R. *Man Adapting*. Yale University Press, New Haven, p. 337 (1965).

Isaacs, A. Interferon. *Sci. Am.* 204: 51–57 (1961).

Isaacs, A. and Lindenmann, J. Virus interference. 1. The interferon. *Proc. Roy. Soc.* B 147: 258–273 (1957).

Lindenmann, J. From interference to interferon: A brief historical introduction. *Phil. Trans. R. Soc. Lond.* B 299: 3–6 (1982).

Lindenmann, J., Burke, D., and Isaacs, A. Studies on the production, mode of action and properties of interferon. *Br. J. Exp. Path.* 38: 551–562 (1957).

2 THE INTERFERON GENE FAMILY

Our knowledge of the IFN genes to a large extent is based on an analysis of these genes in humans and in mice. In both species, multiple genes code for many different IFN-α proteins, but there is only one gene coding for IFN-β and one for IFN-γ.

I. THE HUMAN INTERFERON STRUCTURAL GENES

A. The IFN-α/IFN-β Gene Cluster

1. *The Human IFN-α Genes*

There are at least 24 nonallelic genes or pseudogenes coding for structurally different Hu IFN-α proteins (Weissmann and Weber, 1986). Contrary to the majority of the structural genes that have been characterized in the human genome, including the genes of all the other cytokines, the Hu IFN-α genes lack introns, a feature they share with all other known mammalian IFN-α genes.

TABLE 2.1 The Amino Acid Sequences of Different Hu IFN-α Subtypes Derived from cDNA or Genomic DNA Sequences.[a]

	S1 S10	S20	S23	1 10	20
IFN-α consensus	MALSFSLLMA	VLVLSYKSIC	SLG	CDLPQTHSLG	NRRALILLAQ
IFN-α1	..SP.A...V	LV...C..S.	...	....E....D	...T.M....
IFN-αD	..SP.A...V	LV...C..S.	...	....E....D	...T.M....
IFN-α2	...T.A..V.	L....C..S.	.V.	....	S..T.M....
IFN-αA	...T.A..V.	L....C..S.	.V.	....	S..T.M....
IFN-αK (α6)	...P.A....	LV...C..S.	..D	....	H..TMM....
IFNα5 (G)	...P.V....	LV..NC....	...	.........S	...T.MIM..
IFN-αH1 (αH2)	...P...M..	LV...C..S.	...	.N.S.....N	...T.M.M..
IFN-αB2 (α8)	...T.Y..V.	LV......FS	...	....	....
IFN-αB	...T.Y.MV.	LV......FS	...	....	....
IFN-α4b	....	....	...	....	....
IFN-αC	....	....	...	....	........G.
IFN-αL (ψα10)	....	.........★	...	.......T.R	........G.
IFN-αJ1 (α7)	..R......V	....	...	.........R	....
IFN-αJ2	..R......V	....	...	.........R	....
IFN-αI	....	....	...	....	....
IFN-αF	....	....	...	....	....
IFN-αWA	....	....	...	....	....
IFN-αGx-1	...P...M..	LV...C..S.	...	.N.S.....N	...T.MI...
IFN-α76	....	....	...	....	....

	80	90	100	110
IFN-α consensus	KDSSAAWDES	LLEKFSTELY	QQLNDLEACV	IQEVGVEETP
IFN-α1	.........D	..D..C....	....	M..ER.G...
IFN-αD	.........D	..D..C....	....	M..ER.G...
IFN-α2	.........T	..D..Y....	....	..G...T...
IFN-αA	.........T	..D..Y....	....	..G...T...
IFN-αK (α6)	....V....R	..D.LY....	....	M...W.GG..
IFN-α5 (G)	.....T...T	..D..Y....	.........M	M......D..
IFN-αH1 (αH2)	.N.......T	.....YI..F	..M.......	....
IFN-αB2 (α8)	......L..T	..DE.YI..D	.......S..	M.....I.S.
IFN-αB	......L..T	..DE.YI..D	VLC	D.....I.S.
IFN-α4b	E......EQ.	....	....	....
IFN-αC	E......EQ.	....	....	....
IFN-αL (ψα10)	E......EQ.	........I.	....	....
IFN-αJ1 (α7)	E......EQ.	....	....	....
IFN-αJ2	E......EQ.	....	....	....
IFN-αI	E......EQ.	....	....N.....	.....M....
IFN-αF	.....T.EQ.	.........N	.....M....	....
IFN-αWA	.........T	..D..YI..F	....	T.......IA
IFN-αGx-1	.....T...T	..D..Y....	.........M	M......D..
IFN-α76	E......EQ.	....	....	....

[a] The sequences, including the signal peptide, are presented in comparison with a consensus sequence, and residues are indicated only when they are different from the consensus sequence. In the latter, residues common to all listed sequences are underlined. Sequences with numeric designation are from Weissmann and collaborators, and sequences

IFNs are secreted proteins, synthesized as preIFNs, and membrane insertion is assured by the leader sequence at the N-terminal end. PreIFNs contain a 23-residue leader sequence, cleaved off during maturation, and the mature IFN-α sequence contains 166 amino acids, with the exception of Hu IFN-α2, which contains only 165 amino acids. A comparison of IFN-α coding sequences (Table 2.1) reveals four highly conserved cysteines, which determine the two sulfide bonds that are responsible for the sensitivity of the molecule to reducing agents: Cys 1 is bonded to Cys 99 and Cys 29 is bonded to Cys 139 (Wetzel, 1981). With a few exceptions, most Hu IFN-α genes do not encode typical *N*-glycosylation sequences, which explains why Hu IFN-α, contrary to Mu IFN-α, is usually not glycosylated (Pestka, 1983).

30	40	50	60	70
MGRISPFSCL	KDRHDFGFPQ	EEFDGNQFQK	AQAISVLHEM	IQQTFNLFST
.S....S...	M.........	....	.P.......L	...I....T.
.S....S...	M.........	....	.P.......L	...I....T.
.R...L....	....	...-......	.ET.P.....	...I......
.RK..L....	....	...-......	.ET.P.....	...I......
.R...L....	......R...	....	.E.......V	....
....	....	....	....	....
.R........	......E...	....	....	M.........
.R........	......E...	....DK....	....	....
.R........	......E...	....DK....	....	....
.....H....	.........E	.....H....	T.........	....
....	......RI..	....	....	....
....	......RI..	....	....	....
....	....E.R..E	.....H....	T.........	....
....	....E.R..E	.....H....	T.........	....
....	...P...L..	....	T.........	....
....	....	....	....	....
.....H....	...Y......	.V........	.....AF...	....
....	....	....	....	....
.....H....	.........E	.....H....	....	....

120	130	140	150	160	166
LMNEDSILAV	RKYFQRITLY	LTEKKYSPCA	WEVVRAEIMR	SFSFSTNLQK	RLRRKD
...A......	K...R.....	....	....	.L.L.....E	.....E
...V......	K...R.....	....	....	.L.L.....E	.....E
..K.......	....	.K........	....	...L.....E	S..S.E
..K.......	....	.K........	....	...L.....E	S..S.E
....	....	....	....	...S.R...E	.....E
...V....T.	....	....	....	...L.A...E	.....E
....	K.........	.M........	....	....	......
..Y.......	....	........S..	....	...L.I....	..KS.E
..Y.......	....	........S..	....	...L.I....	..KS.E
...V......	....	....	....	.L........	......
....	....	.I.R......	....	.L........	......
....	....	.I.R......	....	.L........	......
.....F....	....	.M........	....	........K.	G.....
.....F....	....	.M........	....	....	......
....	....	....	....	.L........	I.....
...V......	K.........	....	....	...L.KIF.E	.....E
....	....	.MG.......	....	....	G.....
...V....T.	....	....	....	...L.A...E	.....E
....	....	....	....	.L........	......

A to L are from Pestka, Goeddel, and collaborators. The table is adapted and modified from Pestka (1986). The relevant references can be found in the text, and the amino acid one-letter symbols are explained in Table 2.5.

There are at least 18 Hu IFN-αI nonallelic genes, 4 of which are pseudogenes, and at least 6 Hu IFN-αII genes, 5 of which are pseudogenes. The distinction between the subfamilies I and II is based on significant differences in the overall structure: The IFN-αI genes encode mature polypeptides of 165 or 166 amino acids, whereas the IFN-αII genes encode mature proteins of 172 amino acids. The coding sequences of the genes encoding the members of the Hu IFN-αI subfamily diverge from each other by an average of 8% (upper limit 13%) in the replacement, and an average of 24% (upper limit 35%) in the silent sites. In contrast, the divergence in the coding sequence between the human αI and αII subfamilies is of the order of 30% in the replacement sites and 67% in the silent sites. This significant difference in amino acid sequence results in extensive antigenic differences between human IFN-αI and IFN-αII proteins as shown by lack of cross-neutralization with polyclonal antisera. It has therefore been suggested that the IFN-αII molecules represent a separate family of IFNs that should be called IFN-ω (Adolf, 1987).

There is more homology between the human and bovine IFN-αII genes than between the human IFN-αI and αII genes, and it has been estimated that the class I and class II IFN-α genes diverged more than 100 million years ago, prior to the mammalian radiation (Goeddel et al., 1980; Mantei et al., 1980; Nagata et al., 1980; Streuli et al., 1980; Goeddel et al., 1981; Lawn et al., 1981a,b; Yelverton et al., 1981; Fuke et al., 1984; Gren et al., 1984; Torczynski et al., 1984; Capon et al., 1985; Feinstein et al., 1985; Hauptman and Swetly, 1985; Henco et al., 1985). For the sake of convenience, and since little work has been done with Hu IFN-αII, we will follow common usage and refer to all members of the IFN-αI subfamily as IFN-α.

How can one explain the high degree of homology between the many nonallelic IFN-α genes? Gene conversion as a result of mismatch repair and unequal crossover appear to have contributed extensively to the similarity between different IFN-α genes. Unequal crossover is characterized by nonreciprocal events in the immediate vicinity of the point of recombination, resulting in homogenization of nonallelic genes over a certain length. An extreme example of homology, most likely resulting from gene conversion, is the Hu IFN-α13 gene, whose coding region—but not its 5′ and 3′ flanking regions—is identical to that of the Hu IFN-α1 gene (Todokoro et al., 1984).

The 5′ flanking regions of the IFN-α genes have a highly conserved 42 bp purine-rich region located immediately downstream from position −117. This region is rich in GAA and GAAA repeats, and contains the information that is necessary for the inducible transcription of the gene (see Chapter 3) (Ragg and Weissmann, 1983; Benjamin et al., 1983; Fisher et al., 1983; Weidle and Weissmann, 1983). The 3′ flanking regions are of variable length, up to about 450 nucleotides; some genes have several polyadenylation sites and can give rise to mRNAs of different lengths (Mantei and Weissmann, 1982; Henco et al., 1985). The 3′ noncoding regions also contain the sequence motifs ATTTA or TTATTTAT, which occur many times more than would be expected by chance, and which are also present in the 3′ flanking regions of several other cytokine genes and protooncogenes. Such genes have in common that they are usually only transiently expressed in response to different inducers, and encode proteins that are frequently involved in the inflammatory response. It has been proposed that this region of the gene confers instability upon the transcribed mRNA, and contributes to the short half-life of such mRNAs. Direct experimental evidence for this comes from the observation that introducing a 51 nucleotide AT sequence from the 3′ flanking region of the gene encoding granulocyte-monocyte colony-stimulating factor (the GM-CSF gene, see Chapter 8) into the 3′ flanking region of the rabbit β-globin gene renders the otherwise stable β-globin mRNA highly unstable in vivo (Caput et al., 1986; Shaw and Kamen, 1986).

2. The Human IFN-β Gene

Contrary to the many genes coding for the different IFN-alpha species, only a single gene, coding for what is commonly called "fibroblast IFN," has been fully characterized in humans. The term "fibroblast IFN," however, is misleading, since fibroblasts can produce more than one IFN species, and the more correct designation is Hu IFN-β. Like the IFN-α genes, the IFN-β gene does not contain intervening sequences. The gene codes for a 166-amino-acid-long mature peptide, preceded by a 21-amino-acid signal sequence. The amino acid homology of Hu IFN-β with the Hu IFN-α proteins is about 30%, and the nucleotide homology of the coding sequence is about 45%. There seems to be relatively little allelism in the IFN-β gene, since at least six independently cloned genes or cDNAs have identical coding sequences, and another characterized coding sequence differs from the previous six only in one position—encoding a Leu to Met change at position 47. The Hu IFN-β sequence contains an *N*-glycosylation site at position 180, which explains why Hu IFN-β is glycosylated (Derynck et al., 1980; Houghton et al., 1980; Taniguchi et al., 1980a,c; Derynck et al., 1981; Houghton et al., 1981; Mory et al., 1981; Tavernier et al., 1981; May and Sehgal, 1985).

Hu IFN-β has three cysteine residues at amino acid positions 17, 31, and 141 (Table 2.2). Cys 31 and 141 are involved in disulfide bridging, leaving Cys 17 free. When the latter is replaced by serine, there is no loss in biological activity, whereas replacement of Cys 141 by Tyr does result in the loss of biological activity (Shepard et al., 1981; Mark et al., 1984). Since IFN-α and IFN-β genes can be coordinately induced, it is not surprising that there is extensive sequence homology between the 5′ flanking regions of the Hu IFN-β and the Hu IFN-α genes upstream from the TATA box. As discussed in the chapter on IFN induction, these regions play a role in the inducible expression of the IFN genes (Chapter 3) (Degrave et al., 1981; Ohno and Taniguchi, 1981).

3. The Chromosomal Location of the Hu IFN-α and β Genes

The Hu IFN-α gene family is located on chromosome 9. This has been determined by Southern hybridization assays using a Hu IFN-α cDNA to probe DNA extracted from human–mouse and human–hamster hybrid cell

TABLE 2.2 Comparison of the Deduced Amino Acid Sequences, Including the Signal Peptides, of IFNβ of Human, Murine, and Bovine Origin.[a]

	S10	S20	S21
IFN-β consensus	M T x R C L L Q x A	L L L C F S T T A L	S
Hu-IFN-β	. . N K . . . . I .	. . . .	.
Mu-IFN-β	. N N . W I . H A .	F	.
Bo-IFN-β1	. . Y . . . . . M V	. . . .	.
Bo-IFN-β2	. . H . . . . . M V	. . . .	.
Bo-IFN-β3	. . Y . . . . P M V	. . . .	.
	10	**20**	**30**
IFN-β consensus	x S Y x L L x F Q Q	R x S x x x C Q K L	L x Q L x x x x x x
Hu-IFN-β	M . . N . . G . L .	. S . N F Q	. W . . N G R L E Y
Mu-IFN-β	I N . K Q . Q L . E	. T N I R K . . E .	. E . . N G K I - -
Bo-IFN-β1	R . . S . . R . . .	. Q . L K E	. G . . P S T S Q H
Bo-IFN-β2	R . . S . . R . . .	. R . L A L	. R . . P S T P Q H
Bo-IFN-β3	R . . S . . R . . .	. R . A E V	. G . . H S T P Q H
	40	**50**	**60**
IFN-β consensus	C L x x R M D F x x	P E E M K Q x Q Q F	Q K E D A A L x I Y
Hu-IFN-β	. . K D . . N . D I	. . . I . . L . . .	. . . . T . .
Mu-IFN-β	N . T Y . A . . K I	. . . . T E - K M -	. . S Y T . F A . Q
Bo-IFN-β1	. . E A . . . . Q M	. . . . . . E . . .	. . . . . I . V M .
Bo-IFN-β2	. . E A . . . . Q M	. . . . . . A . . .	. . . . . I . V . .
Bo-IFN-β3	. . E A K . . . Q V	. . . . N . A . . .	R . . . . I . V . .
	70	**80**	**90**
IFN-β consensus	E M L Q N I F x I F	R x D F S S T G W N	E T I V E x L L x E
Hu-IFN-β	. . . . A . .	. Q . S	. . . . . N . . A N
Mu-IFN-β	. . . . . V . L V .	. N N	. . . . V R . . D .
Bo-IFN-β1	. V . . H . . G . L	T R . . . . S	. . . I . D . . K .
Bo-IFN-β2	. . . . Q . . N . L	T R . . . . S	. . . I . D . . E .
Bo-IFN-β3	. . . . Q . . N . L	T R . . . . S	. . . I . D . . V .

lines, each of which had retained a different combination of human chromosomes (Owerbach et al., 1981; Slate et al., 1982). In situ hybridization to human metaphase chromosomes further localizes IFN-α genes on the short arm of chromosome 9 (p21-pter) (Trent et al., 1982). All available evidence suggests that the Hu IFN-α genes are clustered in one chromosomal region. R-looping analysis of human genomic DNA with IFN mRNA from induced human leucocytes, as well as analysis of genomic clones, indicates that the distance between the different Hu IFN-α genes ranges from 4 to 18 kb, with an average of about 10 kb. Subsets of up to three IFN-α genes arranged in tandem have been isolated from one or a few overlapping genomic clones. In the bovine species, members of both the IFN-αI and αII subfamily appear to be interspersed on the chromosome, suggesting that the Hu IFN-αI and αII genes are probably also interspersed on chromosome 9. However, since the exact size and composition of the gene cluster has not been determined, it cannot be excluded that some IFN-α genes are not part of the cluster (Brack

	100	110	120
IFN-β consensus	LYxQxNxLKT	VLEEKxEKEN	xTxGxxMSS--L
Hu-IFN-β	V.H.I.H...	.....L...D	F.R.KL...--.
Mu-IFN-β	.HQ.TVF...	.....Q.-.R	L.WE--...TA.
Bo-IFN-β1	..W.M.R.QP	IQK.IMQ.Q.	S.TEDTIV---P
Bo-IFN-β2	..E.M.H.EP	IQK.IMQ.Q.	S.M.DTTV---.
Bo-IFN-β3	..G.M.R.QP	IQK.IMQEQ.	F.M.DTTV---.
	130	**140**	**150**
IFN-β consensus	HLKxYYxRxx	xYLKxKEYxx	CAWTVVRVEI
Hu-IFN-β	...R..G.IL	H...A...SH	.....I....
Mu-IFN-β	...S..W.VQ	R...LMK.NS	Y..M...A..
Bo-IFN-β1	..GK..FNLM	Q..ES...DR	......Q.Q.
Bo-IFN-β2	..RK..FNLV	Q...S...NR	........Q.
Bo-IFN-β3	...K..FNLV	Q..ES...NR	........Q.
	160	**166**	
IFN-β consensus	LRNFxFIxRL	TGYLRN	
Hu-IFN-β	....Y..N..	......	
Mu-IFN-β	F...LI.R..	.RNFQ.	
Bo-IFN-β1	.T.VS.LM..	...V.D	
Bo-IFN-β2	....S.LT..	.....E	
Bo-IFN-β3	.T..S.LM..	.AS..D	

[a] The sequences are presented as they differ from a consensus sequence, and the amino acids of the consensus sequence that are common to all sequences are underlined. Positions where no clear consensus exists are indicated in the consensus sequence by "x." The table is adapted from Pestka (1986). The relevant references can be found in the text, and the amino acid one-letter code is explained in Table 2.5.

et al., 1981; Lawn et al., 1981a; Ullrich et al., 1982; Lund et al., 1984; Capon et al., 1985).

Like the IFN-α genes, the IFN-β gene is situated on chromosome 9, in the same region as the IFN-α genes (pter-q12) (Shows et al., 1982). Although the IFN-β gene is probably clustered with the IFN-α genes, they may not be adjacent, since a 36-kb fragment, cloned in a human genomic cosmid hybrid, contains the IFN-β gene but no IFN-α genes (Gross et al., 1981). Segregation analysis of restriction fragment length polymorphism of the Hu IFN-α and IFN-β genes in several human families also indicates close linkage of these genes (Ohlsson et al., 1985). Furthermore, the IFN-β gene is distal to the IFN-α cluster, as shown by a translocation in human monocytic leukemia, in which a break in the short arm of chromosome 9 has split the IFN genes and translocated the IFN-β gene, but not the α genes, from chromosome 9 to chromosome 11 (Diaz et al., 1986).

In addition to the Hu IFN-β locus on chromosome 9, the presence of Hu IFN-β-related DNA has been shown on chromosomes 2 and 4 (Sagar et al., 1982, 1984, 1985). The corresponding genes however, or their products, have not been characterized. Interferon induction studies in human–hamster and human–mouse somatic cell hybrids also implicate chromosomes 2 and 5 as coding for antiviral activity, but again, the putative genes have not been isolated or characterized (Slate and Ruddle, 1980).

B. The Hu IFN-γ Gene

1. Structure

The Hu IFN-γ gene, which exists only as a single copy gene, differs significantly from the Hu IFN-α and IFN-β structural genes, with which it shares no obvious sequence homology, although there is evidence suggesting some evolutionary relationship. The gene contains three introns: intron 1 is 1,238 bp long; intron 2 95 bp, and intron 3 2,422 bp. The four exons code for 38, 23, 61, and 44 amino acids, respectively, resulting in a polypeptide of 166 amino acids, 20 of which constitute the signal peptide (Table 2.3) (Devos et al., 1982; Gray and Goeddel, 1982; Gray et al., 1982; Taya et al., 1982).

It is intriguing that the smallest intron is located just in the middle of the Glu-Glu sequence, which is conserved among the three IFN species at virtually the same position: residues 41 and 42 in IFN-α and IFN-γ and residues 42 and 43 in IFN-β. Of the 18 identical amino acids or very conservative replacements between IFN-γ and IFN-α, all except one are located in the third and fourth exon (Epstein, 1982; Taya et al., 1982).

The deduced amino acid sequence reveals two potential *N*-glycosylation sites at positions 28 and 100. This explains the existence of two different molecular weight species of Hu IFN-γ, one of 20 kDa and another of 25 kDa. In the 20 kDa species only one site is glycosylated, whereas in the 25 kDa species both sites are glycosylated (Yip et al., 1982; Rinderknecht et al., 1984). The protein is also characterized by a remarkably high content of basic residues, more particularly two clusters of four residues each, Lys-Lys-Lys-Arg at position 89–92 and Lys-Arg-Lys-Arg at position 131–134. The basic nature of the mature protein is possibly related to the acid lability of its biological activity (Devos et al., 1982).

Although the coding sequences of the IFN-γ cDNAs that are isolated from three different individuals are quite similar, some allelism is observed. Compared to the first published sequence (Gray et al., 1982), one has an arginine instead of a glutamine residue at position 140 (Devos et al., 1982) and another has a lysine to glutamine change at position 9 (Nishi et al., 1985). These changes do not affect the function of the protein.

The 5′ flanking region of the Hu IFN-γ gene contains nucleotide sequences also found in the 5′ flanking regions of other genes that are specifically expressed by T cells, such as the structural gene for IL-2 and the gene encoding the Tac component of the IL-2 receptor. This sequence, about 200 bp long, functions as an inducible transcriptional enhancer and seems to control T-cell specific gene expression (see also Chapter 13) (Fujita et al., 1983, 1986). Another region of potential regulatory importance is the sequence AAGTGTAATTTTTTGAGTTTCTTTT, which is in a DNAse I sen-

TABLE 2.3 Comparison of the Deduced Amino Acid Sequences of Human, Murine, Bovine, and Rat IFN-γ[a]

	S1 S10	S20	1 10	20	30
HuIFN-γ	MKYTSYILAF	QLCIVLGSLG	CYCQDPYVKE	AENLKKYFNA	GHSDVADNGT
MuIFN-γ	.NA.HC...L	..FLMAV.-.	...HGTVIES	L.S.NN...S	SGI..-EEKS
BoIFN-γ	......F..L	L..GL..FS.	S.G.GQFFR.	I....E....	SSP...KG.P
RaIFN-γ	.SA.RRV.VL	...LMAL.-.	....GTLIES	L.S..N...S	SSM.AMEGKS
	40	50	60	70	80
HuIFN-γ	LFLGILKNWK	EESDRKIMQS	QIVSFYFKLF	KNFKDDQSIQ	KSVETIKEDM
MuIFN-γ	...D.WR..Q	KDG.M..L..	..I...LR..	EVL..N.A.S	NNISV.ESHL
BoIFN-γ	..SD......	D...K..I..	....	E.L..N.V..	R.MDI..Q..
RaIFN-γ	.L.D.WR..Q	KDGNT..LE.	..I...LR..	EVL..N.A.S	NNISV.ESHL
	90	100	110	120	130
HuIFN-γ	NVKFFNSNKK	KRDDFEKLTN	YSVTDLNVQR	KAIHELIQVM	AELSPAAKTG
MuIFN-γ	ITT..SNS.A	.K.A.MSIAK	FE.NNPQ...	Q.FN...R.V	HQ.L.ESSLR
BoIFN-γ	FQ..L.GSSE	.LE..K..IQ	IP.D..QI..	...N...K..	ND...KSNLR
RaIFN-γ	ITN..SNS.A	.K.A.MSIAK	FE.NNPQI.H	..VN...R.I	HQ...ESSLR
	140	146			
HuIFN-γ	KRKRSQMLFR	GRRASQ			
MuIFN-γ	.....RC				
BoIFN-γ	......N...	.....M			
RaIFN-γ	.....RC				

[a] Amino acids that are identical to the human sequence are indicated by a dot. The amino acid residues of the human sequence that are common to the three other sequences are underlined. The one-letter amino acid code is explained in Table 2.5.

Source: Adapted from Weissmann and Weber (1986).

sitive site in the first intron of the IFN-γ gene and is 83% homologous to a sequence in the 5′ flanking region of the IL-2 gene. In the latter, this sequence is about 300 bp upstream of the promoter (Hardy et al., 1985).

2. Chromosomal Location

Southern blot analysis of DNA from human–rodent somatic cell hybrids has located the Hu IFN-γ gene on chromosome 12, in the p12.05-qter region (Naylor et al., 1983).

C. Synthetic Genes Can Direct Synthesis of IFN

Synthetic genes encoding biologically active products have been constructed for a number of peptides, including human somatostatin and human insulin A and B chains (Itakura et al., 1977; Goeddel et al., 1979). Similarly, synthetic genes for Hu IFN-α1, IFN-α2, and IFN-γ, consisting of the respective coding regions placed under the control of natural or synthetic promoters active in bacteria, have been expressed into functional IFN proteins (Edge et al., 1981; De Maeyer et al., 1982; Windass et al., 1982; Edge et al., 1983; Tanaka et al., 1983).

The cloning and expression of synthetic IFN genes makes it possible to construct modified genes by introducing specific codon changes at any number of positions along the entire length of the coding sequence, resulting in IFN analogs with defined amino acid changes. Such analogs can then be used for probing structure-function relationships and define regions of IFN molecules that are particularly important for biological activity (Edge et al., 1986). The major conclusions derived from such studies with Hu IFN-α2 analogs are discussed briefly.

1. C- and N-Terminal Truncated Analogs

The 13 amino acid residues at the carboxy terminus are not essential for biological activity, since IFN-α2 (1-155) has normal antiviral and antiproliferative activity. Similarly, IFN-α2 (4-165) is biologically active, indicating that the first three residues, and therefore also the integrity of the Cysl-Cys98 disulfide bridge, are not critical for activity. The same conclusion has been reached by an approach involving chemical modification of the disulfide bonds: Cys1-Cys89 is not critical for antiviral activity, contrary to Cys29-Cys138, which is essential (Wetzel et al., 1983). This probably explains why the corresponding Cys31-141 is the only disulfide bridge conserved in Hu IFN-β.

2. The Region of Residues 98–114

This region has received attention because it is the region of highest divergence (7 out of 12 residues, see Table 2.1) between Hu IFN-α1 and Hu IFN-α2. Hu IFN-α1 differs from Hu IFN-α2 by its lower specific antiviral activity on human cells and its lack of recognition by the monoclonal antibody NK2

(Secher and Burke, 1980). The residues that are critical for lack of recognition by the NK2 antibody are Asn^{112} and Ala^{113}, since (Asn^{112}, Ala^{113}) IFN-α2 is not recognized by the antibody. Thus, the presence of this Asn-Ala sequence in IFN-α1 is sufficient to cause the lack of recognition, whereas the corresponding Lys-Glu sequence in IFN-α2 appears to make an important contribution to the epitope for the monoclonal antibody. Apparently, this part of the IFN-α2 molecule is antigenically very active, since another independently generated, monoclonal antibody (LI-1) also reacts with the Glu 113 residue (Trown et al., 1985). This region of the molecule is hydrophilic, which is consistent with it being an antigenic site. Tertiary structure prediction also suggests that this region is in an exterior, exposed location (Sternberg and Cohen, 1982).

Substitution of completely unrelated sequences from IFN-γ into positions 98–114 of IFN-α2 does not affect biological activity, indicating that the specific amino acid side chains in this region are not important for activity. Moreover, when the decapeptide sequence from positions 101–110 is completely deleted and Ile^{100} is linked directly to Met^{111}, there is no loss of biological activity. A comparable deletion has occurred naturally in Mu IFN-α4, which has an internal deletion of five amino acids at positions 104–108 without loss of biological activity (Kelley and Pitha, 1985a).

3. *Residues Adjacent to the Conserved Cysteine in Position 29*

The residues Leu 30, Asp 32, and Arg 33 are highly conserved in all Hu IFN-α species as well as in Hu IFN-β (see Fig. 2.1), and replacement of these three amino acids by alanine results in complete loss of activity on human cells. However, ($Ala^{30,32,33}$) IFN-α2, although also devoid of antiviral and antiproliferative activity on bovine MDBK cells, does compete for binding and acts as a competitive antagonist of Hu IFN-α2 on these cells. This is discussed in detail in chapter 4 (Camble et al., 1986; Edge et al., 1986; Marcucci and De Maeyer, 1986).

II. THE MURINE IFN STRUCTURAL GENES

A. The Murine IFN-α/IFN-β Gene Cluster

1. *The Mu IFN-α Genes*

At least 12 nonallelic genes or pseudogenes have been identified in the Mu IFN-α gene family, but Southern blot analysis of murine genomic DNA suggests that the list is not complete and that more genes will be identified (Fig. 2.2). Ten different genes have been cloned and expressed into biologically active Mu IFN-α (Table 2.4).

The general structure of the Mu IFN-α genes is comparable to that of the corresponding human genes. The mature proteins contain 166 or 167 amino acids (except for Mu IFN-α4, which has a five-codon deletion between co-

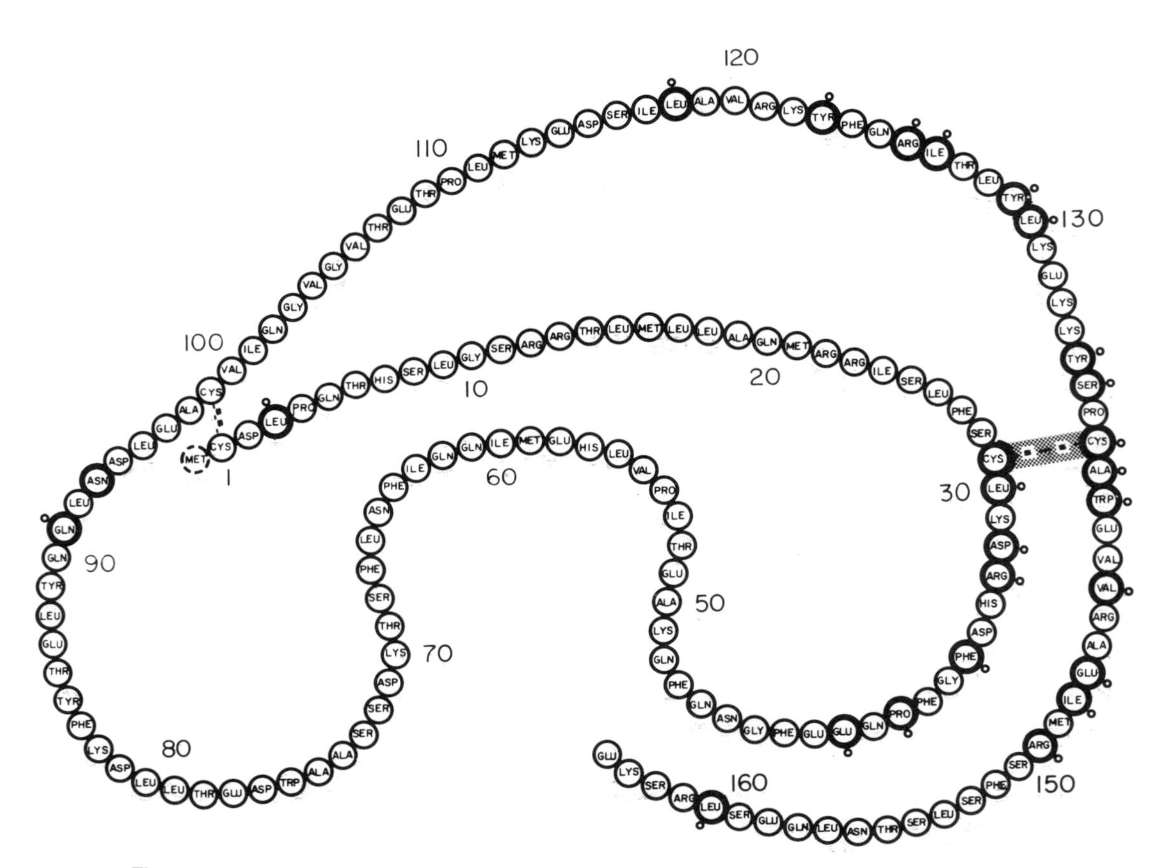

Figure 2.1 The amino acid sequence of Hu IFN-α2, indicating the residues conserved in all Hu IFN-α species and in Hu IFN-β. These conserved residues are indicated by a thick circle and a second, small circle. (Adapted from Edge et al., 1986.)

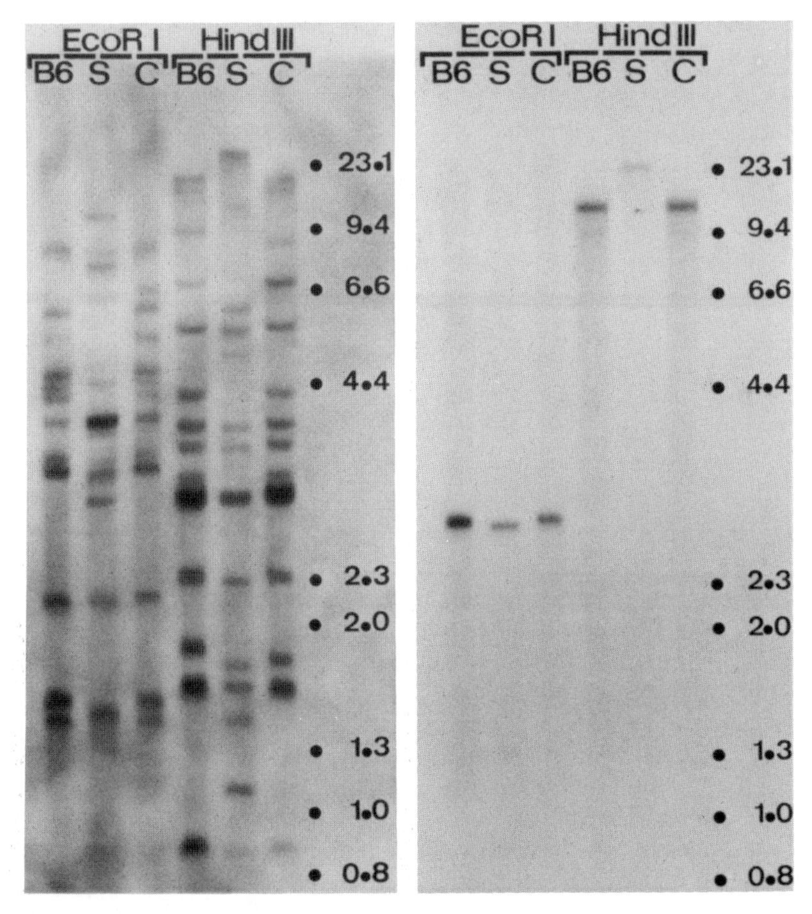

Figure 2.2 Southern blot analysis of genomic DNA from three different mouse strains probed with a Mu IFN-α2 cDNA (left) or Mu IFN-β cDNA (right), illustrating the multiple IFN-α genes, the single IFN-β gene, and the restriction fragment length polymorphism among three different mouse strains. The fragment size markers are expressed in kilobase (kb). The IFN-α genes corresponding to some of the fragments in the BALB/c HindIII digest are the following:

Size of fragment in kb:	Mu IFN-α genes contained in the fragment:
20	α5 and α6k
9	α6
7	α9
5.6	α4
3.1	α2
1	α1

B6: C57BL/6; S: Spretus; C: BALB/c. (Seif and De Maeyer-Guignard, 1986.)

TABLE 2.4 Deduced Amino Acid Sequences of Nine Different Mu IFN-α Species, Compared to a Consensus Sequence.[a]

	S1 S10	S20	S24	1 10	20
Mu-IFN-α consensus	MARLCAFLMV	LAVMSY-WST	CSLG	CDLPQTHNLR	NKRALTLLVQ
Mu-IFN-α1	....	......-.P.	....	....	....
Mu-IFN-α2	........VM	.I....-..I	....	....H.Y...	.....KV.A.
Mu-IFN-α4	I	.VM...Y..A	....	....H.Y..G	......V.EE
Mu-IFN-α5	....	.P.L..-.P.	....	....	.........K
Mu-IFN-α6	....	......-...	....	.......K..	........I.
Mu-IFN-α6k	....	...L..-.P.	....	....	...L.....K
Mu-IFN-αA	.........T	.L....-...	....	....	....
Mu-IFN-α9	...PF.....	.V.I..-...	....	....	..KI....A.

	80	90	100	110	120
Mu-IFN-α consensus	KDSSAAWNAT	LLDSFCNDLH	QQLNDLQACL	MQEVGVQEPP	LTQEDSLLAV
Mu-IFN-α1	....	....	.......G..	..Q.....F.	.....A....
Mu-IFN-α2	.A........	....	.......T..	..Q.......	.....A....
Mu-IFN-α4	..L..T....	....	......K..V	..-----...	....
Mu-IFN-α5	....	.......EV.	......K..V	........S.	....
Mu-IFN-α6	.......D..	...T.....Y	....	V...RL....	....V.....
Mu-IFN-α6k	....	.........H	.L.....G..	..Q.EI.AL.	....
Mu-IFN-αA	.......D.S	....	......K..V	....	.....Y....
Mu-IFN-α9	....	......TG..	.......G..	..L..MK.L.	......Q..M

dons 102 and 108) and the maximum divergence for replacement sites is about 13%. Therefore, the coding sequences of the different Mu IFN-α genes are about as closely related among one another as are the different IFN-α genes in humans. The four cysteines (Cysl-Cys99 and Cys29-Cys139) that are responsible for the disulfide bonds in Hu IFN-α are also present in Mu IFN-α, and an additional cysteine residue is found at position 86. A putative glycosylation site at position 78 (Asn-Ala-Thr) is found in most Mu IFN-α sequences, which explains why natural murine α IFNs are glycosylated. In the 5′ flanking region a stretch of GAAA-rich repeated sequences is likely to be important for virus-inducible transcription, as is the case for the homologous region of Hu IFN-α genes (Kelley et al., 1983; Shaw et al., 1983; Daugherty et al., 1984; Kelley and Pitha, 1985a,b; Le Roscouet et al., 1985; Ryals et al., 1985; Zwarthoff et al., 1985; Seif and De Maeyer-Guignard, 1986).

2. *The Mu IFN-β Gene*

The single copy, intronless Mu IFN-β gene, codes for a preIFN of 182 amino acid residues, containing a 21-amino-acid signal peptide. The amino acid sequence displays 48% homology with that of Hu IFN-β (Table 2.2) and the coding sequence 63% with the positions of identity scattered throughout the sequence, but relatively more frequent in the middle and carboxy terminal portions. The mature protein consists of 161 residues, with three potential *N*-glycosylation sites at positions 29, 69, and 76. Glycosylation explains the large difference between the 17-kDa molecular weight calculated from the amino acid sequence and the apparent molecular weight of about 28 to 35

30	40	50	60	70
MRRLSPLSCL	KDRKDFGFPQ	EKVDAQQIQK	AQAIPVLSEL	TQQILNIFTS
....	....	........K.	....	....
....PF....	...Q.....L	....N.....	.......RD.	...T..L...
....P.....	L	....N.....	....L..RD.	......L...
....	....	...G.....E	....	...V......
....	....	....TLK...	EK.......V	....
....	....	...G.....E	.......T..	.....TL...
....	......R...	.........N	.......Q..	...V......
....	....	.........E	....	.....TL...

130	140	150	160	167 169
RKYFHRITVY	LREKKHSPCA	WEVVRAEVWR	ALSSSANLLA	RLSEEKE
....	....	....	.......V.G	..R....
....	....	....	.....V...P	......E
.T........	..K....L..	...I......	.....T....	....
....	.......L..	....	.....V....	...K.E.
....	....	....	.......V.G	..R....
.T.......F	....	....	......K...	..N.DE
.T........	....	....	.M:...K...	......E
K.........	....	....	.....V....	....

[a] Only differences from the Consensus sequence are listed. Mu IFN-α1 and α2 are from Shaw et al. (1983); Mu IFN-α4, 5, 6, and 6k are from Kelley and Pitha (1985a) and from Zwarthoff et al. (1985); Mu IFN-αA is from Daugherty et al. (1984), and Mu IFN-α9 is from Seif and De Maeyer-Guignard (1986).

Source: Adapted and modified from Pestka (1986). The one-letter amino acid code is explained in Table 2.5.

TABLE 2.5 One-Letter and Three-Letter Codes for the Amino Acids

A	ala	alanine
B	asx	aspartic acid or asparagine
C	cys	cysteine
D	asp	aspartic acid
E	glu	glutamic acid
F	phe	phenylalanine
G	gly	glycine
H	his	histidine
I	ile	isoleucine
K	lys	lysine
L	leu	leucine
M	met	methionine
N	asn	asparagine
P	pro	proline
Q	gln	glutamine
R	arg	arginine
S	ser	serine
T	thr	threonine
V	val	valine
W	trp	tryptophan
Y	tyr	tyrosine

kDa of natural Mu IFN-β. Contrary to Hu IFN-β, only a single cysteine residue is present at position 17, indicating the absence of disulfide bonds. Only the sequence of the Mu IFN-β gene derived from C3H mice is known; there may be allelic differences with mice of other genotypes, but RFLP between different inbred mouse strains is very limited (Higashi et al., 1983).

3. Chromosomal Location of the Mu IFN-α and IFN-β Genes

Like the Hu IFN-α genes, most and maybe all Mu IFN-α genes are clustered. This is evident from the presence of two complete and part of a third Mu IFN-α gene (α1, α4, and α5) in one phage and two other Mu IFN-α genes (α2 and α6T) in another phage belonging to a BALB/c genomic library. A 28-kb region of BALB/c genomic DNA contains four different IFN-α genes (Mu IFN-α1, α4, α5, and α6P), and an intergenic 1,000-nucleotide-long conserved sequence is associated with three of these four genes, indicating that this particular cluster evolved through tandem duplication (Kelley and Pitha, 1985a; Zwarthoff et al., 1985).

The murine IFN-α gene cluster is on chromosome 4. This has been determined by a variety of approaches all giving concordant results. Southern blot hybridization with Mu IFN-α1 or α2 probes of hamster–mouse somatic cell hybrid DNAs shows a correlation of three BamH1 fragments, four EcoRI fragments, and seven HindIII fragments and mouse chromosome 4 (Kelley et al., 1983; Lovett et al., 1984; Van der Korput et al., 1985). Segregation of one polymorphic EcoRI and three polymorphic HindIII restriction fragments in genomic DNA from BALB/c and C57BL/6 mice and recombinant inbred and congenic strains derived from them indicates linkage of these Mu IFN-α restriction fragments to the minor histocompatibility locus *H-15,* near the coat color locus "brown" (Dandoy et al., 1984).

4. The Location of the Ifa Locus on the Linkage Map in a Region of Immunological Interest

According to the nomenclature rules of mouse genetics, the Mu IFN-α gene cluster is designated as the *Ifa* locus. Based on the segregation of several *Ifa* restriction fragments, linkage analysis of the *Ifa* locus, of the major urinary protein locus *Mup-1,* the coat color locus *b* (brown), and the coat color dilution locus *m* (misty), establishes the gene order on chromosome 4, from the centromere, as follows: *Mup-1 - b - Ifa - m.* A different linkage analysis, including additional genes on chromosome 4, gives as the most likely order: *Mup-1 - Lv - b - Orm-1, Orm-2 - Ifa - Bmfr-1* (De Maeyer and Dandoy, 1987; Nadeau et al., 1986). *Lv* is the locus coding for the enzyme δ-aminolevulate dehydratase, which catalyzes the condensation of δ-aminolevulate into porphobilinogen, and *Orm-1* and *Orm-2* are two tightly linked loci coding for α1 acid glycoproteins, two liver-derived acute phase proteins, produced in response to a variety of systemic injuries such as infection or inflammation (see the section on IL-1, in Chapter 13, for a definition of acute phase proteins) (Baumann et al., 1984; Baumann and Berger, 1985). *Bmfr-1* con-

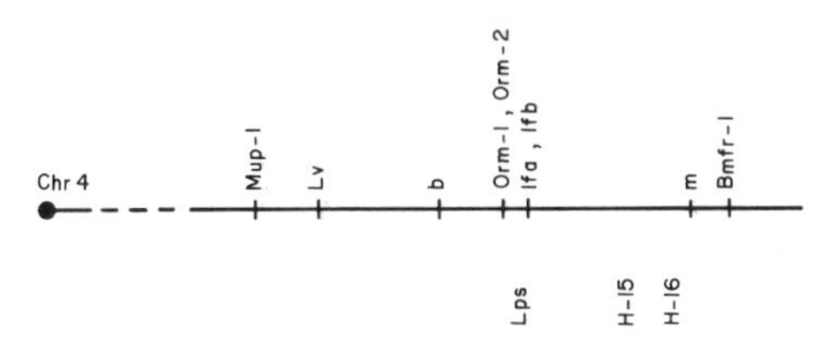

Figure 2.3 The position of the *Ifa* and *Ifb* loci on mouse chromosome 4, relative to other loci of immunological interest. Since the *Lps, H-15,* and *H-16* loci have never been included in a segregational analysis together with *Ifa* or *Ifb,* only the approximate position of these loci is given. (Based on De Maeyer and Dandoy, 1987 and Nadeau et al., 1986.)

trols the responsiveness of B lymphocytes to T-cell derived, B-cell maturation factors, including Mu IFN-γ (Sidman et al., 1986). When the results of these two linkage analyses are combined, the following gene order is obtained: *Mup-1* - *Lv* - *b* - *Orm-1, Orm-2,* - *Ifa* - *m* (the position of *Bmfr-1* relative to *m* is not known, but based on a comparison of different segregational analyses, the two loci appear close). Moreover, the *Ifa* locus is situated in exactly the same chromosomal region as the *Lps* locus, which affects a variety of responses to endotoxin, including the induction of IFN-α/β, IL-1, and TNF-α, and which has been mapped at 6 $\pm$ 2 cM telomeric to *Mup-1*. The exact position of *Lps* with regard to *Ifa* is not known, because these two loci have never been included into one and the same segregation analysis. Since *Ifa* and *Lps* appear to be very close, such an analysis would be the only way of knowing their position relative to each other on the linkage map.

In situ hybridization to chromosome 4 of Mu IFN-α cDNA further confirms the location of *Ifa* on chromosome 4 in the region C3–C6. In view of the relatively loose correlation between physical and recombinational maps, this localization is compatible with the position of *Ifa* on the linkage map (Cahilly et al., 1985; Cheng et al., 1986).

It is noteworthy that the region of chromosome 4 to which *Ifa* has been mapped contains several genes that are involved in the immune response and in systemic response to bacterial and viral infections: *Ifa, Ifb* (see subsequent discussion), *Orm-1* and *Orm-2, Lps, H-15, H-16,* and *Bmfr-1.* Furthermore, the same chromosomal region most likely contains the *Jt* locus, which has been implicated in the murine T-cell-surface I-J antigen expression. The I-J cell-surface structure distinguishes suppressor T cells from other lymphocytes (Hayes et al., 1984; Sachs et al., 1984). Thus, the region of chromosome 4 distal from the *b* locus is characterized by an important grouping of genes that are involved in immune responsiveness and resistance to infectious agents (Fig. 2.3).

5. *Chromosomal Location and Linkage of the* Ifb *Locus*

Like the *Ifa* locus, the Mu IFN-β gene, designated as the *Ifb* locus, is on chromosome 4 (Dandoy et al., 1985; Kelley et al., 1985; Van der Korput et al., 1985).

For the fine mapping of the *Ifa* locus, advantage has been taken of the restriction fragment length polymorphism that is observed when genomic DNA from different inbred mouse strains is probed with Mu IFN-α cDNA. Although there is no extensive polymorphism, it is sufficient to follow segregation analysis of at least one or two Ifa restriction fragments. However, segregation of the *Ifb* locus cannot be followed this way, because the genomic DNAs of the common inbred mouse strains do not display RFLP when probed with Mu IFN-β cDNA. Apparently, in the commonly used laboratory mice, allelism of the Mu IFN-β gene is very limited. This is probably a founder effect, resulting from the peculiar origin of most inbred strains. Indeed, analysis of mitochondrial DNA, which has a strictly maternal mode of inheritance, shows that almost all of these strains possess the same *Mus musculus domesticus* type of mitochondrial DNA. This has been interpreted to suggest that all classical inbred strains could be related by descent from one or a few related females, which would explain the limited polymorphism (Ferris et al., 1982). However, studies on the characterization of the Y chromosome and other genetic markers show that most classical inbred strains are a mosaic of various genetic components, with allelic diversity resulting from crosses between *Mus musculus domesticus* and *Mus musculus musculus* (Bishop et al., 1985).

One way to obtain mice polymorphic at the *Ifb* locus would be to develop new inbred strains from recently captured house mice that belong to the species *Mus musculus domesticus,* from which the laboratory mouse was originally derived. It is also possible to use mice belonging to an altogether different species of the genus *Mus,* such as *Mus spretus* (Lataste). Mice of this species normally live in the wild in the Mediterranean basin, and although they have the same habitat as *Mus domesticus,* they never mate with the latter. *Mus spretus* represents, in fact, a different species, which diverged from *Mus domesticus* about 3 million years ago (Ferris et al., 1983). To give an idea of what this means in terms of evolutionary distance, it is commonly believed that the man–ape split occurred about 4.5 million years ago and the mouse–rat split probably about 10 million years ago.

Like *Mus musculus, Mus spretus* has 40 acrocentric chromosomes, and because of the relatively recent divergence of these two species, certain interspecies crosses are still possible. The extensive genetic polymorphism existing between these two species because of their long separation can thus be exploited for linkage studies. *Mus musculus* females can be mated with *Mus spretus* males but only the female F1 progeny of this cross are fertile, whereas the male F1 progeny are sterile. This is in line with Haldane's rule, which states that the heterogametic sex is sterile in the F1 progeny of an interspecies cross. The fertile F1 females can be backcrossed to the *Mus musculus* parents, which because of the extensive polymorphism, makes it possible to follow segregation of many loci.

When genomic DNA from *spretus* mice is probed with a Mu IFN-α probe and is compared with the DNA of BALB/c, C57BL/6, or DBA/2 mice, a much more extensive restriction fragment polymorphism is observed than when *Mus musculus* DNAs are compared with one another. Likewise, although HindIII or EcoRI restricted DNAs from BALB/c, C57BL/6 of DBA/2 mice display no RFLP when probed with Mu IFN-β cDNA, *Mus spretus* DNA does display RFLP with regard to *Mus musculus* DNA (Fig. 2.2). Segregation analysis in this interspecies cross has established a close linkage of the *Ifa* and *Ifb* loci, since both loci consistently cosegregate in backcross progeny. The recombinational distance between *Ifa* and *Ifb* is estimated between 0 and 7 cM (Dandoy et al., 1985; Nadeau et al., 1986).

B. The Mu IFN-γ Gene

1. Structure

The structure of the Mu IFN-γ gene is comparable to that of the Hu IFN-γ gene. It contains four exons and three introns. The coding part of the gene displays an overall nucleotide homology with the human gene of about 65%; the overall protein homology is 40%. Mature Mu IFN-γ has 136 amino acids, which is ten residues shorter than mature Hu IFN-γ, resulting from the deletion of ten carboxy terminal amino acids (Table 2.3). Like the human form, Mu IFN-γ has 2 *N*-glycosylation sequences and an excess of basic residues. There are three cysteine residues at positions 1, 3, and 136, indicating the potential for an intramolecular disulfide bridge between either of the amino terminal residues and position 136. This is in contrast to Hu-IFN γ, which has only one cysteine residue. The encoded size of the mature protein, based on the amino acid sequence, is 15,894, which is smaller than the size of the natural product based on molecular sieve chromatography, which is 38 kDa. This discrepancy is due to glycosylation, and possibly also to dimerization (Gray and Goeddel, 1983; Dijkmans et al., 1985; Nagata et al., 1986).

2. Chromosomal Location of the Ifg *Locus*

The Mu IFN-γ structural gene (*Ifg* locus) is not on the same chromosome as the *Ifa* and *Ifb* loci. Southern blot analysis of DNA from hamster–mouse cell hybrids localizes the *Ifg* locus to chromosome 10 (Naylor et al., 1984).

III. THE EVOLUTION OF IFN GENES

IFNs or, in the absence of detailed characterization, virus-induced proteins with species-specific antiviral activity, have been induced in cells derived from representatives of all vertebrate classes, except in amphibia. IFNs have been induced in bony fish (Oie and Loh, 1971; De Kinkelin and Dorson, 1973); reptiles (Falcoff and Fauconnier, 1965); birds, in which IFNs were discovered (Isaacs and Lindenman, 1957; Lindenmann et al., 1957); and of course mammals, in which the IFN system has been studied most extensively.

The presence of IFN-γ, however, has only been unambiguously described in mammals.

A. IFN-α and IFN-β Genes Share a Common Ancestry

The significant degree of homology, corresponding to about 28% at the amino acid level and 45% at the nucleotide level, between Hu IFN-α and Hu IFN-β suggests that these genes are derived from a common ancestor by gene duplication. The possibility that IFN-α and IFN-β genes evolved separately after an original gene duplication is supported by the existence of greater homology between Hu IFN-β and Mu IFN-β (48% at the amino acid level) and between Hu IFN-β and bovine IFN-β (55% at the amino acid level) than between Hu IFN-β and the different Hu IFN-αs (Table 2.2) (Leung et al., 1984; Weissmann and Weber, 1986). Opinions are divided as to the timing of the IFN-α/IFN-β gene split, and much depends on how the mutation fixation rates are calculated. The problem is compounded further by our ignorance of the frequency of gene rectification, which, by counteracting the divergence due to nucleotide changes, has played a significant role in the evolution of IFN-α genes. Depending on how the calculation is carried out, it has been estimated that the IFN-α/IFN-β gene split could have happened before the emergence of vertebrates, or at least prior to the divergence of fish and amphibia about 400 million years ago, or even later at about the time of the mammals–birds/reptiles divergence. The minimal value of the time elapsed since the IFN-α/IFN-β divergence corresponds to 250 million years when the calculation of the mutation fixation rates is based on the human IFN-α and -β genes, and 155 million years when the murine genes are taken as a basis for calculation. The difference results from the fact that the calculated mutation fixation rates for the IFN-α genes in mice is about double that in humans. The reason for this difference is not clear, but it may be related to the difference in generation time between both species (Laird et al., 1969; Taniguchi et al., 1980b; Wilson et al., 1983; Miyata et al., 1985; Weissmann and Weber, 1986).

Additional evidence for an early divergence between the α and β genes is provided by a Southern blot analysis of the genomic DNA of a wide range of vertebrates and invertebrates. A complex IFN-α multigene family is detected in all mammals examined with a Hu IFN-α 2 probe, but little or no cross-hybridization is observed with DNA from nonmammalian vertebrates or from invertebrates. Hu IFN-β cDNA, however, successfully cross-hybridizes not only to other mammalian DNAs, but also to nonmammalian vertebrate DNAs, ranging from bony fish to birds. No cross-hybridization with invertebrate DNA is observed with either the α or the β probe (Wilson et al., 1983). Since this conclusion is based on hybridization signals only, and not on sequencing and characterization of the actual genes, some caution must be exerted in its interpretation.

It would be of great interest to have the sequence data of the IFN genes of lower vertebrates and to see whether they already have α and β IFN genes, or only one ancestral type of gene. Whatever the actual time of the α–β split, the IFN-α gene family was certainly established prior to the mammalian radiation, about 85 million years ago, since all mammalian species examined in some detail have large IFN-α gene families (Wilson et al., 1983; Weissmann and Weber, 1986). It has been suggested that the primitive gene, from which the IFN-α and IFN-β genes are derived, itself originated from the duplication of an original segment, giving rise to the amino and the carboxy terminal halves of the present IFN-α and IFN-β genes (Erickson et al., 1984).

Contrary to some highly conserved genes like the histone genes, the limited analysis conducted further down the evolutionary scale has not discovered significant homologies with invertebrate IFN sequences. The Drosophila genome contains no sequences that hybridize with the coding sequence of Mu IFN-α2, although there is some homology between the untranslated but transcribed 3′ end of the *ect* gene in Drosophila and the Mu IFN-α2 gene. The *ect* gene plays a role in the early development of ectodermal structures in Drosophila, and the significance of the homology of the 3′ untranslated region of this gene with the corresponding regions in IFN-α genes is difficult to assess (Nakanishi et al., 1986).

B. Concerted Evolution of IFN-α Genes

IFN-α genes of the same species, be it human or murine, are very closely related to each other over entire gene regions. Gene duplication is the mechanism by which closely related genes are generated, and comparable to the globin genes, this has been the case for the IFN-α genes that undoubtedly arose by a series of gene duplications (Miyata and Hayashida, 1982; Ullrich et al., 1982). Also, as in the globin gene family, gene conversion and unequal crossing-over have contributed to the extensive homology between the different IFN-α genes of the same species. Indeed, because of the existence of a multigene family, the phenomenon of gene conversion and unequal crossing-over can correct the sequence of one gene against that of another (Egel, 1981). As a result, the degree of sequence homology between the two genes changes significantly around the breakpoint where the recombination took place, as illustrated by several genes in the human and murine IFN-α gene family. For example, Hu IFN-α14 is related to Hu IFN-α5 with its 5′ region and to Hu IFN-α7 with its 3′ end. The replacement of a stretch of nucleotides, large or small, with the sequence from another gene corrects the sequence divergence that would normally result from mutations or deletions, with the consequence that such genes are more "homogenized" than would be predicted on the basis of an independent accumulation of mutations in each gene (Gillespie and Carter, 1983). Such concerted evolution of IFN-α genes is most probably responsible for the high degree of similarity between the IFN-α genes within the same species—an extreme example of which is provided by Hu IFN-α1 and α13—two nonallelic genes with identical coding, but not with identical flanking sequences (Todokoro et al., 1984).

In spite of their common, and probably interspersed, chromosomal location, concerted evolution apparently has not occurred between the IFN-αI and IFN-αII genes, since the bovine and the human αII family are more closely related to one another than are either the human αI and αII or the bovine αI and αII family. It has been estimated that class I and class II IFN-α genes diverged about 120 million years ago, which implies that most mammals should have the two distinct IFN-α gene families (Capon et al., 1985).

C. Multiple IFN-β Genes in Some Animal Species

Only one IFN-β gene has been identified unambiguously in humans and mice, and judging from the Southern blot hybridization analysis of genomic DNA from several primates, and from the rabbit, cat, and lion, there is only one IFN-β gene in these species as well. However, in other mammalian species, such as bovines, horses, and pigs, at least five related IFN-β genes are present. Manifestly, the phenomenon of gene duplication did not exclusively give rise to the IFN-α gene family, but it did also create in some animal species an IFN-β gene family (Wilson et al., 1983; Leung et al., 1984).

D. The IFN-γ Gene

Because of the pronounced differences in gene structure and nucleotide sequence, the evolutionary relationship of the IFN-γ gene to the other IFN genes is not clear. Yet, the many similarities in the biological activities of the different gene products would seem to suggest some relationship. It is, therefore, interesting that in both humans and mice sequence homologies have been found between the third exon of the IFN-γ gene, corresponding to the region 41–104 of the mature polypeptide, and the IFN-α and IFN-β genes. When superposed on a tentative three-dimensional model for α/β IFNs, a correlation with a domain encompassing two out of four α-helical structures is found. These two α-helical domains correspond to regions 57–72 and 143–154, respectively (DeGrado et al., 1982; Sternberg and Cohen, 1982; Tavernier and Fiers, 1984).

There is also evidence, independent of the foregoing, that murine IFNs α, β, and γ share at least one epitope. This is based on neutralization studies of the three murine IFN species with a rat monoclonal antibody. The monoclonal antibody binds to and also neutralizes both natural Mu IFN α and Mu IFN-β, which although somewhat unusual, can be explained by the obvious homology between these IFNs. However, the antibody also neutralizes natural Mu IFN-γ. Since all three native Mu IFN species are glycosylated, one cannot draw any definite conclusions as to the structural implications of this finding. Carbohydrate structures are widely shared between different glycoproteins, and an antibody against the carbohydrate moiety of a particular glycoprotein is likely to cross-react with other glycoproteins, even if the polypeptide moieties are unrelated. But the monoclonal antibody also neutralizes unglycosylated, recombinant Mu IFN-α, β, and γ, made by *E. Coli* and therefore devoid of the carbohydrate moieties that are present on natural IFNs. This shows that the epitope against which the antibody is directed is part of the protein domain and not part of the sugar moiety, and that there is at least one common or highly similar domain in the tertiary structure of the three IFN species (De Maeyer-Guignard and De Maeyer, 1985).

Whether or not the IFN-γ and the IFN-α/β genes share a common precursor cannot be decided from the available evidence. Analysis of the IFN genes of more primitive vertebrates may be able to provide some insight into the possible evolutionary relationship of the IFN-γ gene and the other IFN genes.

IV. ARE THERE DIFFERENCES IN THE BIOLOGICAL ACTIVITIES OF THE VARIOUS IFN-α SUBTYPES?

The biological activities of the various IFN-α subtypes seem to be relatively similar, as are the biological activities of IFN-α and β. Does current knowledge provide a rationale for the existence of so many IFN subtypes? This question should be viewed first in the light of our knowledge about gene families in general.

Families of protein-encoding genes are common to vertebrates and invertebrates; examples are the genes for histones, actin and tubulin. Multiple gene copies are sometimes indicative of a great need for transcripts; the genes coding for tRNA, of which there are over a 1,000 copies in the human genome, are one example of this. In the case of protein-encoding genes, the multiple copies of histone genes in the genomes of most eukaryotes are explained by the large amounts of histone proteins that are needed as nuclear structural proteins. Yet, this simple explanation would not seem to apply to the IFN genes, since in a normal, uninfected cell, transcription of IFN genes is limited, and even under optimal conditions of induction in the laboratory and in virus-infected cells, the number of IFN molecules produced per cell is still quite low (see Chapter 3). Furthermore, many genes, encoding proteins that are vital for the functioning of higher vertebrates, do not occur as families of near-duplicate genes but are present as only one or maybe two copies per haploid genome.

Another well-documented example of a gene family is provided by the β-globin-like gene cluster on chromosome 11 in humans. In this case the arrangement of the genes in cluster reflects the order in which they are expressed during the course of the individual's development, from gestation to adult life. The spatial organization of the IFN gene cluster in humans or mice is not known, nor do we have much information concerning the selective induction and expression of the different members of the IFN-α gene cluster.

Although there are some examples of the selective production of IFN-β or of some IFN-α subtypes, there is by and large insufficient information concerning selective induction and transcription of the different IFN-α and IFN-β genes. It is, therefore, not possible to engage in a meaningful discussion about the possibility of different functions for some of the IFN subtypes, and we can only consider the evidence that the biological properties of different IFN-α subtypes can vary. Here again, information is limited, mainly because until very recently, in the most widely used experimental system, which is the mouse, practically all experiments were carried out with Mu IFN-α/β, rather than with the individual molecular IFN species.

One of the best documented differences in biological activity between two members of the Hu IFN-α family is the relatively low antiviral activity of Hu IFN-α1 (or IFN-αD) on human cells, as compared to Hu IFN-α2 (or IFN-αA) (Streuli et al., 1981). It has been estimated that, compared to Hu IFN-α2, 100 times more molecules of Hu IFN-α1 are needed per cell to induce an antiviral effect of comparable magnitude, and about a 1,000 times more molecules to obtain a comparable stimulation of NK cell activity (Rehberg et al., 1982). This difference in specific activity is accompanied by an approximately tenfold higher binding efficiency of Hu IFN-α2 to the high affinity IFN receptors on human cells (Yonehara et al., 1983).

Assuming that the mRNA frequency in induced cells is proportional to the amount of the translation products, Hu IFN-α1 and α2 appear to be the most frequent subtypes in Sendai-virus-induced human buffy coat cells (Nagata et al., 1980; Goeddel et al., 1981). On the basis of mRNA frequency, Hu IFN-α1 is the most predominant subtype in leukocyte IFN preparations, and this is supported by the fact that the spectrum of activity of this IFN on a range of mammalian cells corresponds most closely to that of Hu leukocyte IFN (Weck et al., 1981). Thus, of the Hu IFN-α genes that are predominantly transcribed during a Sendai virus infection of human leukocytes, one codes for a relatively inefficient IFN-α subtype when the antiviral activity is taken as the basis for comparison, but there may be some other advantage in producing this IFN. This possibility is at least suggested by the capacity of Hu IFN-α1 to induce human monocytes to express MHC class II antigens and to produce IL-1, and the inability of Hu IFN-α2 to do the same at comparable concentrations (Rhodes et al., 1986). As discussed in Chapters 9 and 13, an IFN-α subtype that stimulates MHC class II expression and IL-1 production in monocytes offers a definite advantage for coping with infection, which would more than make up for the relative disadvantage of a lower specific antiviral activity. In this respect it is also relevant that Hu IFN-α1, although it has much less specific antiviral activity than IFN-α2 in monkey cells in vitro, is significantly more efficient in protecting squirrel monkeys against an EMC virus infection in vivo. This demonstrates that properties other than direct antiviral activity in vitro are also important for host defense against virus infections (Stebbing et al., 1983).

The comparison of the biological activities of different IFN-α subtypes shows that they can display different functional efficacies, for example, also for the antiproliferative effect (Fig. 2.4) (Fish et al., 1983).

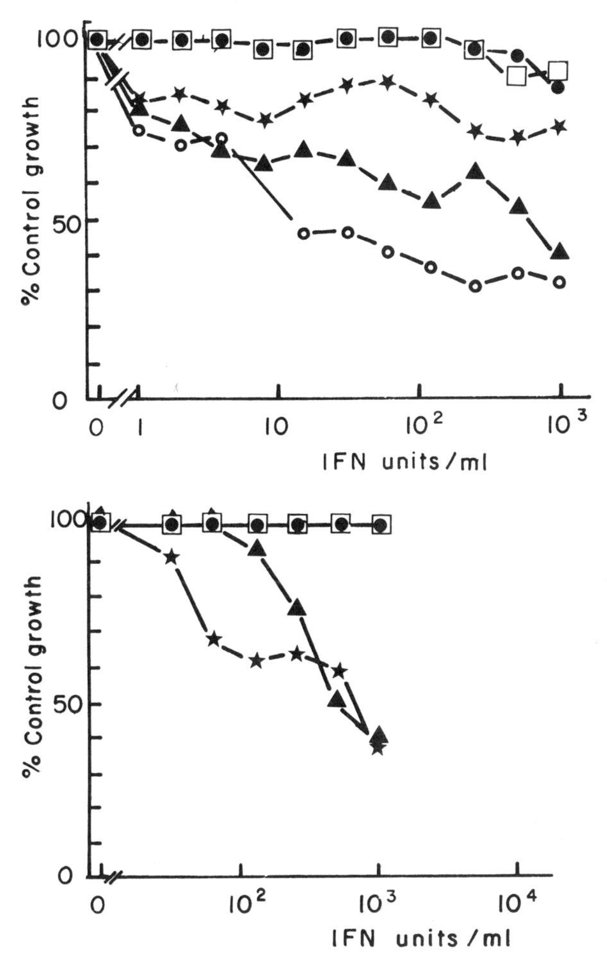

Figure 2.4 There can be differences in the biological activities of various IFN-α subtypes, as demonstrated, for example, by different antiproliferative efficacies. Both on HL60 and on T98G cells, IFNαA and αD are without effect at the concentrations tested, whereas IFN-αC, B, and F express different degrees of activity. (*a*) Effect on myeloid leukemic HL60 cells in suspension cultures. (*b*) Effect on T98G cells in monolayer cultures. IFN-αA: □; IFN-αB: ○; IFN-αC: ▲; IFN-αD: ●; IFN-αF: ★. (Adapted from Fish et al., 1983.)

Another example of a qualitative difference is provided by Hu IFN-αJ, which, contrary to other IFN-α subtypes, lacks the ability to boost human natural killer cell activity (NK cells, see Chapter 11), and can in fact interfere with the boosting of these cells by other IFN subtypes, probably by acting as a competitive antagonist through binding to the NK cell receptor (Ortaldo et al., 1984). It will be necessary to explore in much greater detail the possible differences in the functions of IFN-β and of the various IFN-α subtypes before the reason for so many IFN subtypes become clear. It may also turn out that there is no other reason than nature's own way of making multiple gene copies to play around with different possibilities until something new and useful comes up. After all, we are relatively well off with only one IFN-β gene, whereas bovines have at least five.

REFERENCES

Adolf, G. R. Antigenic structure of human interferon ω1 (Interferon alpha II): Comparison with other human interferons. *J. Gen. Virol.* 68: 1669–1676 (1987).

Baumann, H. and Berger, F. G. Genetics and evolution of the acute phase proteins in mice. *Mol. Gen. Genet.* 201: 505–512 (1985).

Baumann, H., Held, W. A., and Berger, F. G. The acute phase response of mouse liver. Genetic analysis of the major acute phase reactants. *J. Biol. Chem.* 259: 566–573 (1984).

Benjamin, W. R., Steeg, P. S., and Farrar, J. J. Production of murine immune interferon by a T lymphocyte cell line: Regulation by an EL4 thymoma-derived factor(s). In *Interleukins, Lymphokines and Cytokines,* Oppenheim, J. J., Cohen, S. and Landy, M. (eds.) Academic Press, New York, pp. 609–615 (1983).

Bishop, C. E., Boursot, P., Baron, B., Bonhomme, F., and Hatat, D. Most classical *Mus musculus domesticus* laboratory mouse strains carry a *Mus musculus musculus* Y chromosome. *Nature* 315: 70–72 (1985).

Brack, C., Nagata, S., Mantei, N., and Weissmann, C. Molecular analysis of the human interferon-α gene family. Gene 15: 379–394 (1981).

Cahilly, L. A., George, D., Daugherty, B. L., and Pestka, S. Subchromosomal localization of mouse IFN-α genes by in situ hybridization. *J. Interferon Res.* 5: 391–395 (1985).

Camble, R., Petter, N. N., Trueman, P., Newton, C. R., Carr, F. J., Hockney, R. C., Moore, V. E., Greene, A. R., Holland, D., and Edge, M. D. Functionally important conserved amino-acids in interferon-α2 identified with analogues produced from synthetic genes. *Biochem. Biophys. Res. Comm.* 134: 1404–1411 (1986).

Capon, D. J., Shepard, H. M., and Goeddel, D. V. Two distinct families of human and bovine interferon-α genes are coordinately expressed and encode functional polypeptides. *Mol. Cell Biol.* 5: 768–779 (1985).

Caput, D., Beutler, B., Hartog, K., Thayer, R., Brown-Shimer, S., and Cerami, A. Identification of a common nucleotide sequence in the 3′-untranslated region of mRNA molecules specifying inflammatory mediators. *Proc. Natl. Acad. Sci.* (USA) 83: 1670–1674 (1986).

Cheng, Z. Y., Lovett, M., Epstein, L. B., and Epstein, C. J. The mouse IFN-α (Ifa) locus: Correlation of physical and linkage maps by in situ hybridization. *Cytogenet. Cell. Genet.* 41: 101–106 (1986).

Dandoy, F., De Maeyer, E., Bonhomme, F., Guenet, J. L., and De Maeyer-Guignard, J. Segregation of restriction fragment length polymorphism in an interspecies cross of laboratory and wild mice indicates tight linkage of the murine IFN-β gene to the murine IFN-α genes. *J. Virol.* 56: 216–220 (1985).

Dandoy, F., Kelley, K. A., De Maeyer-Guignard, J., De Maeyer, E., and Pitha, P.

M. Linkage analysis of the murine interferon-α locus on chromosome 4. *J. Exp. Med.* 160: 294–302 (1984).

Daugherty, B., Martin-Zanca, D., Kelder, B., Collier, K., Seamans, T. C., Hotta, K., and Pestka, S. Isolation and bacterial expression of a murine α leukocyte interferon gene. *J. Interferon Res.* 4: 635–643 (1984).

De Grado, W. F., Wasserman, Z. R., and Chowdhry, V. Sequence and structural homologies among type I and type II interferons. *Nature* 300: 379–381 (1982).

Degrave, W., Derynck, R., Tavernier, J., Haegeman, G., and Fiers, W. Nucleotide sequence of the chromosomal gene for human fibroblast (β1) interferon and of the flanking regions. *Gene* 14:137–143 (1981).

De Kinkelin, P. and Dorson, M. Interferon production in rainbow trout (*salmo gairdneri*) experimentally infected with Egtved virus. *J. Gen. Virol.* 19: 125–127 (1973).

De Maeyer, E. and Dandoy, F. Linkage analysis of the murine interferon-α locus (*Ifa*) on chromosome 4. *J. Heredity* 78: 143–146 (1987).

De Maeyer, E., Skup, D., Prasad, K. S. N., De Maeyer-Guignard, J., Williams, B., Meacock, P., Sharpe, G., Pioli, D., Hennam, J., Schuch, W., and Atherton, K. Expression of a chemically synthesized human α1 interferon gene. *Proc. Natl. Acad. Sci.* (USA) 79: 4256–4259 (1982).

De Maeyer-Guignard, J. and De Maeyer, E. Immunomodulation by interferons: Recent developments. In: *Interferon 6,* I, Gresser, Ed., Academic Press, London, pp. 69–91 (1985).

Derynck, R., Content, J., De Clercq, E., Volckaert, G., Tavernier, J., Devos, R., and Fiers, W. Isolation and structure of a human fibroblast interferon gene. *Nature* 285: 542–547 (1980).

Derynck, R., Devos, R., Remaut, E., Saman, E., Stanssens, P., Tavernier, J., Volckaert, G., Content, J., De Clercq, E., and Fiers, W. Isolation and characterization of a human fibroblast interferon gene and its expression in *Escherichia coli. Rev. Infect. Dis.* 3: 1186–1195 (1981).

Devos, R., Cheroutre, H., Taya, Y., Degrave, W., Van Heuverswyn, H., and Fiers, W. Molecular cloning of human immune interferon cDNA and its expression in eukaryotic cells. *Nucl. Acids Res.* 10: 2487–2501 (1982).

Diaz, M. O., Le Beau, M. M., Pitha, P., and Rowley, J. D. Interferon and c-ets-1 genes in the translocation (9;11) (p22; q23) in human acute monocytic leukemia. *Science* 231: 265–267 (1986).

Dijkmans, R., Volckaert, G., Van Damme, J., De Ley, M., Billiau, A., and De Somer, P. Molecular cloning of murine interferon γ (Mu IFN-γ) cDNA and its expression in heterologous mammalian cells. *J. Interferon Res.* 5: 511–520 (1985).

Edge, M. D., Camble, R., Moore, V. E., Hockney, R. C., Carr, F. J., and Fitton, J. E. Interferon analogues from synthetic genes: An approach to protein structure-activity studies. In: *Interferon 7,* I. Gresser, Ed., Academic Press, London, pp. 2–46 (1986).

Edge, M. D., Greene, A. R., Heathcliffe, G. R., Meacock, P. A., Schuch, W., Scanlon, D. B., Atkinson, T. C., Newton, C. R., and Markham, A. F. Total synthesis of a human leukocyte interferon gene. *Nature* 292: 756–762 (1981).

Edge, M. D., Greene, A. R., Heathcliffe, G. R., Moore, V. E., Faulkner, N. J.,

Camble, R., Petter, N. N., Trueman, P., Schuch, W., Hennam, J., Atkinson, T. C., Newton, C. R., and Markham, A. F. Chemical synthesis of a human interferon-α2 gene and its expression in *Escherichia coli*. *Nucl. Acids Res.* 11: 6419–6435 (1983).

Egel, R. Intergenic conversion and reiterated genes. *Nature* 290: 191–192 (1981).

Epstein, L. B. Interferon-γ: Success, structure and speculation. *Nature* 295: 453–454 (1982).

Erickson, B. W., May, L. T., and Sehgal, P. B. Internal duplication in human α1 and β1 interferons. *Proc. Natl. Acad. Sci.* (USA) 81: 7171–7175 (1984).

Falcoff, E. and Fauconnier, B. In vitro production of an interferon-like inhibitor of viral multiplication by a poikilothermic animal cell, the tortoise (*Tetsudo greca*). *Proc. Soc. Exp. Biol. Med.* 118: 609 (1965).

Feinstein, S. I., Mory, Y., Chernajovsky, Y., Maroteaux, L., Nir, U., Lavie, V., and Revel, M. Family of human α-interferon-like sequences. *Mol. Cell. Biol.* 5: 510–517 (1985).

Ferris, S. D., Sage, R. D., Prager, E. M., Ritte, U., and Wilson, A. C. Mitochondrial DNA evolution in mice. *Genetics* 105: 681–721 (1983).

Ferris, S. D., Sage, R. D., and Wilson, A. C. Evidence from mtDNA sequences that common laboratory strains of inbred mice are descended from a single female. *Nature* 295: 163–165 (1982).

Fish, E. N., Banerjee, K., and Stebbing, N. Human leukocyte interferon subtypes have different antiproliferative and antiviral activities on human cells. *Biochem. Biophys. Res. Comm.* 112: 537–545 (1983).

Fisher, P. B., Miranda, A. F., Babiss, L. E., Pestka, S., and Weinstein, I. B. Opposing effects of interferon produced in bacteria and of tumor promoters on myogenesis in human myoblast cultures. *Proc. Natl. Acad. Sci.* (USA) 80: 2961–2965 (1983).

Fujita, T., Shibuya, H., Ohashi, T., Yamanishi, K., and Taniguchi, T. Regulation of human interleukin-2 gene: Functional DNA sequences in the 5′ flanking region for the gene expression in activated T lymphocytes. *Cell* 46: 401–407 (1986).

Fujita, T., Takaoka, C., Matsui, H., and Taniguchi, T. Structure of the human interleukin 2 gene. *Proc. Natl. Acad. Sci.* (USA) 80: 7437–7441 (1983).

Fuke, M., Hendrix, L. C., and Bollon, A. P. Pseudogene IFN-αL: Removal of the step codon in the signal sequence permits expression of active human interferon. *Gene* 32: 135–140 (1984).

Gillespie, D. and Carter, W. Concerted evolution of human interferon α genes. *J. Interferon Res.* 3: 83–88 (1983).

Goeddel, D. V., Kleid, D. G., Bolivar, F., Heyneker, H. L., Yansura, D. G., Crea, R., Hirose, T., Kraszewski, A., Itakura, K., and Riggs, A. D. Expression in *Escherichia coli* of chemically synthesized genes for human insulin. *Proc. Natl. Acad. Sci.* (USA) 76: 106–110 (1979).

Goeddel, D. V., Leung, D. W., Dull, T. J., Gross, M., Lawn, R. M., McCandliss, R., Seeburg, P. H., Ullrich, A., Yelverton, E., and Gray, P. W. The structure of eight distinct cloned human leukocyte interferon cDNAs. *Nature* 290: 20–26 (1981).

Goeddel, D. V., Yelverton, E., Ullrich, A., Heyneker, H. L., Miozzari, G., Holmes, W., Seeburg, P. H., Dull, T., May, L., Stebbing, N., Crea, R., Maeda, S.,

McCandliss, R., Sloma, A., Tabor, J. M., Gross, M., Familletti, P. C., and Pestka, S. Human leukocyte interferon produced by *E. coli* is biologically active. *Nature* 287: 411–416 (1980).

Gray, P. W. and Goeddel, D. V. Structure of the human immune interferon gene. *Nature* 298: 859–863 (1982).

Gray, P. W. and Goeddel, D. V. Cloning and expression of murine immune interferon cDNA. *Proc. Natl. Acad. Sci.* (USA) 80: 5842–5846 (1983).

Gray, P. W., Leung, D. W., Pennica, D., Yelverton, E., Najarian, R., Simonsen, C. C., Derynck, R., Sherwood, P. J., Wallace, D. M., Berger, S. L., Levinson, A. D., and Goeddel, D. V. Expression of human immune interferon cDNA in *E. coli* and monkey cells. *Nature* 295: 503–508 (1982).

Gren, E., Berzin, V., Jansone, I., Tsimanis, A., Vishnevsky, Y., and Apsalons, U. Novel human leukocyte interferon subtype and structural comparison of α interferon genes. *J. Interferon Res.* 4: 609–617 (1984).

Gross, G., Mayr, U., Bruns, W., Grosveld, F., Dahl, H. H. M., and Collins, J. The structure of a thirty-six kilobase region of the human chromosome including the fibroblast interferon gene IFN-β. *Nucl. Acids Res.* 9: 2495–2507 (1981).

Hardy, K. J., Peterlin, B. M., Atchison, R. E., and Stobo, J. D. Regulation of expression of the human interferon γ gene. *Proc. Natl. Acad. Sci* (USA) 82: 8173–8177 (1985).

Hauptmann, R. and Swetly, P. A novel class of human type I interferons. *Nucl. Acids Res.* 13: 4739–4749 (1985).

Hayes, C. E., Klyczek, K. K., Krum, D. P., Whitcomb, R. M., Hullett, D. A., and Cantor, H. Chromosome 4Jt gene controls murine T cell surface I-J expression. *Science* 223: 559–563 (1984).

Henco, K., Brosius, J., Fujisawa, A., Fujisawa, J. I., Haynes, J. R., Hochstadt, J., Kovacic, T., Pasek, M., Schambock, A., Schmid, J., Todokoro, K., Walchli, M., Nagata, S., and Weissmann, C. Structural relationship of human interferon-α genes and pseudogenes. *J. Mol. Biol.* 185: 227–260 (1985).

Higashi, Y., Sokawa, Y., Watanabe, Y., Kawade, Y., Ohno, S., Takaoka, C. and Taniguchi, T. Structure and expression of a cloned cDNA for mouse interferon-β. *J. Biol. Chem.* 258: 9522–9529 (1983).

Houghton, M., Eaton, M. A. W., Stewart, A. G., Smith, J. C., Doel, S. M., Catlin, G. H., Lewis, H. M., Patel, T. P., Emstage, J. S., Carey, N. H., and Porter, A. G. The complete amino acid sequence of human fibroblast interferon as deduced using synthetic oligodeoxyribonucleotide primers of reverse transcriptase. *Nucl. Acids Res.* 8: 2885–2894 (1980).

Houghton, M., Jackson, I. J., Porter, A. G., Doel, S. M., Catlin, G. H., Barber, C., and Carey, N. H. The absence of introns within a human fibroblast interferon gene. *Nucl. Acids Res.* 9: 247–266 (1981).

Isaacs, A. and Lindenmann, J. Virus interference. II. Some properties of interferon. *Proc. Roy. Soc. B* (London) 147: 268–273 (1957).

Itakura, K., Hirose, T., Crea, R., Riggs, A. D., Heyneker, H. L., Bolivar, F., and Boyer, H. W. Expression in *Escherichia coli* of a chemically synthesized gene for the hormone somatostatin. *Science* 198: 1056–1063 (1977).

Kelley, K. A., Kozak, C. A., Dandoy, F., Sor, F., Skup, D., Windass, J. D., De

Maeyer-Guignard, J., Pitha, P. M., and De Maeyer, E. Mapping of murine interferon-α genes to chromosome 4. Gene 26: 181–188 (1983).

Kelley, K. A., Kozak, C. A., and Pitha, P. M. Localization of the mouse interferon-β1 gene to chromosome 4. *J. Interferon Res.* 5: 409–413 (1985).

Kelley, K. A. and Pitha, P. M. Characterization of a mouse interferon gene locus I. Isolation of a cluster of four α interferon genes. *Nucl. Acids Res.* 13: 805–823 (1985a).

Kelley, K. A. and Pitha, P. M. Characterization of a mouse interferon gene locus II. Differential expression of α-interferon genes. *Nucl. Acids. Res.* 13: 825–839 (1985b).

Laird, C. D., McConaughy, B. L., and McCarthy, B. J. Rate of fixation of nucleotide substitutions in evolution. *Nature* 224: 149–154 (1969).

Lawn, R. M., Adelman, J., Dull, T. J., Gross, M., Goeddel, D., and Ullrich, A. DNA sequence of two closely linked human leukocyte interferon genes. *Science* 212: 1159–1162 (1981a).

Lawn, R. M., Gross, M., Houck, C. M., Franke, A. E., Gray, P. V., and Goeddel, D. V. DNA sequence of a major human leukocyte interferon gene. *Proc. Natl. Acad. Sci.* (USA) 78: 5435–5439 (1981b).

Le Roscouet, D., Vodjdani, G., Lemaigre-DuBreuil, Y., Tovey, M. G., Latta, M., and Doly, J. Structure of a murine α interferon pseudogene with a repetitive R-type sequence in the 3′ flanking region. *Mol. Cell. Biol.* 5: 1343–1348 (1985).

Leung, D. W., Capon, D. J., and Goeddel, D. V. The structure and bacterial expression of three distinct bovine interferon-β genes. *Biotechnology* 2: 458–464 (1984).

Lindenmann, J., Burke, D. C., and Isaacs, A. Studies on the production, mode of action and properties of interferon. *Br. J. Exp. Pathol.* 38: 551–562 (1957).

Lovett, M., Cox, D. R., Yee, D., Boll, W., Weissmann, C., Epstein, C. J., and Epstein, L. B. The chromosomal location of mouse interferon α genes. *EMBO J.* 3: 1643–1646 (1984).

Lund, B., Edlund, T., Lindermaier, W., Ny, T., Collins, J., Lundgren, E., and Von Gabain, A. Novel cluster of α-interferon gene sequences in a placental cosmid DNA library. *Proc. Natl. Acad. Sci. (USA) 81: 2435–2439 (1984).*

Mantei, N., Schwarzstein, M., Streuli, M., Panem, S., Nagata, S., and Weissmann, C. The nucleotide sequence of a cloned human leukocyte interferon cDNA. *Gene* 10: 1–10 (1980).

Mantei, N. and Weissmann, C. Controlled transcription of a human α-interferon gene introduced into mouse L cells. *Nature* 297: 128–132 (1982).

Marcucci, F. and De Maeyer, E. An interferon analogue, [$Ala^{30,32,33}$] HuIFN-α2, acting as a HuIFN-α2 antagonist on bovine cells. *Biochem. Biophys. Res. Comm.* 134: 1412–1418 (1986).

Mark, D. F., Lu, S. D., Creasey, A. A., Yamamoto, R., and Lin, L. S. Site-specific mutagenesis of the human fibroblast interferon gene. *Proc. Natl. Acad. Sci.* (USA) 81: 5662–5666 (1984).

May, L. T. and Sehgal, P. B. On the relationship between human interferon α1 and β1 genes. *J. Interferon Res.* 5: 521–526 (1985).

Miyata, T. and Hayashida, H. Recent divergence from a common ancestor of human IFN-α genes. *Nature* 295: 165–168 (1982).

Miyata, T., Hayashida, H., Kukuno, R., Toh, H., and Kawade, Y. Evolution of interferon genes. In: *Interferon 6,* I. Gresser, Ed., Academic Press, London, pp. 1–30 (1985).

Mory, Y., Chernajovsky, Y., Feinstein, S. I., Chen, L., Nir, U., Weissenbach, J., Malpiece, Y., Tiollais, P., Marks, D., Ladner, M., Colby, C., and Revel, M. Synthesis of human interferon β1 in *Escherichia coli* infected by a λ phage recombinant containing a human genomic fragment. *Eur. J. Biochem.* 120: 197–202 (1981).

Nadeau, J. H., Berger, F. G., Kelley, K. A., Pitha, P. M., Sidman, C. L., and Worrall, N. Rearrangement of genes located on homologous chromosomal segments in mouse and man: The location of α- and β-interferon, α-1 acid glycoprotein-1 and -2, and aminolevulinate dehydratase on mouse chromosome 4. *Genetics* 104: 1239–1255 (1986).

Nagata, K., Kikuchi, N., Ohara, O., Teraoka, H., Yoshida, N., and Kawade, Y. Purification and characterization of recombinant murine immune interferon. *FEBS Lett.* 205: 200–204 (1986).

Nagata, S., Mantei, N., and Weissmann, C. The structure of one of the eight or more distinct chromosomal genes for human interferon-α. *Nature* 287: 401–408 (1980).

Nakanishi, Y., Paco-Larson, M. L., and Garen, A. DNA homology between the 3′-untranslated regions of a developmentally regulated *Drosophila* gene and a mouse α-interferon gene. *Gene* 46: 79–88 (1986).

Naylor, S. L., Gray, P. W., and Lalley, P. A. Mouse immune interferon (IFN-γ) gene is on chromosome 10. *Somatic Cell Mol. Genet.* 10: 531–534 (1984).

Naylor, S. L., Sakaguchi, A. Y., Shows, T. B., Law, M. L., Goeddel, D. V., and Gray, P. W. Human immune interferon gene is located on chromosome 12. *J. Exp. Med.* 57: 1020–1027 (1983).

Nishi, T., Fujita, T., Nishi-Takaoka, C., Saito, A., Matsumoto, T., Sato, M., Oka, T., Itoh, S., Yip, Y. K., Vilcek, J., and Taniguchi, T. Cloning and expression of a novel variant of human interferon-γ. *J. Biochem.* 97: 153–159 (1985).

Ohlsson, M., Feder, J., Cavalli-Sforza, L. L., and von Gabain, A. Close linkage of α and β interferons and infrequent duplication of β interferon in humans. *Proc. Natl. Acad. Sci.* (USA) 82: 4473–4476 (1985).

Ohno, S. and Taniguchi, T. Structure of a chromosomal gene for human interferon β. *Proc. Natl. Acad. Sci.* (USA) 78: 5305–5309 (1981).

Oie, H. K. and Loh, P. C. Reovirus type 2: Induction of viral resistance and interferon production in fathead minnow cells. *Proc. Soc. Exp. Biol. Med.* 136: 369–373 (1971).

Ortaldo, J. R., Herberman, R. B., Harvey, C., Osheroff, P., Pan, Y. C. E., Kelder, B., and Pestka, S. A species of human α interferon that lacks the ability to boost human natural killer activity. *Proc. Natl. Acad. Sci.* (USA) 81: 4926–4929 (1984).

Owerbach, D., Rutter, W. J., Shows, T. B., Gray, P., Goeddel, D. V., and Lawn, R. M. Leucocyte and fibroblast interferon genes are located on human chromosome 9. *Proc. Natl. Acad. Sci.* (USA) 78: 3123–3127 (1981).

Pestka, S. The human interferons: From protein purification and sequence to cloning and expression in bacteria: Before, between and beyond. *Arch. Biochem. Biophys.* 221: 1–37 (1983).

Pestka, S. Interferon from 1981 to 1986. In: *Methods in Enzymology,* Vol. 119, Part C, pp. 3–14, S. Pestka, Ed., Academic Press, Orlando (1986).

Pestka, S., Kelder, B., Rehberg, E., Ortaldo, J. R., Herberman, R. B., Kempner, E. S., Moschera, J. A., and Tarnowski, S. J. The specific molecular activities of interferons differ for antiviral, antiproliferative, and natural killer cell activities. In: *The Biology of the Interferon System 1983,* E. De Maeyer and H. Schellekens, Eds., Elsevier Science Publisher, Amsterdam, pp. 535–549 (1983).

Ragg, H. and Weissmann, C. Not more than 117 base pairs of 5′-flanking sequence are required for inducible expression of a human IFN-α gene. *Nature* 303: 439–442 (1983).

Rehberg, E., Kelder, B., Hoal, E. G., and Pestka, S. Specific molecular activities of recombinant and hybrid leukocyte interferons. *J. Biol. Chem.* 257: 11497–11502 (1982).

Rhodes, J., Ivanyi, J., and Cozens, P. Antigen presentation by human monocytes: Effects of modifying major histocompatibility complex class II antigen expression and interleukin 1 production by using recombinant interferons and corticosteroids. *Eur. J. Immunol.* 16: 370–375 (1986).

Rinderknecht, E., O'Connor, B. H., and Rodriguez, H. Natural human interferon-γ. Complete amino acid sequence and determination of sites of glycosylation. *J. Biol. Chem.* 259: 6790–6797 (1984).

Ryals, J., Dierks, P., Ragg, H., and Weissmann, C. A 46-nucleotide promoter segment from an IFN-α gene renders an unrelated promoter inducible by virus. *Cell* 41: 497–507 (1985).

Sachs, D. H., Lynch, D. H., and Epstein, S. L. The I-J dilemma: New developments. *Immunol. Today* 5: 94–95 (1984).

Sagar, A. D., Sehgal, P. B., May, L. T., Inouye, M., Slate, D. L., Shulman, L. and Ruddle, F. H. Interferon-β related DNA is dispersed in the human genome. *Science* 223: 1312–1315 (1984).

Sagar, A. D., Sehgal, P. B., May, L. T., Slate, D. L., Shulman, L., Baker, P. E., and Ruddle, D. H. Interferon-β-related DNA on human chromosome 4. *Somatic Cell Mol. Genet.* 11: 403–408 (1985).

Sagar, A. D., Sehgal, P. B., Slate, D. L., and Ruddle, F. H. Multiple human β interferon genes. *J. Exp. Med.* 156: 744–755 (1982).

Secher, D. S. and Burke, D. C. A monoclonal antibody for large-scale purification of human leukocyte interferon. *Nature* 285: 446–450 (1980).

Seif, I. and De Maeyer-Guignard, J. Structure and expression of a new murine interferon-α gene: MuIFN-αI9. *Gene* 43: 111–121 (1986).

Shaw, G. D., Boll, W., Taira, H., Mantei, N., Lengyel, P., and Weissmann, C. Structure and expression of cloned murine IFN-α genes. *Nucl. Acids. Res.* 11: 555–573 (1983).

Shaw, G. D. and Kamen, R. A conserved AU sequence from the 3′ untranslated region of GM-CSF mRNA mediates selective mRNA-degradation. *Cell* 46: 659–667 (1986).

Shepard, H. M., Leung, D., Stebbing, N., and Goedel, D. V. A single amino acid change in IFN-β1 abolishes its antiviral activity. *Nature* 294: 563–565 (1981).

Shows, T. B., Sakaguchi, A. Y., Naylor, S. L., Goeddel, D. G., and Lawn, R. M.

Clustering of leucocyte and fibroblast interferon genes on human chromosome 9. *Science* 218: 373–374 (1982).

Sidman, C. L., Marshall, J. D., Beamer, W. G., Nadeau, J. H., and Unanue, E. R. Two loci affecting B cell responses to B cell maturation factors. *J. Exp. Med.* 163: 116–128 (1986).

Slate, D. L., D'Eustachio, P., Pravtcheva, D., Cunningham, A. C., Nagata, S., Weissmann, C., and Ruddle, F. H. Chromosomal location of a human α interferon gene family. *J. Exp. Med.* 155: 1019–1024 (1982).

Slate, D. L. and Ruddle, F. H. Somatic cell genetic analysis of interferon production and response. *Ann. N.Y. Acad. Sci.* 350: 174–178 (1980).

Stebbing, N., Weck, P. K., Fenno, J. T., Estell, D. A., and Rinderknecht, E. Antiviral effects of bacteria-derived human leukocyte interferons against encephalomyocarditis virus infection of squirrel monkeys. *Arch. Virol.* 76: 365–372 (1983).

Sternberg, M. J. E. and Cohen, F. E. Prediction of the secondary and tertiary structures of interferon from four homologous amino acid sequences. *Int. J. Biol. Macromol.* 4: 137–144 (1982).

Streuli, M., Hall, A., Boll, W., Stewart II, W. E., Nagata, S., and Weissmann, C. Target cell specificity of two species of human interferon-α produced in *Escherichia coli* and of hybrid molecules derived from them. *Proc. Natl. Acad. Sci.* (USA) 78: 2848–2852 (1981).

Streuli, M., Nagata, S., and Weissmann, C. At least three human type α interferons: Structure of $\alpha 2$. *Science* 209: 1343–1346 (1980).

Tanaka, S., Oshima, T., Ohsuye, K., Ono, T., Mizono, A., Ueno, A., Nakazato, H., Tsujimoto, M., Higashi, N., and Noguchi, T. Expression in *Escherichia coli* of chemically synthesized gene for the human immune interferon. *Nucl. Acids Res.* 11: 1707–1722 (1983).

Taniguchi, T., Fujii-Kuriyama, Y., and Muramatsu, M. Molecular cloning of human interferon cDNA. *Proc. Natl. Acad. Sci.* (USA) 77: 4003–4006 (1980a).

Taniguchi, T., Mantei, N., Schwarzstein, M., Nagata, S., Muramatsu, M., and Weissmann, C. Human leukocyte and fibroblast interferons are structurally related. *Nature* 285: 547–549 (1980b).

Taniguchi, T., Ohno, S., Fujii-Kuriyama, Y., and Muramatsu, M. The nucleotide sequence of human fibroblast interferon cDNA. *Gene* 10: 11–15 (1980c).

Tavernier, J., Derynck, R., and Fiers, W. Evidence for a unique human fibroblast interferon (IFN-$\beta 1$) chromosomal gene devoid of interfering sequences. *Nucl. Acids Res.* 9: 461–471 (1981).

Tavernier, J. and Fiers, W. The presence of homologous regions between interferon sequences. *Carlsberg Res. Commun.* 49: 359–364 (1984).

Taya, Y., Devos, R., Tavernier, J., Cheroutre, H., Engler, G., and Fiers, W. Cloning and structure of the human immune interferon-γ chromosomal gene. *EMBO J.* 1: 953–959 (1982).

Todokoro, K., Kioussis, D., and Weissmann, C. Two non-allelic human interferon α genes with identical coding regions. *EMBO J.* 3: 1809–1812 (1984).

Torczynski, R. M., Fuke, M., and Bollon, A. P. Human genomic library screened with 17-base oligonucleotide probes yields a novel interferon gene. *Proc. Natl. Acad. Sci.* (USA) 81: 6451–6455 (1984).

Trent, J. M., Olson, S., and Lawn, R. M. Chromosomal localization of human leukocyte, fibroblast, and immune interferon genes by means of in situ hybridization. *Proc. Natl. Acad. Sci.* (USA) 79: 7809–7813 (1982).

Trown, P. W., Heimer, E. P., Felix, A. M., and Bohoslawec, O. Localization of the epitopes for binding of the monoclonal antibodies LI-1 and LI-8 to leukocyte interferons. In: *The Biology of the Interferon System 1984,* H. Kirchner and H. Schellekens, Eds., Elsevier Science Publishers, Amsterdam, pp. 69–76 (1985).

Ullrich, A., Gray, A., Goeddel, D. V., and Dull, T. J. Nucleotide sequence of a portion of human chromosome 9 containing a leukocyte interferon gene cluster. *J. Mol. Biol.* 156: 467–486 (1982).

Van der Korput, J. A. G. M., Hilkens, J., Kroezen, V., Zwarthoff, E. C., and Trapman, J. Mouse interferon α and β genes are linked at the centromere proximal region of chromosome 4. *J. Gen. Virol.* 66: 493–502 (1985).

Weck, P. K., Apperson, S., May, L., and Stebbing, N. Comparison of the antiviral activities of various cloned human interferon-α subtypes in mammalian cell cultures. *J. Gen. Virol.* 57: 233–237 (1981).

Weidle, U. and Weissmann, C. The 5′-flanking region of a human IFN-α gene mediates viral induction of transcription. *Nature* 303: 442–446 (1983).

Weissmann, C. and Weber, H. The interferon genes. *Progr. Nucl. Acid Res.* Vol. 33, pp. 251–302, W. E. Cohn and K. Moldave, Eds., Academic Press, Orlando (1986).

Wetzel, R. Assignment of the disulphide bonds of leukocyte interferon. *Nature* 289: 606–607 (1981).

Wetzel, R., Johnston, P. D., and Czarniecki, C. W. Roles of the disulfide bonds in a human α interferon. In: *The Biology of the Interferon System 1983,* E. De Maeyer and H. Schellekens, Eds., Elsevier Science Publishers, Amsterdam, pp. 101–112 (1983).

Wilson, V., Jeffreys, A. J., Barrie, P. A., Boseley, P. G., Slocombe, P. M., Easton, A., and Burke, D. C. A comparison of vertebrate interferon gene families detected by hybridization with human interferon DNA. *J. Mol. Biol.* 166: 457–475 (1983).

Windass, J. D., Newton, C. R., De Maeyer-Guignard, J., Moore, V. E., Markham, A. F., and Edge, M. D. The construction of a synthetic *Escherichia coli* trp promoter and its use in the expression of a synthetic interferon gene. *Nucl. Acids Res.* 10: 6639–6657 (1982).

Yelverton, E., Leung, D., Weck, P., Gray, P. W., and Goeddel, D. V. Bacterial synthesis of a novel human leucocyte interferon. *Nucl. Acids. Res.* 9: 731–741 (1981).

Yip, Y. K., Barrowclough, B. S., Urban, C., and Vilcek, J. Purification of two species of human γ (immune) interferon. *Proc. Natl. Acad. Sci.* (USA) 79: 1820–1824 (1982).

Yonehara, S., Yonehara-Takahashi, M., Ishii, A., and Nagata, S. Different binding of human interferon $\alpha 1$ and $\alpha 2$ to common receptors on human and bovine cells. *J. Biol. Chem.* 258: 9046–9049 (1983).

Zwarthoff, E. C., Mooren, A. T. A., and Trapman, J. Organization, structure and expression of murine interferon α genes. *Nucl. Acids Res.* 13: 791–804 (1985).

3 INDUCTION OF IFN-α AND IFN-β

To understand IFN induction we have to learn how the expression of IFN genes is controlled. In prokaryotes the sudden switching on of genes by some substances is known as induction, and the mechanisms that control the expression of several inducible genes are now well understood. Genetic regulation of inducible genes in eukaryotes, however, is considerably more complex and poorly understood. It is, therefore, not surprising that for a long time we have known less about the mechanism of IFN induction than about many other aspects of the IFN system. This is now changing through the use of expression vectors containing different IFN genes and their various deletion mutants in the upstream regulatory regions, and we are well en route to a better understanding of how IFN gene expression is regulated. Our knowledge of the functioning of inducible genes in vertebrates in general will benefit from this work.

The production of IFN-α and IFN-β is not a specialized cell-function, and probably all cells of the organism are capable of producing these IFNs. This has never been systematically investigated for each of the 200 or so different cell types that make up the vertebrate organism. However, whenever cells of a specific differentiated type have been examined, it has been possible to induce production of either IFN-α or IFN-β, and usually of both. In fact, there is a good possibility that IFN genes become inducible as a result of differentiation, since undifferentiated mouse embryonal carcinoma cells become inducible only upon differentiation, and mouse embryo tissues cannot be induced to produce IFN before day 7, whereas by day 13 all the major embryonic tissues have become competent (Burke et al., 1978; Barlow et al., 1984).

Production of IFN-γ is a specialized function of antigen-activated T cells, and since the expression of this gene is induced by a different mechanism, we will discuss its induction in Chapter 11.

Representatives of all classes of vertebrates except amphibia have been shown to produce IFN, but the distinction between the α and the β species—as well as the existence of IFN-γ—has only been unambiguously demonstrated in mammals. One of the problems encountered in studies on induction is that they have been performed in cell populations in which the percentage of IFN-producing cells was an unknown. Reproducible methods for studying IFN production at the single cell level using in situ hybridization methods have only become recently available, and the few studies performed using these techniques so far indicate that in some cell cultures only a fraction of the population may be inducible at any given time (Fig. 3.1) (Zawatzky et al., 1985; Enoch et al., 1986). This makes the interpretation of some experiments more difficult. For example, although frequently one can demonstrate a correlation between the amount of inducer and the amount of IFN produced, it is not always possible to decide whether this happens because each cell makes more IFN or because more cells are induced.

We should add that theoretical extrapolations, based on an analysis of correlation between the amount of viral inducer and IFN yield, suggest that in some in vitro systems virtually all cells in the population respond to IFN induction. This is probably a function of the particular inducer-cell combination, as well as the physiological condition of the cells (Sekellick and Marcus, 1982, 1986). For example, contrary to what is observed in cultures of mouse C-243 cells, in which only a fraction of the cells is inducible, in situ hybridization reveals that virtually all cells are IFN-β producers in cultures of murine peritoneal macrophages that are infected with Newcastle disease virus (see Fig. 10.1, Chapter 10).

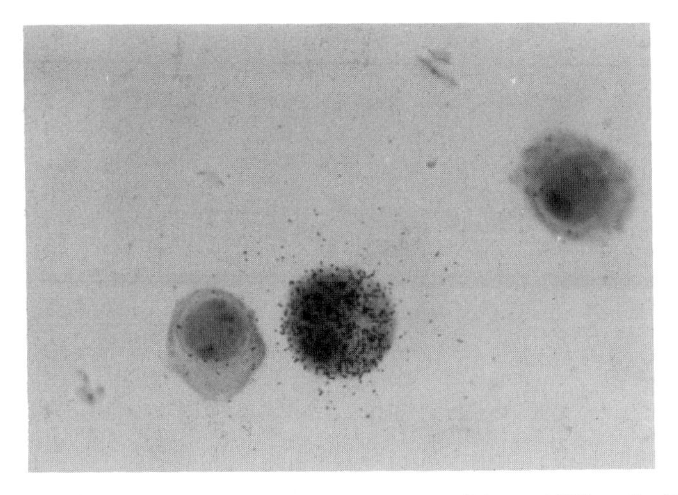

Figure 3.1 In some cell cultures, even under optimal conditions of IFN production, only a fraction of the cells appear to be producers. This is demonstrated, for example, when cells of the high IFN-producing mouse C-243 cell line are induced with Newcastle disease virus and analyzed by in situ hybridization 5 hours after the onset of induction. Since these cells produce both Mu IFN-α and β, a mixture of ^{35}S-labeled Mu IFN-α2 and IFN-β cDNA has been used to probe for the presence of IFN mRNA. The picture shows a producer cell, surrounded by two nonproducers. (From Zawatzky et al., 1985.)

Most cells, whether in the organism or in culture, do not release measurable amounts of IFN-α/β, nor can IFN be extracted from them. There are, however, exceptions, and spontaneous IFN production—meaning in the absence of any added inducer—can occur in cultures of macrophages, of lymphoblastoid cells, and of leukemic and normal peripheral leukocytes (Northrop and Deinhardt, 1967; Smith and Wagner, 1967; Swart and Young, 1969; Zajac et al., 1969; De Maeyer et al., 1971b; Talas et al., 1972; Adams et al., 1975; Pickering et al., 1980; Ablashi et al., 1982; Pickering et al., 1983). These cells are all derived from the hemopoietic system, but spontaneous, low-level IFN production can also occur in cells of other origins, for example, in continuous lines of mouse L or C-243 cells. This type of spontaneous production is characterized by the appearance of low but measurable levels of IFN in the extracellular milieu, and therefore IFN production can be directly demonstrated. It is, however, not known whether this is the result of production of significant amounts of IFN by a few cells in the population or of low production by many cells. Low levels of Hu IFN-α1 and IFN-α2 mRNA—but not of several other IFN-α subtypes or of Hu IFN-β—are constitutively transcribed in spleen, liver, kidney, and peripheral blood lymphocytes of normal individuals (Tovey et al., 1987). Again, whether this represents a substantial level of IFN transcription in some cells or a very low level in all cells is not known.

There are also indirect demonstrations of spontaneous IFN-α/β production, in which one cannot isolate and directly measure IFN, but nevertheless can obtain proof of its presence and activity by showing that certain effects can only occur in the presence of anti-IFN antibodies. For example, the high resistance to virus infection that characterizes a mutant line of murine 3T6 cells can be reversed by the treatment of the cells with anti-IFN-α/β serum. These cells are apparently engaged in constitutive IFN synthesis but do not secrete enough to be measured in an IFN assay (Jarvis and Colby, 1978). Similarly, some macrophages can be rendered permissive for vesicular stomatitis virus replication by previously treating the mice from which they are derived with anti-IFN-α/β serum (Gresser et al., 1985).

In view of the mounting evidence that IFNs play a role in the normal regulation of cell growth and replication and that growth factors and other cytokines can induce IFN-α and IFN-β, it is quite possible that very low levels of some IFN species are made by all cells during a certain period of the cell cycle (Friedman-Einat et al., 1982; Wells and Mallucci, 1985). This aspect of IFN production is related to the effect of IFN on cell replication and is discussed in Chapter 7.

If normally, then, most cells either do not display a measurable or at the most a very low level of IFN-α/β production, exposure to a variety of agents results in the production and secretion of IFN-α or IFN-β, or, more frequently, of a mixture of both. We will first consider the many agents that can induce IFN.

I. THERE ARE MANY IFN INDUCERS

A. Viruses

IFN was discovered during an investigation of the phenomenon of viral interference (Isaacs and Lindenmann, 1957), and viruses are still the most widely studied natural inducers of IFN-α/β. All animal viruses, irrespective of their structure and mode of replication, can induce IFN, and the number of virus-cell combinations used to study virus-induced IFN-α/β production is almost limitless. An extensive list of these can be found in Ho (1984).

B. Other Microorganisms

In addition to viruses, many other infectious agents can induce IFN production, but usually do so in vivo only and at lower levels than those produced during acute viral infections. Frequently, the IFN induced in this way has been shown to be an acid stable type I IFN, without further characterization. The following are the most important nonviral infectious agents that have been shown to induce IFN-α/β.

a. Rickettsia

These intracellular parasites induce acid stable type I IFN in murine and chick embryo cell cultures (Kazar, 1966; Kohno et al., 1970).

b. Bacteria and Endotoxin

Many bacteria, especially those that replicate inside animal cells, such as *Listeria monocytogenes,* can induce IFN-α/β. This occurs usually only after systemic infection of the host, or in cultures of cells derived from the hemopoietic system. Endotoxin, the lipopolysaccharide derived from the cell walls of gram-negative bacteria, and the M proteins of group A streptococci are also IFN-inducers (Youngner and Stinebring, 1964; De Maeyer et al., 1971b; Kirchner et al., 1977; Nagane and Minagawa, 1983; Havell, 1986; Weigent et al., 1986). IFN induction by bacterial endotoxin occurs mainly in macrophages and is discussed in Chapter 10.

c. Mycoplasma

Infection of cell cultures with mycoplasma can result in a decreased capacity to produce IFN upon stimulation by other agents, but mycoplasma under certain conditions can also be IFN inducers. Contamination of some tumor cell lines with Mycoplasma orale 1 is responsible for the induction by these cells of IFN-α in lymphocytes; when the tumor cells are cured from the infection, they no longer induce IFN (Rinaldo et al., 1974; Lombardi and Cole, 1978; Birke et al., 1981).

d. Protozoa

Infection of mice with several protozoa, including *Toxoplasma gondii* and *Plasmodium berghei* results in the production of IFN-α/β (Freshman et al., 1966; Huang et al., 1968). Infection with different strains of *Trypanosoma Cruzi* similarly induces low levels of IFN-α/β in mice (Sonnenfeld and Kierszenbaum, 1981). Children acutely ill with Plasmodium falciparum infection produce IFN of the α type, which can be isolated from their serum (Ojo-Amaize et al., 1981).

C. Growth Factors and Other Cytokines

Several growth factors and cytokines have been shown to induce the synthesis of IFN-α or IFN-β.

Colony stimulating factor 1 (CSF-1), also known as macrophage growth factor, is a proliferative signal for committed macrophage precursors and a stimulatory signal for more differentiated cells. It acts through a specific receptor at the surface of these cells. This receptor is related to a protooncogene c-*fms* (Sherr et al., 1985). Stimulation of murine bone-marrow cells with CSF-1 results in the production of Mu IFN-α/β (Moore et al., 1984). This is further discussed in Chapter 8.

Platelet-derived growth factor (PDGF) is one of the major mitogenic signals for connective tissue cells. It induces Mu IFN-β mRNA in confluent monolayers of murine BALB/c 3T3 cells (see Chapter 7) (Zullo et al., 1985).

Interleukin-1 (IL-1) and tumor necrosis factor (TNF) are regulatory cytokines with pleiotropic activities. In human diploid fibroblasts, they induce the synthesis of Hu IFN-β and of Hu IFN-β2 (see Chapter 13).

Interleukin-2 (IL-2), a lymphokine that is vital for T-cell function and IFN-γ induction, can induce the production of Mu IFN-α/β in mouse bone-marrow cells (see Chapter 13).

IFN-γ sometimes acts as an inducer of IFN-α. Part of the antiviral activity induced in mouse L-929 cells by Mu IFN-γ can be neutralized by monoclonal antibody to Mu IFN-α (Hughes and Baron, 1987).

Since IFN induction by growth factors and other cytokines is discussed in Chapters 7, 8, and 13, we will limit the discussion in the present chapter to IFN induction by viruses and double-stranded RNA.

II. INDUCTION BY DOUBLE-STRANDED RNA AND VIRUSES

Both natural and synthetic double-stranded RNAs (dsRNA) induce IFN-α/β with high efficiency (Field et al., 1967; Lampson et al., 1967). The origin of the dsRNA is not important for this function, and even cellular dsRNA from normal cells can induce IFN in cells identical to those from which it was derived (De Maeyer et al., 1971a; Kimball and Duesberg, 1971; Stern and Friedman, 1971). A stable secondary structure and a high molecular weight are prime requisites for induction.

Of the synthetic polynucleotides, the homopolymer pair polyrIrC is the most active and also the most widely used for induction studies. The structural requirements of dsRNAs for optimal IFN induction and for optimal activation of some enzymes that are activated in IFN-treated cells (the P1 protein kinase and the different 2-5A synthetases) are not quite identical, but it is, nevertheless, intriguing that dsRNA can act as both an IFN inducer and an activator of enzymes involved in IFN activity (see Chapters 5 and 6).

In spite of their tremendous differences in structure and mode of replication, all animal viruses induce IFN production under appropriate conditions. Does this point to the existence of different modes of induction, or is there evidence for a common mechanism? This question is far from settled; the best hypothesis at present is that for many viruses there is a common pathway for induction, going through the formation of dsRNA.

There are indeed many arguments in favor of dsRNA as an essential intermediate for IFN induction with RNA viruses. Reovirus, which contains a dsRNA genome, induces IFN-α/β even in the absence of virus replication (Long and Burke, 1971; Lai and Joklik, 1973; Winship and Marcus, 1980). Other RNA viruses, without a double-stranded genome, require formation of replicative intermediates or at least some RNA synthesis to trigger IFN induction (Falcoff and Falcoff, 1970; Lomniczi and Burke, 1970; Clavell and Bratt, 1971; Marcus and Fuller, 1979; Marcus and Sekellick, 1980). One of the most compelling arguments for dsRNA as an inducer during a virus infection is the demonstration with defective interfering vesicular stomatitis virus particles that viral dsRNA generated within the cell is the actual trigger for IFN production. Theoretical considerations suggest that in this particular system a single molecule of dsRNA is sufficient for induction (Fig. 3.2) (Marcus and Sekellick, 1977, 1980; Sekellick and Marcus, 1982).

Can the formation of dsRNA explain IFN induction by DNA viruses? In cell culture, but not always in the animal, DNA viruses as a rule are less efficient IFN inducers than are RNA viruses. The formation of dsRNA has been demonstrated during the replication of vaccinia virus and of adenovirus (Colby and Duesberg, 1969; Petterson and Philipson, 1974), and more as a result of "guilt by association" than of a formal demonstration, it has been assumed that this provides the trigger for IFN-α/β synthesis. For other DNA viruses, however, there is no direct evidence for the formation of dsRNA during their replicative cycle, and one cannot exclude that other molecules

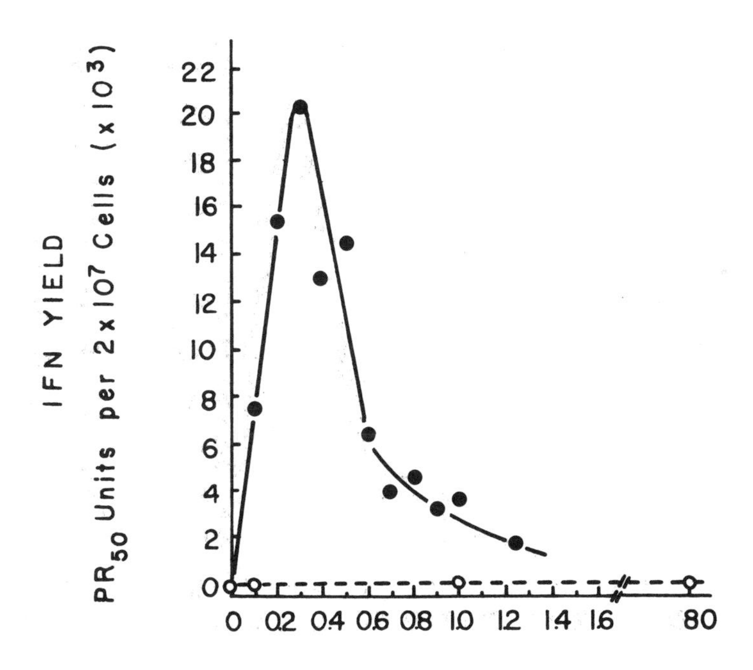

Figure 3.2 Dose (viral multiplicity)—response (IFN yield) curve showing the IFN-inducing capacity of vesicular stomatitis virus defective-interfering (DI) particles (●—●). Statistical analysis of the results suggests that a single viral particle per cell can be sufficient to trigger IFN production in chick embryo cell cultures. (Adapted from Marcus and Sekellick, 1977.)

generated during virus replication or metabolic changes resulting from the presence of viruses in cells may also be involved in stimulating the transcription of the IFN genes. For many viruses, such as *herpetoviridae, papovaviridae,* ând *parvoviridae,* we are completely in the dark as to how they trigger IFN production. In DNA viruses with a double-stranded genome, there is always the theoretical possibility of the transcription of opposite strands, which could provide the potential for dsRNA formation and IFN induction. For example, RNA transcripts derived from the DNA strand opposite to that encoding the mRNA for the ICP-O viral polypeptide have been detected in neurons that are latently infected with the herpes simplex virus (Stevens et al., 1987).

In view of the experimental evidence available, it is relatively easy to make at least a theoretical case for dsRNA as a common denominator for IFN induction when RNA viruses are concerned. Nevertheless, a rigid demonstration that the formation of dsRNA during an RNA virus infection is the only IFN inducer is lacking even for some RNA viruses, and indeed, there are indications that other mechanisms may also be involved. One can, for example, induce Hu IFN-α in peripheral blood mononuclear cells by exposing them to glutaraldehyde fixed cells expressing at their surface herpes simplex or Newcastle disease virus antigens (Capobianchi et al., 1985). In this particular instance an interaction beween a virus surface component and the mononuclear cell membrane is sufficient to induce an IFN-α synthesis. The glycoproteins of Sendai virus are capable of inducing acid stable type I IFN production in murine spleen cells, but are unable to do so in mouse L fibroblasts (Ito et al., 1978). Although this mode of IFN induction is probably limited to cells derived from the hemopoietic system and is not operative in fibroblast-like cells, it is, however, a demonstration that IFN-α and IFN-β production can be triggered by inducers other than dsRNA. In addition, polypeptides such as growth factors can induce IFN-β in macrophages or fibroblast-like cells, in the absence of any viral inducer (Moore et al., 1984; Van Damme et al., 1985a,b; Zullo et al., 1985; Kohase et al., 1986).

Nevertheless, even for IFN induction by growth factors and polypeptides in general, one could propose the involvement of dsRNA as the proximal inducer as a working hypothesis. The various growth factors such as PDGF, CSF-1, IL-1β, and IL-2, which have been shown to induce IFN-α or IFN-β, also induce expression of the protooncogene c-*myc*. The latter belongs to the competence genes, whose expression is correlated with proliferative activity of cells, and mitogenic stimulation of lymphocytes or fibroblasts leads to a fast increase in c-*myc* mRNA levels (Kelly et al., 1983; Armelin et al., 1984). The nucleotide sequence of the c-*myc* gene shows that a region of exon 1 is highly complementary to a region of exon 2, which means that the mRNA could form a stable stem-loop structure, including a double-stranded region spanning about 75 bp, which theoretically is sufficient to act as an IFN inducer (Saito et al., 1983). Moreover, transcription of the normal c-*myc* gene can occur at the same time in the sense and antisense direction (Nepveu and Marcu, 1986; Kindy et al., 1987). Gene transcription of coding and noncoding strands in an overlapping fashion provides the possibility of dsRNA formation, and dsRNA resulting from symmetrical transcription could theoretically serve as an intracellular IFN-α or IFN-β inducer.

This hypothesis is compatible with the observation that in growth-factor-stimulated cells, c-*myc* mRNA usually appears before IFN mRNA. It could also explain why the IFN genes belong to the late activated genes after growth factor stimulation, since IFN induction would be the direct consequence of the enhanced transcription of the c-*myc* gene. Derepression of IFN genes by cellular dsRNA, resulting from either hairpin structures in some mRNAs or duplex formation of mRNAs derived from genes transcribed from opposite strands of the same DNA locus (Nepveu and Marcu, 1986; Adelman et al., 1987, Kindy et al., 1987) appears as a unifying working hypothesis for dsRNA as the universal, proximal inducer of IFN-α/β.

For the time being, however, no hard facts indicate that IFN induction can or should be explained by a "universal inducer," which would be a common denominator of the many agents, viral and others, shown to stimulate IFN-α/β synthesis. In addition, no experimental evidence unambiguously points to the existence of a unique "proximal inducer," acting as the last link in the chain of events leading to induction and being directly responsible for derepression of the IFN genes. These questions are still unanswered.

III. MANY OTHER PROTEINS ARE COINDUCED WITH IFNs

Unlike induction of genes in prokaryotes, whereby one inducer usually activates one operon, IFN inducers activate several other genes at the same time. In human fibroblasts, polyrIrC induces, in addition to Hu IFN-β, the synthesis of at least 13 and maybe as many as 23 other proteins, which have not been further characterized (Raj and Pitha, 1980; Content et al., 1982). Some of these coinduced genes are probably linked, but unrelated to the Hu IFN-β gene on chromosome 9 (Gross et al., 1981). In murine 3T3 cells, polyrIrC as well as vesicular stomatitis virus, induce the expression not only of the Mu IFN-β gene, but also of the "competence" gene family, consisting among others of the c-*myc* and c-*fos* genes, which can also be induced by the platelet-derived growth factor (Zullo et al., 1985). In mouse cells, Newcastle disease virus induces, in addition to the IFN-α and IFN-β genes on chromosome 4, two other genes situated on chromosome 12 and on the X chromosome, respectively. The gene on the X chromosome, whose expression is actually enhanced rather than induced, codes for a protein that is involved in stimulation of erythropoietic activity and that is also an inhibitor of metalloproteinases and of collagenase. This gene is the murine homologue of the human EPA/TIMP gene (Skup et al., 1982; Docherty et al., 1985; Gasson et al., 1985; Kelley and Pitha, 1985; Kelley et al., 1986; Gewert et al., 1987; Jackson et al., 1987).

Induction of the IFN genes is manifestly not a highly specific event, restricted to the IFN structural genes only, and the regions that confer inducibility seem to be shared by many other genes. The reason for coordinate induction of the expression of so many dispersed genes is unknown. One can speculate that they are all involved in regulatory circuits that still have to be defined.

IV. CONTROL OF IFN-α AND IFN-β SYNTHESIS

Production of IFN-α and IFN-β is controlled at both the transcriptional and the posttranscriptional levels. For many studies on the molecular mechanism of IFN induction polyrIrC has been used; it has the advantage that no viral replication is required and that the amount of inducer can be more easily controlled.

A. Control at the Transcriptional Level

Gene expression in eukaryotic cells is controlled at several levels. To allow for transcription, the gene must be in an active state, which requires changes in the chromatin structure to permit access of RNA polymerase. Histones and DNA, the most important components of chromatin, are held together in the nucleosomes. Acquisition of the "active" state is the first requirement for gene activation and has in fact been shown to occur in IFN genes upon induction. Induction-specific changes in nucleosome structure can be demonstrated in mouse L cells that are transfected with cosmid containing 36 kb of human genomic DNA including and surrounding the Hu IFN-β gene. These transfected cells contain multiple copies of the gene and are inducible by polyrIrC. Soon after induction with polyrIrC, structural changes occur in the nucleosomes upstream of the gene, and the DNA becomes more accessible as demonstrated by the fact that it is no longer protected from nuclease activity (Bode et al., 1986).

The second step of gene activation consists of initiation of transcription and is controlled by the promoter region of the structural gene. The leftward boundary of the promoter is always upstream of the TATA or Hogness box, a ubiquitous sequence of 7 or 8 bp that lies usually about 20 to 30 bp upstream from the starting point for transcription. The activity of a promoter can be significantly increased by the presence of other DNA regions, known as enhancers. These do not have to be in a fixed location, but they can act at different positions in the gene; some enhancers are tissue specific.

The first indication suggesting that IFN induction involves a transcriptional event came from studies with Actinomycin D (Wagner, 1964). Later it was shown that functional IFN mRNA cannot be extracted from uninduced cells, and the absence of mRNA is borne out in Northern blot analysis by the lack of hybridization of mRNA from uninduced cells that are probed with IFN-α or IFN-β cDNA. Transcription starts early after induction, the actual time being a function of the inducer-cell system studied. For example, in human fibroblasts induced with polyrIrC a significant level of Hu IFN-β mRNA is already present 2 hours after the inducer has been added; it increases until about 7 hours after induction, and then declines (Content et al., 1983). When human lymphoblastoid Namalwa cells are treated with the Sendai virus, IFN-β and IFN-α are produced (Havell et al., 1977). One can thus study the expression of both β and α genes and show that Sendai virus induces the coordinate synthesis of IFN-α and IFN-β mRNAs in Namalwa cells. Detectable levels of mRNA are present 3 hours after the onset of induction, reach a maximum at 9 hours, and then decline (Shuttleworth et al., 1983).

There are many other experimental variations on this theme, but these two examples suffice to make the point that transcription starts rapidly upon induction, reaches a peak, and is then terminated, despite the continuous presence of the inducer or the addition of more inducer. The cause of the shutoff of IFN synthesis is unknown; it occurs in cell cultures and in the animal and is usually followed by a period of hyporesponsiveness to renewed induction. It is unlikely that a negative feedback by IFN itself is responsible for the arrest of transcription, since treatment of cells, even with very high doses (up to 100,000 units/ml) of IFN-α/β increases rather than decreases polyrIrC-induced IFN synthesis in murine C-243 cells (De Maeyer-Guignard et al., 1980). More likely candidates for the shutoff function are provided by the many proteins whose synthesis is coinduced with IFN by viruses or polyrIrC. Crude IFN preparations often contain one or several substances that can decrease IFN production (Friedman, 1966).

Additional evidence that a transcriptional event is involved in IFN induction is provided by the observation that manipulations resulting in enhanced transcription also result in increased IFN production. This can be done, among others, by treatment of cells with butyric acid. This compound changes the structure of nucleosomes through hyperacetylation of histones, which results in a better accessibility of the gene for RNA polymerase (Candido et al., 1978; Bertrand et al., 1984). In Sendai virus-induced Namalwa cells, for example, pretreatment with sodium butyrate results in a 15-fold increase of IFN-α mRNA and a fourfold increase of IFN-β mRNA (Shuttleworth et al., 1983). An increased rate of transcription of IFN-α mRNA could be due to either the synthesis of more mRNA copies per gene or an increased number of IFN-α genes transcribed. Until recently it was not easy to distinguish between these two possibilities. However, since there is only one IFN-β gene in human cells, an increased rate of transcription can clearly be attributed to an increased number of mRNA copies per gene, provided one works under conditions in which all cells are induced.

Demethylation of cytosine at CpG sites plays a role in the control of eukaryotic gene expression (Felsenfeld and McGhee, 1982), and hypomethylation of DNA could be another factor contributing to enhanced transcription of IFN genes. For example, treatment of Namalwa cells with the demethylating agent 5-azacytidine is sufficient by itself to induce IFN production, and furthermore, causes a 50-fold increase of NDV-induced IFN-α production (Tovey et al., 1983). A direct demonstration that this effect is actually due to hypomethylation however, is lacking.

Cycloheximide, an inhibitor of protein synthesis, can markedly increase the polyrIrC-induced production of Hu IFN-β, and in fact cycloheximide alone induces very low levels of IFN-β and significant levels of IFN-β2 (Tan and Berthold, 1977; Weissenbach et al., 1980). It is believed that cycloheximide treatment can both stabilize IFN mRNA (see subsequent discussion) and increase the rate of transcription. When Chinese hamster ovary cells are transfected with a plasmid carrying the Hu IFN-β gene, cycloheximide, in the absence of any other inducer, stimulates the production of IFN-β mRNA. Nuclear runoff transcription assays show that this is a direct effect on transcription (Ringold et al., 1984). Similar results are obtained in mouse cells that are transformed with a bovine papilloma virus-Hu IFN-β gene recombinant: Cycloheximide treatment results in transcription of the IFN gene (Maroteaux et al., 1983). The most straightforward conclusion of these observations is that one or several proteins are involved in preventing the transcription of the IFN-β gene in uninduced cells, and that transcription is stimulated by inhibiting the synthesis of these proteins. Induction of IFN-β by inhibitors of macromolecular synthesis has prompted the "repressor-depletion" hypothesis, which explains IFN induction by the inhibition of short-lived repressor molecules that normally repress expression of the IFN-β gene (Tan and Berthold, 1977).

The rate of transcription of the IFN genes can also be influenced by IFN itself. Pretreatment of cells with low doses of IFN just before induction frequently results in enhanced IFN production, a phenomenon known as "priming" (Isaacs and Burke, 1958; Stewart et al., 1971). The priming of polyrIrC-induced human fibroblasts with Hu IFN-β leads to a tenfold increased transcription rate of the Hu IFN-β gene. This is not an effect of IFN treatment on gene transcription in general, since in the same cells the transcription rate of the IFN-β2 gene is not affected by priming. It is not an exclusive effect on the IFN genes either, because the rate of transcription of other polyrIrC activated genes, clustered near the IFN-β gene, is also increased by priming. This shows that different genes are differentially affected by priming. It is possible that for structural reasons, for example, the changes in chromatin structure discussed in the first paragraph, some adjacent genes are coinfluenced by priming with IFNs (Nir et al., 1985).

B. Control of IFN Synthesis at the Posttranscriptional Level: The Stability of IFN mRNA

In eukaryotes, mRNA is synthesized as a large precursor molecule in the nucleus. After maturation, consisting of reduction in size, capping at the 5′ and polyadenylation at the 3′ end, the mature mRNA moves to the cytoplasm for translation. For many eukaryotic mRNAs the functional half-life in the cytoplasm can be relatively long, up to 24 hours. Some mRNAs, however, have much shorter half-lives, of 30 minutes or less. Manifestly, mRNA stability is one of the factors intervening in the control of protein synthesis, and rapid degradation of mRNA in the absence of further transcription is one way of quickly terminating the synthesis of a protein. This appears to be the case for IFN, since IFN-α/β mRNAs in general are rapidly degrading mRNAs. In this respect they are comparable to mRNAs of other cytokines and lymphokines and of some protooncogenes. These mRNAs all share an AU-rich region in their 3′ noncoding region, and the possibility has been raised that this common structural feature is a recognition signal for rapid mRNA degradation (Caput et al., 1986; Shaw and Kamen, 1986).

When the 3′ noncoding regions of the Mu IFN-α2 and Mu IFN-α4 genes are replaced by the corresponding region of the rabbit globin gene, under the control of the SV40 early promoter, a fourfold higher IFN production is observed in monkey COS cells (van Heuvel et al., 1986). Features of the 3′ and 5′ untranslated regions of the Hu IFN-mRNA can affect the translational efficacy of the mRNA, independently of their effect on stability. This has been demonstrated by replacing the 3′ and 5′ untranslated regions of Hu IFN-β mRNA by the corresponding regions of Xenopus β-globin mRNA; such chimeric mRNA displays greatly increased translational efficacy in *Xenopus* oocytes (Kruys et al., 1987).

IFN production can be modulated by treatments that influence the stability of IFN mRNA. When virus-induced human lymphoblastoid cells are kept at 28°C instead of the usual 37°C, the overall IFN-α and IFN-β yield is increased. This is due to an increased half-life of IFN mRNA, which remains associated for a longer time with ribosomes resulting in a prolonged translation (Morser and Shuttleworth, 1981).

Superinduction is the enhancement of specific protein expression by inhibitors of translation or transcription, or by a combination of both. The possible mechanisms are enhanced transcription and increased stability of mRNA. Stabilization of cytoplasmic IFN mRNA probably contributes to the enhanced and prolonged synthesis of Hu IFN-β in polyrIrC-induced human fibroblasts that have been treated with cycloheximide and Actinomycin D in a superinduction scheme. In cells, that are not exposed to these metabolic inhibitors, active degradation of IFN mRNA starts a few hours after induction by polyrIrC, whereas in superinduced cells IFN synthesis goes on for several more hours (Raj and Pitha, 1983). Since superinduction can also stimulate IFN mRNA transcription (Dinter and Hauser, 1987), the relative contribution of stabilization and enhanced transcription to increased IFN synthesis is difficult to assess.

In polyrIrC-induced human fibroblasts the shutoff of IFN synthesis is accompanied by the degradation of IFN mRNA, which is thus the cause of arrested synthesis. In Newcastle disease virus-induced Namalwa cells however, IFN-α mRNA continues to be present after IFN synthesis has virtually ceased. Up to 3 hours after IFN production has stopped, stable, in vitro translatable, IFN mRNA can still be isolated from the induced cells (Berger et al., 1980). What prevents this mRNA from being further translated in the cell is not known, but it shows that, in addition to factors influencing the stability of IFN mRNA, other as yet undefined mechanisms of translational control regulate IFN synthesis.

V. THE STRUCTURAL FEATURES OF IFN-α/β GENES INVOLVED IN THE REGULATION OF EXPRESSION

A. Cell-Specific Factors Influence IFN Induction

Before discussing the molecular aspects of IFN gene regulation, we should review the evidence that cell-specific factors influence the expression of IFN genes. In human fibroblast-like cells in vitro, polyrIrC induces the production of IFN-β only. This is not because fibroblast cannot express the IFN-α genes, since Newcastle disease virus induces both IFN-α and IFN-β and there are other examples of simultaneous induction of both IFN-α and IFN-β in diploid fibroblasts (Havell et al., 1978; Hayes et al., 1979). It is also not because polyIrC is incapable of derepressing IFN-α genes in fibroblasts in general, since in murine fibroblast-like cells polyIrC, as well as Newcastle disease virus, consistently induce the synthesis of both IFN-β and IFN-α. In murine peritoneal macrophages little or no IFN-α is induced by Newcastle disease virus, and the major IFN species induced is IFN-β. In macrophages derived from bone-marrow progenitor cells in culture, however, Newcastle disease virus induces both Mu IFN-α and Mu IFN-β.

Such cell-determined selective induction could be due to either differential transcription of the IFN genes or differential posttranscriptional processing of mRNA. The latter possibility has not received much attention to date, whereas it has been shown that differential transcription can be involved. The ratio of IFN-α to IFN-β mRNA transcripts varies significantly with cell type as well as with the inducer when one compares IFN induction in human peripheral blood leukocytes, human lymphoblastoid cells, HeLa cells, and human fibroblasts. Moreover, the proportion of individual IFN-α mRNA species is also dependent on cell type (Hiscott et al., 1984). This is a good indication that cell or tissue-specific factors influence IFN induction, by provoking preferential activation of certain IFN genes, depending on the inducer. It is, therefore, theoretically possible that differences in the regulatory regions of the IFN genes are responsible for these cell-specific induction effects. However, since frequently IFN-α and IFN-β genes are coordinately induced, no compelling reason exists for supposing that the regions conferring inducibility to these genes are grossly different.

It is clear that the induction of IFN synthesis by viruses, dsRNAs, or other inducers is mainly caused by an activation of transcription, and much less or not at all by transient stabilization of rapidly turning over mRNA. Which features of the IFN genes are involved in induction, and by what mechanism is transcription activated upon induction? The different experimental approaches to these questions will be considered separately for the IFN-α and IFN-β genes.

B. IFN-α Genes

1. Inducibility Is a Characteristic of the IFN-α Gene Itself

The Hu IFN-α1 gene contains the necessary information for inducibility, since when it is introduced into murine cells by a cotransformation procedure with the herpes thymidine kinase gene, the transfected cells can be induced by Newcastle disease virus to synthesize Hu IFN-α mRNA, initiated at the correct start site (Mantei and Weissmann, 1982). Thus although the transfected human IFN-α gene is integrated at multiple sites of the mouse genome, it remains inducible by viruses. Inducibility is, therefore, not the result of the chromosomal environment, but rather, is directly under control of the gene itself.

Which part of the gene controls inducibility? The Hu IFN-α1 gene promoter region contains the necessary information: when the 675 bp upstream of the IFN-α1 coding region are joined to the transcription unit of the rabbit globin gene, and mouse cells are then transformed by this hybrid gene, correctly initiated globin mRNA only appears after viral induction. If, on the other hand, the globin promoter is fused to the IFN-α1 coding sequence, correct transcription no longer depends on viral induction, but instead, has become constitutive. The sequences of the IFN-α1 gene promoter region that are responsible for induction can be further delineated by examining the inducibility of a series of constructed 5′ deletion mutants. This reveals that the region required for inducible transcription is located between positions -117 and -74. This region contains a purine-rich stretch of 42 bp, located immediately downstream of position -117, which is highly conserved in all Hu IFN-α genes (Benjamin et al., 1983; Fisher et al., 1983).

Can this promoter region be separated further in a region mediating induction and a region required for initiation of transcription per se? The boundaries of the IFN-α1 promoter region required for virus inducibility have been further delineated by joining various segments of the 5′ flanking region of the IFN-α1 gene to truncated segments of the rabbit β globin gene, containing the part of the promoter that is involved in initiation of transcription. The hybrid gene is then inserted into a vector containing the HSV thymidine kinase gene, which makes it possible to transform thymidine kinase negative murine cells and then to select for thymidine kinase positive transformed cells. The transformed cells, containing the various test genes, can then be assayed for virus inducibility of β globin synthesis. This procedure shows that the region of the Hu IFN-α1 gene required to confer virus inducibility is located between positions -109 and -64 of the promoter.

The region required for maximal inducibility has two sets of repeats, a perfect repeat of the pentamer CAGAA and an imperfect repeat of the octamer A(A/T)GGAAAG. Thirteen out of 20 nucleotides conserved in all IFN-α and IFN-β gene promoters are located within these repeats, suggesting their importance for inducibility (Ryals et al., 1985).

C. The IFN-β Gene

1. The Interferon Gene Regulatory Element (IRE)

The region required for induction of the Hu IFN-β gene has been determined by analyzing the induced expression of various deletion mutants that are introduced into murine C127 cells on a bovine papilloma virus vector (Zinn et al., 1982). Under these conditions, the sequences required for efficient induction by Sendai virus are located within the 77 base pairs immediately upstream from the mRNA cap site (Chernajovsky et al., 1983). The regulatory sequence between position -77 and -37 corresponds to the region required for maximal inducibility between positions -109 and -64 of the Hu IFN-α1 gene described in the previous section. This sequence has been called the IFN gene regulatory element (IRE) (Goodbourn et al., 1985). The IRE regulates inducibility in an orientation independent manner, since its reinsertion in either direction into an IFN-β gene deletion mutant restores inducibility. Furthermore, the IRE confers inducibility upon a heterologous promoter, also in an orientation independent manner. In addition, even when the IRE is placed 800 bp upstream of the -73 position or 360 bp downstream of the polyA addition site of the poorly inducible -73 deletion mutant of the Hu IFN-β gene, inducibility is enhanced. It is typical of enhancer elements that they influence transcription in an orientation independent manner at a considerable distance from a promoter, which furthermore, can be heterologous. The IRE, therefore, displays the properties of an enhancer. A minimal regulatory element of 14 bp has been identified within this enhancer; sequences closely related to this minimal element can be found five times within the 5′ flanking regions of the Hu IFN-beta and IFN-α2 genes (Goodbourn et al., 1985). Inducible enhancer elements have been described for other genes, such as the metallothionein genes and the human c-*fos* gene (Karin et al., 1984; Treisman, 1985).

2. Organization of the IRE

How is transcription of the IFN gene controlled by the inducible enhancer element? In prokaryotes, transcription of those structural genes that are not constitutively expressed is controlled by regulatory proteins, which can be either negative-acting or positive-acting. Positive-acting proteins bind to DNA near a promoter site and increase the efficiency of the binding of RNA polymerase. Negative-acting proteins repress a gene by binding to DNA near a promoter site, thereby physically preventing the access of RNA polymerase and transcription. When repressors bind to certain specific molecules, called inducers, their binding affinity for DNA decreases and the gene can be transcribed (Ptashne, 1986).

Results from the laboratory of T. Maniatis (Goodbourn et al., 1986; Zinn and Maniatis, 1986) indicate that the IFN-β enhancer is under negative control, and that the IRE contains two functionally distinct sequences—a constitutive transcription element (-77 to -55) and a negative regulatory element (-55 to -36). Deletion of the negative regulatory domain results in a high level of mRNA synthesis in the absence of inducer. This indicates that the factors that are necessary for transcription are already present in uninduced cells, but they are blocked by the negative regulatory region, which is adjacent to or maybe overlapping with the constitutive transcription element. It is possible that this constitutive transcription element contains binding sites for more than one transacting, transcription factor, and that the binding of these factors is blocked by another factor bound to the negative regulatory element.

Direct evidence for interaction between cellular factors and the regulatory regions of the IFN-β gene is obtained from DNAse I genomic footprinting analysis of this region. In uninduced cells, factors that bind to DNA are present in a region located between positions -167 and -94 from the cap site, and are in a second region located between -68 and -38. This second region corresponds to the negative regulatory element of the IRE. After induction with polyrIrC, these factors disappear and another factor binds to a region located between -77 and -64, which corresponds to the transcription regulatory region of the IRE. The best hypothesis to explain these observations is that there are two proteins bound to the 5′ flanking region of the IFN-β gene before induction. Protein I interacts with the region between −55 and −38, and protein II interacts with a large region extending from -167 to -94 (Fig. 3.3). Both these proteins act as repressors of the IFN gene. After induction, these proteins dissociate from the gene and protein III can then bind to the constitutive enhancer element in the IRE. Protein III is, therefore, probably the IFN transcription factor (ITF), whose binding is blocked by repressor protein I in uninduced cells. It is interesting that the ITF binding site (-76 to -64) is homologous to several other viral and cellular constitutive enhancer elements, found in a variety of other genes (Table 3.1). This could explain why other cellular proteins are coinduced with IFN when cells are exposed to viruses or to polyrIrC.

Repressor protein II binds to a second negative regulatory element, upstream of the IRE, in uninduced cells. Its dissociation results in a five- to tenfold increase in the constitutive level of the IFN-β gene expression,

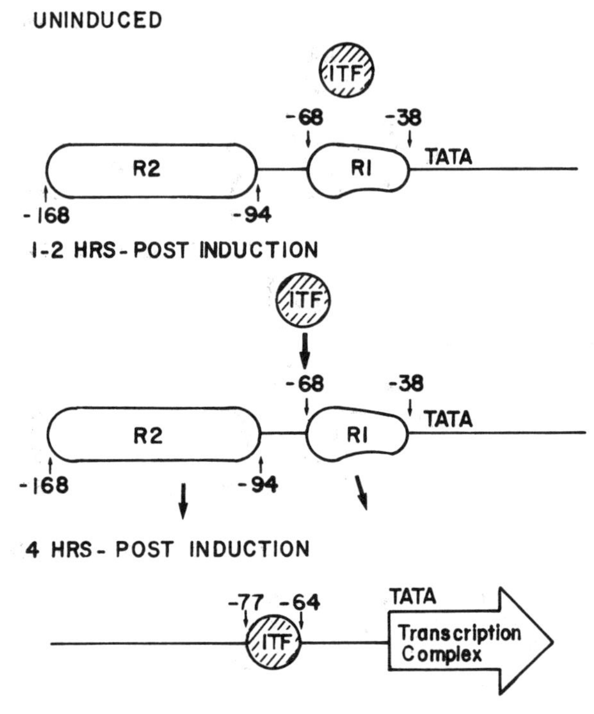

Figure 3.3 A proposed mechanism for Hu IFN-β gene induction. In the uninduced state the repressor proteins, R1 and R2, interact with their corresponding regions. After induction, the repressor proteins dissociate from the gene, and another protein, ITF (for Interferon Transcription Factor) binds to the IRE (for Interferon Regulatory Element). (Adapted from Zinn and Maniatis, 1986.)

TABLE 3.1 Sequence Homologies to the Constitutive Transcription Element in the IRE of the Hu IFN-β Gene[ab]

	−76									−64						−59
Hu IFN-β	A	G	A	A	G	T	G	A	A	A	G	T	G	G	G	A	A	A
Hu HSP70 promoter	A	G	A	A	G	G	G	A	A	A								
Adenovirus 5 E1a enhancer	G	G	A	A	G	T	G	A	A	A								
Immunoglobulin heavy chain gene enhancer											G	T	G	G	C	A	A	G
c-fos gene regulatory element											G	T	G	G	A	A	A	C
Adenovirus 2 E1a enhancer											G	T	G	G	T	A	A	A

Source: Adapted from Goodbourn et al. 1986.

[a] A comparison of the Hu IFN-β gene constitutive transcription element to functionally important sequence elements from enhancer and promoter regions of several other genes.

[b] references: Banerji et al. 1983; Hearing and Shenk, 1983; Hen et al., 1983; Treisman, 1985; Wu et al. 1985.

without affecting induced levels (Goodbourn et al., 1986; Zinn and Maniatis, 1986).

The results from other groups are somewhat at variance with regard to the upstream boundary of the 5′ region conferring inducibility to the Hu IFN-β gene. When Murine L929 or Ltk$^-$ cells are transfected with various 5′ deletion mutants of the Hu IFN-β gene and induced with either Newcastle disease virus or polyrIrC, the upstream boundary of DNA sequences required to support the maximum level of induction lies between -117 and -67 from the cap site, rather than between -77 and -37 (Dinter et al., 1983; Fujita et al., 1985). Since in this case expression is measured in L cells and not in C127 cells, the difference could be caused by cell-specific factors influencing IFN induction. A definite answer to this question can only be provided by comparative studies in different transfected cells with deletion mutants carried by the same vector and using the same inducer.

A possible candidate for a repressor of the Hu IFN-β gene is a nuclear protein that has been isolated from human leukemic KG-1 cells. This protein binds selectively to a 163 bp fragment situated between positions -135 and -202 of the upstream region of the Hu IFN-β gene, and does not bind to the corresponding region of the Hu IFN-α1 gene (Xanthoudakis and Hiscott, 1987).

3. *An Hexamer Sequence Unit Acting As An Inducible Enhancer*

Repeated sequence units that are present in the upstream region between -65 and -109 relative to the cap site of the Hu IFN-β gene have also been described as being important for conferring virus inducibility to the gene. This region consists of quasi-tandem repeats of homologous hexamer sequences. Chemically synthesized "consensus" hexamer sequence units, placed in tandem repeats upstream of cognate and noncognate promoters, confer virus inducibility upon these promoters. The most efficient hexamer sequence unit, AAGTGA, contributes incrementally in virus-induced activation of transcription and manifests properties of an inducible, rather than of a constitutive enhancer (Fujita et al., 1987).

4. *Evidence for a Transacting Activation Factor*

Activation of the Hu IFN-β gene by a preexisting, transacting factor that is normally present in cells is suggested by experiments showing that the induction of Hu IFN-β mRNA in poorly inducible human cells can be stimulated 200-fold by fusing these cells to highly inducible mouse cells, or by a combined treatment with IFN and cycloheximide before induction. The latter observation suggests that the transacting activation factor is also IFN-inducible (Enoch et al., 1986). It has been the general experience that inducibility of IFN genes is dominant over uninducibility in cell hybrids and heterokaryons (Carver et al., 1968; Guggenheim et al., 1968; Burke et al., 1980).

REFERENCES

Ablashi, D. V., Baron, S., Armstrong, G., Faggioni, A., Viza, D., Levine, P. H., and Pizza, G. Spontaneous production of high levels of leukocyte (α) interferon by a human lymphoblastoid B-cell line (LDV/7). *Proc. Soc. Exp. Biol. Med.* 171:114–119 (1982).

Adams, A., Lidin, E., Strander, H., and Cantell, K. Spontaneous interferon production and Epstein-Barr virus antigen expression in human lymphoid cell lines. *J. Gen. Virol.* 28: 219–223 (1975).

Adelman, J. P., Bond, C. T., Douglass, J., and Herbert, E. Two mammalian genes transcribed from opposite strands of the same locus. *Science* 235: 1514–1517 (1987).

Armelin, I. A., Armelin, M. C. S., Kelly, K., Stewart, T., Leder, P., Cochran, B. H., and Stiles, C. D. Functional role for c-*myc* in mitogenic response to platelet-derived growth factor. *Nature* 310: 655–660 (1984).

Banerji, J., Olson, L., and Schaffner, W. A lymphocyte specific cellular enhancer is located downstream of the joining region in immunoglobulin heavy chain genes. *Cell* 33: 729–740 (1983).

Barlow, D. P., Randle, B. J., and Burke, D. C. Interferon synthesis in the early post-implantation mouse embryo. *Differentiation* 27: 229–235 (1984).

Benjamin, R., Steeg, P. S., and Farrar, J. J. Production of murine immune interferon by a T lymphocyte cell line: Regulation by an EL4 thymoma-derived factor(s). In: *Interleukins, Lymphokines, and Cytokines,* Oppenheim, J. J., Cohen, S., and Landy, M., Eds., Academic Press, New York, pp. 609–615 (1983).

Berger, S. L., Hitchcock, M. J. M., Zoon, K. C., Birkenmeier, C. S., Friedman, R. M., and Chang, E. H. Characterization of interferon messenger RNA synthesis in Namalva cells. *J. Biol. Chem.* 255: 2955–2961 (1980).

Bertrand, E., Erard, M., Gomez-Lira, M., and Bode, J. Influence of histone hyperacetylation on nucleosomal particles as visualized by electron microscopy. *Arch. Biochem. Biophys.* 229: 395–398 (1984).

Birke, C., Peter, H. H., Langenberg, U., Muller-Hermes, W. J. P., Peters, J. H., Heitmann, J., Leibold, W., Dalluge, H., Krapf, E., and Kirchner, H. Mycoplasma contamination in human tumor cell lines: Effect on interferon induction and susceptibility to natural killing. *J. Immunol.* 127: 94–98 (1981).

Bode, J., Pucher, H. J., and Maass, K. Chromatin structure and induction-dependent conformational changes of human interferon-β genes in a mouse host cell. *Eur. J. Biochem.* 158: 393–401 (1986).

Burke, D. C., Ege, T., and Ringertz, N. R. Production of chick interferon by reactivating chick erythrocytes. *J. Gen. Virol.* 50: 437–440 (1980).

Burke, D. C., Graham, C. F., and Lehman, J. M. Appearance of interferon inducibility and sensitivity during differentiation of murine teratocarcinoma cells in vitro. *Cell* 13: 243–248 (1978).

Candido, E. P. M., Reeves, R., and Davie, J. Sodium butyrate inhibits histone deacetylation in cultured cells. *Cell* 14: 105–113 (1978).

Capobianchi, M. R., Facchini, J., Di Marco, P., Antonelli, G., and Dianzani, F. Induction of α interferon by membrane interaction between viral surface and peripheral blood mononuclear cells. *Proc. Soc. Exp. Biol. Med.* 178: 551–556 (1985).

Caput, D., Beutler, B., Hartog, K., Thayer, R., Brown-Shimer, S., and Cerami, A. Identification of a common nucleotide sequence in the 3′-untranslated region of mRNA molecules specifying inflammatory mediators. *Proc. Natl. Acad. Sci.* (USA) 83: 1670–1674 (1986).

Carver, D. H., Seto, D. S. Y., and Migeon, B. R. Interferon production and action in mouse, hamster and somatic hybrid mouse–hamster cells. *Science* 160: 558–559 (1968).

Chernajovsky, Y., Mory, Y., Vaks, B., Feinstein, S. I., Segev, D., and Revel, M. Production of human interferon in *E. coli* under lac and tryplac promoter control. *Ann. N.Y. Acad. Sci.,* pp. 88–96 (1983).

Clavell, L. A. and Bratt, M. A. Relationship between the ribonucleic acid-synthesizing capacity of ultraviolet-irradiated Newcastle disease virus and its ability to induce interferon. *J. Virol.* 8: 500–508 (1971).

Colby, C. and Duesberg, P. H. Double-stranded RNA in vaccinia virus infected cells. *Nature* 222: 940–944 (1969).

Content, J., De Wit, L., Pierard, D., Derynck, R., De Clercq, E., and Fiers, W. Secretory proteins induced in human fibroblasts under conditions used for the production of interferon. *Proc. Natl. Acad. Sci.* (USA) 79: 2768–2772 (1982).

Content, J., De Wit, L., Tavernier, J., and Fiers, W. Human fibroblast interferon RNA transcripts of different sizes in poly(I).poly(C) induced cells. *Nucl. Acids Res.* 11: 2627–2638 (1983).

De Maeyer, E. and De Maeyer-Guignard, J. Interferon structural and regulatory genes in the mouse. In: *Interferons as Cell Growth Inhibitors and Antitumor Factors,* UCLA Symposia on Molecular and Cellular Biology, New Series, Vol. 50, R. Friedman, T. Merigan, and T. Sreevalsan, Eds., Alan Liss, Inc. New York, pp. 435–445 (1986).

De Maeyer, E., De Maeyer-Guignard, J., and Montagnier, L. Double-stranded RNA from rat liver induces interferon in rat cells. *Nature New Biol.* 229: 109–110 (1971a).

De Maeyer, E., Fauve, R. M., and De Maeyer-Guignard, J. Production d'interféron au niveau du macrophage. *Ann. Inst. Pasteur* 120: 438–446 (1971b).

De Maeyer-Guignard, J., Cachard, A., and De Maeyer, E. Electrophoretically pure mouse interferon has priming but no blocking activity in poly(I.C)-induced cells. *Virology* 102: 222–225 (1980).

Dinter, H. and Hauser, H. Superinduction of the human interferon-β promoter. *EMBO J.* 6: 599–604 (1987).

Dinter, H., Hauser, H., Mayr, U., Lammers, R., Bruns, W., Gross, G., and Collins, J. Human interferon-β and co-induced genes: Molecular studies. In: *The Biology of the IFN System 1983,* E. De Maeyer and H. Schellekens, Eds., Elsevier, Amsterdam, pp. 33–44 (1983).

Docherty, A. J. P., Lyons, A., Smith, B. J., Wright, E. M., Stephens, P. E., Harris, T. J. R., Murphy, G., and Reynolds, J. J. Sequence of human tissue inhibitor of metalloproteinases and its identity to erythroid-potentiating activity. *Nature* 318: 66–69 (1985).

Enoch, T., Zinn, K., and Maniatis, T. Activation of the human β-interferon gene requires an interferon-inducible factor. *Mol. Cell. Biol.* 6: 801–810 (1986).

Falcoff, R. and Falcoff, E. Induction de la synthèse d'interféron par des RNA bicaténaires. II. Etude de la forme intermédiaire de réplication du virus Mengo. *Biochim. Biophys. Acta* 199:147–158 (1970).

Felsenfeld, G. and McGhee, J. Methylation and gene control. *Nature* 296: 602–603 (1982).

Field, A. K., Tytell, A. A., Lampson, G. P., and Hilleman, M. R. Inducers of interferon and host resistance. II. Multistranded synthetic polynucleotide complexes. *Proc. Natl. Acad. Sci.* (USA) 58: 1004–1010 (1967).

Fisher, P. B., Miranda, A. F., Babiss, L. E., Pestka, S., and Weinstein, I. B. Opposing effects of interferon produced in bacteria and of tumor promoters on myogenesis in human myoblast cultures. *Proc. Natl. Acad. Sci.* (USA) 80: 2961–2965 (1983).

Freshman, M. M., Merigan, T. C., Remington, J. S., and Brownlee, I. E. In vitro and in vivo antiviral action of an interferon-like substance induced by toxoplasma gondii. *Proc. Soc. Exp. Biol. Med.* 123: 862–866 (1966).

Friedman, R. M. Effect of interferon treatment on interferon production. *J. Immunol.* 96: 872–877 (1966).

Friedman-Einat, M., Revel, M., and Kimchi, A. Initial characterization of a spontaneous interferon secreted during growth and differentiation of Friend erythroleukemia cells. *Mol. Cell. Biol.* 2: 1472–1480 (1982).

Fujita, T., Ohno, S., Yasumitsu, H., and Taniguchi, T. Delimitation and properties of DNA sequences required for the regulated expression of human interferon-β gene. *Cell* 41: 489–496 (1985).

Fujita, T., Shibuya, H., Hotta, H., Yamanishi, K., and Taniguchi, T. Interferon-β gene regulation: Tandemly repeated sequences of a synthetic 6 bp oligomer function as a virus-inducible enhancer. *Cell* 49: 357–367 (1987).

Gasson, J. C., Golde, D. W., Kaufman, S. E., Westbrook, C. A., Hewick, R. M., Kaufman, R. J., Wong, G. G., Temple, P. A., Leary, A. C., Brown, E. L., Orr, E. C., and Clark, S. C. Molecular characterization and expression of the gene encoding human erythroidpotentiating activity. *Nature* 315: 768–771 (1985).

Gewert, D. R., Coulombe, B., Castelino, M., Skup, D., and Williams, B. R. G. Characterization and expression of a murine gene homologous to human EPA/TIMP: a virus-induced gene in the mouse. *EMBO J.* 6: 651–657 (1987).

Goodbourn, S., Burnstein, H., and Maniatis, T. The human β-interferon gene enhancer is under negative control. *Cell* 45: 601–610 (1986).

Goodbourn, S., Zinn, K., and Maniatis, T. Human β-interferon gene expression is regulated by an inducible enhancer element. *Cell* 41: 509–520 (1985).

Gresser, I., Vignaux, F., Belardelli, F., Tovey, M. G., and Maunoury, M. T. Injection of mice with antibody to mouse interferon α/β decreases the level of 2′-5′oligoadenylate synthetase in peritoneal macrophages. *J. Virol.* 53: 221–227 (1985).

Gross, G., Mayr, U., and Collins, J. New poly I-C inducible transcribed regions are linked to the human IFN-β gene in a genomic clone. In: *The Biology of the*

Interferon System 1981, E. De Maeyer, G. Gallasso, and H. Schellekens, Eds., Elsevier, Amsterdam, pp. 85–90 (1981).

Guggenheim, M. A., Friedman, R. M., and Rabson, A. S. Interferon: Production by chick erythrocytes activated by cell fusion. *Science* 159: 542–543 (1968).

Havell, E. A. Augmented induction of interferons during Listeria monocytogenes. *J. Infect. Dis.* 153: 960–969 (1986).

Havell, E. A., Hayes, T. G., and Vilcek, J. Synthesis of two distinct interferons by human fibroblasts. *Virology* 89: 330–334 (1978).

Havell, E. A., Yip, Y. K., and Vilcek, J. Characteristics of human lymphoblastoid (Namalva) interferon. *J. Gen. Virol.* 38: 51–59 (1977).

Hayes, T. S., Yip, Y. K., and Vilcek, J. Le interferon production by human fibroblasts. *Virology* 98: 351–363 (1979).

Hearing, P. and Shenk, T. The adenovirus type 5E1A transcriptional control region contains a duplicated enhancer sequence. *Cell* 33: 695–703 (1983).

Hen, R., Borrelli, E., and Chambon, P. An enhancer element is located 340 base pairs upstream from the adenovirus-2 Ela cap site. *Nucl. Acids Res.* 11: 8747–8760 (1983).

Hiscott, J., Cantell, K., and Weissmann, C. Differential expression of human interferon genes. *Nucl. Acids Res.* 12: 3727–3746 (1984).

Ho, M. Induction and inducers of interferon. In: *Interferon 1. General and Applied Aspects,* A. Billiau, Ed., Elsevier, Amsterdam, pp. 79–124 (1984).

Huang, K. Y., Schultz, W., and Gordon, F. B. Interferon induced by plasmodium berghei. *Science* 162: 123–124 (1968).

Hughes, T. K. and Baron, S. A large component of the antiviral activity of mouse interferon-γ may be due to its induction of interferon-alpha. *J. Biol. Regul. Homeost. Agents* 1: 29–32 (1987).

Isaacs, A. and Burke, D. C. Mode of action of interferon. *Nature* 182: 1073–1074 (1958).

Isaacs, A. and Lindenmann, J. Virus interference. I. The interferon. *Proc. Roy. Soc.* B147: 258–267 (1957).

Ito, Y., Nishiyama, Y., Shimokata, K., Nagata, I., Takeyama, H., and Kunii, A. The mechanism of interferon induction in mouse spleen cells stimulated with HVJ. *Virology* 88: 128–137 (1978).

Jackson, I. J., LeCras, T. D., and Docherty, A. J. P. Assignment of the TIMP gene to the murine X-chromosome using an inter-species cross. *Nucl. Acids Res.* 15: 4357 (1987).

Jarvis, A. P. and Colby, C. Murine interferon system regulation: Isolation and chaacterization of a mutant 3T6 cell engaged in the semiconstitutive synthesis of interferon. *Cell* 14: 355–363 (1978).

Karin, M., Haslinger, A., Holtgreve, H., Cathala, G., Slater, E., and Baxter, J. D. Activation of a heterologous promoter in response to dexamethasone and cadmium by metallothionein gene 5′-flanking DNA. *Cell* 36: 371–379 (1984).

Kazar, J. Interferon-like inhibitor in mouse sera induced by Rickettsiae. *Acta Virol.* 10: 277 (1966).

Kelley, K. A. and Pitha, P. M. Differential effect of poly rI.rC and Newcastle disease virus on the expression of interferon and cellular genes in mouse cells. *Virology* 147: 382–393 (1985).

Kelley, K. A., Pitha, P. M., De Maeyer-Guignard, J., De Maeyer, E., and Kozak, C. Assignment of two mouse genes coinduced with interferon to chromosomes 12 and X. *J. Interferon Res.* 6: 51–57 (1986).

Kelly, K., Cochran, B. H., Stiles, C., and Leder, P. Cell-specific regulation of the c-*myc* gene by lymphocyte mitogens and platelet-derived growth factor. *Cell* 35: 603–610 (1983).

Kimball, P. C. and Duesberg, P. H. Virus interference by cellular double-stranded RNA. *J. Virol.* 7: 697–706 (1971).

Kindy, M. S., McCormack, J. E., Buckler, A. J., Levine, R. A., and Sonenshein, G. E. Independent regulation of transcription of the two strands of the c-myc gene. *Mol. Cell Biol.* 7: 2857–2862 (1987).

Kirchner, H., Hirt, H. M., Becker, H., and Munk, K. Production of an antiviral factor by murine spleen cells after treatment with Corynebacterium parvum. *Cell. Immunol.* 31: 172–176 (1977).

Kohase, M., Henriksen-Destefano, D., May, L. T., Vilcek, J., and Sehgal, P. B. Induction of β-interferon by tumor necrosis factor: A homeostatic mechanism in the control of cell proliferation. *Cell* 45: 659–666 (1986).

Kohno, S., Kohase, M., Sakata, H., Shimizu, Y., Hikita, M., and Shishido, A. Production of interferon in primary chick embryonic cells infected with Rickettsia Mooseri. *J. Immunol.* 105: 1553–1558 (1970).

Kruys, V., Wathelet, M., Poupart, P., Contreras, R., Fiers, W., Content, J., and Huez, G. The 3′ untranslated region of the human interferon-beta mRNA has an inhibitory effect on translation. *Proc. Natl. Acad. Sci.* (USA) 84: 6030–6034 (1987).

Lai, M. H. T. and Joklik, K. The induction of interferon by temperature-sensitive mutants of reovirus, UV-irradiated reovirus, and subviral reovirus particles. *Virology* 51: 191–204 (1973).

Lampson, G. P., Tytell, A. A., Field, A. K., Nemes, M. M., and Hilleman, M. R. Inducers of interferon and host resistance. I. Double-stranded RNA from extracts of *penicillium funiculosum*. *Proc. Natl. Acad. Sci.* (USA) 58: 782–789 (1967).

Lombardi, P. S. and Cole, B. C. Induction of a pH-stable interferon in sheep lymphocytes by mycoplasmatales virus MVL2. *Infect. Immun.* 20: 209–214 (1978).

Lomniczi, B. and Burke, D. C. Interferon production by temperature sensitive mutants of Semliki forest virus. *J. Gen. Virol.* 8: 55–68 (1970).

Long, W. F. and Burke, D. C. Interferon production by double-stranded RNA: A comparison of induction by reovirus to that by a synthetic double-stranded polynucleotide. *J. Gen. Virol.* 12: 1–11 (1971).

Mantei, N. and Weissmann, C. Controlled transcription of a human α-interferon gene introduced into mouse L cells. *Nature* 297: 128–132 (1982).

Marcus, P. and Fuller, F. J. Interferon induction by viruses. II. Sindbis virus: Interferon induction requires one-quarter of the genome-genes G and A. *J. Gen. Virol.* 44: 169–177 (1979).

Marcus, P. I. and Sekellick, M. J. Defective interfering particles with covalently linked—RNA induce interferon. *Nature* 266: 815–819 (1977).

Marcus, P. I. and Sekellick, M. J. Interferon induction by viruses. III. Vesicular stomatitis virus: Interferon-inducing particle activity requires partial transcription of gene N. *J. Gen. Virol.* 47: 89–96 (1980).

Maroteaux, L., Chen, L., Mitrani-Rosenbaum, S., Howley, P. M., and Revel, M. Cycloheximide induces expression of the human interferon β1 gene in mouse cells transformed by bovine papillomavirus-interferon β1 recombinants. *J. Virol.* 47: 89–95 (1983).

Moore, R. N., Larsen, H. S., Horohov, D. W., and Rouse, B. T. Endogenous regulation of macophage proliferative expansion by colony-stimulating factor-induced interferon. *Science* 223: 178–181 (1984).

Morser, J. and Shuttleworth, J. Low temperature treatment of Namalwa cells causes superproduction of interferon. *J. Gen. Virol.* 56: 163–174 (1981).

Nakane, A. and Minagawa, T. Alternative induction of α/β interferons and γ interferon by listeria monocytogenes in mouse spleen cell cultures. *Cell. Immunol.* 75: 283–291 (1983).

Nepveu, A. and Marcu, K. B. Intragenic pausing and anti-sense transcription within the murine c-myc locus. *EMBO J.* 5: 2859–2865 (1986).

Nir, U., Maroteaux, L., Cohen, B., and Mory, I. Priming affects the transcription rate of human interferon-β1 gene. *J. Biol. Chem.* 260: 14242–14247 (1985).

Northrop, R. L. and Deinhardt, F. Production of interferon-like substances by human bone marrow tissues in vitro. *J. Nat. Cancer Inst.* 39: 685–689 (1967).

Ojo-Amaize, E. A., Salimonu, L. S., Williams, A. I. O., Akinwolere, O. A. O., Shabo, R., Alm, G., and Wigzell, H. Positive correlation between degree of parasitemia, interferon titres, and natural killer cell activity in plasmodium falciparum-infected children. *J. Immunol.* 127: 2296–2300 (1981).

Petterson, U. and Philipson, L. Synthesis of complementary RNA sequences during productive adenovirus infection. *Proc. Natl. Acad. Sci.* (USA) 71: 4887–4889 (1974).

Pickering, L. A., Kronenberg, L. H., and Stewart, W. E. II. Spontaneous production of human interferon. *Proc. Natl Acad. Sci.* (USA) 77: 5938–5942 (1980).

Pickering, L. A., Lin, L. S., and Sarkar, F. H. Characteristics of spontaneously produced and of virus-induced human LuKII cell interferons. *Arch. Virol.* 75: 201–211 (1983).

Ptashne, M. *A Genetic Switch. Gene Control and Phage Lambda.* Blackwell Scientific Publications, Palo Alto (USA) (1986).

Raj, N. B. K. and Pitha, P. M. Synthesis of new proteins associated with the induction of interferon in human fibroblast cells. *Proc. Natl. Acad. Sci.* (USA) 77: 4918–4922 (1980).

Raj, N. B. K. and Pitha, P. M. Two levels of regulation of β interferon gene expression in human cells. *Proc. Natl. Acad. Sci.* (USA) 80: 3923–3927 (1983).

Rinaldo, C. R., Jr., Cole, B. C., Overall, J. C., Jr., and Glasgow, L. A. Induction of interferon in mice by mycoplasmas. *Infect. Immun.* 10: 1296–1301 (1974).

Ringold, G. M., Dieckmann, B., Vannice, J. L., Trahey, M., and McCormick, F. Inhibition of protein synthesis stimulates the transcription of human β-interferon genes in Chinese hamster ovary cells. *Proc. Natl. Acad. Sci.* (USA) 81: 3964–3968 (1984).

Ryals, J., Dierks, P., Ragg, H., and Weissmann, C. A 46-nucleotide promoter segment from an IFN-α gene renders an unrelated promoter inducible by virus. *Cell* 41: 497–507 (1985).

Saito, H., Hayday, A. C., Wiman, K., Hayward, W. S., and Tonegawa, S. Activation of the c-*myc* gene by translocation: A model for translational control. *Proc. Natl. Acad. Sci.* (USA) 80: 7476–7480 (1983).

Sekellick, M. J. and Marcus, P. I. Interferon induction by viruses. VIII. Vesicular stomatitis virus: (+)D1-011 particles induce interferon in the absence of standard virions. *Virology* 117: 280–285 (1982).

Sekellick, M. J. and Marcus, P. I. Induction of high titer chicken interferon. In: *Methods in Enzymology,* Interferons, Part C, S. Pestka, Ed., Vol. 119, Academic Press, Orlando, pp. 115–125 (1986).

Shaw, G. and Kamen, R. A conserved AU sequence from the 3′ untranslated region of GM-CSF mRNA mediates selective mRNA-degradation. *Cell* 46: 659–667 (1986).

Sherr, C. J., Rettenmier, C. W., Sacca, R., Roussel, M. F., Look, A. T., and Stanley, E. R. The c-*fms* proto-oncogene product is related to the receptor for the mononuclear phagocyte growth factor, CSF-1. *Cell* 41: 665–676 (1985).

Shuttleworth, J., Morser, J., and Burke, D. C. Expression of interferon-α and interferon-β genes in human lymphoblastoid (Namalwa) cells. *Eur. J. Biochem.* 133: 399–404 (1983).

Skup, D., Windass, J. D., Sor, F., George, H., Williams, B. R. G., Fukuhara, H., De Maeyer-Guignard, J., and De Maeyer, E. Molecular cloning of partial cDNA copies of two distinct mouse IFN-β mRNAs. *Nucl. Acids. Res.* 10: 3069–3084 (1982).

Smith, T. J. and Wagner, R. R. Rabbit macrophages interferons. I. Conditions for biosynthesis by virus-infected and uninfected cells. *J. Exp. Med.* 125: 559–577 (1967).

Sonnenfeld, G. and Kierszenbaum, F. Increased serum levels of an interferon-like activity during the acute period of experimental infection with different strain of trypanosoma cruzi. *Am. J. Trop. Med. Hyg.* 30: 1189–1191 (1981).

Stern, R. and Friedman, R. M. Ribonucleic acid synthesis in animal cells in the presence of actinomycin. *Biochemistry* 10: 3635–3645 (1971).

Stevens, J. G., Wagner, E. K., Devi-Rao, G. B., Cook, M. L., and Feldman, L. T. RNA complementary to a herpervirus α gene mRNA is prominent in latently infected neurons. *Science* 235: 1056–1059 (1987).

Stewart II, W. E., Gosser, L. B., and Lockart, R. Z., Jr. Priming: a nonantiviral function of interferon. *J. Virol.* 7: 792–801 (1971).

Swart, B. E. and Young, B. G. Inverse relationship of interferon production and virus content in cell lines from Burkitt's lymphoma and acute leukemias. *J. Natl. Cancer Inst.* 42: 941–944 (1969).

Talas, M., Szolgay, E., and Rozsa, K. Further study of spontaneous interferon produced by hamster peritoneal cells. *Arch. Ges. Virusforsch.* 38: 149–158 (1972).

Tan, Y. H. and Berthold, W. A mechanism for the induction and regulation of human fibroblastoid interferon genetic expression. *J. Gen. Virol.* 34: 401–411 (1977).

Tovey, M. G., Streuli, M., Gresser, I., Gugenheim, J., Blanchard, B., Guymarho, J., Vignaux, F., and Gigou, M. Interferon messenger RNA is produced constitutively in the organs of normal individuals. *Proc. Natl. Acad. Sci.* (USA) 84: 5038–5042 (1987).

Tovey, M. G., Vincent, C., Gresser, I., and Revel, M. The effect of DNA methylation on the expression of human interferon-α genes. In: *The Biology of the Interferon System 1983,* E. De Maeyer and H. Schellekens, Eds., Elsevier, Amsterdam, pp. 45–50 (1983).

Treisman, R. Transient accumulation of c-*fos* RNA following serum stimulation requires a conserved 5′ element and c-*fos* 3′ sequences. *Cell* 42: 889–902 (1985).

Van Damme, J., De Ley, M., Opdenakker, G., Billiau, A., De Somer, P., and Van Beeumen, J. Homogeneous interferon-inducing 22K factor is related to endogenous pyrogen and interleukin-1. *Nature* 314: 266–268 (1985a).

Van Damme, J., Opdenakker, G., Billiau, A., De Somer, P., De Wit, L., Poupart, P., and Content, J. Stimulation of fibroblast interferon production by a 22K protein from human leukocytes. *J. Gen. Virol.* 66: 693–700 (1985b).

van Heuvel, M., Bosveld, I. J., Luyten, W., Trapman, J., and Zwarthoff, E. C. Transient expression of murine interferon-alpha genes in mouse and monkey cells. *Gene* 45: 159–165 (1986).

Wagner, R. R. Inhibition of interferon biosynthesis by actinomycin D. *Nature* 204: 49–51 (1964).

Weigent, D. A., Huff, T. L., Peterson, J. W., Stanton, G. J., and Baron, S. Role of interferon in streptococcal infection in the mouse. *Microbial Pathogenesis* 1: 399–407 (1986).

Weissenbach, J., Chernajovsky, Y., Zeevi, M., Shulman, L., Soreq, H., Nir, U., Wallach, D., Perricaudet, M., Tiollais, P., and Revel, M. Two interferon mRNAs in human fibroblasts: In vitro translation and *E. coli* cloning studies. *Proc. Natl Acad. Sci.* (USA) 77: 7152–7156 (1980).

Wells, V. and Mallucci, L. Expression of the 2-5A system during the cell cycle. *Exp. Cell. Res.* 159: 27–36 (1985).

Winship, T. R. and Marcus, P. I. Interferon induction by viruses. VI. Reovirus: Virion genome dsRNA as the interferon inducer in aged chick embryo cells. *J. Interferon Res.* 1: 155–167 (1980).

Wu, B. J., Kingston, R. E., and Morimoto, R. Human HSP70 promoter contains at least two distinct regulatory domains. *Proc. Natl. Acad. Sci.* (USA) 83: 629–633 (1986).

Xanthoudakis, S. and Hiscott, J. Identification of a nuclear DNA binding protein associated with the interferon-beta upstream regulatory region. *J. Biol. Chem.* 262: 8298–8302 (1987).

Youngner, J. S. and Stinebring, W. R. Interferon production in chickens injected with *Brucella abortus*. *Science* 144: 1022–1023 (1964).

Zajac, B. A., Henle, W., and Henle, G. Autogenous and virus-induced interferons from lines of lymphoblastoid cells. *Cancer Res.* 29: 1467–1475 (1969).

Zawatzky, R., De Maeyer, E., and De Maeyer-Guignard, J. Identification of individual interferon-producing cells by in situ hybridization. *Proc. Natl. Acad. Sci.* (USA) 82: 1136–1140 (1985).

Zinn, K. and Maniatis, T. Detection of factors that interact with the human β-interferon regulatory region in vivo by DNAse I footprinting. *Cell* 45: 611–618 (1986).

Zinn, K., Mellon, P., Ptashne, M., and Maniatis, T. Regulated expression of an extrachromosomal human β-interferon gene in mouse cells. *Proc. Natl. Acad. Sci.* (USA) 79: 4897–4901 (1982).

Zullo, J. N., Cochran, B. H., Huang, A. S., and Stiles, C. D. Platelet-derived growth factor and double-stranded ribonucleic acids stimulate expression of the same gene in 3T3 cells. *Cell* 43: 793–800 (1985).

4 INTERFERON RECEPTORS

Protein effector molecules, such as hormones, growth factors, and interferons, bind to specific receptors at the cell surface. These receptors frequently consist of glycoproteins, sometimes associated as dimers or tetramers of homologous or heterologous subunits. Ligand-receptor binding is characterized by its high affinity, which clearly distinguishes it from the nonspecific, nonreceptor mediated forms of binding to the cell surface. As a result of specific ligand-receptor interaction, the generation of intracellular signals elicits the appropriate physiological response to the ligand.

In the case of many protein signaling molecules the ligand-receptor complex enters the cell by receptor-mediated endocytosis, a selective concentration mechanism that enables cells to ingest ligands. During the process of endocytosis the part of the plasma membrane that contains the ligand-receptor complex invaginates and is pinched off to enter the cell as a vesicle. Some of these endocytotic vesicles are coated on the inside with bristlelike structures, originating from regions of the plasma membrane called coated pits. These coated regions are characterized on their cytoplasmic side by a hexagonal lattice, which consists principally of clathrin, a fibrous protein. Many ligands initially bind to receptors located outside clathrin-coated pits, but the ligand-receptor complexes then diffuse in the plasma membrane to cluster in coated pits, from which they enter the cell, moving to other vesicles called receptosomes. It is from the receptosomes that the ligand is delivered to various intracellular organelles such as the Golgi apparatus or the nucleus. Very often endocytotic vesicles fuse with lysosomes, resulting in a rapid breakdown of the macromolecules of the ligand-receptor complex.

Whether or not the uptake of the ligand-receptor complex is a necessary requirement for biological activity is a highly debated point for many ligands. The binding of the signaling molecule to the receptor at the cell surface is often assumed to be sufficient to start the chain of events leading to the specific ligand activity. It has indeed been shown for some receptors that the binding of molecules other than ligands can also deliver specific signals. Antibodies that mimic the receptor-binding site of a given ligand can bind to the corresponding receptor, for example, the insulin receptor or the thyroid-stimulating hormone receptor, and induce to some extent events that normally would result from ligand-receptor binding (Smith et al., 1977; Jacobs et al., 1978). This shows that at least for some effects ligand is not required, provided the receptor site is occupied. However, since the antibody-receptor complex is usually internalized, it still could mean that internalization of the receptor is necessary. We will see that the same question comes up for the IFN receptors.

Receptor binding generally results in intracellular activation of second messengers such as calcium ions (Ca^{2+}) and cyclic AMP. The calcium messenger system is a very common way by which extracellular signals regulate cell function. There are two branches in this system: a calmodulin branch, involved in brief responses or in the initial phase of sustained response, and the protein kinase C branch, involved in the sustained phase of cellular response. The major transducing event in this system is the turnover of plasma membrane phosphatidylinositides, giving rise to two intracellular messengers: inositol triphosphate and diacylglycerol, which initiate the flow of information in the calmodulin and protein kinase C branches, respectively (Rasmussen, 1986). These pathways are very likely involved in signal transduction after IFN-receptor binding, since a rapid and transient increase in diacyglycerol concentrations occurs in human lymphoblastoid Daudi cells and in human fibroblasts within seconds of their exposure to Hu IFN-α or Hu IFN-β. This effect is proportional to the number of IFN receptors present on the cell membrane (Yap et al., 1986a,b).

Cyclic AMP exerts its effects by activating the cyclic AMP-dependent protein kinases, which represent only a fraction of the number of protein kinases that can be activated in a cell. In some cases there is an association of cyclic AMP-dependent protein kinases, which shows that the cyclic AMP messenger system can influence the calcium messenger system (Rasmussen, 1986).

As a result of endocytosis, the receptor-ligand complex can end up in lysosomal vesicles, where it is degraded. If degradation occurs more rapidly than synthesis and replacement, the number of receptors at the cell surface decreases, with a concomitant decrease in target cell sensitivity to the ligand, an event known as receptor down-regulation. Some receptors however, such as low-density lipoprotein receptors, are not degraded, but after dissociation of the ligand they are most probably recycled back to the cell surface to serve again (Goldstein et al., 1979).

I. BINDING OF IFNS TO THEIR RECEPTORS

To qualify as "receptor binding," ligand binding has to be specific. This means that the number of binding sites must be finite and that the binding of the ligand should be inhibited by competition with the same or a structurally related ligand. Usually, in these tests the first-bound ligand is radioactively labeled, whereas the competitive ligand is unlabeled. In addition to binding, the appropriate physiological response should be obtained, which implies that the labeling procedure should not inactivate the ligand. It is also desirable that the ligand consist of a single molecular species, but this requirement has not always been met with IFN because of the nature of many IFN preparations (see appendix). In this chapter we will emphasize as much as possible studies that were performed with single molecular species.

In accordance with the preceding criteria, one can demonstrate specific binding of radiolabeled IFN molecules to cell-surface receptors. Apparent dissociation constants (K_d) or the number of binding sites per cell can be calculated by mathematical models, such as the Scatchard plot, originally devised to measure the interaction of proteins and small molecules (Scatchard, 1949). The Scatchard analysis of various IFN-binding data gives apparent dissociation constants ranging from 2×10^{-9} to 2.5×10^{-11} M. The average number of binding sites per cell is within the range of a few thousand (Table 4.1), which is a low figure when compared to many other protein activator molecules, for which typical values of receptor sites vary from 10^4 to 10^5 per cell. The reason for the low receptor number reported for IFNs is not clear. There are a few reports of higher values, such as 3.5×10^4 sites on

Table 4.1. IFN Binding Sites and Affinity Constants

IFN	Cell Type	Binding Sites per Cell	Apparent K_d (M)	References
Human				
Leukocyte	Daudi	4,200	2.0×10^{-10}	Mogensen et al., 1981
	Raji	900	5.0×10^{-10}	
Rec αA	T cells	230	1.8×10^{-10}	Faltynek et al., 1986
	Daudi	7,300	1.6×10^{-10}	
	LGL[a]	500	2.2×10^{-10}	
Rec αA	Daudi	3,550	0.9×10^{-10}	Langer et al., 1986a
	LGL	185	2.3×10^{-10}	
Rec αJ	Daudi	3,550	2.3×10^{-9}	
	LGL	185	4.3×10^{-9}	
Rec α2	B cells	444	5.0×10^{-10}	Hannigan et al., 1986
Rec γ	Monocytes	4,000	4.0×10^{-9}	Finbloom et al., 1985
	Lymphoma	1,800	1.5×10^{-10}	Rashidbaigi et al., 1985
	T cells	520	4.5×10^{-11}	Faltynek et al., 1986
	Daudi	7,900	5.1×10^{-11}	
	LGL	760	3.3×10^{-11}	
	Platelets	200	2.0×10^{-10}	Molinas et al., 1987
Mouse				
α/β	L1210 Leukemia cells	1,000	1.0×10^{-10}	Aguet and Blanchard, 1981
β	L cells	5,000	9.8×10^{-10}	Kushnaryov et al., 1985
Rec γ	L cells	11,000	5.0×10^{-10}	Langer et al., 1986b
	T cells	9,100	9.6×10^{-10}	
	L1210 Leukemia cells	4,000	5.0×10^{-10}	Wietzerbin et al., 1986

[a] LGL: Large granular lymphocytes.

Daudi cells. These represent, however, low affinity binding sites, with a dissociation constant of 1×10^{-8} M, which should be further characterized (Hannigan et al., 1984b).

A. Binding of IFN-α and IFN-β

The existence of a common receptor for IFN-α and IFN-β is demonstrated by the competition of Hu IFN-β for binding sites on human lymphoblastoid cells occupied by ^{125}I-labeled recombinant Hu IFN-αA. Contrary to Hu IFN-β, Hu IFN-γ does not displace the bound IFN-α (Branca and Baglioni, 1981). Murine IFNs display the same behavior: IFN-α and IFN-β compete for the same binding sites, but IFN-γ has a separate binding site (Aguet and Blanchard, 1981; Aguet et al., 1982). There are sufficient structural homologies between IFN-α and IFN-β to allow for the possibility of common receptor-binding domains, and nothing is unusual in the fact that they share a common receptor since these IFN species evolved from the same ancestral gene (see Chapter 2). Competitive receptor-binding experiments using several Hu IFN-α subspecies indicate that Hu IFN-α1, −2, −4, −5, −6, −7, and −8 all share the same receptors when assayed on human cells of various origins. However, the IFN-α subspecies can have significantly different binding affinities: Hu IFN-α1 has a much lower binding affinity for human cells than does Hu IFN-α2 (Aguet et al., 1984). This raises the possibility that some of the differential effects, observed when the biological activities of different IFN-α subspecies are compared, are due to variations in affinity for the receptor. For the time being, however, there is no proof that affinity for the receptor and different biological activities are related.

Many receptor studies have been performed on lymphoblastoid Daudi cells, originally derived from a patient with Burkitt's lymphoma, and these cells can hardly be called normal. However, when one compares the number and affinity of receptors for Hu IFN-αA and of Hu IFN-γ on Daudi cells and on freshly isolated normal lymphocytes, no significant differences are found. We can, therefore, be confident that studies in Daudi cells yield valuable information of a general nature (Faltynek et al., 1986).

B. Do IFN-Resistant Cells Lack Receptors?

Is there a correlation between receptor binding or lack thereof and sensitivity or resistance to IFN action? This has been tested in cell lines that display complete or partial resistance to the antiviral and anticellular action of IFNs.

A subline of mouse lymphoid L1210 cells, resistant to the antiviral and anticellular action of Mu IFN-α/β, lacks functional cell-surface receptors, and two human cell lines responding poorly to Hu IFN-α have very low IFN-binding activity (Baglioni et al., 1982). This is a clear case of cause and effect: In the absence of IFN receptors there is no IFN activity. Undifferentiated murine embryo carcinoma cells, however, are resistant to all IFN effects, yet have specific receptors and IFN-resistant clones of Friend leukemia cells bind Mu IFN-α/β with the same affinity as IFN-sensitive cells (Aguet et al., 1981; Affabris et al., 1983).

Human lymphoblastoid Daudi cells are exquisitely sensitive to the antiviral and antiproliferative effects of Hu IFN-α, which they bind with high affinity (Mogensen et al., 1981), but one can isolate mutant Daudi cells that have become IFN resistant. One of the resistant lines binds Hu IFN-α2 and Hu IFN-α8 with the same affinity as the wild-type sensitive line. Other resistant lines, however, display different binding characteristics: Although they also bind the two subtypes of IFN-α, the kinetics of binding differ by the absence of an initial peak of binding, observed in wild-type Daudi cells (Dron et al., 1986). The functional meaning of these altered binding kinetics is not understood, but the fact that one of the resistant lines has the same binding kinetics as the wild-type cells is a clear indication that resistance can occur in the presence of apparently normal receptor binding.

In conclusion, although some resistant cells do lack IFN receptors, others bind IFN just as efficiently as sensitive cells. The binding of IFN to high affinity receptors is a necessary first step, which, however, is not sufficient to obtain a physiological response.

C. Binding of IFN-γ

The receptors for Hu IFN-γ are different from the other IFN receptors, since neither Hu IFN-α nor IFN-β compete for binding sites with Hu IFN-γ (Branca and Baglioni, 1981; Rashidbaigi et al., 1985). The number of receptor sites on human monocytes has been estimated to be about 4,000 and on human histiocytic lymphoma cells about 1,800, and the dissociation constant is 8×10^{-9} M and 5.4×10^{-10}, respectively (Finbloom et al., 1985; Rashidbaigi et al., 1985). Like IFN-α and IFN-β, IFN-γ is degraded after endocytosis.

The presence of specific and separate receptors for IFN-γ explains the synergistic antiviral and antiproliferative effect when IFN-γ is combined with IFN-α or with IFN-β, in contrast to the purely additive effects when only IFN-α and IFN-β are combined (Fleischmann et al., 1979; Czarniecki et al., 1984).

The receptor for Hu IFN-γ on monocytes appears to be different from the one on fibroblasts. Hu IFN-γ specifically binds with high affinity to both types of cells, but with significantly different binding characteristics. Moreover, acid-treated IFN-γ, in spite of the inactivation of its antiviral activity, still effectively competes for binding to the fibroblast receptor, but no longer competes for binding to the monocyte receptor. These binding differences are accompanied by differences in biological activity, since although it induces an antiviral state in fibroblasts, Hu IFN-γ does not do so in monocytes (Orchansky et al., 1986).

The Scatchard analysis indicates about 4,000 binding sites for rec Mu IFN-γ on lymphoid L1210 cells and an apparent K_d of 5×10^{-10} M. In these cells the binding of IFN-γ results in down-regulation of binding sites, and reappearance of the receptor can be completely blocked by cycloheximide and also by tunicamycin. This is a strong indication that the IFN-γ receptor is not recycled, and furthermore, that it is a glycoprotein (Wietzerbin et al., 1986). The Hu IFN-γ receptor on human fibroblasts, however, does appear to be either recycled to some extent or else recruited from a large intracellular pool (Anderson et al., 1982).

Interestingly, human platelets have specific high affinity receptors for Hu IFN-γ, with an apparent K_d of 2×10^{-10} M, which is of the same order of magnitude as the affinity constant of the other IFN-γ receptors. The number of receptors on the platelet membrane is about 200, and one can calculate from this that the total number of IFN-γ receptors available on platelets is about 4.5×10^{10} per milliliter of blood, which is higher than the total number of IFN-γ lymphocyte receptors for the same volume. Platelets have no receptors for IFN-α (Molinas et al., 1987).

D. Different Genes Code for the IFN-α/β and the IFN-γ Receptors

The structure of the IFN receptors has not yet been determined and no DNA probes are available for mapping the structural genes by segregation analysis of the restriction fragment length polymorphism. The evidence for the chromosomal localization of IFN receptor genes is, therefore, solely derived from measurements of biological activity and of IFN binding.

1. The Gene for the Human IFN-α/β Receptor (IFRC Locus) Is on Chromosome 21

A locus determining the sensitivity of human cells to the antiviral and antiproliferative action of Hu IFN-α and Hu IFN-β has been assigned to chromosome 21, using mouse–human somatic cell hybrids. In these hybrids the absence of chromosome 21 results in total insensitivity to IFN (Tan et al., 1973). However, in spite of their sensitivity to the other human IFN species, mouse–human somatic cell hybrids containing chromosome 21 as the only human chromosome are not sensitive to Hu IFN-γ and the cell membranes of such hybrids have high affinity binding sites for Hu IFN-α and Hu IFN-β, but not for Hu IFN-γ (Raziuddin et al., 1984; Fournier et al., 1985). The exact position of the *IFRC* locus is on the distal part of the long arm of chromosome 21 in band q21-qter (Epstein and Epstein, 1976).

Down's syndrome is a genetic disease that is due to an extra copy of chromosome 21; such patients, therefore, have an extra copy of the *IFRC* locus. Cells derived from individuals with trisomy 21 are more sensitive to the antiviral and antiproliferative actions of IFN-α/β, indicating a gene dosage effect. This is confirmed by the observation that the extra copy of the *IFRC* locus results in an increased number of receptors, as shown by the direct proportionality between the number of chromosomes 21 in monosomic, disomic (diploid), and trisomic human fibroblasts and the number of specific Hu IFN-α receptors at the surface of these cells (Epstein et al., 1982). Moreover, the relative quantities of the 2-5A synthetase and of three other IFN-induced proteins correspond closely to the ratios expected on the basis of a gene dosage relationship (Fig. 4.1) (Weil et al., 1980, 1983). The increased sensitivity of trisomic fibroblasts to the antiviral and anticellular action of IFN shows that there can be a direct correlation between the number of high affinity receptors and the intensity of IFN-induced effects.

Lymphocytes from patients with trisomy 21 have three times more binding sites for Hu IFN-α than lymphocytes from normal individuals (Mogensen et al., 1982). This increase is greater than what one would expect on the basis of a strict gene dosage effect and therefore is different from what one sees in fibroblasts. Apparently other factors influence the binding efficiency of IFN in lymphocytes of these patients.

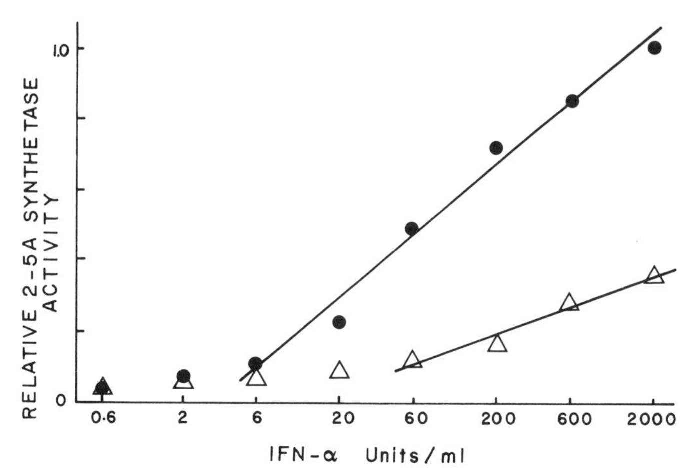

Figure 4.1 In human cells trisomic for chromosome 21, the extra copy of the *IFRC* locus, coding for the IFN-α and IFN-β receptor, results in an increased number of receptors and an increased sensitivity of the cells to IFN action. This is shown by the induction of 2-5A synthetase in diploid and trisomic cells by Hu IFN-α. Trisomic strain: ●; diploid strain: △. (Adapted from Weil et al., 1983.)

2. The Gene for the Mu IFN-α/β Receptor (Ifrec Locus) Is on Chromosome 16

The somatic cell hybrid approach, successfully used for the chromosomal assignment of the human *IFRC* locus, has also revealed the chromosomal assignment of the *Ifrec* locus. The sensitivity of hamster–mouse somatic cell hybrids to Mu IFN-α/β correlates with the presence of murine chromosome 16 (Cox et al., 1980). In humans the *IFRC* locus is linked to the superoxide dismutase locus (*SOD*-1), and a similar linkage has been found in the mouse between *Ifrec* and *Sod*-1 (Lin et al., 1980). This demonstration of evolutionary conserved linkage between the IFN receptor locus and the superoxide dismutase locus in humans and mice provides at the same time indirect evidence for the correspondence between *IFRC* and *Ifrec*.

The *Ifrec* locus has not been tested for sensitivity to Mu IFN-γ in somatic cell hybrids, and therefore one can not make a definite statement concerning the exclusive affinity of the gene product of this locus for Mu IFN-α/β. Yet, since receptor-binding studies on murine cells show that the receptor for Mu IFN-γ is different, we can infer that *Ifrec* only codes for Mu IFN-α/β receptors.

3. The Gene Encoding the Human IFN-γ Receptor or Its Binding Component Is on Chromosome 6

We have seen how in the absence of molecular probes the *IFRC* locus in humans and the *Ifrec* locus in mice have been mapped in somatic cell hybrids by correlating biological activity with the presence of a given chromosome, and in the case of *IFRC*, by correlating specific binding as well. The approach for mapping the Hu IFN-γ receptor has similarly made use of somatic hamster–human or mouse–human cell hybrids. Radiolabeled recombinant Hu IFN-γ can be crosslinked with disuccinimidyl suberate to its receptor on human cells, and an IFN-receptor complex of a molecular size of 117,000 dalton can then be isolated. Such a complex is not formed when Hu IFN-γ interacts with normal hamster or mouse cells, but it can be isolated after the interaction of IFN-γ with those somatic cell hybrids that carry human chromosome 6. The presence of the long arm of chromosome 6 is necessary and sufficient for the formation of the complex. This is a good indication that the IFN-γ receptor, or at least its binding subunit, is encoded by a gene on the long arm of chromosome 6. There is, however, an intriguing difference with what one observes when the IFN-α/β receptor is studied in somatic cell hybrids. Although the binding of IFN-γ correlates with the presence of chromosome 6, biological activity, as measured by the induction of the antiviral state, does not (Rashidbaigi et al., 1986). This suggests that, like the IL-2 receptor (see Chapter 14), the IFN-γ receptor could consist of different subunits, of which only the binding component would be encoded by chromosome 6. Alternatively, it is also possible that in order to exert its biological activity, IFN-γ needs other species-specific proteins, coming into play after receptor binding and internalization, and encoded by genes on other chromosomes. A gene on chromosome 21 appears to be involved, since the presence of human chromosome 21, in addition to human chromosome 6, in hamster-human or mouse-human somatic cell hybrids is necessary and sufficient to cause enhanced HLA cell surface antigen expression after IFN-γ treatment of the cells (Jung et al., 1987).

4. *The Gene Encoding the Murine IFN-γ Receptor (Ifgr Locus) is on Mouse Chromosome 10*

Binding and cross-linking reactions using recombinant Mu IFN-γ on mouse-hamster cell hybrids reveal that the presence of mouse chromosome 10 is necessary and sufficient for the formation of the major 95-125 kDa cross-linked complex that characterizes Mu IFN-γ binding. The formation of this complex, however, is not sufficient to induce the antiviral state; for binding to result in biological activity, genes on other mouse chromosomes appear to be required. This provides an exact parallel to the situation in human cells. The *Ifgr* locus on murine chromosome 10 is probably closely linked to the gene encoding the murine protooncogene c-*myb* (Mariano et al., 1987).

II. INTERNALIZATION OF THE IFN-RECEPTOR COMPLEX

A. Evidence for Internalization

Ligand binding is usually measured at 4°C, and to follow the fate of the ligand-receptor complex, one has to bring the cells back to 37°C. When this is done, part of the cell-bound IFN is internalized and subsequently degraded. Chemical compounds, that are known to inhibit degradation of receptor-bound ligands, such as chloroquine, which inhibits lysosomal activity, are also able to inhibit the degradation of IFN. This suggests that at least some of the receptor-bound IFN ends up in lysosomes where it is degraded (Branca et al., 1982). One can also follow the fate of the receptor-bound IFN visually, although the low number of receptor sites per cell does not make this easy. By coupling them to colloidal gold, one can confer an electron dense label to IFN molecules to follow their fate with the electron microscope. This technique has shown that recombinant Hu IFN-α2, bound to MDBK cells at 4°C, is clustered in coated pits as soon as 1 minute after the cells have been brought back to 37°C. Several minutes later IFN is found in receptosomes, and finally in lysosomes (Fig. 4.2). What one actually observes is, of course, only the colloidal gold particles, and their presence could mean either the presence of IFN or of IFN degradation products (Zoon et al., 1983). Electron microscopic observation of the fate of Mu IFN-β bound to mouse L-cell receptors, as visualized by ferritin labeled IgG, also

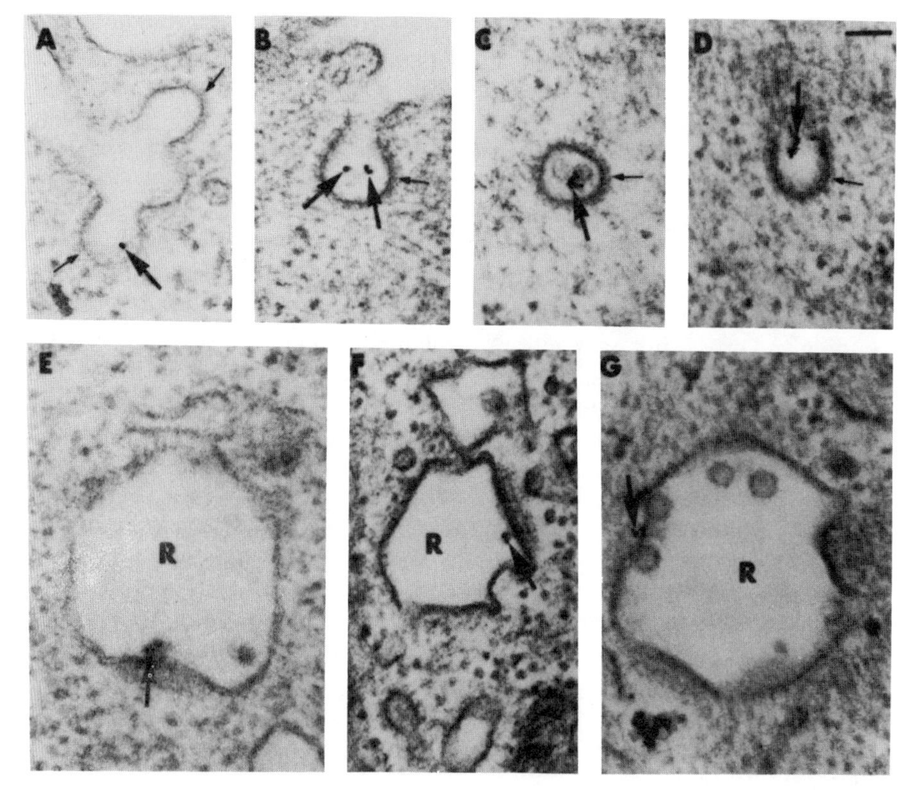

Figure 4.2 Electron microscopic visualization of the internalization of Hu IFN-αA labeled with colloidal gold. Labeled IFN (arrows) clusters in clathrin-coated pits on the cell surface (A, B, C, and D) and in receptosomes inside the cell (E, F, and G). Magnification: ×90.000. (From Zoon *et al.,* 1983, with permission.)

shows the presence of IFN in coated pits a few minutes after binding (Kushnaryov et al., 1983, 1985). Evidently, at least part, if not all, of receptor-bound IFN-α and IFN-β enters the cell by receptor-mediated endocytosis via the coated pit pathway and is eventually degraded by lysosomal enzymes.

Are the IFN receptors degraded as well or are they recycled back to the cell surface like some other polypeptide receptors? This question can only be approached indirectly, since there is presently no way for specifically labeling the IFN receptor except by complex formation with labeled IFN. When recombinant Hu IFN-α2 binds to receptors of the human T-cell line Molt 4, the internalization of IFN, followed by its degradation, is accompanied by a marked decline in the ability of the cells to bind more IFN. This suggests that the IFN-α receptors do not return to the cell surface and are down-regulated as a result of degradation (Evans and Secher, 1984). Down-regulation of the IFN-α/β receptor occurs also on human Daudi cells as a result of interaction with Hu IFN-αA or Hu IFN-β and on normal lymphocytes after the binding of Hu IFN-α2. Regeneration of receptors depends on protein synthesis, indicating that at least some of the "used" receptors are not recycled back to the cell membrane (Branca and Baglioni, 1982; Lau et al., 1986).

The plasma membrane is in a state of continuous turnover, and surface molecules, including receptors, are constantly internalized and either recycled back to the surface or degraded and replaced by newly synthesized ones. Is the internalization of the IFN receptor activated because it is occupied by an IFN molecule or is internalization of the ligand-receptor complex a random event, not resulting from receptor occupancy, but from rather, the normal flow of surface molecules? To put it more colorfully, Does IFN push the down button of an elevator, or does it just jump onto a running conveyor belt? This relates to the general question of the initiation of endocytosis: Do endocytotic vacuoles form constitutively or are they induced by the ligand-receptor complex? The events that link ligand binding and endocytosis are largely unknown, and there is evidence for both constitutive and induced endocytosis. Apparently, the clustering of the ligand and receptor on the cell membrane can in some instances trigger endocytosis: This seems to occur in phagocytosis, which is, however, a highly specialized form of endocytosis from which one can probably not extrapolate to other systems (Steinman et al., 1983).

In the case of IFN-α, analysis of steady-state binding of rec Hu IFN-α2 to MDBK cells, and of subsequent internalization and degradation, indicates that occupied and unoccupied receptors are cleared from the cell surface at approximately the same rate. For this particular IFN-cell combination this rate has been calculated to range from 38 to 50 receptors per cell per min, and the half-life of internalized ligand has been calculated to be 48 minutes (Zoon et al., 1986b). This is in contrast to observations with EGF or insulin, in which the ligand does induce an increase in the rate of receptor internalization (Krupp and Lane, 1981; Wiley and Cunningham, 1981).

If the binding of IFN to its receptor does not increase the rate of receptor endocytosis, what then could be the mechanism of down-regulation? We know that the number of receptors for growth factors or polypeptide hormones can be modulated not only by the proper ligand, but also by heterologous ligands: This means that a decrease in receptors or receptor affinity can have more causes than direct ligand binding. For example, binding of the platelet-derived growth factor to its own receptor can decrease the binding of the epidermal growth factor (EGF) to the corresponding receptor, and the fibroblast-derived growth factor can modulate EGF receptors (Rozengurt et al., 1982; Bowen-Pope et al., 1983). IFNs can do the same: rec IFN-α2 can decrease the number of insulin-binding sites on human lymphoblastoid cells and the number of EGF-binding sites on bovine MDBK cells, and Hu IFN-γ can modify the expression of the Hu IFN-α receptor and decrease the specific binding of Hu IFN-α by increasing the apparent K_d of the IFN-α-receptor interaction (Faltynek et al., 1984; Hannigan et al., 1984a; Zoon et al, 1986a). It is, therefore, theoretically possible that the decrease in IFN receptors is not a direct result of ligand binding, that is, of bona fide down-regulation, but results indirectly from other IFN-induced cell-surface changes. Steady-state studies on other IFN-cell combinations in addition to the Hu IFN-α-MDBK system, which is somewhat artificial since human IFN is made to interact with the bovine IFN-α receptor, should tell us whether the lack of accelerated ligand-receptor endocytosis is a general property of IFN receptors.

B. Is Internalization of IFN or of the IFN-Receptor Complex Necessary for IFN action?

This is a highly debated issue, not only for IFNs but also for many other ligands, for which there is presently no definite answer.

When Mu IFN-α/β is covalently coupled to Sepharose beads, it can still induce the antiviral state in cells (Ankel et al., 1973). The straightforward interpretation of this observation is that IFNs do not have to enter the cell to exert their antiviral action. However, as those working with antibody-sepharose columns well know, there is always leakage from such systems, and one cannot exclude that some free IFN molecules, after binding to the receptor, do get into the cell and generate the signals that are necessary for their action.

When the receptor is bypassed by microinjecting IFNs α or β directly into the cell, no antiviral state is induced (Higashi and Sokawa, 1982; Huez et al., 1983). Again, this is an incomplete demonstration, since receptor-mediated endocytosis may introduce IFN into cellular compartments that are different from those reached by microinjection, and furthermore, we cannot exclude the possibility that IFN molecules are processed when they enter cells the physiological way.

Instead of microinjecting IFN, one can microinject anti-IFN antibodies and then treat the cells with IFN, with the aim of neutralizing any IFN that enters the cell, and thus preventing its action. Microinjection into nucleus or cytoplasm of anti-Hu IFN-α antibodies does not prevent the establishment of the antiviral state, suggesting that the intracellular presence of IFN is not required (Arnheiter and Zoon, 1984). Such an experiment, however, is only unambiguously interpretable in the case of a positive result, since again, one cannot exclude the processing of IFN molecules after they enter the cell, the protection of IFN epitopes by binding to the receptor, or the differential compartimentalization of endocytosed IFN and injected antibody.

Genetic engineering offers another way for bypassing the receptor, and some experiments indicate that IFN-γ can induce an antiviral state without having to bind to its receptor first. Mouse cells can be transformed with a truncated cDNA encoding mature Hu IFN-γ protein lacking the signal peptide. The Hu IFN-γ produced in this way accumulates to high levels in the murine cells, stimulates the 2-5A synthetase system, and induces an antiviral state in these cells. Since Hu IFN-γ is normally not active on murine cells, the species-specific block appears to reside at the receptor level and not at any subsequent intracellular stage (Sanceau et al., 1987).

C. Nuclear Membranes Contain High Affinity Sites for Mu IFN-β and Mu IFN-γ

Several peptide hormones, particularly gonadotrophins, insulin, and EGF, become associated with the target cell nucleus after internalization. Distinct nuclear-binding sites have been proposed for insulin and EGF, with the implication that these sites are important for the action of these polypeptides. This is a highly debated issue, since when doing direct binding studies with isolated nuclei, it is very difficult to exclude the fact that the nuclei are not contaminated with other cellular components that can have receptors. This uncertainty compromises to some extent the interpretation of receptor-binding studies with isolated nuclei (King and Cuatrecasas, 1981).

The same issue arises when the interesting question of nuclear receptors for IFNs is approached. Immunoferritin labeled Mu IFN-β, after binding to the plasma membrane receptor and internalization via coated pits, can be found around the nucleus and Golgi apparatus by about 3 minutes and in lysosomes after about 5 minutes. Isolated nuclear membranes from murine L929 fibroblasts bind both Mu IFN-β and Mu IFN-γ in a receptor-specific way, with an apparent Kd of about 5×10^{-10}, and the IFN-γ-binding sites are distinct from those that bind IFN-β (Figs. 4.3 and 4.4). The function of these sites in gene activation by IFNs is not known (Kushnaryov et al., 1985; MacDonald et al., 1986).

D. Inhibition of the Intracellular Breakdown of IFN Does Not Seem to Affect Gene Activation by IFN

After ligand-receptor complexes have entered the cell via coated pits and have been taken up by endosomes, many polypeptide ligands move further

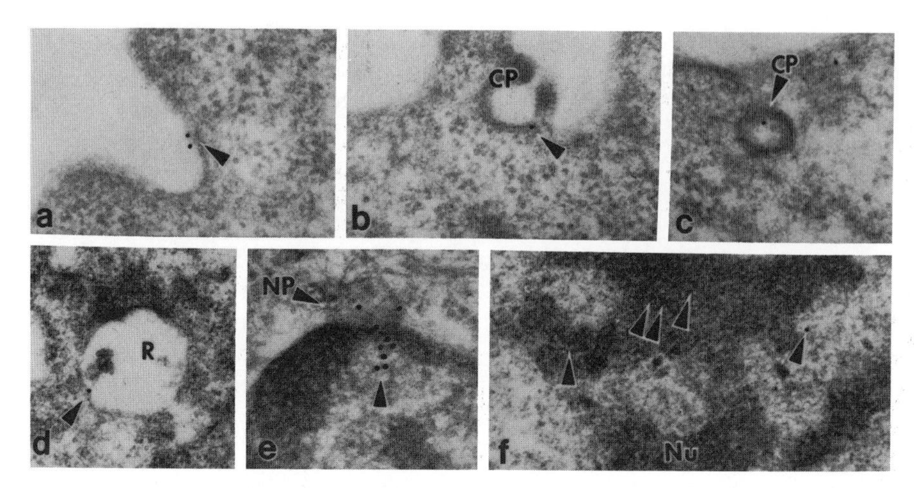

Figure 4.3 Electron microscopic visualization by a postembedding immunogold technique of receptor-mediated endocytotic movement of Mu IFN-γ through the cytoplasm and into the nucleus of murine L929 fibroblasts. IFN-γ is first located in coated pits (a and b), then in coated vesicles (c) and in receptosomes (d), in a nuclear pore (e) and inside the nucleus (f). In a time interval not exceeding 2 minutes the majority of the label is located inside the nucleus. (From MacDonald et al., 1986, with permission).

on to lysosomes, where they are degraded, sometimes together with their receptor. The lysosomotropic amine, methylamine, has been shown to inhibit the degradation of polypeptide hormones. It inhibits, for example, the clustering and internalization of EGF-receptor complexes in fibroblasts. The mechanism of this is obscure; it could be due to the raising of the intracellular pH, interfering with protein-protein interactions, or by interfering with transglutaminase activity by binding calcium (Maxfield et al., 1979).

Is intracellular breakdown of IFNs necessary for activity? Methylamine inhibits the internalization of Hu IFN-α, subsequent to receptor binding on Daudi cells (Branca et al., 1982). Treatment of human glioblastoma T98G cells with methylamine, followed by transcriptional analysis of IFN-induced genes by nuclear runoff assays, provides evidence against the necessity of intracellular breakdown of IFN in the transcriptional response to Hu IFN-α2. Transcription of several IFN-induced genes, including the 2-5A synthetase gene, is activated as early as 5 minutes following the exposure of human glioblastoma cells to IFN-α (Fig. 4.5). By correlating kinetics of binding with the nuclear runoff assays, one can establish a direct temporal relationship between receptor occupancy and transcriptional activation of the IFN-induced genes. The prevention by methylamine of degradation of internalized IFN does not affect the transcriptional response of these IFN-induced genes, suggesting that receptor occupancy and not degradation is the determining factor in the induction of transcription (Hannigan and Williams, 1986). One can, however, still not completely exclude a role for internalized, nondegraded, IFN receptor complexes, since it is virtually impossible to

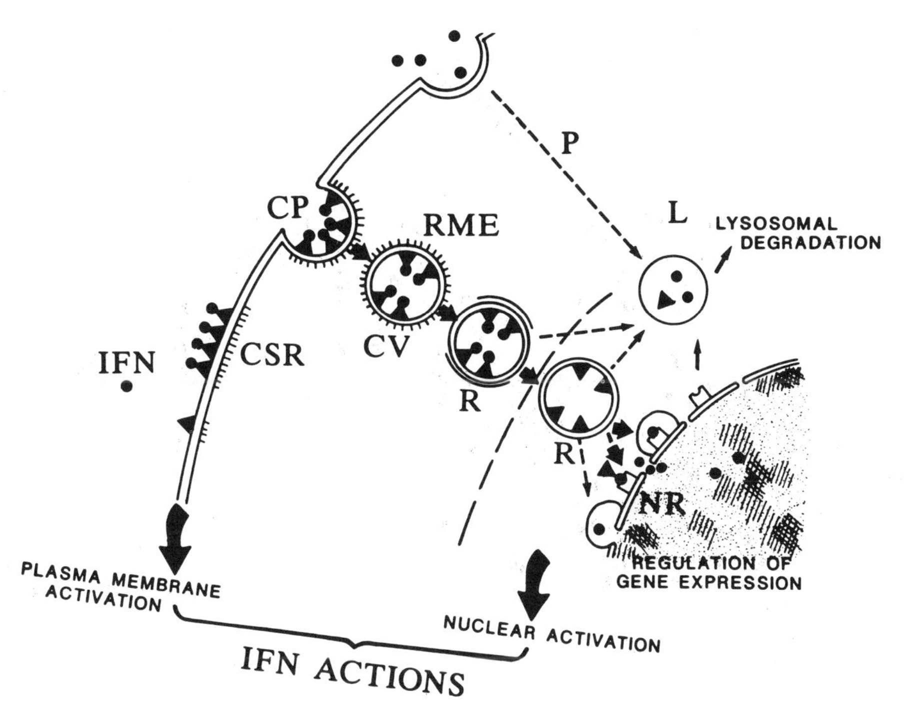

Figure 4.4 Scheme of internalization of Mu IFN-β and IFN-γ by mouse L929 fibroblasts. IFN enters the cells in two ways: by pinocytosis (P) for delivery mostly to lysosomes (L) and by receptor-mediated endocytosis (RME). IFN binds to cytoplasmic surface receptors (CSR) and in coated pits (CP) and is translocated by coated vesicles (CV) and receptosomes to nuclear receptors and to nuclear pores. (Diagram provided by and derived from the data of V. Kushnaryov, H. MacDonald, J. J. Sedmak, and S. E. Grossberg, The Medical College of Wisconsin). As discussed in section II C, the significance of binding to nuclear receptors is a controversial issue.

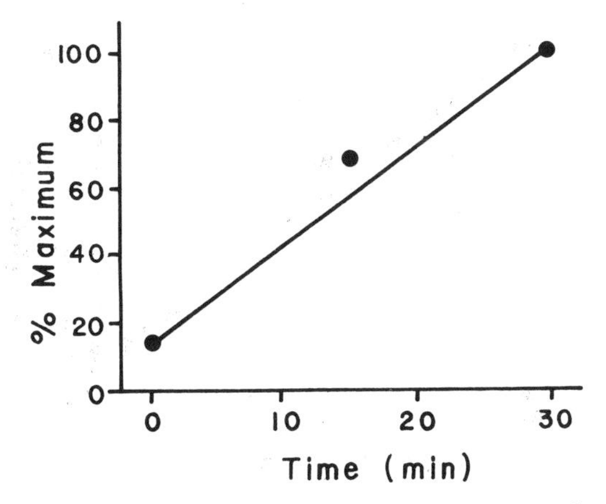

Figure 4.5 Transcription of several IFN-induced genes starts as early as 5 minutes after receptor binding. The figure shows rapid transcriptional induction of 2-5A synthetase by Hu IFN-α2 in T98G cells as measured by nuclear runoff reactions. (Adapted from Hannigan and Williams, 1986.)

ascertain that the prevention of internalization by methylamine is 100% efficient.

III. CHARACTERIZATION OF IFN RECEPTORS

Competitive binding studies with the different IFN-α subspecies and IFN-β show that they all compete for the same binding sites. This does not, however, tell us that there is only one receptor, and theoretically there could be several different types of receptors with comparable affinities. In addition, evidence indicates the existence of receptors with different affinities, whose expression is influenced by the physiological state of the cell. Nonproliferating human lymphoid cells display only one high affinity binding component, whereas proliferating cells have in addition a lower affinity component, which is expressed to at least ten times the copy number, per cell, of the high affinity component (Hannigan et al., 1986).

It is, therefore, possible that more than one IFN receptor will have to be characterized.

A. The IFN-α/β Receptor

Receptor-bound Hu IFN-α2 can be covalently crosslinked to the binding part of the receptor, and one can then isolate the ligand-receptor complex for further characterization. The size of the complex isolated in this way, as measured by SDS gel electrophoresis, is approximately 140 to 150 kDa. If one molecule of IFN-α is present in such a complex, the molecular mass of the receptor, or of its binding subunit, should be around 125,000 (Joshi, 1982; Faltynek et al., 1983).

B. The IFN-γ Receptor

Crosslinking of rec Mu IFN-γ to its receptor on mouse L1210 lymphoid cells yields a complex with a MW of approximately 110 kDa. The molecular weight of the binding unit is probably around 95 kDa, assuming that one molecule of IFN-γ is bound per receptor (Wietzerbin et al., 1986). Other estimations range from 75 to 85 kDa. These values are lower than those reported for the human IFN-γ receptor, which could reflect differences in the size of the receptor polypeptides or in the degree of the posttranslational modification of the proteins (Langer et al., 1986b).

IV. WHICH DOMAINS OF THE IFN MOLECULES BIND TO THE RECEPTOR?

Receptor binding by IFNs is by and large species-specific, and therefore, this question corresponds to asking at the same time, which structural features of the IFN molecule are responsible for species specificity? Like antigenic determinants, these features could be continuous or discontinuous. Discontinuous determinants are residues from different parts of the sequence, brought together by the folding of the polypeptide chain to its native structure.

Rec Hu IFN-α, truncated at either the amino or carboxy terminal, retains its antiviral activity. We know, therefore, that the first 15 amino-terminal or last 13 carboxy-terminal residues can be dispensed with for receptor binding (Levy et al., 1981; Streuli et al., 1981; Franke et al., 1982). The 10–16 carboxy terminal residues of receptor-bound Hu IFN-αA remain accessible to monoclonal antibodies that are specifically directed against them: This is additional evidence that this part of the molecule is not directly interacting with the receptor (Arnheiter et al., 1983).

Since all Hu IFN-α subtypes, as well as Hu IFN-β bind to the same receptors, one can search for common structural elements by looking for positions at which residues are strictly conserved. The 165 or 166 amino acid sequence of these IFNs contains 35 such positions (see Chapter 2, Table 2.1). The synthesis, cloning, and expression of synthetic IFN-α genes make it possible to introduce changes into these conserved positions and thus to investigate structure-function relationships. When the conserved amino acids Leu30, Asp32, and Arg33 of Hu IFN-α2 are all replaced with alanine residues, the molecule loses its antiviral and antiproliferative activity in human and bovine cells. When one introduces single amino acid changes at each of these three conserved positions, the molecules retain activity, but less than the unchanged protein. At position 30, replacement of leucine with the approximately isosteric but hydrophylic asparagine leads to a 100-fold drop in activity. The acidic side chain of aspartic acid 32 can be extended by replacement with glutamic acid, neutralized by replacement with asparagine or deleted by replacement with alanine without any effect on activity. The replacement of Arg 33 with alanine results in a 1,000-fold drop in specific activity, whereas replacement by methionine gives an analog with no detectable antiviral activity. It is clear that the biological activity of Hu IFN-α2 is very sensitive to changes at the conserved residues in positions 30, 32, and 33 (Camble et al., 1986). The presence of the side chains of these three amino acids is an absolute requirement for the induction of biological activity on human and bovine cells. It is also an absolute requirement for binding to human cells, since the analog (Ala 30, 32, 33) Hu IFN-α2 no longer binds to the IFN-α receptor on human cells. This analog, however, in spite of its lack of biological activity on bovine MDBK cells, still binds to the IFN-α receptor on these cells, with an affinity only slightly lower than that of normal Hu IFN-α2. As a result, it competes for binding with Hu IFN-α2 and acts as a competitive antagonist by inhibiting both antiviral and antiproliferative activity of native Hu IFN-α (Marcucci and De Maeyer, 1986).

This shows that the highly conserved cluster of amino acids 30, 32, and 33 is necessary for receptor binding on human cells; these amino acids are most probably part of the domain that interacts directly with the receptor. In this respect it may be revealing that Hu IFN-α1, which is the only IFN-α subtype to have a methionine residue in position 31 instead of a lysine, binds with less affinity than the other IFN-α species (Uzé et al., 1985).

Since the IFN analog (Ala 30, 32, 33) Hu IFN-α2 binds in a specific way and with high affinity to the IFN-α receptor on bovine cells without inducing the antiviral state or without inhibiting cell proliferation, it is clear that, at least in the heterologous system of human IFN and bovine cells, receptor binding is not sufficient to trigger activity. Something additional, not provided by the structural features of this analog, is apparently required. We do not know what this is; one possibility would be that the analog has enough steric resemblance for binding, but that it lacks the features to induce the activation step following binding, which has been postulated for Hu IFN-α (Mogensen and Bandu, 1983).

The conserved amino acids at positions 30, 32, and 33 are obviously very important for the receptor binding of Hu IFN-α, since when replaced by alanine residues, receptor binding is abolished. However, although this domain of the Hu IFN-α molecule is necessary for binding, it is apparently not sufficient, since the same amino acids (leucine, aspartic acid, and arginine at positions 30, 32, and 33, respectively) occur at identical positions in the Mu IFN-α molecules, which display much less binding affinity for the human receptor. It is, nevertheless, revealing that Mu IFN-β, which contrary to Mu IFN-α has not conserved these three amino acids, has no activity on human cells.

Additional regions of the Hu IFN-α molecules are believed to be important for receptor binding. A binding site located somewhere within amino acids 1-60 (and which may or may not correspond to positions 30, 32, and 33), and a second site within amino acids 93 to 150, has been suggested by the use of two rec Hu IFN-α hybrids (Meister et al., 1986). The presence of two residues in Hu IFN-α, Lysine at position 121, and Arginine at position 125, greatly increases activity in murine cells (Weber et al., 1987). These residues are naturally already present in Hu IFN-α1, which has a higher activity on murine cells than the other Hu IFN-α subtypes. Interestingly, the same residues are found in almost identical positions (122 and 126) in most Mu IFN-α subtypes. This is in line with the suggestion that the N-terminal part of the IFN-α molecule strongly influences activity on human cells, whereas the C-terminal moiety is relatively more important for activity on murine cells (Streuli et al., 1981).

Another example, albeit somewhat different, of binding without inducing biological activity is provided by rec Hu IFN-αJ. This IFN binds to IFN-α receptors of natural killer cells, without boosting killer activity, and it competes with Hu IFN-αA for binding sites on these cells. One could say that it acts as an analog competitor, but only on NK cells. It has normal antiviral and antiproliferative activities on human cells (Langer et al, 1986a). Obviously, binding to the receptor, even at the specific, high affinity sites occuped by IFNs, does not always start the chain of events resulting in gene activation.

REFERENCES

Affabris, E., Romeo, G., Belardelli, F., Jemma, C., Mechti, N., Gresser, I., and Rossi, G. B. 2-5A synthetase activity does not increase in interferon-resistant Friend leukemia cell variants treated with α/β interferon despite the presence of high-affinity interferon receptor sites. *Virology* 125: 508–512 (1983).

Aguet, M., Belardelli, F., Blanchard, B., Marcucci, F., and Gresser, I. High affinity binding of ^{125}I-labeled mouse interferon to a specific cell surface receptor. IV. Mouse γ interferon and cholera toxin do not compete for the common receptor site of α/β interferon. *Virology* 117: 541–544 (1982).

Aguet, M. and Blanchard, B. High affinity binding of ^{125}I-labeled mouse interferon to a specific cell surface receptor. II. Analysis of binding properties. *Virology* 115: 249–261 (1981).

Aguet, M., Gresser, I., Hovanessian, A. G., Bandu, M. T., Blanchard, B., and Blangy, D. Specific high-affinity binding of ^{125}I-labeled mouse interferon to interferon resistant embryonal carcinoma cells in vitro. *Virology* 114: 585–588 (1981).

Aguet, M., Grobke, M., and Dreiding, P. Various human interferon-α subclasses cross-react with common receptors: Their binding affinities correlate with their specific biological activities. *Virology* 132: 211–216 (1984).

Anderson, P., Yip, Y. K., and Vilcek, J. Human interferon-γ is internalized and degraded by cultured fibroblasts. *J. Biol. Chem.* 258: 6497–6502 (1982).

Ankel, H., Chany, C., Galliot, B., Chevalier, M. J., and Robert M. Antiviral effect of interferon covalently bound to sepharose. *Proc. Natl. Acad. Sci.* (USA) 70: 2360–2363 (1973).

Arnheiter, H., Ohno, M., Smith, M., Gutte, B., and Zoon, K. C. Orientation of a human leukocyte interferon molecule on its cell surface receptor: Carboxyl terminus remains accessible to a monoclonal antibody made against a synthetic interferon fragment. *Proc. Natl. Acad. Sci.* (USA) 80: 2539–2543 (1983).

Arnheiter, H. and Zoon, K. C. Microinjection of anti-interferon antibodies into cells does not inhibit the induction of an antiviral state by interferon. *J. Virol.* 52: 284–287 (1984).

Baglioni, C., Branca, A. A., D'Alessandro, S. B., Hossenlopp, D., and Chadha, K. C. Low interferon binding activity of two human cell lines which respond poorly to the antiviral and antiproliferative activity of interferon. *Virology* 122: 202–206 (1982).

Bowen-Pope, D. F., Dicorleto, P. E., and Ross, R. Interactions between the receptors for platelet-derived growth factor and epidermal growth factor. *J. Cell Biol.* 96: 679–683 (1983).

Branca, A. A. and Baglioni, C. Evidence that types I and II interferons have different receptors. *Nature* 294: 768–770 (1981).

Branca, A. A. and Baglioni, C. Down-regulation of the interferon receptor. *J. Biol. Chem.* 257: 13197–13200 (1982).

Branca, A. A., Faltynek, C. R., D'Alessandro, S. B., and Baglioni, C. Interaction of interferon with cellular receptors. Internalization and degradation of cell-bound interferon. *J. Biol. Chem.* 257: 13291–13296 (1982).

Camble, R., Petter, N. N., Trueman, P., Newton, C. R., Carr, F. J., Hockney, R. C., Moore, V. E., Greene, A. R., Holland, D., and Edge, M. D. Functionally important conserved amino-acids in interferon-α2 identified with analogues produced from synthetic genes. *Biochem. Biophys. Res. Commun.* 134: 1404–1411 (1986).

Cox, D. R., Epstein, L. B., and Epstein, C. J. Genes coding for sensitivity to interferon (IfRec) and soluble superoxide dismutase (SOD-1) are linked in mouse and man and map to mouse chromosome 16. *Proc. Natl. Acad. Sci.* (USA) 77: 2168–2172 (1980).

Czarniecki, C. W., Fennie, C. W., Powers, D. B., and Estell, D. A. Synergistic antiviral and antiproliferative activities of *Escherichia coli* derived human α, β and γ interferons. *J. Virol.* 49: 490–496 (1984).

Dron, M., Tovey, M. G., and Uze, G. Isolation of Daudi cells with reduced sensitivity to interferon. IV. Characterization of clones with altered binding of human interferon α subspecies. *J. Gen. Virol.* 67: 663–669 (1986).

Epstein, L. B. and Epstein, C. J. Localization of the gene AVG for the antiviral expression of immune and classical interferon to the distal portion of the long arm of chromosome 21. *J. Infect. Dis.* 133: A56–A62 (1976).

Epstein, C. J., McManus, N. H., Epstein, L. B., Branca, A. A., D'Alessandro, S. B., and Baglioni, C. Direct evidence that the gene product of the human chromosome 21 locus, *IFRC,* is the interferon-α receptor. *Biochem. Biophys. Res. Commun.* 107: 1060–1066 (1982).

Evans, T. and Secher, D. Kinetics of internalization and degradation of surface-bound interferon in human lymphoblastoid cells. *EMBO J.* 3: 2975–2978 (1984).

Faltynek, C. R., Branca, A. A., McCandless, S., and Baglioni, C. Characterization of an interferon receptor on human lymphoblastoid cell lines. *Proc. Natl. Acad. Sci.* (USA) 80: 3269–3273 (1983).

Faltynek, C. R., McCandless, S., and Baglioni, C. Treatment of lymphoblastoid cells with interferon decreases insulin binding. *J. Cell Physiol.* 121: 437–441 (1984).

Faltynek, C. R., Princler, G. L., and Ortaldo, J. R. Expression of IFN-α and IFN-γ receptors on normal human small resting T lymphocytes and large granular lymphocytes. *J. Immunol.* 136: 4134–4139 (1986).

Finbloom, D. S., Hoover, D. L., and Wahl, L. M. The characteristics of binding of human recombinant interferon-γ to its receptor on human monocytes and human monocyte-like cell lines. *J. Immunol.* 135: 300–305 (1985).

Fleischmann, W. R., Jr., Georgiades, J. A., Osborne, L. C., and Johnson, H. M. Potentiation of interferon activity by mixed preparations of fibroblast and immune interferon. *Infect. Immun.* 26: 248–253 (1979).

Fournier, A., Zhang, Z. Q., and Tan, Y. H. Human $\beta:\alpha$ but not α interferon binding site is a product of the chromosome 21 interferon action gene. *Somat. Cell Mol. Genet.* 11: 291–295 (1985).

Franke, A. E., Shepard, H. M., Houck, C. M., Leung, D. W., Goeddel, D. V., and Lawn, R. M. Carboxyterminal region of hybrid leukocyte interferons affects antiviral specificity. *DNA* 1: 223–230 (1982).

Goldstein, J. L., Anderson, R. G. W., and Brown, M. S. Coated pits, coated vesicles, and receptor-mediated endocytosis. *Nature* 279: 679–685 (1979).

Hannigan, G. E., Fish, E. N., and Williams, B. R. G. Modulation of human interferon-α receptor expression by human interferon-γ. *J. Biol. Chem.* 259: 8084–8086 (1984a).

Hannigan, G. E., Gewert, D. R., and Williams, B. R. G. Characterization and regulation of α-interferon receptor expression in interferon-sensitive and -resistant human lymphoblastoid cells. *J. Biol. Chem.* 259: 9456–9460 (1984b).

Hannigan, G. E., Lau, A. S., and Williams, B. R. G. Differential human interferon-α receptor expression on proliferating and nonproliferating cells. *Eur. J. Biochem.* 157: 187–193 (1986).

Hannigan, G. E. and Williams, B. R. G. Transcriptional regulation of interferon-responsive genes is closely linked to interferon receptor occupancy. *EMBO J.* 5: 1607–1613 (1986).

Higashi, Y. and Sokawa, Y. Microinjection of interferon and 2′, 5′-oligoadenylate into mouse L cells and their effects on virus growth. *J. Biochem.* 91: 2021–2028 (1982).

Huez, G., Silhol, M., and Lebleu, B. Microinjected interferon does not promote an antiviral response in HeLa cells. *Biochem. Biophys. Res. Commun.* 110: 155–160 (1983).

Jacobs, S., Chang, K.-J., and Cuatrecasas, P. Antibodies to purified insulin receptor have insulin-like activity. *Science* 200: 1283–1284 (1978).

Joshi, A. R., Sarkar, F. H., and Gupta, S. L. Interferon receptors. Cross-linking of human leukocyte interferon $\alpha 2$ to its receptor on human cells. *J. Biol. Chem.* 257: 13884–13887 (1982).

Jung, V., Rashidbaigi, A., Jones, C., Tischfield, J. A., Shows, T. B., and Pestka, S. Human chromosomes 6 and 21 are required for sensitivity to human interferon gamma. *Proc. Natl. Acad. Sci.* (USA) 84: 4151–4155 (1987).

King, A. C. and Cuatrecasas, P. Peptide hormone-induced receptor mobility, aggregation, and internalization. *N. Engl. J. Med.* 305: 77–88 (1981).

Krupp, M. and Lane, D. On the mechanism of ligand-induced down-regulation of insulin receptor level in the liver cell. *J. Biol. Chem.* 256: 1689–1694 (1981).

Kushnaryov, V. M., MacDonald, H. S., Sedmak, J. J., and Grossberg, S. E. Ultras tructural distribution of interferon receptor sites on mouse L fibroblasts grown in suspension: Ganglioside blockade of ligand binding. *Infect. Immun.* 40: 320–329 (1983).

Kushnaryov, V. M., MacDonald, H. S., Sedmak, J. J., and Grossberg, S. E. Murine interferon-β receptor-mediated endocytosis and nuclear membrane binding. *Proc. Natl. Acad. Sci.* (USA) 82: 3281–3285 (1985).

Langer, J. A., Ortaldo, J. R., and Pestka, S. Binding of human α-interferons to natural killer cells. *J. Interferon Res.* 6: 97–105 (1986a).

Langer, J. A., Rashidbaigi, A., and Pestka, S. Preparation of ^{32}P-labeled murine immune interferon and its binding to the mouse immune interferon receptor. *J. Biol. Chem.* 261: 9801–9804 (1986b).

Lau, A. S., Hannigan, G. E., Freedman, M. H., and Williams, B. R. G. Regulation of interferon receptor expression in human blood lymphocytes in vitro and during interferon therapy. *J. Clin. Invest.* 77: 1632–1638 (1986).

Levy, W. P., Rubinstein, M., Shively, J., Del Valle, U., Lai, C. Y., Moschera, J.,

Brink, L., Gerber, L., Stein, S., and Pestka, S. Amino acid sequence of a human leukocyte interferon. *Proc. Natl. Acad. Sci.* (USA) 78: 6186–6190 (1981).

Lin, P. F., Slate, D. L., Lawyer, F. C., and Ruddle, F. H. Assignment of the murine interferon sensitivity and cytoplasmic superoxide dismutase genes to chromosome 16. *Science* 209: 285–287 (1980).

MacDonald, H. S., Kushnaryov, V. M., Sedmak, J. J., and Grossberg, S. E. Transport of γ-interferon into the cell nucleus may be mediated by nuclear membrane receptors. *Biochem. Biophys. Res. Commun.* 138: 254–260 (1986).

Marcucci, F. and De Maeyer, E. An interferon analogue $Ala^{30,32,33}$ Hu IFN-α2 acting as a Hu IFN-α2 antagonist on bovine cells. *Biochem. Biophys. Res. Commun.* 134: 1412–1418 (1986).

Mariano, T. M., Kozak, C. A., Langer, J. A., and Pestka, S. The mouse immune interferon receptor gene is located on chromosome 10. *J. Biol. Chem.* 262: 5812–5814 (1987).

Maxfield, F. R., Willingham, M. C., Davies, P. J. A., and Pastan, I. Amines inhibit the clustering of α2-macroglobulin and EGF on the fibroblast cell surface. *Nature* 277: 661–663 (1979).

Meister, A., Uze, G., Mogensen, K. E., Gresser, I., Tovey, M. G., Grutter, M., and Meyer, F. Biological activities and receptor binding of two human recombinant interferons and their hybrids. *J. Gen. Virol.* 67: 1633–1643 (1986).

Mogensen, K. E. and Bandu, M. T. Kinetic evidence for an activation step following binding of human interferon-α2 to the membrane receptors of Daudi cells. *Eur. J. Biochem.* 134: 355–364 (1983).

Mogensen, K. E., Bandu, M. T., Vignaux, F., Aguet, M., and Gresser, I. Binding of ^{125}I-labeled human α-interferon to human lymphoid cells. *Int. J. Cancer* 28: 575–582 (1981).

Mogensen, K. E., Vignaux, F., and Gresser, I. Enhanced expression of cellular receptors for human interferon-α on peripheral lymphocytes from patients with Down's syndrome. *FEBS Lett.* 140: 285–287 (1982).

Molinas, F. C., Wietzerbin, J., and Falcoff, E. Human platelets possess receptors for a lymphokine: Demonstration of high specific receptors for Hu IFN-γ. *J. Immunol.* 138: 802–806 (1987).

Orchansky, P., Rubinstein, M., and Fischer, D. G. The interferon-γ receptor in human monocytes is different from the one in nonhematopoietic cells. *J. Immunol.* 136: 169–173 (1986).

Rashidbaigi, A., Kung, H. F., and Pestka, S. Characterization of receptors for immune interferon in U937 cells with ^{32}P-labeled human recombinant immune interferon. *J. Biol. Chem.* 260: 8514–8519 (1985).

Rashidbaigi, A., Langer, J. A., Jung, V., Jones, C., Morse, H. G., Tischfield, J. A., Trill, J. J., Kung, H. F., and Pestka, S. The gene for the human immune interferon receptor is located on chromosome 6. *Proc. Natl. Acad. Sci.* (USA) 83: 384–388 (1986).

Rasmussen, H. The calcium messenger system (First of two parts.) *New Engl. J. Med.* 314: 1094–1101 (1986).

Raziuddin, A., Sarkar, F. H., Dutkowski, R., Shulman, L., Ruddle, F. H., and Gupta, S. L. Receptors for human α and β interferon but not for γ interferon are specified by human chromosome 21. *Proc. Natl. Acad. Sci.* (USA) 81: 5504–5508 (1984).

Rozengurt, E., Collins, M., Brown, K. D., and Pettican, P. Inhibition of epidermal growth factor binding to mouse cultured cells by fibroblast-derived growth factor. Evidence for an indirect mechanism. *J. Biol. Chem.* 257: 3680–3686 (1982).

Sanceau, J., Sondermeyer, P., Beranger, F., Falcoff, R., and Vaquero, C. Intracellular human γ-interferon triggers an antiviral state in transformed murine L cells. *Proc. Natl. Acad. Sci.* (USA) 84: 2906–2910 (1987).

Scatchard, G. The attraction of proteins for small molecules and ions. *Ann. N.Y. Acad. Sci.* 51: 660–672 (1949).

Smith, B. R., Pyle, G. A., Peterson, V. B., and Hall, R. Interaction of thyroid stimulating antibodies with the human thyrotropin receptor. *J. Endocrinol.* 75: 401–407 (1977).

Steinman, R. M., Mellman, I. S., Muller, W. A., and Cohn, Z. A. Endocytosis and the recycling of plasma membrane. *J. Cell. Biol.* 96: 1–27 (1983).

Streuli, M., Hall, A., Boll, W., Stewart, W., II, Nagata, S., and Weissmann, C. Target cell specificity of two species of human interferon-α produced in *Escherichia coli* and of hybrid molecules derived from them. *Proc. Natl. Acad. Sci.* (USA) 78: 2848–2852 (1981).

Tan, Y. H., Tischfield, J., and Ruddle, F. H. The linkage of genes for the human interferon-induced antiviral protein and indophenol oxidase-B traits to chromosome G-21. *J. Exp. Med.* 137: 317–330 (1973).

Uze, G., Mogensen, K. E., and Aguet, M. Receptor dynamics of closely related ligands: "fast" and "slow" interferons. *EMBO J.* 4: 65–70 (1985).

Weber, H., Valenzuela, D., Lujber, G., Gubler, M., and Weissmann, C. Single amino acid changes that render human IFN-α2 biologically active on mouse cells. *EMBO J.* 6: 591–598 (1987).

Weil, J., Epstein, L. B., and Epstein, C. J. Synthesis of interferon-induced polypeptides in normal and chromosome 21-aneuploid human fibroblasts: Relationship to relative sensitivities in antiviral assays. *J. Interferon Res.* 1: 111–124 (1980).

Weil, J., Tucker, G., Epstein, L. B., and Epstein, C. J. Interferon induction of (2′-5′)oligoisoadenylate synthetase in diploid and trisomy 21 human fibroblasts: Relation to dosage of the interferon receptor gene (IRFC). *Hum. Genet.* 65: 108–111 (1983).

Wietzerbin, J., Gaudelet, C., Aguet, M., and Falcoff, E. Binding and cross-linking of recombinant mouse interferon-γ to receptors in mouse leukemic L1210 cells: interferon-γ internalization and receptor down-regulation. *J. Immunol.* 136: 2451–2455 (1986).

Wiley, H. S. and Cunningham, D. D. A steady state model for analyzing the cellular binding, internalization and degradation of polypeptide ligands. Cell 25: 433–440 (1981).

Yap, W. H., Teo, T. S., McCoy, E., and Tan, Y. H. Rapid and transient rise in diacylglycerol concentration in Daudi cells exposed to interferon. *Proc. Natl. Acad. Sci.* (USA) 83: 7765–7769 (1986a).

Yap, W. H., Teo, T. S., and Tan, Y. H. An early event in the interferon-induced transmembrane signaling process. *Science* 234: 355–358 (1986b).

Zoon, K. C., Arnheiter, H., Zur Nedden, D., Fitzgerald, D. J. P., and Willingham, M. C. Human interferon α enters cells by receptor-mediated endocytosis. *Virology* 130: 195–203 (1983).

Zoon, K. C., Karasaki, Y., Zur Nedden, D. L., Hu, R., and Arnheiter, H. Modulation of epidermal growth factor receptors by human α interferon. *Proc. Natl. Acad. Sci.* (USA) 83: 8226–8230 (1986a).

Zoon, K. C., Zur Nedden, D., Hu, R., and Arnheiter, H. Analysis of the steady state binding, internalization, and degradation of human interferon-$\alpha 2$. *J. Biol. Chem.* 261: 4993–4996 (1986b).

5 INTERFERONS AS MULTIFUNCTIONAL GENE ACTIVATORS

The diversity of IFN effects is of such magnitude that a comprehensive description of the changes taking place in cells as a result of exposure to IFNs is still far away. Since it took a long time before IFN was completely purified, and since recombinant, single species IFNs have only recently become available, mixtures of natural IFNs, often hardly purified, have been used for many studies on the mechanisms of IFN action. It is now evident that subtypes of IFN-α can differ by their antiviral, anticellular, and immunomodulating activities, and this is reflected by differences in the biochemical changes that are induced. For example, a mixture of natural Hu IFN-α subspecies, derived from lymphoblastoid cells, enhances the synthesis of eight specific peptides in human peripheral blood T lymphocytes from every donor out of 18 tested, whereas the single subspecies Hu IFN-α2 produces this effect in T cells from some donors only (Cooper, 1982; Cooper et al., 1982). This is illustrative of both the variation observed when different IFN-α subspecies are used and the variation in the response to IFN due to genetic differences between individuals.

The total number of genes whose expression can either be induced or enhanced by all the different subspecies of IFN-α or by IFN-β is probably the same, since these IFNs share common receptors. We do not know what that number is: in one system—human fetal lung fibroblasts treated with Hu leukocyte IFN—the enhanced synthesis or the de novo induction of 17 polypeptides has been identified by two-dimensional electrophoresis. In the same cells, Hu IFN-γ, which binds to a different receptor, induces the synthesis of 24 polypeptides, consisting of the 17 that are induced by IFN-α, plus another 7 that are not found in IFN-α-treated cells (Weil et al., 1983a,b,c). Many, but probably not all, of these proteins appear in IFN-treated cells as the result of a primary response of the transcribed genes, and not as a secondary, cascadelike response, resulting from the induction of one or a few intermediate proteins, which then induce other genes. For example, in Hu IFN-α-treated T98G neuroblastoma cells, the onset of transcription is very rapid for seven different Hu IFN-α inducible genes for which cDNA probes are available, including the metallothionein II and a class I HLA gene, and synthesis of new proteins is not required for induction of these genes (Friedman et al., 1984). However, another gene, coding for a 56 kDa protein, which is transcribed in Hu IFN-α-induced HeLa cells, does appear to require protein synthesis before its transcription can take place, since its induction can be blocked by cycloheximide (Kusari and Sen, 1986). This is, therefore, most likely an example of a secondary induction, resulting from the primary induction of another protein. To complicate matters somewhat more, in other cells or even in different lines of HeLa cells, the same gene seems to be directly inducible (Wathelet et al., 1986b).

To exert their different activities, IFNs first have to bind to specific receptors at the cell surface. IFN-α and IFN-β share the same receptors and IFN-γ has its own receptors; the average number of receptor sites at the cell surface is of the order of a few thousands, which is considerably lower than the usual number of receptors for biologically active peptides. A detailed discussion of IFN receptors and the interaction of IFNs and their receptors can be found in Chapter 4.

Receptor binding is a necessary condition for IFN-α/β action; if the receptor is bypassed by microinjecting IFN-α or IFN-β molecules directly into the cell, IFN is not active, at least not as judged by its antiviral effect (Higashi and Sokawa, 1982; Huez et al., 1983). After IFN binding, the receptor-IFN complex is taken into the cell by endocytosis, via receptosomes, but whether or not internalization of IFN is necessary for activity is a matter of debate (Yonehara et al., 1983; Zoon et al., 1983). We do not know what happens between the interaction with the receptor and the induction of the specific genes that are either derepressed or whose expression is enhanced by IFNs. Transcription and translation are required for IFN action, since Actinomycin D or inhibitors of protein synthesis inhibit the establishment of the antiviral state (Taylor, 1964; Friedman and Sonnabend, 1965). This is confirmed by studies showing that the establishment, but not the maintenance, of the antiviral state requires the presence of the nucleus (Radke et al., 1974). In general, the binding of IFNs to their receptors is species specific, which explains the general species specificity of IFNs.

The molecular characterization of all the IFN-induced polypeptides, either primary or secondary, the elucidation of their function, and the regulation of their induction is the major object of the study of IFN action.

I. IFN-INDUCED PROTEINS WITH KNOWN RELATION TO THE ANTIVIRAL STATE

Of the many IFN activities, the antiviral state has been best characterized at the biochemical level, because it was the first known and for a long time the most easily measured biological activity of IFNs. But even the mechanisms of the establishment of the antiviral state are still not well understood. A complete understanding of all metabolic changes that are responsible for the antiviral state will also be relevant to other IFN activities, such as the antiproliferative effect, since the IFN-induced enzymes that are responsible for the antiviral activity can do more than just inhibit viral replication.

Several IFN-induced proteins involved in antiviral activity have been described, and some have been characterized at the molecular level.

A. Oligo-Adenylate Synthetase (2-5A Synthetase)

This enzyme, also called $(2'\text{-}5')A_n$ synthetase or (2′,5′)oligo(A) adenyltransferase, is constitutively present in tissues and in cultured cells, although usually at very low levels; its concentration increases by one to several orders of magnitude in IFN-treated cells. When activated by double-stranded RNA, it polymerizes ATP into a series of 2′-5′ linked oligomers $(ppp(A2'p)_n)$, of which the trimer is the most abundant. These oligomers, collectively called 2-5A, activate a latent cellular endoribonuclease (Fig. 5.1). The kinetics of induction of 2-5A synthetase vary between different cells and induction is usually very transient, disappearing after cessation of IFN treatment (Shulman and Revel, 1980). IFN-γ is much less efficient an inducer of 2-5A synthetase than IFN-α or IFN-β (Baglioni and Maroney, 1980; Verhaegen-Lewalle et al., 1982). Functional mRNAs of different sizes, 1.65 and 1.85 kb in human and 1.8 and 3.6kb in murine cells, have been isolated. The human 1.6 and 1.8 kb mRNAs are transcribed from the same gene but are spliced differently (Merlin et al., 1983; St Laurent et al., 1983; Benech et al., 1985a,b; Saunders et al., 1985). The nucleotide sequence corresponding to the 1.6 and 1.8 kb human mRNAs is known, and the 40 and

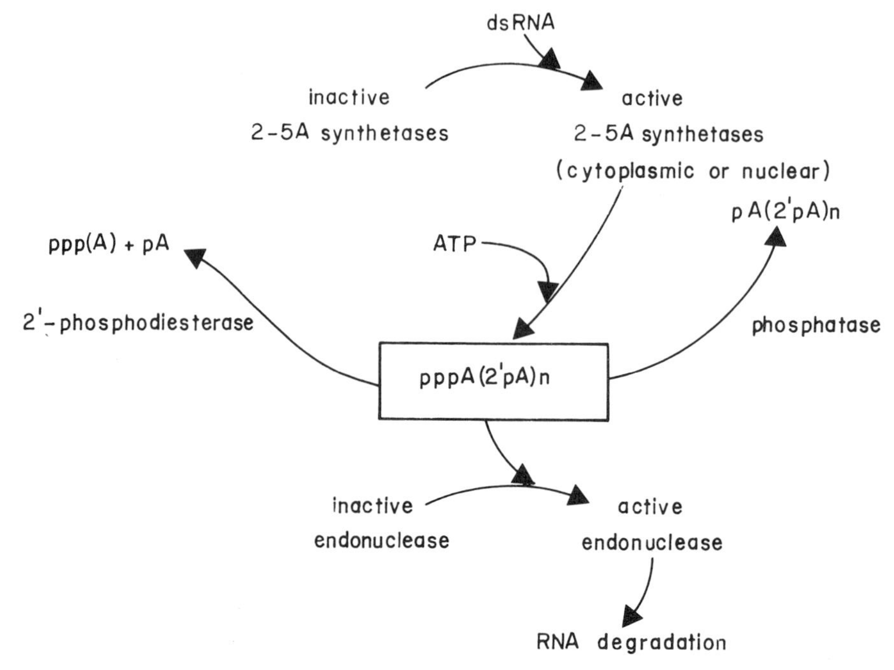

Figure 5.1 The 2-5A synthetase pathway.

46 kDa synthetases derived from them are identical in their first 346 residues, but they are different at their C-terminal ends (Benech et al., 1985b; Wathelet et al., 1986a). The structural gene for these polypeptides is located on human chromosome 12; it consists of six exons, contained within 12 kb of DNA (Fig. 5.2) (Williams et al., 1986; Wathelet et al., 1987). In human cells a

Figure 5.2 The genomic organization of the human 2-5A synthetase gene, as determined by electron microscopy.

Exons (E, in bp)	Intron (I, in bp)
E_1: 256 ± 33 (6)	I_1: 1.425 ± 159 (24)
E_2: 300 ± 25 (8)	I_2: 2.286 ± 256 (8)
E_3: 197 ± 24 (24)	I_3: 5.550 ± 232 (17)
E_4: 225 ± 18 (24)	I_4: 835 ± 43 (5)
E_5: 421 ± 36 (6)	
( ): number of samples	

Differential splicing of transcripts of this gene results in an mRNA of either 1.6 or 1.8 kb. The full length cDNA corresponding to the 1.6 kb mRNA (encoding the 40 kDa 2-5A synthetase) was hybridized to a 33-kb genomic fragment containing the complete 2-5A synthetase gene. After denaturation, reannealing took place for 3 hours at room temperature. The cDNA annealed to the corresponding coding sequences of the gene, leaving the five introns corresponding to this particular mRNA transcript as free loops. (Courtesy of M. Wathelet and J. Content, University of Brussels.)

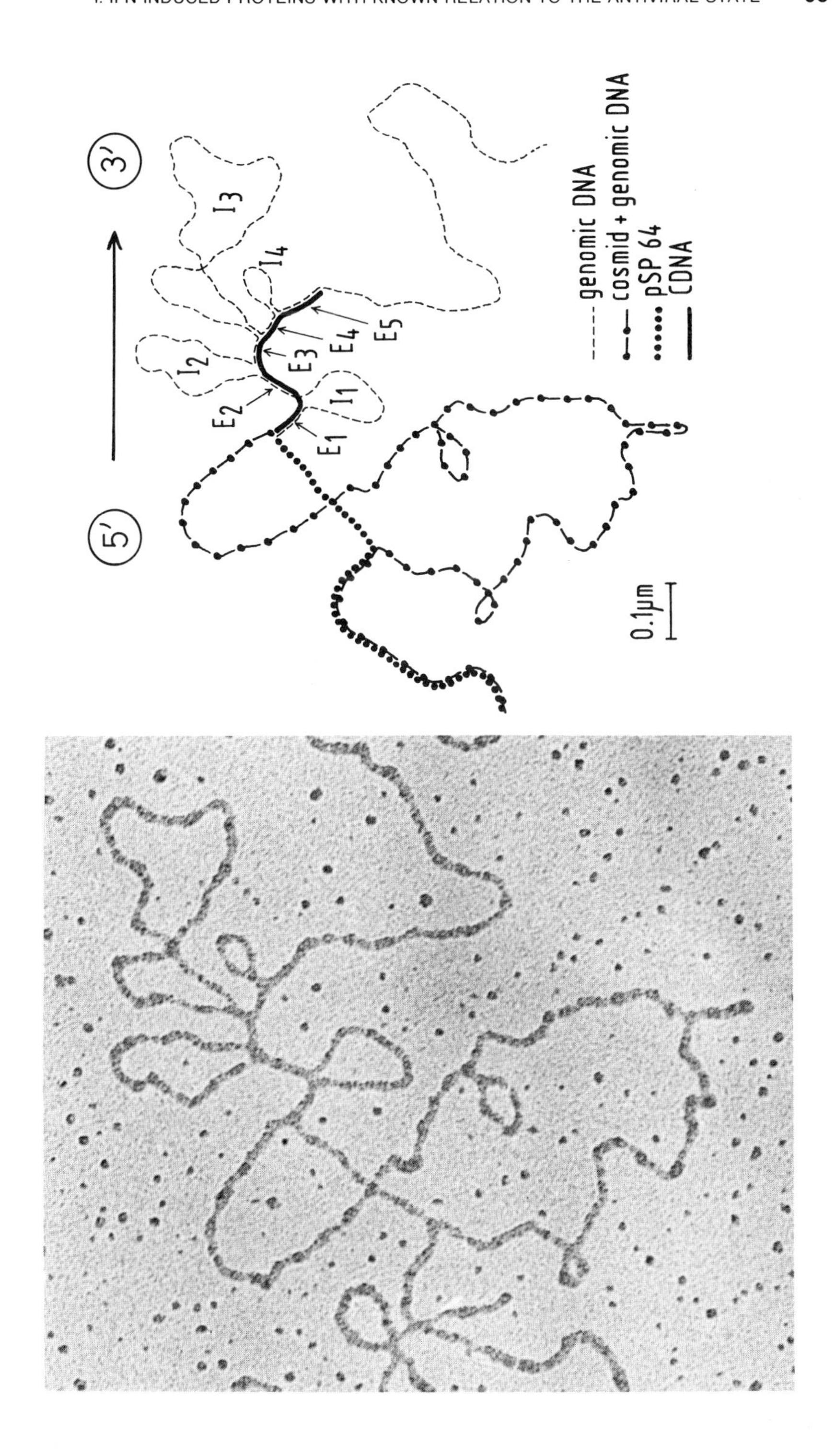
5'
3'
I1
I2
I3
I4
E1
E2
E3
E4
E5
0.1μm
genomic DNA
cosmid + genomic DNA
pSP 64
CDNA

TABLE 5.1 Different MW Forms of 2-5A Synthetase

	mRNA Size (Kb)	MW Form (kDa)	Localization
Mouse	1.5	42	Nucleus
	3.5	100	Cytoplasm
Human	1.6	40	Nucleus
	1.8	46	Nucleus
	3.6[a]	69	Nucleus and membrane structure
		100	Microsomes

[a] There is no formal evidence that the 3.6 kb mRNA codes for the 69 and 100 kDa synthetases.

larger 2-5A synthetase with a molecular mass of 80–105 kDa is frequently present (Yang et al., 1981; Wells et al., 1984), and this form could theoretically be derived from the 3.60 kb mRNA. Although the relationship of this larger enzyme to those coded for by the 1.6 and 1.8 kb mRNAs is not completely elucidated, it is antigenically related to the lower MW forms. Indeed, in human cells one can detect four different 2-5A synthetases, all antigenically related, with sizes corresponding to 40 and 46 kDa and to two larger proteins of 67 and 100 kDa. The four enzymes differ by their preferential intracellular localization, by their dsRNA requirements, and by their optimal conditions for activity (Table 5.1) (Chebath et al., 1987; Hovanessian et al., 1987). The activation requirements of the high MW 2-5A synthetase are significantly different from those of the low molecular weight enzyme, in that it needs a thousand times less dsRNA and furthermore, is activated by some types of dsRNA, for example, polyrArU, which is unable

TABLE 5.2 Homology of Aminoacid Sequence of the Murine 42 kDa and the Human 46 kDa 2-5A Synthetase

```
                1                                   20
MOUSE           M E H G L R S I P A W T L D K F I E D Y L L P D T T F G A D V K S A V N V
                        - -     - -     - - - - - - - - - - - - -
HUMAN 1.8 Kb        M D L R N T P A K S L D K F I E D Y L L P D T C F R M Q I N H A I D I
                    100                                       120
S F E D Q L N R R G E F I K E I K K Q L Y E V Q H E R R F R V K F E V Q S S W W P N A R S L S F K L
  -   - - - - - - - - - -   - -     - -       -   - -   -   - - - - - -       -   -   -   - - -   -
T F Q D Q L N R R G E F I Q E I R R Q L E A C Q R E R A F S V K F E V Q A P R W G N P R A L S F V L
                  200                                   220                                       240
S T C F T E L Q R N F L K Q R P T K L K S L I R L V K H W Y Q L C K E K L G K P L P P Q Y A L E L L
- - - - - - - - -   - - - - - - - - - - - - - - - - - - - - -   - -   - - - -   - - - - - - - - - -
S T C F T E L Q R D F L K Q R P T K L K S L I R L V K H W Y Q N C K K K L G K - L P P Q Y A L E L L
                      300                             320                                       340
Y L H R Q L R K A R P V I L D P A D P T G N V A G G N P E G W R R L A E E A D V W L W Y P C F I K K
- -   - - -   -   - - - - - - - - - - - - -     - -   -   - - -   - -   - -     - -   - - - -
Y L R R Q L T K P R P V I L D P A D P T G N L G G G D P K G W R Q L A Q E A E A W L N Y P C F K N W
          360
V P F E Q V E E N W T C I L L
        -   - -   - - -     -
A S T P Q A E E D W T C T I L
```

Reference: Ichii et al., 1986

to activate the lower molecular weight enzyme (Ilson et al., 1986). There is manifestly more than one functional form of 2-5A synthetase, and we are dealing with a complex multienzyme system.

The murine 42 kDa 2-5A synthetase displays 69% homology at the amino acid level and 73% homology at the nucleotide level with the human 46 kDa enzyme (Table 5.2) (Ichii et al., 1986).

The presence of 2-5A synthetase has been demonstrated in mammals, in reptiles, and probably in amphibia, but not in the lower eukaryotes and in prokaryotes (Cayley et al., 1982). Low amounts of 2-5A synthetase activity can normally be found in a variety of animal cells and tissues, and the level fluctuates with growth and hormone status. For example, in regenerating rat liver, the intracellular concentration of 2-5A synthetase activity and the amount of 2-5A oligonucleotides is inversely related to the proliferative activity of the cells (Etienne-Smekens et al., 1983; Smekens-Etienne et al., 1983).

What is the role of these enzymes in the absence of viral infection? It is very difficult to decide whether the low levels of 2-5A synthetase activity found in different tissues and cells are the result of a constitutive production, or are induced by the very low amounts of "spontaneous" IFN that are present in many cell lines and also in vivo in the absence of manifest virus infection.

Apparently constitutive 2-5A synthetase is present in several organs of normal mice, with maximum levels in the spleen (Krust et al., 1982). The enzyme activity decreases, without entirely disappearing, if the animals are treated with anti-IFN-α/β globulin. Manifestly, at least part of the base

```
    40                                    60                                      80
V C D F L K E R C F Q G A A H P V R V S K V V K G G S S G K G T T L K G R S D A D L V V F L N N L T
  -   - - - - - - -   -       - -   - - - - - - - - - - - - - - - -   - - - - - - - - - - -     - -
I C G F L K E R C F R G S S Y P V C V S K V V K G G S S G K G T T L R G R S D A D L V V F L S P L T
        140                                     160                                           180
S A P H L H Q E V E F D V L P A F D V L G H V N T S S K P D P R I Y A I L I E E C T S L G K D G E F
-       -       - - - - - - - - - -   - -         -   - -   -   - -     - - - - - -   -   -   - - -
S S L Q L G E G V E F D V L P A F D A L G Q L T G S Y K P N P Q I Y V K L I E E C T D L Q K E G E F
                                       260                                     280
T V F A W E Q G N G C Y E F N T A Q G F R T V L E L V I N Y Q H L R I Y W T K Y Y D F Q H Q E V S K
- -   - - -   -           - - - - - - - - - - - - - - - - - -   -   - - - - - - - - -             -
T V Y A W E R G S M K T H F N T A Q G F R T V L E L V I N Y Q Q L C I Y W T K Y Y D F K N P I I E K
                 348                                                                       349
D G S R V S S W D V P - - - - - - - - - - - - - - - - - - - - - - - - - - - - - - - - - - - T V V P
- - -   - - - -                                                                             -
D G S P V S S W I L L A E S N S T D D E T D D P R T Y Q K Y G Y I G T H E Y P H F S H R P S T L Q A
```

levels of 2-5A synthetase do not result from constitutive synthesis, but rather, from induction by endogenous IFN. Where does the endogenous IFN come from? Macrophages from conventionally reared mice release low levels of spontaneous IFN-α/β whereas macrophages from germ-free animals are devoid of spontaneous IFN production (De Maeyer et al., 1971). One possible source of endogenous IFN production in mice is macrophages being continuously induced by LPS (see Chapter 10). This possibility receives experimental support from the observation that germ-free and pathogen-free mice have tenfold lower base levels of synthetase in their spleens (Galabru et al., 1985; Hearl and Johnston, 1987).

Another link between the presence of 2-5A synthetase and IFN production is suggested by the significantly higher levels of 2-5A synthetase in the spleens from female mice. There is an influence of sex on IFN production in mice, and during virus infection female mice in general produce more IFN-α/β than do male mice (Zawatzky et al., 1982; De Maeyer-Guignard et al., 1983). It is, therefore, entirely possible that female mice may also have higher levels of "spontaneous" IFN production, and consequently, have more 2-5A synthetase. Hormonal influence is also a distinct possibility.

The 2-5A synthetase activity that is present in mononuclear cells of healthy humans is relatively high and constant, and during acute viral infections the enzyme activity goes up. It is also elevated in patients with a variety of autoimmune diseases, such as lupus erythematosus, vasculitis, scleroderma, and dermatomyositis, and in patients with Epstein-Barr-virus-related malignancies (Schattner et al., 1981). The elevated levels of 2-5A synthetase in many autoimmune diseases may be related to the presence of acid-labile IFN-α which often accompanies this type of illness. There is, however, no formal evidence for a causal relationship.

The 2-5A synthetase is only activated by dsRNA, usually present as a result of infection with RNA viruses, and this raises the question of its activation and role in uninfected cells. It is technically very difficult to look for the presence of low amounts of oligoisoadenylates or of the endonuclease (RNase L) activated by these oligomers. In spite of this difficulty, it has been possible to detect the presence of RNase L in several mammalian cells and tissue extracts and the presence of 2-5A in the nuclei of a continuous line of human tumor cells (Nilsen et al., 1981, 1982; Hearl and Johnston, 1985). Possibly certain cellular RNAs, such as heterogeneous nuclear RNAs that contain double-stranded regions, can be activators, and maybe other structural features could replace the requirement for double-strandedness.

One can imagine that the mere presence of a basal 2-5A synthetase level could provide some protection to viral infection, since in the case of infection with many RNA viruses the first round of virus replication provides the dsRNA structures that are necessary to activate the enzyme. Why then are cells in general susceptible to virus infection, in spite of the presence of base levels of synthetase? We will see in Chapter 6 that this is not the only antiviral mechanism coming into play in IFN-treated cells, but that there are several other mechanisms. For example, an antiviral state can be induced by Hu IFN-α and Hu IFNβ in diploid MRC5 cells in the absence of any detectable 2-5A synthetase. Moreover, high levels of 2-5 synthetase can be induced by IFN in cells that do not, nevertheless, become resistant to virus infection (Verhaegen et al., 1980; Meurs et al., 1981).

B. Ribonuclease L

This latent endoribonuclease is activated by the 2-5A oligomers made by the 2-5A synthetase; it is, therefore, frequently not an IFN-induced enzyme, but an IFN-activated enzyme, although under certain conditions it does appear to be induced by IFN (Lengyel, 1982; Krause et al., 1985). It plays an essential role in the 2-5A pathway and is responsible for some important IFN activities. The molecular weight of RNase L has been estimated to be about 77 to 85 kDa (Floyd-Smith et al., 1982; Wreschner et al., 1982). The activation of the enzyme by 2-5A is completely reversible, and it probably involves the binding of the oligomer. Once activated, RNase L is able to cleave single-stranded viral RNAs, cellular mRNAs, and ribosomal RNAs, but not dsRNAs (Ratner et al., 1977). The cleavage occurs preferentially at UpNp sequences, leaving 3′ phosphate termini (Floyd-Smith et al., 1981; Hovanessian et al., 1981; Wreschner et al., 1981).

C. Protein Kinase P1

In cells treated with IFN one observes an increase of a dsRNA-dependent protein kinase activity. This kinase catalyzes the phosphorylation of serine and threonine residues of the α subunit of the eukaryotic protein synthesis initiation factor 2 protein with a molecular mass of 67 kDa in mice and of 68 kDa in humans (Lebleu et al., 1976; Roberts et al., 1976; Zilberstein et al., 1976; Farrel et al., 1977). It is virtually certain that protein P1, which acts as a substrate for the kinase activity, becomes in fact the kinase itself once it is phosphorylated (Galabru and Hovanessian, 1985). Good evidence exists that the P1 protein has a dsRNA-dependent, ATP-binding site, which is conceivably the catalytic site of the protein kinase (Bischoff and Samuel, 1985). The apparent native molecular weight of the murine kinase has been estimated to be 62 kDa, rather than the value of 67 kDa obtained with denatured and phosphorylated enzyme (Berry et al., 1985). The increased enzyme activity resulting from IFN treatment is due to de novo synthesis of enzyme and to enhanced activity resulting from autophosphorylation (Vandenbussche et al., 1978).

The P1 kinase is distinct from another serine-threonine kinase, protein kinase C, a Ca^{2+}- and phospholipid-dependent enzyme, which is, among others, the site of action of biologically active phorbol diesters. IFN-γ stimulates the activity of protein kinase C in macrophages, but there is no evidence that this stimulation has anything to do with antiviral activity (Becton et al., 1985). The IFN-γ-enhanced activity of protein kinase C probably does not result from de novo synthesis, but rather, from a modulation of an existing enzyme to a state of higher catalytic activity (Hamilton et al., 1985).

D. The Mx Protein

This 72 kDa protein, induced in mouse cells by Mu IFN-α and IFN-β but not by IFN-γ, renders mice specifically resistant to infection with the influenza virus, but not with other viruses. In this the protein is unique, since all other enzymes and proteins with antiviral activity induced by IFNs do not display such narrow virus specificity, but instead, are capable of inhibiting a broader range of viruses. The structural gene for the *Mx* protein in mice (*Mx* locus) is on chromosome 16; mice with the *Mx*+ allele are inducible by IFN-α/β to express the *Mx* protein, whereas mice having the Mx^- allele cannot be induced to express the Mx protein (Horisberger et al., 1983). In humans, a comparable protein has been isolated using monoclonal antibodies raised against the murine protein; the function of the human protein is not yet known (Staeheli and Haller, 1985). The *Mx* locus is discussed in detail in Chapter 15.

II. OTHER IFN-INDUCED PROTEINS

A. Guanylate-Binding Proteins

Several cellular proteins bind guanine nucleotides with high affinity. The exact function of many of these proteins is not known, but some guanylate-binding proteins are implicated in the control of the hormone-sensitive adenylate cyclase system, controlling different protein kinase activities (Gilman, 1984). Guanylate-binding proteins in general probably serve as branch points for the transduction of information across the cellular membrane. In both humans and mice, IFNs augment the synthesis of several guanylate-binding proteins.

In mice, Mu IFN-α/β or IFN-γ induce the synthesis of a family of guanylate-binding proteins (GBP). The minor GBPs are synthesized equally well in cells from many mouse strains, but the major protein, GBP-1, with a molecular mass of 65 kDa, is induced by IFN in some mouse strains, but not in others (Staeheli et al., 1984). The inducibility of GBP-1 is controlled by a single autosomal locus, situated on the distal part of chromosome 3, near the alcohol dehydrogenase gene complex (Prochazka et al., 1985). The *Gbp*-1 locus has two alleles, *Gbp*-1^a for inducibility and *Gbp*-1^b for the lack of inducibility (Staeheli et al., 1984) and it does not influence the inducibility of the other, minor GBPs. The function of GBP-1 is unknown; its presence or absence in IFN-treated cells does not correlate with the induction of the antiviral state directed against VSV or the influenza virus (Staeheli *et al.*, 1983).

In humans, treatment of diploid fibroblast cells with Hu IFN-α, β, or γ induces the synthesis of a GBP with a molecular mass of 67 kDa. After induction by IFN-γ, the 67 kDa GBP appears in the cytoplasm of diploid human fibroblasts and lymphocytes. However, for reasons not understood, the protein is not induced in transformed cells (Cheng *et al.*, 1983, 1985).

B. Tubulin

Microtubules, together with actin filaments, are important components of the cytoskeleton. The major protein constituent of microtubules is tubulin, a dimer consisting of two 50 kDa subunits, α and β tubulin, with a closely related amino acid sequence. Treatment with Hu IFN-α or Hu IFN-β results in a marked increase of α- and β-tubulin mRNA in human lymphoblastoid cells (Fellous et al., 1982). This effect on tubulin synthesis represents probably one of the ways by which IFNs influence cell membrane components and cell shape (Tovey et al., 1975; Pfeffer et al., 1980).

An intact tubulin network is required for some actions of IFN, since 3T3 cells treated with tubulin disrupting agents, such as colchicine, become less sensitive to the antiproliferative activity of Hu IFN-β (Ebsworth et al., 1986). The stimulation of tubulin by IFN may, therefore, be part of a positive feedback loop-enhancing IFN activity.

C. Metallothionein II

Metallothioneins are heavy metal-binding proteins of high cysteine content, which are ubiquitously distributed in animals and plants. They are encoded by a large gene family, and their synthesis is usually induced by heavy metals and glucocorticoids. These proteins function in the regulation of trace-metal metabolism, in the storage of metal ions in the liver, and as a protective mechanism against heavy metal toxicity. Metallothionein II is one of the major metallothionein proteins to be expressed in human liver; its structural gene is on chromosome 16 (Karin and Richards, 1982; Schmidt et al., 1984).

Transcription of the gene for metallothionein II is induced in human T98G neuroblastoma cells by Hu leukocyte or lymphoblastoid IFN. The rate of transcription is comparable to that observed after induction with dexamethasone, but the mRNA accumulates more slowly, due to differences in post-transcriptional events between induction by IFN and by the glucocorticoid. The IFN-induced transcription of the metallothionein II gene is clearly dose-dependent: About 30 times more IFN is required to trigger the transcription of this gene than is necessary to induce the transcription of HLA genes. This is an example of the dose-dependent initiation of transcription, which appears as one of the ways by which the action of IFNs is regulated (Friedman and Stark, 1985). How the induction of metallothionein II is in any way related to the biological activity of IFN is not known.

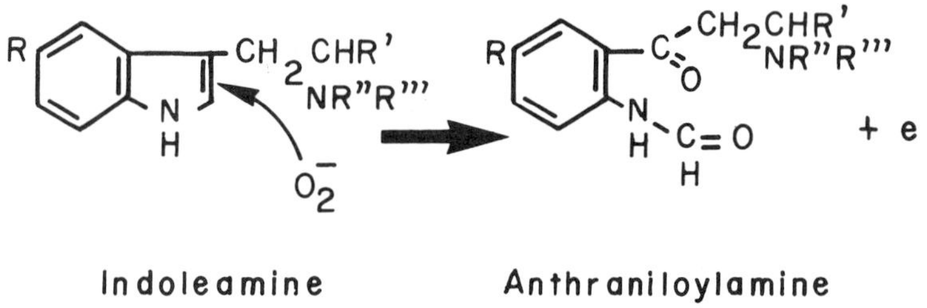

Indoleamine Anthraniloylamine

Figure 5.3 Activity of indoleamine 2,3-dioxygenase, which acts upon various indoleamine derivatives, including tryptophan. (Adapted from Hayashi, 1985.)

D. Indoleamine 2,3-Dioxygenase

This enzyme, found in many tissues in the organism, is a 42 kDa heme-containing dioxygenase that plays a crucial role in tryptophan degradation (Fig. 5.3). It appears in large amounts in mouse lung tissue as a result of an influenza virus infection, and its synthesis is also stimulated in LPS-treated mice. Murine IFN-α and IFN-β induce high levels of indoleamine 2,3-dioxygenase in mouse lung cells (Fig. 5.4). Hu IFN-γ, and not Hu IFN-α or Hu IFN-β, induces a tryptophan-degrading enzyme in human fibroblasts, which is also indoleamine 2,3-dioxygenase. As a result of tryptophan degradation, the growth of the obligate intracellular protozoon *Toxoplasma gondii* is inhibited (Yoshida et al., 1981; Pfefferkorn, 1984; Hayashi, 1985; Pfefferkorn et al., 1986).

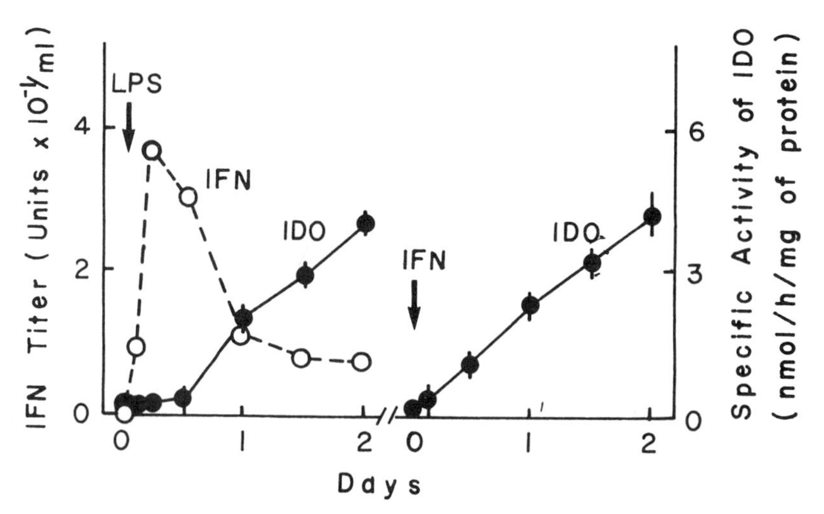

Figure 5.4 Induction of indoleamine 2,3 dioxygenase (IDO) and IFN-α/β (IFN) by endotoxin (LPS) (left panel) and by Mu IFN-α/β (right panel) in mouse lung slices. (Adapted from Hayashi, 1985.)

Hu IFN also stimulates tryptophan degradation in normal and malignant human cells in vitro, and moreover, plasma tryptophan levels in cancer patients undergoing Hu IFN-γ treatment are lowered as a result of the treatment (Byrne et al., 1986). Since tryptophan is a precursor of physiologically important coenzymes, it is conceivable that the enhanced degradation of this amino acid contributes to the antiproliferative effect of IFN.

E. A 12 kDa Protein, Probably a Mediator of Inflammation

Platelet factor 4 and β-thromboglobulin are two chemotactic proteins, most likely mediators of inflammation and wound healing, stored in the α granules of platelets and released on degranulation. In human mononuclear cells, fibroblasts, and cndothelial cells, Hu IFN-γ and, to a much lesser extent, Hu IFN-β induce the transcription of a gene coding for a protein of molecular mass 12,378, which has a significant amino acid homology to these two platelet proteins (Luster et al., 1985).

F. A 20 kDa Membrane Protein Induced in Human Cells

Hu IFN-α and Hu IFN-β enhance the synthesis of a 20 kDa protein, which is a component of the membranes of several lymphoblastoid and diploid human cells. The induction of this protein may be related to the antiproliferative action of IFN-β, since the protein is only induced in cells whose replication is inhibited by IFN-β and not in cells that are resistant to the antiproliferative action of IFN-β. The lack of induction of the 20 kDa protein in the latter does not result from a general resistance to IFN action, because a 15-kDa cytoplasmic protein is induced by IFN-β in these cells, and furthermore, they can become resistant to viral infection after IFN treatment (Knight et al., 1985).

G. The 56 kDa Protein

A cDNA corresponding to a 56-kDa protein induced by IFN-β but not by IFN-γ in human cells has been isolated; the protein encoded by the full length cDNA has 478 amino acid residues, and its normal location is most likely the cytoplasm. Its function is not known (Chebath et al., 1983; Larner et al., 1984; Wathelet et al., 1986b).

The gene encoding the 56-kDa protein is situated on human chromosome 10, most likely at the junction of 10q25-q26 (Kusari et al., 1987).

H. Other Proteins of Unknown Function

A number of other proteins and partial or complete cDNAs corresponding to IFN-inducible genes have been described, but we will have to wait for further information concerning the structure and function of these proteins

TABLE 5.3 Chromosomal Location of Some IFN-Inducible Genes[a]

Chromosome	Gene	Protein(s)	Function
		In Mice	
1	202	56 kDa	?
2	β2 microglobulin	12 kDa	Cell surface antigen
3	Gbp-1	65 kDa	?
16	Mx	72 kDa	Resistance to influenza virus
17	MHC class I and class II	Several	Cell surface antigens
		In Humans	
6	MHC class I and class II	Several	Cell surface antigens
10	561	56 kDa	?
12	2-5A Synthetase	40 and 46 kDa	Antiviral, plus other less well-defined functions (cell growth and differentiation?)
15	β 2 microglobulin	11.6 kDa	Cell surface antigen
16	Metallothionein II		Storage and treatment of metal ions

[a] The relevant references can be found in the text.

before a meaningful discussion is possible. One of these is a 56-kDa murine protein induced by Mu IFN-β. The structural gene for this protein, referred to as gene 202, is on mouse chromosome 1 (Samanta et al., 1984). Another IFN-induced protein of unknown function is a 15-kDa cytoplasmic protein induced by Hu IFN-β in Daudi cells (Korant et al., 1984; Blomstrom et al., 1986). IFN-inducible genes are evidently on many different chromosomes, and Table 5.3 lists some human and murine IFN-inducible genes for which the chromosomal location is known.

I. MHC Class I and Class II Cell-Surface Antigens

IFNs exert important immunomodulating effects by inducing or enhancing the expression of the structural genes of the MHC locus in humans and mice. This aspect of IFN action is discussed extensively in Chapter 9.

III. REGULATORY FEATURES OF GENE ACTIVATION BY IFNs

The majority of the IFN-inducible proteins discussed in this chapter can be induced by all three IFN species, although frequently with different efficacies. There are, however, several exceptions to this. Some genes, for example, the Mx locus in the mouse cannot be induced by IFNγ. Others, on the contrary, such as class II MHC antigens or the 12 kDa protein with homology to β-thromboglobulin are preferentially—although not exclusively—induced by IFN-γ. How is the transcriptional response to IFN treatment regulated? What are the features of these genes that determine inducibility by IFNs? Different approaches to obtaining answers to these important questions are emerging.

A. Different Pathways for Gene Activation?

The great variety and number of genes activated by IFNs is rather striking, and we know next to nothing about the links between receptor occupancy and the onset of transcription. There is probably more than one pathway leading to gene activation by IFNs, since Hu lymphoblastoid IFN-α induces a set of 15 proteins in wild-type Daudi cells, only 8 of which are induced in Daudi cells that are resistant to the antiproliferative effect (Dron et al., 1985). This differential control in wild-type and resistant cells shows that there is at least one branch point in the pathway leading to IFN-induced gene transcription. It cannot be excluded that this branch point is at the earliest possible site of IFN action and that it is the reflection of functional heterogeneity among IFN receptors on the plasma membrane (McMahon et al., 1986).

B. Evidence for IFN-Regulated Transcriptional Factors

IFN-inducible genes are activated with varying rates and duration of transcription, and modulation of transcription appears as an important control mechanism for IFN action. For some genes it has been shown that transcriptional induction is followed by a programmed decline in transcription, coupled with the inability of further IFN treatment to restimulate transcription. This regulatory event has been called "desensitization"; it requires one or several proteins and is not merely the result of receptor down-regulation (Kusari and Sen, 1986; Larner et al., 1986).

C. A Consensus Sequence Influencing IFN-α/β-Induced Transcription

The 5′ regions of several IFN-α inducible human and mouse genes contain a consensus sequence that may be involved in the regulation of transcription of at least some IFN-induced genes, and that has been called the interferon regulatory sequence or IRS (Friedman and Stark, 1985). This consensus sequence has been found in the 5′ flanking region of class I HLA genes, of the Metallothionein II gene, and of the murine class I H-2K^b gene (Table 5.4). In the murine β-2-microglobulin gene the consensus sequence is not found upstream of the cap site, but rather, is in the 3′ noncoding region. In the murine 202 gene a sequence that is partially homologous with the consensus sequence is part of the coding sequence of the first exon (Friedman and Stark, 1985; Israel et al., 1986; Kimura et al., 1986; Samanta et al., 1986). The gene coding for the human 56 kDa protein (see Section II G) has

TABLE 5.4 Homologous 5′ Flanking Sequences in Some IFN Inducible Genes (IRS Region)

HLA-DR		G T G T T G A A C C T C - A G A G T T T C T C C T C T - C A T	
HLA-A3		A T T C C C C A A C T C C G C A G T T T C T T T T C T - C T C	
HLA		A T T C G C C A C C T C C G C A G T T T C T C T T C T T C T C	
MT-II		T C T C T G G A C C T - - G C A G T T T C T C C T C T - C T A	
H2-K^b	−164	T T C C C C A T C T C C A C A G T T T C A C T T C T G C A	−135
Consensus		T T C N $^{G}_{C}$ N A C C T C N G C A G T T T C T C $^{G}_{C}$ T C T - C T	

References: Friedman and Stark, 1985; Israel et al., 1986; Kimura et al., 1986.

sequences that are similar to the IRS at three different locations: one on the sense strand and two on the antisense strand (Wathelet et al., 1986).

In the murine class I H-2K and H-2L genes the IFN response sequence overlaps with an enhancer. Fusion of the promoter region, containing the IFN response sequence, to an unrelated gene, confers inducibility by IFN-α on this gene. Moreover, when such an engineered gene, containing the murine promoter and IFN-response sequence, is transfected into human cells, it becomes inducible by Hu IFN-α, -β, and -γ. This shows that although the receptors for IFN-α/β and IFN-γ are different, the final target at the level of this gene is the same (Israel et al., 1986). Conversely, one can transfect a human IFN-inducible gene into murine cells and show that it becomes inducible by Mu IFN-α or Mu IFN-β. The sequences that are necessary for IFN-induced gene expression, and the mechanisms governing such an expression, appear to be conserved between humans and mice (Kelly et al., 1986).

The IRS is not present in all IFN-inducible genes; it is, for example, absent in the 2-5A synthetase gene. The latter, however, contains two DNA segments on the antisense strand that are homologous to the IRE region of the Hu IFN-β gene promoter (see Chapter 3 for a definition of the IRE region). Another IFN-inducible gene, coding for the human 56 kDa protein,

TABLE 5.5 Region of the Human 2-5A Synthetase Gene and the 56 kDa Gene Homologous with the IRE of Hu IFN-β

Locally homologous regions between the 2-5A synthetase gene and the IRE

```
      A A T T G A A A G G G A A A A   +815 to +829
      * *   * * * * * * * * *   * *
      A A C T G A A A G G G A G A A   -87 to -73 (IRE region)

A A A A G G A   A A G T G C A A A G   +1266 to +1282
* * *   * * *   * * * * *   * * * *
A A A G G G A G A A G T G   A A A G   -82 to -66 (IRE region)
```

Locally homologous regions between the 56 K gene and the IRE

```
                        A G A A G G G A A A   T G G G A A A   +340 to +356 (last intron)
                        * * * * *   * * * *   * * * * * * *
                        A G A A G T G A A A G T G G G A A A   -76 to -59 (IRE region)
A C C T A G G G A A A C C G A A A G G G G A A A G T G A A A   -116 to -87 (sense strand)
* *   * * * *   * * * *   * * * * * * *     * * * * * * * *
A C A T A G G A A A A C T G A A A G G G A G A A G T G A A A   -96 to -67 (IRE region)
```

Reference: Wathelet et al., 1987.

which was mentioned previously, contains two longer segments that are homologous to the IRE, one on the sense strand at position −116 and one on the antisense strand at position +340 in the first intron (Table 5.5). These regions could constitute the target for induction by viruses or dsRNA, which raises the very intriguing possibility that some genes can be activated by both IFN and IFN inducers (Wathelet et al., 1987).

REFERENCES

Baglioni, C. and Maroney, P. A. Mechanism of action of human interferons. Induction of 2′-5′-oligo(A) polymerase. *J. Biol. Chem.* 255: 8390–8393 (1980).

Becton, D. L., Adams, D. O., and Hamilton, T. A. Characterization of protein kinase C activity in interferon-γ treated murine peritoneal macrophages. *J. Cell Physiol.* 125: 485–491 (1985).

Benech, P., Merlin, G., Revel, M., and Chebath, J. 3′ end structure of the human (2′-5′) oligo A synthetase gene: Prediction of two distinct proteins with cell type-specific expression. *Nucl. Acids Res.* 13: 1267–1281 (1985a).

Benech, P., Mory, Y., Revel, M., and Chebath, J. Structure of two forms of the interferon-induced (2′-5′)-oligo A synthetase of human cells based on cDNAs and gene sequences. *EMBO J.* 4: 2249–2256 (1985b).

Berry, M. J., Knutson, G. S., Lasky, S. R., Munemitsu, S. M., and Samuel, C. E. Mechanism of interferon action. Purification and substrate specificities of the double-stranded RNA-dependent protein kinase from untreated and interferon-treated mouse fibroblasts. *J. Biol. Chem.* 260: 11240–11247 (1985).

Bischoff, J. R. and Samuel, C. E. Mechanism of interferon action. The interferon-induced phosphoprotein P1 possesses a double-stranded RNA-dependent ATP-binding site. *J. Biol. Chem.* 260: 8237–8239 (1985).

Blomstrom, D. C., Fahey, D., Kutny, R., Korant, B. D., and Knight, E., Jr. Molecular characterization of the interferon-induced 15-kDa protein. Molecular characterization of the interferon-induced 15-kDa protein. Molecular cloning and nucleotide and amino acid sequence. *J. Biol. Chem.* 261: 8811–8816 (1986).

Byrne, G. I., Lehmann, L. K., Kirschbaum, J. G., Borden, E. C., Lee, C. M., and Brown, R. R. Induction of tryptophan degradation in vitro and in vivo: A γ-interferon-stimulated activity. *J. Interferon Res.* 6: 389–396 (1986).

Cayley, P. J., White, R. F., Antoniw, J. F., Walesby, N. J., and Kerr, I. M. Distribution of the ppp(A2′p)$_n$A-binding protein and interferon-related enzymes in animals, plants, and lower organisms. *Biochem. Biophys. Res. Commun.* 108: 1243–1250 (1982).

Chebath, J., Benech, P., Hovanessian, A., Galabru, J., and Revel, M. Four different forms of interferon-induced 2′,5′-oligo(a) synthetase identified by immunoblotting in human cells. *J. Biol. Chem.* 262: 3852–3857 (1987).

Chebath, J., Merlin, G., Metz, R., Benech, P., and Revel, M. Interferon-induced 56,000 Mr protein and its mRNA in human cells: Molecular cloning and partial sequence of the cDNA. *Nucl. Acids Res.* 11: 1213–1226 (1983).

Cheng, Y. S. E., Becker-Manley, M. F., Chow, T. P., and Horan, D. C. Affinity purification of an interferon-induced human guanylate-binding protein and its characterization. *J. Biol. Chem.* 260: 15834–15839 (1985).

Cheng, Y. S. E., Colonnot, R. J., and Yin, F. H. Interferon induction of fibroblast proteins with guanylate binding activity. *J. Biol. Chem.* 258: 7746–7750 (1983).

Cooper, H. L. Effect of bacterially produced interferon-$\alpha 2$ on synthesis of specific peptides in human peripheral lymphocytes. *FEBS Lett.* 140: 109–112 (1982).

Cooper, H. L., Fagnani, R., London, J., Trepel, J., and Lester, E. P. Effect of interferons on protein synthesis in human lymphocytes: Enhanced synthesis of eight specific peptides in T cells and activation-dependent inhibition of overall protein synthesis. *J. Immunol.* 128: 828–833 (1982).

De Maeyer, E., Fauve, R. M., and De Maeyer-Guignard, J. Production d'interféron au niveau du macrophage. *Ann. Inst. Pasteur* 120: 438–446 (1971).

De Maeyer-Guignard, J., Zawatzky, R., Dandoy, F., and De Maeyer, E. An X-linked locus influences early serum interferon levels in the mouse. *J. Interferon Res.* 3: 241–252 (1983).

Dron, M., Tovey, M. G., and Eid, P. Isolation of Daudi cells with reduced sensitivity to interferon. III. Interferon-induced proteins in relation to the phenotype of interferon resistance. *J. Gen. Virol.* 66: 787–795 (1985).

Ebsworth, N., Rozengurt, R., and Taylor-Papadimitriou, J. Microtubule-disrupting agents reverse the inhibitory effect of interferon on mitogenesis in 3T3 cells. *Exp. Cell. Res.* 165: 255–259 (1986).

Etienne-Smekens, M., Vandenbussche, P., Content, J., and Dumont, J. E. (2′-5′)oligoadenylate in rat liver: Modulation after partial hepatectomy. *Proc. Natl. Acad. Sci.* (USA) 80: 4609–4613 (1983).

Farrell, P. J., Balkow, K., Hunt, T., Jackson, R. J., and Trachsel, H. Phosphorylation of initiation factor eIF-2 and the control of reticulocyte protein synthesis. *Cell* 11: 187–200 (1977).

Fellous, A., Ginzburg, I., and Littauer, U. Z. Modulation of tubulin mRNA levels by interferon in human lymphoblastoid cells. *EMBO J.* 1: 835–839 (1982).

Floyd-Smith, G., Slattery, E., and Lengyel, P. Interferon action: RNA cleavage pattern of a (2′-5′) oligoadenylate-dependent endonuclease. *Science* 212: 1030–1032 (1981).

Floyd-Smith, G., Yoshie, O., and Lengyel, P. Interferon action. Covalent linkage of (2′-5′)pppApApA (^{32}P)pCp to (2′-5′(A)$_n$-dependent ribonucleases in cell extracts by ultraviolet irradiation. *J. Biol. Chem.* 257: 8584–8587 (1982).

Friedman, R. L., Manly, S. P., McMahon, M., Kerr, I. M., and Stark, G. R. Transcriptional and posttranscriptional regulation of interferon-induced gene expression in human cells. *Cell* 38: 745–755 (1984).

Friedman, R. L. and Stark, G. R. α-interferon-induced transcription of HLA and metallothionein genes containing homologous upstream sequences. *Nature* 314: 637–639 (1985).

Friedman, R. M. and Sonnabend, J. A. Inhibition of interferon action by puromycin. *J. Immunol.* 95: 696–703 (1965).

Galabru, J. and Hovanessian, A. G. Two interferon-induced proteins are involved in the protein kinase complex dependent on double-stranded RNA. *Cell* 43: 685–694 (1985).

Galabru, J., Robert, N., Buffet-Janvresse, C., Riviere, Y., and Hovanessian, A. G. Continuous production of interferon in normal mice: effect of anti-interferon globulin, sex, age, strain and environment on the levels of 2-5A synthetase and p67K kinase. *J. Gen. Virol.* 66: 711–718 (1985).

Gilman, A. G. G proteins and dual control of adenylate cyclase. *Cell* 36: 577–579 (1984).

Hamilton, T. A., Becton, D. L., Somers, S. D., Gray, P. W., and Adams, D. O. Interferon-γ modulates protein kinase C activity in murine peritoneal macrophages. *J. Biol. Chem.* 260: 1378–1381 (1985).

Hayaishi, O. Indoleamine 2,3-dioxygenase with special reference to the mechanism of interferon action. *Biken J.* 28: 39–49 (1985).

Hearl, W. G. and Johnston, M. I. Levels of 2′,5′-oligoadenylates in the tissues of pathogen-free mice. In: *The 2-5A System: Molecular and Clinical Aspects of the Interferon-Regulated Pathway,* Alan R. Liss, Inc., pp. 19–24 (1985).

Hearl, W. G. and Johnston, M. I. Accumulation of 2′,5′-oligoadenylates in encephalomyocarditis virus-infected mice. *J. Virol.* 61: 1586–1592 (1987).

Higashi, Y. and Sokawa, Y. Microinjection of interferon and 2′-5′-oligoadenylate into mouse L cells and their effects on virus growth. *J. Biochem.* 91: 2021–2028 (1982).

Horisberger, M. A., Staeheli, P., and Haller, O. Interferon induces a unique protein in mouse cells bearing a gene for resistance to influenza virus. *Proc. Natl. Acad. Sci.* (USA) 80: 1910–1914 (1983).

Hovanessian, A. G., Laurent, A. G., Chebath, J., Galabru, J., Robert, N., and Svab, J. Identification of 69-kd and 100-kd forms of 2-5A synthetase in interferon-treated human cells by specific monoclonal antibodies. *EMBO J.* 6: 1273–1280 (1987).

Hovanessian, A. G., Rollin, P., Riviere, Y., Pouillart, P., Sureau, P., and Montagnier, L. Protein kinase in human plasma analogous to that present in control and interferon-treated HeLa cells. *Biochem. Biophys. Res. Commun.* 103: 1371–1377 (1981).

Huez, G., Silhol, M., and Lebleu, B. Microinjected interferon does not promote an antiviral response in HeLa cells. *Biochem. Biophys. Res. Commun.* 110: 155–160 (1983).

Ichii, Y., Fukunaga, R., Shiojiri, S., and Sokawa, Y. Mouse 2-5A synthetase cDNA: Nucleotide sequence and comparison to human 2-5A synthetase. *Nucl. Acids Res.* 14: 10117 (1986).

Ilson, D. H., Torrence, P. F., and Vilcek, J. Two molecular weight forms of human 2′-5′-oligoadenylate synthetase have different activation requirements. *J. Interferon Res.* 6: 5–12 (1986).

Israel, A., Kimura, A., Fournier, A., Fellous, M., and Kourilsky, P. Interferon response sequence potentiates activity of an enhancer in the promoter region of a mouse H-2 gene. *Nature* 322: 743–746 (1986).

Karin, M. and Richards, R. I. Human metallothionein genes—Primary structure of the metallothionein-II gene and a related processed gene. *Nature* 299: 797–803 (1982).

Kelly, J. M., Porter, A. C. G., Chernajovsky, Y., Gilbert, C. S., Stark, G. R., and

Kerr, I. M. Characterization of a human gene inducible by α and β-interferons and its expression in mouse cells. *EMBO J.* 5: 1601–1606 (1986).

Kimura, A., Israel, A., Le Bail, O., and Kourilsky, P. Detailed analysis of the mouse H-2K^b promoter: Enhancer-like sequences and their role in the regulation of class I gene expression. *Cell* 44: 261–272 (1986).

Knight, E., Fahey, D., and Blomstrom, D. C. Interferon-β enhances the synthesis of a 20,000 dalton membrane protein: A correlation with the cessation of cell growth. *J. Interferon Res.* 5: 305–313 (1985).

Korant, B. D., Blomstrom, D. C., Jonak, G. J., and Knight, E., Jr. Interferon-induced proteins. Purification and characterization of a 15,000 dalton protein from human and bovine cells induced by interferon. *J. Biol. Chem.* 259: 14835–14839 (1984).

Krause, D., Silverman, R. H., Jacobsen, H., Leisy, S. A., Dieffenbach, C. W., and Friedman, R. M. Regulation of ppp $(A2'p)_nA$-dependent RNase levels during interferon treatment and cell differentiation. *Eur. J. Biochem.* 146: 611–618 (1985).

Krust, B., Riviere, Y., and Hovanessian, A. G. p67K kinase in different tissues and plasma of control and interferon-treated mice. *Virology* 120: 240–246 (1982).

Kusari, J. and Sen, G. C. Regulation of synthesis and turnover of an interferon-inducible mRNA. *Mol. Cell. Biol.* 6: 2062–2067 (1986).

Kusari, J., Szabo, P., Grzeschik, K. H., and Sen, G. C. Chromosomal localization of the interferon-inducible human gene encoding mRNA 561. *J. Interferon Res.* 7: 53–59 (1987).

Larner, A. C., Chaudhuri, A., and Darnell, J. E., Jr. Transcriptional induction by interferon. New protein(s) determine the extent and length of the induction. *J. Biol. Chem.* 261: 453–459 (1986).

Larner, A. C., Jonak, G., Cheng, Y. S. E., Korant, B., Knight, E., and Darnell, J. R. Transcriptional induction of two genes in human cells by β-interferon. *Proc. Natl. Acad. Sci.* (USA) 81: 6733–6737 (1984).

Lebleu, B., Sen, G. C., Shaila, S., Cabrer, B., and Lengyel, P. Interferon, double-stranded RNA, and protein phosphorylation. *Proc. Natl. Acad. Sci.* (USA) 73: 3107–3111 (1976).

Lengyel, P. Biochemistry of interferons and their actions. *Ann. Rev. Biochem.* 51: 251–282 (1982).

Luster, A. D., Unkeless, J. C., and Ravetch, J. V. γ-interferon transcriptionally regulates an early-response gene containing homology to platelet proteins. *Nature* 315: 672–676 (1985).

McMahon, M., Stark, G. R., and Kerr, I. M. Interferon-induced gene expression in wild-type and interferon-resistant human lymphoblastoid (Daudi) cells. *J. Virol.* 57: 362–366 (1986).

Merlin, G., Chebath, J., Benech, P., Metz, R., and Revel, M. Molecular cloning and sequence of partial cDNA for interferon-induced (2′-5′oligo(A)synthetase mRNA from human cells. *Proc. Nat. Acad. Sci.* (USA) 80: 4904–4908 (1983).

Meurs, E., Hovanessian, A. G., and Montagnier, L. Interferon mediated antiviral state in human MRC5 cells in the absence of detectable levels of 2-5A synthetase and protein kinase. *J. Interferon Res.* 1: 219–232 (1981).

Nilsen, T. W., Maroney, P. A., and Baglioni, C. Double-stranded RNA causes synthesis of 2′-5′-oligo(A) and degradation of messenger RNA in interferon-treated cells. *J. Biol. Chem.* 256: 7806–7811 (1981).

Nilsen, T. W., Wood, D. L., and Baglioni, C. Presence of 2′-5′-oligo(A) and of enzymes that synthesize, bind, and degrade 2′,5′-oligo(A) in HeLa cell nuclei. *J. Biol. Chem.* 257: 1602–1605 (1982).

Pfeffer, L. M., Wang, E., and Tamm, I. Interferon effects on microfilament organization, cellular fibronectin distribution, and cell motility in human fibroblasts. *J. Cell. Biol.* 85: 9–17 (1980).

Pfefferkorn, E. R. Interferon-γ blocks the growth of Toxoplasma gondii in human fibroblasts by inducing the host cells to degrade tryptophan. *Proc. Natl. Acad. Sci.* (USA) 81: 908–912 (1984).

Pfefferkorn, E. R., Rebhun, S., and Eckel, M. Characterization of an indoleamine 2,3-dioxygenase induced by γ-interferon in cultured human fibroblasts. *J. Interferon Res.* 6: 267–279 (1986).

Prochazka, M., Staeheli, P., Holmes, R., and Haller, O. Interferon-induced guanylate-binding proteins: Mapping of the murine *Gbp-1* locus to chromosome 3. *Virology* 145: 273–279 (1985).

Radke, K. L., Colby, C., Kates, J. R., Krider, H. M., and Prescott, D. M. Establishment and maintenance of the interferon-induced antiviral state: Studies in enucleated cells. *J. Virol.* 13: 623–630 (1974).

Ratner, L., Sen, G. C., Brown, G. E., Lebleu, B., Kawakita, M., Cabrer, B., Slattery, E., and Lengyel, P. Interferon, double-stranded RNA and RNA degradation. Characteristics of an enconuclease activity. *Eur. J. Biochem.* 79: 565–577 (1977).

Roberts, W. K., Hovanessian, A., Brown, R. E., Clemens, M. J., and Kerr, I. M. Interferon-mediated protein kinase and low-molecular weight inhibitor of protein synthesis. *Nature* 264: 477–480 (1976).

Samanta, H., Engel, D. A., Chao, H. M., Thakur, A., Garcia-Blanco, M. A., and Lengyel, P. Interferons as gene activators. Cloning of the 5′ terminus and the control segment of an interferon activated gene. *J. Biol. Chem.* 261: 11849–11858 (1986).

Samanta, H., Pravtcheva, D. D., Ruddle, F. H., and Lengyel, P. Chromosomal location of mouse gene 202 which is induced by interferons and specify a 56.5 KD protein. *J. Interferon Res.* 4: 295–300 (1984).

Saunders, M. E., Gewert, D. R., Tugwell, M. E., McMahon, M., and Williams, B. R. G. Human 2-5a synthetase: Characterization of a novel cDNA and corresponding gene structure. *EMBO J.* 4: 1761–1768 (1985).

Schattner, A., Merlin, G., Levin, S., Wallach, D., Hahn, T., and Revel, M. Assay of an interferon-induced enzyme in white blood cells as a diagnostic aid in viral diseases. *The Lancet* ii, 497–499 (1981).

Schmidt, C. J., Hamer, D. H., and McBride, O. W. Chromosomal location of human metallothionein genes: Implications for Menke's disease. *Science* 224: 1104–1106 (1984).

Shulman, L. and Revel, M. Interferon-dependent induction of mRNA activity for (2′-5′)-oligoisoadenylate synthetase. *Nature* 298: 98–100 (1980).

Smekens-Etienne, M., Goldstein, J., Ooms, H. A., and Dumont, J. E. Variation of (2′-5′)oligo(adenylate) synthetase activity during rat-liver regeneration. *Eur. J. Biochem.* 130: 269–273 (1983).

Staeheli, P., Colonno, R. J., and Cheng, Y. S. E. Different mRNAs induced by interferon in cells from inbred mouse strains A/J and A2G. *J. Virol.* 47: 563–567 (1983).

Staeheli, P. and Haller, O. Interferon-induced human protein with homology to protein Mx of influenza virus-resistant mice. *Mol. Cell. Biol.* 5: 2150–2153 (1985).

Staeheli, P., Prochazka, M., Steigmeier, P. A., and Haller, O. Genetic control of interferon action: Mouse strain distribution and inheritance of an induced protein with guanylate-binding property. *Virology* 137: 135–142 (1984).

St-Laurent, G., Yoshie, O., Floyd-Smith, G., Samanta, H., Sehgal, P., and Lengyel, P. Interferon action: Two $(2'\text{-}5')(A)_n$ synthetases specified by distinct mRNAs in Ehrlich ascites tumor cells treated with interferon. *Cell* 33: 95–102 (1983).

Taylor, J. Inhibition of interferon action by actinomycin. *Biochem. Biophys. Res. Commun.* 14: 447–451 (1964).

Tovey, M., Brouty-Boye, D., and Gresser, I. Early effect of interferon on mouse leukemia cells cultivated in a chemostat. *Proc. Natl. Acad. Sci.* (USA) 72: 2265–2269 (1975).

Vandenbussche, P., Content, J., Lebleu, B., and Werenne, J. Comparison of interferon action in interferon resistant and sensitive L1210 cells. *J. Gen. Virol.* 41: 161–166 (1978).

Verhaegen, M., Divizia, M., Vandenbussche, P., Kuwata, T., and Content, J. Abnormal behavior of interferon-induced enzymatic activities in an interferon-resistant cell line. *Proc. Natl. Acad. Sci.* (USA) 77: 4479–4483 (1980).

Verhaegen-Lewalle, M., Kuwata, T., Zhang, Z. X., Declercq, E., Cantell, K., and Content, J. 2-5A synthetase activity induced by interferon α, β, and γ in human cell lines differing in their sensitivity to the anticellular and antiviral activities of these interferons. *Virology* 117: 425–434 (1982).

Wathelet, M., De Vries, L., Defilippi, P., Nols, C., Vandenbussche, P., Pohl, V., Huez, G., and Content, J. Comparative studies of the genomic structure of two human interferon induced genes. In: *The Biology of the Interferon System 1986*, K. Cantell and H. Schellekens, Eds., Martimus Nyhoff, The Hague pp. 85–91 (1987).

Wathelet, M., Moutschen, S., Cravador, A., Dewit, L., Defilippi, P., Huez, G., and Content, J. Full-length sequence and expression of the 42 kDa 2-5A synthetase induced by human interferon. *FEBS Lett.* 196: 113–120 (1986a).

Wathelet, M., Moutschen, S., Defilippi, P., Cravador, A., Collet, M., Huez, G., and Content, J. Molecular cloning, full-length sequence and preliminary characterization of a 56 kDa protein induced by human interferons. *Eur. J. Biochem.* 155: 11–17 (1986b).

Weil, J., Epstein, C. J., and Epstein, L. B. Polypeptide induction by recombinant human IFN-γ. *Nat. Immun. Cell Growth Regul.* 3: 51–60 (1983a).

Weil, J., Epstein, C. J., Epstein, L. B., Semak, J. J., Sabran, J. L., and Grossberg, S. E. A unique set of polypeptides is induced by γ interferon in addition to those induced in common with α and β interferons. *Nature* 301: 437–439 (1983b).

Weil, J., Epstein, C. J., Epstein, L. B., Van Blerkom, J., and Xuong, N. H. Computer-assisted analysis demonstrates that polypeptides induced by natural and recombinant human interferon-γ are the same and that some have related primary structures. *Antiv. Res.* 3: 303–314 (1983c).

Wells, J. A., Swyryd, E. A., and Stark, G. R. An improved method for purifying 2′-5′-oligoadenylate synthetases. *J. Biol. Chem.* 259: 1363–1370 (1984).

Williams, B. R. G., Saunders, M. E., and Williard, H. F. Interferon-regulated human 2-5A synthetase gene maps to chromosome 12. *Somat. Cell Mol. Genet.* 12: 403–408 (1986).

Wreschner, D. H., McCauley, J. W., Skehel, J. J., and Kerr, I. M. Interferon action: Sequence specificity of the ppp(A2′p)$_n$-dependent ribonuclease. *Nature* 289: 414–417 (1981).

Wreschner, D. H., Silverman, R. H., James, T. C., Gilbert, C. S., and Kerr, I. M. Affinity labelling and characterization of the ppp(A2′p)$_n$A-dependent endoribonuclease from different mammalian sources. *Eur. J. Biochem.* 124: 261–268 (1982).

Yang, K., Samanta, H., Dougherty, J., Jayaram, B., Broeze, R., and Lengyel, P. Interferons, double-stranded RNA, and RNA degradation. Isolation and characterization of homogenous human (2′-5′)(A)$_n$ synthetase. *J. Biol. Chem.* 256: 9324–9328 (1981).

Yonehara, S., Ishii, A., and Yonehara-Takahashi, M. Cell surface receptor-mediated internalization of interferon: Its regulation to the antiviral activity of interferon. *J. Gen. Virol.* 64: 249–2418 (1983).

Yoshida, R., Imanishi, J., Oku, T., Kishida, T., and Hayaishi, O. Induction of pulmonary indoleamine 2,3-dioxygenase by interferon. *Proc. Natl. Acad. Sci.* 78: 129–132 (1981).

Zawatzky, R., Kirchner, H., De Maeyer-Guignard, J., and De Maeyer, E. An X-linked locus influences the amount of circulating interferon induced in the mouse by Herpes simplex virus type 1. *J. Gen. Virol.* 63: 325–332 (1982).

Zilberstein, A., Federman, P., Shulman, L., and Revel, M. Specific phosphorylation in vitro of a protein associated with ribosomes of interferon-treated mouse L cells. *FEBS Lett.* 68: 119–124 (1976).

Zoon, K. C., Arnheiter, H., Zur Nedden, D., Fitzgerald, D. J. P., and Willingham, M. C. Human interferon-α enters cells by receptor-mediated endocytosis. *Virology* 130: 195–203 (1983).

6 THE ANTIVIRAL ACTIVITY OF INTERFERONS

When discussing the antiviral activity of IFNs, we need to make a clear distinction between IFN-α and IFN-β on the one hand, and IFN-γ on the other. There is sufficient evidence to advance the thesis that IFN-α and IFN-β play an important role in host resistance to at least some virus infections, and furthermore, that these IFNs occupy the first line of defense before antibodies or cell-mediated immunity come into play. The role of IFN-γ is different, since in primary infections this IFN is only produced after T cells have been sensitized to viral antigens. During viral infections the key function of IFN-γ resides more in the activation of immune reactions than in its direct antiviral effect.

IFNs of all three species can render cells resistant to viral infection by inducing the antiviral state. However, the induction of the antiviral state in cells is only one of the ways by which IFNs participate in the defense of the organism against virus infections. Furthermore, they do so by stimulating NK cells, activating macrophages, augmenting the expression of MHC antigens involved in T-cell recognition of virus-infected cells, and stimulating cell-mediated immunity in general.

I. THE MULTIPLE MECHANISMS OF THE ANTIVIRAL STATE

Following exposure of cells to IFNs, a wide spectrum of changes, designated as the antiviral state, results in cellular resistance to virus infections. We are still far from a complete understanding of the different mechanisms that are activated during the IFN-induced antiviral state; this complexity is a reflection of the different strategies adopted by viruses to ensure their replication.

The diversity of systems in which the antiviral action of IFNs has been studied practically reflects every possible virus-host cell combination, and we will only give some examples to highlight the different stages of the viral replicative cycle that are susceptible to being influenced by IFNs. A few general comments may be useful before going into the details. The replication of all viruses, without exception, can be to a greater or lesser extent inhibited by IFNs. Moreover, there is no example of a complete IFN-resistant viral mutant, although there are many examples of IFN-resistant cells. Within a given virus strain the most one can do is isolate variants whose replication is to a greater or lesser extent inhibited by the antiviral state. The degree of virus inhibition by IFNs is, furthermore, significantly influenced by the cell in which the effect is measured and even by differences in culture conditions such as monolayer or suspension cultures. In addition, in some cells the antiviral state can be selectively directed against certain viruses without affecting others (Herz et al., 1983; Sen and Herz, 1983; Vandenbussche et al., 1983; Esteban et al., 1986; Mistchenko and Falcoff, 1987).

All viruses, in spite of their widely divergent mechanisms of penetration, replication, and assembly, are sensitive to the antiviral action of IFNs, which suggests that the effect of IFNs on viral replication is pleiotropic, as borne out by experimental evidence. In many cells the antiviral state results in the inhibition of virus replication without unduly affecting cellular protein synthesis: This points to the activation of mechanisms that are capable of discriminatory regulation of viral and cellular protein synthesis.

The different stages of viral infection and replication that are theoretically susceptible to being affected in IFN-treated cells are the following: attachment of virions to the cell membrane followed by penetration, uncoating, transcription and translation, virus assembly, maturation, and egress from the cell. Interestingly enough, there are indications from different virus–host cell systems that, except for attachment, each of these steps can be influenced during the antiviral state, depending on the nature of the virus and of the host cell. The major effect, however, is believed to be on translation.

A. Attachment

Attachment of virions to cell-surface structures, frequently glycoproteins, is a necessary prerequisite of infection. The presence or absence of receptor molecules is instrumental in determining whether or not a cell will be susceptible to virus infection, and cells can be permissive for a virus, albeit resistant to infection because suitable receptors are lacking. In view of the many cell-surface changes induced in IFN-treated cells, the theoretical possibility that IFNs influence virus attachment cannot be excluded. For example, some togaviruses, when coated with nonneutralizing antibodies, utilize Fc receptors for their entry into cells (Prires and Porterfield, 1979). Since IFNs enhance Fc receptor expression, they could thereby theoretically stimulate the attachment of antibody coated virions, but this has not been directly demonstrated.

B. Penetration

Viruses like myxoviruses and herpes viruses penetrate into the cell after the fusion of their envelope with the plasma membrane, and their internal constituents are then released into the cytoplasm. In view of the cell-surface changes induced by IFNs, this process could theoretically be influenced in an IFN-treated cell, but there is no evidence for this to occur.

Nonenveloped viruses penetrate into cells as a result of endocytosis. Although this step has generally been found not to be inhibited in IFN-treated cells, there is one example of IFN-mediated inhibition of virus penetration by endocytosis. Penetration, but not adsorption, of the vesicular stomatitis virus (VSV) is impeded by IFN-α/β treatment of mouse cells, human cells, and chicken cells (in the latter case crude chicken IFN, of unknown composition was used). This, for the time being, is a unique example of IFN action, which may well be applicable to other viruses, since, for example, adenovirus and the Rous sarcoma virus are, like VSV, internalized by endocytosis via coated pits and receptosomes (Whitaker-Dowling et al., 1983). Relatively high doses of IFN are required to inhibit the penetration of VSV into cells, and the importance of this phenomenon in the natural defense against virus infection in vivo is difficult to evaluate. In theory it appears as the best possible way to protect cells, since obviously the most efficient way of preventing damage is to keep the virus out of the cell. It leaves virus available at the cell surface to interact with antibodies and to prime antigen-specific T cells. The molecular mechanism that is responsible for reduced endocytosis is not known; it may be related to the various effects of IFN on the cell surface and the cytoskeleton. In any case, reduction of endocytosis does not appear to contribute significantly to the antiviral state, since in most systems where this has been specifically investigated no effect on virus penetration has been observed. To cite two examples, vaccinia or EMC virus penetration into mouse L cells is not impeded by previous IFN treatment of the cells (Benavente et al., 1984; Meurs et al., 1986).

C. Uncoating

Before the expression of viral functions can occur, the viral capsid has to be removed, either entirely or at least partly. This is usually done by host enzymes, already present in the cell, which probably explains why there is very limited evidence for an IFN effect on the uncoating of virus particles, although there are some indications that uncoating can be delayed by IFN-α/β in retrovirus infected cells (Aboud et al. 1980). It has also been suggested that the inhibition of transcription of the SV40 virus in IFN-treated cells could be the result of the defective uncoating of the viral genome (Brennan and Stark, 1983).

D. Transcription

Transcription of the viral genome occurs at different stages of the replicative cycle, depending on the nature of the virus. The genomes of some single-stranded RNA viruses, such as picorna or togaviruses, which consist of the positive or plus (+) strand, are not transcribed immediately after uptake by the cell; they first bind to ribosomes to be translated into a polyprotein, which is subsequently cleaved into the different viral proteins. One of these, the viral polymerase, then transcribes (−) strands from the (+) strand, and the (−) strands then serve as templates for further synthesis of (+) strands. Inhibition of transcription of (+) strand RNA viruses in IFN-treated cells is therefore not due to a direct effect on transcription, but instead, results from the lack of viral polymerase when primary translation of the input (+) strand is inhibited.

The genomes of other single-stranded RNA viruses, like myxoviruses, arena viruses, and rhabdoviruses, consist of negative (−) strands that must first be transcribed into (+) strands to serve as mRNAs. All (−) strand RNA viruses possess their own polymerase, packed in the virion, and start their infectious cycle by transcribing the viral genome into monocistronic mRNAs. Double-stranded RNA viruses like reovirus also bring their polymerase packaged into the virion and start their infectious cycle by a first round of primary transcription into monocistronic messages.

In general, it appears that primary transcription, that is, transcription of the genomes of input virions, is not inhibited in the case of RNA viruses containing their own transcriptase, such as the vesicular stomatitis virus and reovirus (Repik et al., 1974; Wiebe and Joklik, 1975; Baxt et al., 1977; Ball and White, 1978). Later transcription can be inhibited, but this is then an indirect effect, resulting from the lack of synthesis of new polymerase due to the inhibition of viral mRNA translation. There is also, however, some evidence for a direct effect on the primary transcription of VSV genomic RNA in cells treated with either IFN-α or IFN-β (Belkowski and Sen, 1987). Evidence for such an effect has also been obtained from IFN-γ-treated human cells, but in this case the inhibition of primary transcription is modest and cannot account for the total inhibition of virus replication in these cells (Ulker and Samuel, 1985).

One clear-cut example of the inhibition of primary transcription that is responsible for the antiviral state does exist. In influenza virus-infected *Mx*$^{+}$ cells treated with Mu IFN-α/β, primary transcription of the genomic negative strand RNA by the virion associated polymerase is inhibited, as a result of the induction of the Mx protein (Krug et al., 1985). This is a specific effect of the Mx protein, absent in Mu IFN-α/β-treated Mx^- cells (see Chapter 15). A protein similar to the Mx protein is possibly also induced by IFN in cells of other species since in Hu IFN-α-treated bovine cells, primary transcription of influenza virus RNA is inhibited, as well as in IFN-treated chick embryo cells (Bean and Simpson, 1973; Ransohoff et al., 1985).

Poxviruses are DNA viruses that replicate in the cytoplasm and use their own virion-associated polymerase to initiate transcription since they do not have access to the cellular polymerases in the nucleus. Primary transcription of vaccinia virus DNA is not inhibited in IFN-α/β-treated cells, but it appears rather enhanced (Joklik and Merigan, 1966; Esteban et al., 1984).

Most DNA viruses, however, are transcribed and replicate in the nucleus and use the cellular polymerases for transcription. This is the case for herpes viruses, adenoviruses, and papovaviruses. Papovaviruses are small, nonenveloped, double-stranded DNA viruses, and considerable attention has been paid to the effect of IFNs on the transcription of one of them, SV40 virus, which can transform human and murine cells. In the transformed cells the SV40 genome is integrated into the cellular DNA. IFN-α and IFN-β inhibit the accumulation of early viral mRNA in SV40-infected human or mouse cells, but they have no effect on virus-specific early mRNA once the viral DNA has become integrated into the cellular genome (Oxman et al., 1967; Oxman and Levin, 1971; Mozes and Defendi, 1979). The inhibition of SV40 early mRNA accumulation almost certainly results from decreased transcription, but one cannot exclude a concomitant accelerated degradation of viral RNA transcripts (Kingsman and Samuel, 1980). In cells that are transformed by SV40 and subsequently lytically infected with the virus, the accumulation of virion-derived early mRNA is inhibited by IFN treatment, whereas the accumulation of mRNA transcribed from integrated SV40 DNA is not (Garcia-Blanco et al., 1985). The IFN-induced antiviral state manifestly can discriminate between the expression of a gene situated in the viral genome and the expression of the same gene integrated into the cellular genome.

Possibly other examples of transcriptional inhibition will be found as the techniques for analyzing virus replication become more refined.

E. Translation

All viruses, regardless of their different modes of uncoating, transcription, or assembly, have one stage in common during their replicative cycle: use of the cell machinery to express their proteins. This implies that viral mRNAs must be recognizable and translatable by eucaryotic cells. Because this requirement is a common denominator of all virus infections, it makes sense that inhibition of translation has evolved as the major block of virus replication during the antiviral state.

Two pathways that are responsible for the inhibition of translation have been described: the 2-5A pathway and the protein kinase pathway. The induction and activation of the enzymes involved in these pathways is discussed in detail in the preceding chapter on IFNs as multiple gene activators (Chapter 5).

1. The 2-5A Pathway

One of the mechanisms that are responsible for the inhibition of translation is the activation of the 2-5A pathway, by which IFN induces the dsRNA-dependent 2-5A synthetases, which activate RNase L through the synthesis of the 2-5A oligomers. As a result of the presence of this activated endonuclease, viral mRNA is degraded. In vitro, however, the degradation is not limited to viral RNA, and cellular mRNA as well as ribosomal RNA can also be cleaved by RNAse L (Nilsen et al., 1981; Wreschner et al., 1981; Silverman et al., 1983).

The antiviral potential of 2-5A is directly demonstrated by adding extraneous 2-5A oligomers to intact cells. Under conditions favoring the uptake of the oligomers by the cells, endonuclease activity is stimulated and virus production is decreased (Williams and Kerr, 1978; Hovanessian et al., 1979; Bayard et al., 1985). Moreover, direct injection into individual HeLa cells of 2-5A, and especially of phosphodiesterase-resistant 2-5A analogs, renders the cells resistant to vesicular stomatitis virus, and to a lesser extent to mengovirus, in the absence of any IFN treatment (Fig. 6.1) (Defilippi et al., 1986). This is at the same time a demonstration that endoribonuclease L does not have to be induced by IFN, but rather, is directly activated by 2-5A (see Chapter 5). A further demonstration that the 2-5A system specifically contributes to the antiviral state is provided by the partial restoration of EMC

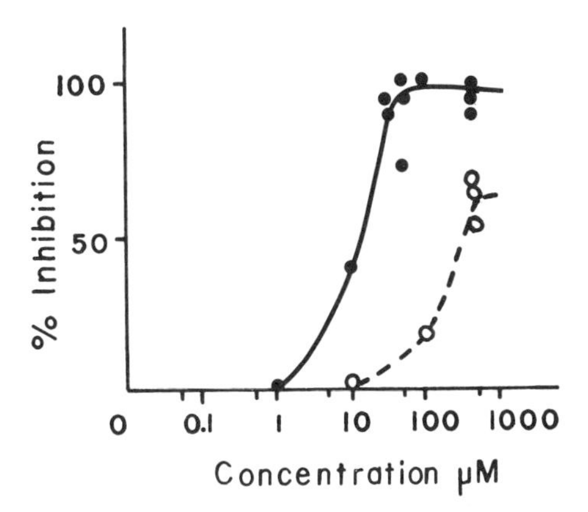

Figure 6.1 A direct demonstration that 2-5A oligonucleotides can suppress virus replication in intact cells. This is done by measuring inhibition of Mengovirus replication after microinjection of natural 2-5A (○—○) or its phosphodiesterase-resistant analog (●—●) into HeLa cells. (Adapted from Defilippi et al., 1986.)

virus RNA synthesis and replication in IFN-treated cells to which an analog inhibitor of 2-5A has been added (Watling et al., 1985).

A further demonstration for an involvement of the 2-5A pathway in the selective establishment of resistance against some, but not all, virus infections is provided by transfecting chinese hamster ovary cells with a cDNA that encodes the human 40 kDa 2-5A synthetase form. Transformed cell clones that constitutively express high levels of the enzyme become resistant to the replication of a picorna virus such as mengovirus, but retain their original sensitivity to vesicular stomatitis virus. This clearly shows that the 2-5A synthetase is sufficient by itself, in the absence of IFN, to cause resistance to some viruses (Chebath et al. 1987).

Contrary to what is observed in vitro, in intact IFN-treated cells viral RNA is preferentially degraded, whereas cellular RNA in general is protected from RNAse L. How does the enzyme discriminate between cellular and viral RNA in the absence of substrate specificity? One suggested mechanism is that of "localized activation." The essential element for preferential degradation of viral RNA could be provided by the presence of double-stranded regions, for example, in replicative intermediates of some RNA viruses. Such regions could cause local activation of the 2-5A synthetase resulting in the presence of activated endonuclease only in the immediate vicinity of the viral RNA, so that the bulk of cellular RNA would remain unaffected (Nilsen and Baglioni, 1979; Baglioni and Nilsen, 1983). Although it will be very difficult to prove unequivocally that this mechanism is indeed operative in intact cells, it appears as the most plausible hypothesis to explain differential degradation.

Because the 2-5A pathway is activated by dsRNA, one would expect it to be most active during infection with dsRNA viruses, since in this case there is no need to wait for the presence of replicative intermediates to activate the system. This is indeed borne out by the rapid rise of 2-5A levels in IFN-treated reovirus-infected Hela cells, followed by a rapid degradation of viral RNA. At high input multiplicities, however, the levels of 2-5A increase so much that, due to the indiscriminate action of ribonuclease L, cellular mRNA and rRNA are also degraded to some extent. Yet, in spite of the extensive mRNA breakdown, the cells manage to maintain a relatively constant level of cellular protein synthesis, because they compensate for the degradation by enhancing cellular mRNA synthesis (Nilsen et al., 1982a, 1983).

The only known function of the 2-5A oligomers is activation of the latent endonuclease, ribonuclease L. Activation of the 2-5A pathway is one of the mechanisms by which IFNs inhibit viruses that replicate by forming dsRNA-containing structures, but this does not imply that the only role of the 2-5A system is antiviral. It is probably also implicated in other cellular functions, such as the regulation of replication and differentiation (Kimchi et al., 1981a,b; Krause et al., 1985). This aspect of the 2-5A system is treated in Chapter 7.

It is evident, furthermore, that the 2-5A pathway is not the only mechanism by which viral mRNA translation is inhibited during the antiviral state. For example, activated RNase L does not play a role in the inhibition of VSV replication by Hu IFN-αA in human cells. In these cells translation of VSV mRNA is greatly inhibited, but this does not result from cleavage or other degradative changes of the mRNA, since viral mRNA isolated from IFN-treated cells is fully intact and can be translated with high efficiency in cell-free systems. In the IFN-treated cells the translational machinery somehow recognizes and discriminates against VSV mRNA, since cellular protein synthesis is hardly affected (Masters and Samuel, 1983a,b).

The appearance of high levels of 2-5A oligomers in IFN-treated cells during virus infection does not necessarily result in high antiviral activity. In human and murine cells treated with IFN-α or IFN-β, vaccinia virus infection causes the appearance of high levels of 2-5A, due to the activation of 2-5A synthetase by the late viral dsRNA. Yet, in spite of these high levels of 2-5A oligomers, vaccinia virus replication is relatively little affected (Rice et al., 1984).

2. The Protein Kinase Pathway

This is the other pathway of translational control that is dependent on the presence of dsRNA structures in IFN-treated cells. The kinase, induced by IFN, is activated in IFN-treated cells by low levels of dsRNA; high levels of dsRNA, however, inhibit activation (Miyamoto et al., 1983). In cell-free systems the dsRNA-dependent protein kinase phosphorylates the α subunit of eIF2, the eukaryotic protein synthesis initiation factor (Fig. 6.2). This initiation factor is composed of three subunits and functions by forming a complex with GTP and met-tRNA$_i$. The GTP moiety is hydrolyzed to GDP

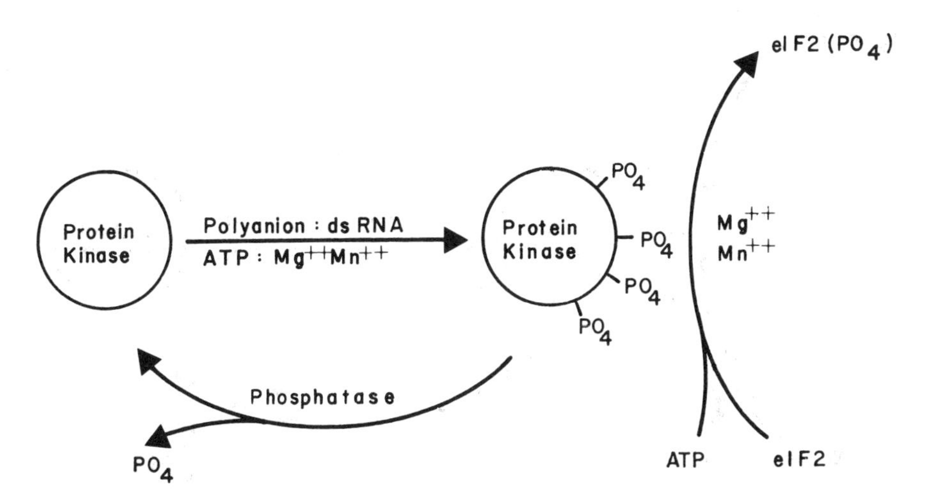

Figure 6.2 The protein kinase P_1 pathway.

as the 80S complex is formed and initiation factors are released to serve in a new round. For this to occur the eIF2-associated GDP must be replaced again by GTP, an exchange catalyzed by a GTP recycling factor. The recycling factor, however, is irreversibly bound by the phosphorylated eIF-2 and is therefore no longer available to recycle GTP, with the result that initiation of translation is inhibited. This sequence of events is very clear in vitro when protein synthesis is measured in cell-free extracts from cells treated with IFN in the presence of dsRNA (Lebleu et al., 1976; Farrel et al., 1977; Lengyel, 1982; Safer, 1983).

Can it be said that the kinase-mediated inhibition of protein synthesis in cell-free systems is relevant to what goes on in virus-infected intact cells? An increased phosphorylation of the 67-kDa protein kinase, due to autophosphorylation (see preceding chapter), occurs in reovirus-infected Mu IFN-α/β-treated mouse cells, and extracts from the same cells are capable of phosphorylating the initiation factor eIF2 in vitro, without addition of dsRNA, normally required in extracts from uninfected cells (Miyamoto and Samuel, 1980; Gupta et al., 1982). This shows that the virus provides enough dsRNA moieties to activate the kinase, but still does not prove that the initiation factor is actually phosphorylated in vivo nor that this is the cause of the inhibition of virus replication. However, one can actually isolate the phosphorylated α subunit of eIF-2 from IFN-treated reovirus-infected cells, thereby showing that the inhibition of translation as a result of kinase activation is at least a theoretical possibility in such cells (Nilsen et al., 1982b; Samuel et al., 1984). A direct demonstration that this can actually occur is provided by the observation that the binding of reovirus mRNA to functional initiation complexes is inhibited in IFN-α/β-treated mouse L cells in which the α subunit of eIF-2 has been phosphorylated (De Benedetti et al., 1985).

Phosphorylation of the eIF-2 initiation factor, as well as activation of the 2-5A pathway has also been shown in intact IFN-treated human cells infected with the encephalomyocarditis virus. We can be confident that both pathways can be active in intact cells, although we do not know the importance of their contribution to the antiviral state (Rice et al., 1985).

Further evidence for an involvement of the protein kinase pathway in the inhibition of virus replication comes from a comparison of wild-type and mutant adenovirus infection in human cells treated with Hu IFN-αA. Adenoviruses on the whole are relatively resistant to the antiviral action of IFNs. This resistance is caused by the presence of VA RNAs: These are small viral RNAs of about 160 nucleotides that accumulate in the cytoplasm of infected cells and become a dominant cytoplasmic RNA in the late phase of infection. Adenovirus VA RNA blocks the activation of the protein kinase that would normally occur in adenovirus-infected cells as a result of dsRNA formation. DsRNA formation takes place late in infection as a result of symmetrical transcription of the adenoviral genome (Petterson and Philipson, 1974). VA RNA most likely competes with dsRNA for binding sites of the protein kinase, and because of its abundancy can occupy a majority of sites and thus prevent activation by dsRNA. Why can VA RNA bind to the kinase without activating it? Its predicted secondary structure contains a number of stable duplex regions of up to 13 bp; this is enough to bind, but not enough to activate, since it has been estimated that a duplex region of at least 50 bp is required for activation (Minks et al., 1979).

Some adenovirus mutants have deletions resulting in the lack of VA RNA formation. When IFN-treated cells are infected with such mutants, the kinase system is activated, and virus replication is much more inhibited than that of the wild-type virus. This combination of genetic and biochemical evidence is a very good demonstration that the protein kinase pathway is involved in the antiviral action of IFNs (Kitajewski et al., 1986; O'Malley et al., 1986).

The inhibition of the protein kinase pathway in IFN-treated cells is not a unique property of adenoviruses, since vaccinia virus is also capable of doing so, but in a different way. Early after a vaccinia virus infection, a protein designated as SKIF (specific kinase inhibitory factor) appears in the infected cells. SKIF inhibits the activation of the protein kinase and the ensuing phosphorylation of the α subunit of eIF2, with the result that protein synthesis can go on. This inhibition is probably the result of competition between SKIF and the kinase for dsRNA, since the effect can be overcome by an excess of dsRNA. Some vaccinia virus mutants, having a deleted section of their genomes, do not cause the formation of SKIF, suggesting that SKIF is either a viral protein, or, less likely, a cellular product induced by a viral protein (Whitaker-Dowling and Youngner, 1984). As a result of the blocked kinase activity, other viruses, for example, the EMC virus, poliovirus, or VSV, are partially rescued from inhibition by IFN if the IFN-treated cells are superinfected with vaccinia virus (Whitaker-Dowling and Youngner, 1983; Paez and Esteban, 1984; Rice and Kerr, 1984; Whitaker-Dowling and Youngner, 1986). The production of SKIF in vaccinia virus-infected cells offers an interesting example of a viral strategy to overcome one mechanism of the antiviral state. However, since the antiviral state has multiple mechanisms, vaccinia virus replication can, nevertheless, be inhibited by IFN.

F. Inhibition of Virion Assembly, Synthesis of Anomalous Particles, and Impeded Release from Infected Cells

Picorna viruses, reoviruses, papovaviruses, and poxviruses are assembled intracellularly, in either the cytoplasm or the nucleus, and are usually released as a result of disintegration of the infected cell. Several enveloped viruses, negative-strand RNA viruses, and retroviruses, are assembled, however, while they are released from the cell, and during this process the viral nucleocapsids are coated by the viral envelope proteins that have been inserted into the plasma membrane. The different stages of assembly and release can be affected during the antiviral state.

The inhbition of retrovirus replication can occur at several levels, but one mechanism is clearly different from what is observed with many other viruses. In chronically infected cells, IFN affects various late steps, such as the postranslational processing of viral precursor proteins, virus assembly, and maturation, leading to the "trapping" of virions at the cell surface. When virions are released from IFN-treated cells, they frequently have reduced infectivity. The inhibition of virus release from an infected cell is usually accompanied by the accumulation of intracellular virions, some of which are RNA deficient, probably as a result of defective virus assembly. Retrovirus assembly requires the proper temporal and spatial cleavage of three major precursor proteins: Gag, which is cleaved to core proteins; gag-pol, which leads to the virion-associated polymerase; and env, which gives rise to the envelope glycoprotein and virion matrix protein. Anomalies in the incorporation of these different proteins into mature virions have been observed in IFN-treated cells, but the exact molecular mechanism has not been elucidated (Billiau et al., 1976; Chang and Friedman, 1977; Chang et al., 1977; Bandyopadhyay et al., 1979; Pitha et al., 1980; Sen and Sarkar, 1980; Aboud et al., 1982; Bilello et al., 1982; Riggin and Pitha, 1982; Aboud and Hassan, 1983; Sen and Pinter, 1983).

Synthesis of anomalous virus particles and the inhibition of virion release from IFN-treated cells are not limited to retroviruses, but they also occur in vaccinia virus and herpes simplex virus type 1 infected cells. Defective vaccinia virus particles are produced in Mu IFN-α/β-treated mouse cells: Virus released from such cells adsorbs, penetrates, and is uncoated less efficiently than normal vaccinia virus. The defect is probably due to modifications of some viral proteins, such as decreased phosphorylation and glycosylation (Esteban, 1984). Rec Hu IFN-α2 or rec Hu IFN-β can block HSV-1 morphogenesis at a late stage of the infectious cycle. Early stages of infection are much less affected, and the major capsid polypeptides are synthesized normally with the typical assembly of viral nucleocapsid cores inside the cell nucleus. It is not clear what causes the block in viral assembly; a significant reduction in the synthesis of two viral glycoproteins may be involved, but there is no proof that this is the major cause of defective assembly (Chatterjee et al., 1985).

An effect on glycosylation of viral proteins in IFN-treated cells has been noticed in several virus-host cell systems, with a concomitant reduction of infectivity of virus particles. Vesicular stomatitis virus particles, released from Mu IFN-α/β-treated mouse fibroblasts, are characterized by a greatly reduced infectivity: Many more noninfectious virions are present. The main reason for this deficiency is in all likelihood a reduced synthesis of VSV virion G glycoprotein; the importance of this particular glycoprotein for VSV infectivity is well known (Printz and Wagner, 1971; Maheswari et al., 1983). In human cells, however, Hu IFN-α does not selectively inhibit the synthesis of the VSV G glycoprotein, nor does Hu IFN-γ (Masters and Samuel, 1984; Ulker and Samuel, 1985). This is one more example of the impossibility to generalize about the mechanism of antiviral activity in IFN-treated cells; the only generalization that can be made is that every virus–host cell system constitutes a special case.

G. Conclusion

The study of the antiviral state in specific virus–host cell systems has provided examples of interference with virus production at virtually every stage of the infectious cycle, but the molecular mechanisms of this pleiotropic antiviral activity are largely unknown. In cells treated with IFN-α and IFN-β the major mechanism appears to be the inhibition of translation, and although the 2-5A and the protein kinase pathways can be involved in this phenomenon, other mechanisms remain to be resolved.

The molecular study of the antiviral action of IFN-γ has received less attention, but we know that although to a lesser extent than IFN-α or IFN-β, it can activate the 2-5A pathway and the dsRNA-dependent protein kinase (Hovanessian et al., 1980).

II. THE ROLE OF IFN-α AND IFN-β IN NATURAL RESISTANCE TO VIRAL INFECTIONS

"IFNs are induced by viruses and IFNs induce an antiviral state in cells, therefore IFNs contribute to host defense against viral infections." This simple and straightforward working hypothesis has turned out to be very hard to test in vivo, principally because of the enormous complexity of the virus–host relationship.

In vitro, in a simple monolayer cell-culture system, one can indeed set up conditions demonstrating that endogenous IFN, produced during virus infection, plays an inhibitory role and limits the spread of infection. This is most easily done by carrying out plaque assays in the presence and absence of anti-IFN-α/β antibodies in the overlay and following the spreading of the infection under the overlay. In many virus-cell combinations the presence of anti-IFN antibodies results in a marked increase of plaque size, meaning that more cells are infected and destroyed when endogenous IFN-α/β is neutralized (Fig. 6.3). This shows that local IFN production, followed by the induction of the antiviral state in the absence of any other effector mechanism that is liable to be stimulated by IFN, can limit the spread of infection.

Because of the complexity of the phenomena involved in viral pathogenesis, it is very hard to single out the role of IFN in vivo—as it is hard to single out the role of any other factor—and much of the evidence for a contribution of IFN in resistance to viral infections is circumstantial. Practically all experiments that are relevant to this problem have been performed in mice infected with many different types of viruses.

The first requirement for IFN to contribute to host defense is obviously

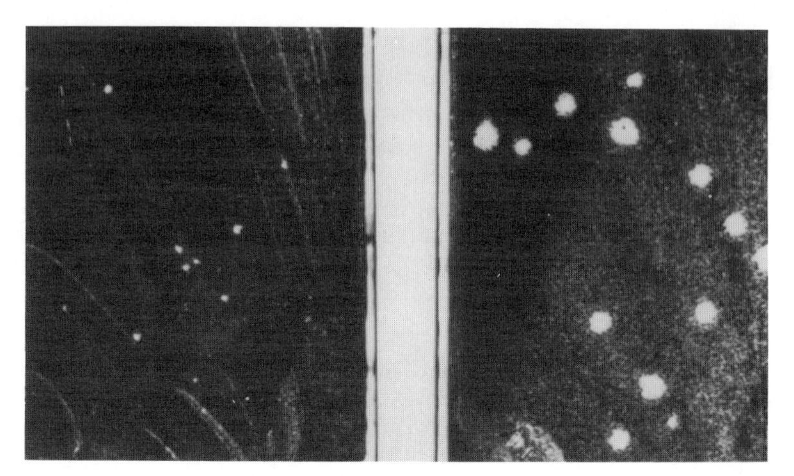

Figure 6.3 The spectacular difference in virus plaque size in the presence or absence of anti-IFN serum offers a direct demonstration that endogenous IFN-α/β, induced by Semliki Forest virus replicating in BALB/c 3T3 cells, inhibits the further replication of the virus and the spreading of the infection under the agarose overlay. Left panel: cells infected with Semliki Forest virus in the absence of anti-Mu IFN-α/β antibodies in the overlay. Right panel: Cells infected with the same amount of virus, but with antibodies to Mu IFN-α/β incorporated into the overlay. The cultures have been fixed and stained with Coomassie brilliant blue 48 hours after infection. (Courtesy of I. Seif.)

that IFN be first produced. In vivo production of IFN has been demonstrated in humans during viral disease and in mouse models using a wide variety of different viruses. In mouse models the kinetics of the appearance of IFN-α/β stand out as one important factor in determining the efficacy of endogenous IFN action. As a rule, during acute infections, early IFN production appears as potentially important for limiting infection, whereas late IFN production generally has no apparent protective effect, but merely reflects the extent of virus replication. Early IFN production is important because of the time required to mount a specific host immune response against the infecting virus; during this interval, IFN production is the only known active defense mechanism.

Convincing evidence that endogenous IFN-α/β can play a role in limiting the spread of infection is provided by the effect of the treatment of virus-infected mice with anti-IFN globulin. Such treatment markedly enhances the severity of infection with viruses as different as herpes simplex, vesicular stomatitis, encephalomyocarditis, or Semliki Forest virus (Gresser, 1984). Moreover, mice that normally display natural resistance toward infection with certain viruses can become susceptible as a result of treatment with anti-IFN globulin. C3H/He mice are resistant to infection with mouse hepatitis virus (MHV-3), and only an exceptional animal dies in the first weeks after infection. When treated with anti-IFN globulin, all mice become fully susceptible and die a few days after infection (Virelizier and Gresser, 1978).

This does not imply the presence of an IFN component in all cases of resistance to virus infection, and there are examples, also from mouse models, where anti-IFN globulin does not affect resistance (Gresser, 1984).

In some virus–host models the virus is administered intranasally, which can be a natural route of infection. However, in the majority of mouse models used for studying the role of IFN, the virus is administered intraperitoneally, intramuscularly, or intracerebrally, which, except for the intramuscular route when the rabies virus is concerned, are hardly natural routes for viral penetration. Caution is, therefore, indicated in extrapolating from these studies to natural viral infections, and there is little information on the relative contribution of endogenous IFN production and action in determining the outcome of viral infections in humans.

There is no reason to doubt that at least part of the protective effect of endogenous IFN results from the direct induction of the antiviral state in different cells in vivo, and, for example, elevated 2-5A synthetase levels are present in peripheral blood mononuclear cells of patients with many different viral diseases (Schattner et al., 1981). The effect of the *Mx* locus in mice infected with influenza virus offers a clear-cut example of the protective effect of endogenous IFN resulting from the induction of an antiviral state in cells (see Chapter 16).

It is, however, clear from mouse-model studies that the antiviral activity of IFN-α/β in vivo not only—and maybe sometimes not principally—results from an induction of the antiviral state in potential target cells, but also from a wide variety of other IFN effects on host defense mechanisms. The best known of these are stimulation by IFN of NK cells, stimulation of cell-mediated immunity, for example, cytotoxic T cells or delayed hypersensitivity, and activation of macrophages. The relevance of these different activities to the antiviral action of IFNs in vivo is discussed at length in the chapters dealing with these topics.

REFERENCES

Aboud, M. and Hassan, Y. Accumulation and breakdown of RNA-deficient intracellular virus particles in interferon-treated NIH 3T3 cells chronically producing Moloney murine leukemia virus. *J. Virol.* 45:489–495 (1983).

Aboud, M., Shoor, R., Bari, S., Hasan, Y., Shurtz, R., Malik, Z. and Salzberg, S. Biochemical analysis and electron microscopic study on intracellular virions in NIH/3T3 mouse cells chronically infected with Moloney murine leukemia virus: Effect of interferon. *J. Gen. Virol.* 62:219–225 (1982).

Aboud, M., Shoor, R., and Salzberg, S. An effect of interferon on the uncoating on murine leukemia virus not related to the antiviral state. *J. Gen. Virol.* 51:425–429 (1980).

Baglioni, C. and Nilsen, T. W. Mechanisms of antiviral action of interferon. In: *Interferon 5,* I. Gresser, Ed., Academic Press, London, pp. 23–42 (1983).

Ball, L. A. and White, C. N. Effect of interferon pretreatment on coupled transcription and translation in cell-free extracts of primary chick embryo cells. *Virology* 84:496–508 (1978).

Bandyopadhyay, A. K., Chang, E. H., Levy, C. C., and Friedman, R. M. Structural abnormalities in murine leukemia viruses produced by interferon-treated cells. *Biochem. Biophys. Res. Commun.* 87:983–988 (1979).

Baxt, B., Sonnabend, J. A., and Bablanian, R. Effects of interferon on vesicular stomatitis virus transcription. *J. Gen. Virol.* 35:325–334 (1977).

Bayard, B., Leserman, L. D., Bisbal, C., and Lebleu, B. Antiviral activity in L1210 cells of liposome-encapsulated (2′-5′) oligo(adenylate) analogues. *Eur. J. Biochem.* 151:319–325 (1985).

Bean, W. J., Jr. and Simpson, R. W. Primary transcription of the influenza virus genome in permissive cells. *Virology* 56:646–651 (1973).

Belkowski, L. S. and Sen, G. C. Inhibition of vesicular stomatitis viral mRNA synthesis by interferons. *J. Virol.* 61:653–660 (1987).

Benavente, J., Paez, E., and Esteban, M. Indiscriminate degradation of RNAs in interferon-treated, vaccinia virus-infected mouse L cells. *J. Virol.* 51:866–871 (1984).

Bilello, J. A., Wivel, N. A., and Pitha, P. M. Effect of interferon on the replication of mink cell focus-inducing virus in murine cells: Synthesis, processing, assembly, and release of viral proteins. *J. Virol.* 43:213–222 (1982).

Billiau, A., Heremans, H., Allen, P. T., De Maeyer-Guignard, J., and De Somer, P. Trapping of oncornavirus particles at the surface of interferon-treated cells. *Virology* 73:537–542 (1976).

Brennan, M. B. and Stark, G. R. Interferon pretreatment inhibits simian virus 40 infections by blocking the onset of early transcription. *Cell* 33:811–816 (1983).

Chang, E. H. and Friedman, R. M. A large glycoprotein of Moloney leukemia virus derived from interferon-treated cells. *Biochem. Biophys. Res. Commun.* 77:392–398 (1977).

Chang, E. H., Myers, M. W., Wong, P. K. Y., and Friedman, R. M. The inhibitory effect of interferon on a temperature sensitive mutant of Moloney murine leukemia virus. *Virology* 77:625–636 (1977).

Chatterjee, S., Hunter, E., and Whitley, R. Effect of cloned human interferons on protein synthesis and morphogenesis of herpes simplex virus. *J. Virol.* 56:419–425 (1985).

Chebath, J., Benech, P., Revel, M., and Vigneron, M. Constitutive expression of (2′-5′) oligo A synthetase confers resistance to picornavirus infection. *Nature* 330:587–588 (1987).

De Benedetti, A., Williams, G. J., Comeau, L., and Baglioni, C. Inhibition of viral mRNA translation in interferon-treated L cells infected with reovirus. *J. Virol.* 55:588–593 (1985).

DeFilippi, P., Huez, G., Verhaegen-Lewalle, M., De Clercq, E., Imai, J., Torrence, P., and Content, J. Antiviral activity of a chemically stabilized 2-5A analog upon microinjection into HeLa cells. *FEBS Lett.* 198:326–332 (1986).

Esteban, M. Defective vaccinia virus particles in interferon-treated cells. *Virology* 133:220–227 (1984).

Esteban, M., Benavente, J., and Paez, E. Effect of interferon on integrity of vaccinia virus and ribosomal RNA in infected cells. *Virology* 134: 40–51 (1984).

Esteban, M., Benavente, J., and Paez, E. Mode of sensitivity and resistance of vaccinia virus replication to interferon. *J. Gen. Virol.* 67:801–808 (1986).

Farrell, P. J., Balkow, K., Hunt, T., Jackson, R. J., and Trachsel, H. Phosphorylation of initiation factor eIF-2 and the control of reticulocyte protein synthesis. *Cell* 11:187–200 (1977).

Garcia-Blanco, M. A., Ghosh, P. K., Jayaram, B. M., Ivory, S., Lebowitz, P., and Lengyel, P. Selectivity of interferon action in simian virus 40-transformed cells superinfected with simian virus 40. *J. Virol.* 53:893–898 (1985).

Gresser, I. Role of interferon in resistance to viral infection in vivo. In *Interferon 2: Interferons and the Immune System,* J. Vilcek and E. De Maeyer, Eds., Elsevier, Amsterdam, pp. 221–247 (1984).

Gupta, S. L., Holmes, S. L., and Mehra, L. L. Interferon action against reovirus: Activation of interferon-induced protein kinase in mouse L929 cells upon reovirus infection. *Virology* 120:495–499 (1982).

Herz, R. E., Rubin, B. Y., and Sen, G. C. Human interferon-α and γ-mediated inhibition of retrovirus production in the absence of an inhibitory effect on vesicular stomatitis virus and encephalomyocarditis virus replication in RD-114 cells. *Virology* 125:246–250 (1983).

Hovanessian, A. G., La Bonnardiere, C., and Falcoff, E. Action of murine γ (immune)-interferon on β (fibroblast)-interferon resistant L1210 and embryonal carcinoma cells. *J. Interferon Res.* 1:125–135 (1980).

Hovanessian, A. G., Wood, J., Meurs, E., and Montagnier, L. Increased nuclease activity in cells treated with pppA2′p5′A2′p5′A. *Proc. Nat. Acad. Sci.* (USA) 76:3261–3265 (1979).

Joklik, W. K. and Merigan, T. C. Concerning the mechanism of action of interferon. *Proc. Natl. Acad. Sci.* (USA) 56:558–565 (1966).

Kimchi, A., Shure, H., Lapidot, Y., Rapoport, S., Panet, A., and Revel, M. Antimitogenic effects of interferon and (2′-5′)-oligoadenylate in synchronized 3T3 fibroblasts. *FEBS Lett.* 134:212–216 (1981a).

Kimchi, A., Shure, H., and Revel, M. Anti-mitogenic function of interferon-induced (2′-5′) oligo(adenylate) and growth-related variations in enzymes that synthesize and degrade this oligonucleotide. *Eur. J. Biochem.* 114:5–10 (1981b).

Kingsman, S. M. and Samuel, C. E. Mechanism of interferon action. Interferon-mediated inhibition of Simian virus 40 early RNA accumulation. *Virology* 101:458–465 (1980).

Kitajewski, J., Schneider, R. J., Safer, B., Munemitsu, S. M., Samuel, C. E., Thimmappaya, B., and Shenk, T. Adenovirus VAI RNA antagonizes the antiviral action of interferon by preventing activation of the interferon-induced eIF-2α kinase. *Cell* 45:195–200 (1986).

Krause, D., Silverman, R. H., Jacobsen, H., Leisy, S. A., Dieffenbach, C. W., and Friedman, R. M. Regulation of ppp $(A2'p)_n$-dependent RNase levels during interferon treatment and cell differentiation. *Eur. J. Biochem.* 146:611–618 (1985).

Krug, R. M., Shaw, M., Broni, B., Shapiro, G., and Haller, O. Inhibition of influenza viral mRNA synthesis in cells expressing the interferon-induced Mx gene product. *J. Virol.* 56:201–206 (1985).

Lebleu, B., Sen, G. C., Shaila, S., Cabrer, B., and Lengyel, P. Interferon, double-stranded RNA, and protein phosphorylation. *Proc. Natl. Acad. Sci.* (USA) 73:3107–3111 (1976).

Lengyel, P. Biochemistry of interferons and their actions. *Ann. Rev. Biochem.* 51:251–282 (1982).

Maheswari, R. K., Husain, M. M., and Friedman, R. M. Low infectivity of vesicular stomatitis virus (VSV) particles released from interferon-treated cells is related to glycoprotein deficiency. *Biochem. Biophys. Commun.* 117:161–168 (1983).

Masters, P. S. and Samuel, C. E. Mechanism of interferon action: Inhibition of vesicular stomatitis virus replication in human amnion U cells by cloned human leucocyte interferon. I. Effect on early and late stages of the viral multiplication cycle. *J. Biol. Chem.* 258:12019–12025 (1983a).

Masters, P. S. and Samuel, C. E. Mechanism of interferon action: Inhibition of vesicular stomatitis virus replication in human amnion U cells by cloned human leucocyte interferon. II. Effect on viral macromolecular synthesis. *J. Biol. Chem.* 258:12026–12033 (1983b).

Masters, P. S. and Samuel, C. E. Mechanism of interferon action. Inhibition of vesicular stomatitis virus in human amnion U cells by cloned human leukocyte interferon. *Biochem. Biophys. Res. Commun.* 119:326–334 (1984).

Meurs, E., Krause, D., Robert, N., Silverman, R. H., and Hovanessian, A. G. The 2-5A system in control and interferon-treated K/BALB cells infected with encephalomyocarditis virus. *Ann. Inst. Pasteur/Virol.* 137E:251–272 (1986).

Minks, M. A., West, D. K., Benvin, S., and Baglioni, C. Structural requirements of double-stranded RNA for the activation of 2′,5′-oligo (A) polymerase and protein kinase of interferon-treated HeLa cells. *J. Biol. Chem.* 254:10180–10183 (1979).

Mistchenko, A. S. and Falcoff., R. Recombinant human interferon-γ inhibits adenovirus multiplication in vitro. *J. Gen. Virol.* 68:941–944 (1987).

Miyamoto, N. G., Jacobs, B. L., and Samuel, C. E. Mechanism of interferon action. Effect of double-stranded RNA and the 5′-0-monophosphate form of 2′,5′-oligoadenylate on the inhibition of reovirus mRNA translation in vitro. *J. Biol. Chem.* 258:15232–15237 (1983).

Miyamoto, N. G. and Samuel, C. E. Mechanism of interferon action. Interferon-mediated inhibition of reovirus mRNA translation in the absence of detectable mRNA degradation but in the presence of protein phosphorylation. *Virology* 107:461–475 (1980).

Mozes, L. W. and Defendi, V. The differential effect of interferon on T antigen production in simian virus 40-infected or transformed cells. *Virology* 93:558–568 (1979).

Nilsen, T. W. and Baglioni, C. Mechanism for discrimination between viral and host mRNA in interferon-treated cells. *Proc. Natl. Acad. Sci.* (USA) 76:2600–2604 (1979).

Nilsen, T. W., Maroney, P. A., and Baglioni, C. Double-stranded RNA causes synthesis of 2′,5′-oligo (A) and degradation of messenger RNA in interferon-treated cells. *J. Biol. Chem.* 256:7806–7811 (1981).

Nilsen, T. W., Maroney, P. A., and Baglioni, C. Synthesis of (2′-5′) oligoadenylate and activation of an endoribonuclease in interferon-treated HeLa cells infected with reovirus. *J. Virol.* 42:1039–1045 (1982a).

Nilsen, T. W., Maroney, P. A., and Baglioni, C. Inhibition of protein synthesis in reovirus-infected HeLa cells with elevated levels of interferon-induced protein kinase activity. *J. Biol. Chem.* 257:14593–14596 (1982b).

Nilsen, T. W., Maroney, P. A., and Baglioni, C. Maintenance of protein synthesis in spite of mRNA breakdown in interferon-treated HeLa cells infected with reovirus. *Mol. Cell. Biol.* 3:64–69 (1983).

O'Malley, R. P., Mariano, T. M., Siekierka, J., and Mathews, M. B. A mechanism for the control of protein synthesis by adenovirus VA RNA1. *Cell* 44:391–400 (1986).

Oxman, M. N., Baron, S., Black, P. H., Takemoto, K. K., Habel, K., and Rowe, W. P. The effect of interferon on SV40 T antigen production in SV40-transformed cells. Virology 32:122–127 (1967).

Oxman, M. N. and Levin, M. J. Interferon and transcription of early virus-specific RNA in cells infected with Simian virus 40. *Proc. Natl. Acad. Sci.* (USA) 68:299–302 (1971).

Paez, E. and Esteban, M. Resistance of vaccinia virus to interferon is related to an interference phenomenon between the virus and the interferon system. *Virology* 134:12–28 (1984).

Petterson, U. and Philipson, L. Synthesis of complementary RNA sequences during productive adenovirus infection. *Proc. Natl. Acad. Sci.* (USA) 71:4887–4891 (1974).

Pitha, P. M., Fernie, B., Maldarelli, F., Hattman, T., and Wivel, N. A. Effect of interferon on mouse leukaemia virus (MuLV). V. Abnormal proteins in virions of Rauscher MuLV produced in the presence of interferon. *J. Gen. Virol.* 46:97–110 (1980).

Printz, P. and Wagner, R. R. Temperature-sensitive mutants of vesicular stomatitis virus: Synthesis of virus-specific proteins. *J. Virol.* 7:651–662 (1971).

Prires, J. S. M. and Porterfield, S. S. Antibody modified enhancement in flavivirus replication in macrophage cell lines. *Nature* 282:509–511 (1979).

Ransohoff, R. M., Maroney, P. A., Nayak, D. P., Chambers, T. M., and Nilsen, T. W. Effect of human αA interferon on influenza virus replication in MDBK cells. *J. Virol.* 56:1049–1052 (1985).

Repik, P., Flamand, A., and Bishop, D. H. L. Effect of interferon upon the primary and secondary transcription of vesicular stomatitis and influenza viruses. *J. Virol.* 14:1169–1178 (1974).

Rice, A. P., Duncan, R., Hershey, J. W. B., and Kerr, I. M. Double-stranded RNA-dependent protein kinase and 2-5A system are both activated in interferon-treated, encephalomyocarditis virus-infected HeLa cells. *J. Virol.* 54:894–898 (1985).

Rice, A. P. and Kerr, I. M. Interferon-mediated, double-stranded RNA-dependent protein kinase is inhibited in extracts from vaccinia virus-infected cells. *J. Virol.* 50:229–236 (1984).

Rice, A. P., Roberts, W. K., and Kerr, I. M. 2-5A accumulated to high levels in interferon-treated, vaccinia virus-infected cells in the absence of any inhibition of virus replication. *J. Virol.* 50:220–228 (1984).

Riggin, C. H. and Pitha, P. M. Effect of interferon on the exogenous Friend murine leukemia virus infection. *Virology* 118:202–213 (1982).

Safer, B. 2B or not 2B: Regulation of the catalytic utilization of eIF-2. *Cell* 33:7–8 (1983).

Samuel, C. E., Duncan, R., Knutson, G. S., and Hershey, J. W. B. Mechanism of interferon action. Increased phosphorylation of protein synthesis initiation factor eIF-2α in interferon-treated, reovirus-infected mouse L929 fibroblasts in vitro and in vivo. *J. Biol. Chem.* 259:13451–13457 (1984).

Schattner, A., Wallach, D., Merlin, G., Hahn, T., Levin, S., and Revel, M. Assay of an interferon-induced enzyme in white blood cells as a diagnostic aid in viral diseases. *The Lancet* ii:497–500 (1981).

Sen, G. C. and Herz, R. E. Differential antiviral effects of interferon in three murine cell lines. *J. Virol.* 45:1017–1027 (1983).

Sen, G. C. and Pinter, A. Interferon-mediated inhibition of production of Gazdar murine sarcoma virus, a retrovirus lacking env proteins and containing an uncleaved gag precursor. *Virology* 126:403–407 (1983).

Sen, G. C. and Sarkar, N. H. Effects of interferon on the production of murine mammary tumor virus by mammary tumor cells in culture. *Virology* 102:431–443 (1980).

Silverman, R. H., Skehel, J. J., James, T. C., Wreschner, D. H., and Kerr, I. M. rRNA cleavage as an index of ppp $(A2'p)_nA$ activity in interferon-treated encephalomyocarditis virus-infected cells. *J. Virol.* 46:1051–1055 (1983).

Ulker, N. and Samuel, C. E. Mechanism of interferon action: Inhibition of vesicular stomatitis virus replication in human amnion U cells by cloned human γ-interferon. II. Effect on viral macromolecular synthesis. *J. Biol. Chem.* 260:4324–4329 (1985).

Vandenbussche, P., Kuwata, T., Verhaegen-Lewalle, M., and Content, J. Effect of interferon on two human choriocarcinoma-derived cell lines. *Virology* 128:474–479 (1983).

Virelizier, J. L. and Gresser, I. Role of interferon in the pathogenesis of viral diseases of mice as demonstrated by the use of anti-interferon serum. V. Protective role in mouse hepatitis virus type 3 infection of susceptible and resistant strains of mice. *J. Immunol.* 120:1616–1619 (1978).

Watling, D., Serafinowska, H. T., Reese, C. B., and Kerr, I. M. Analogue inhibitor of 2-5A action: Effect on the interferon-mediated inhibition of encephalomyocarditis virus replication. *EMBO J.* 4:431–436 (1985).

Whitaker-Dowling, P. and Youngner, J. S. Vaccinia rescue of VSV from interferon-induced resistance: Reversal of translation block and inhibition of protein kinase activity. *Virology* 131:128–136 (1983).

Whitaker-Dowling, P. and Youngner, J. S. Characterization of a specific kinase inhibitory factor produced by vaccinia virus which inhibits the interferon-induced protein kinase. *Virology* 137:171–181 (1984).

Whitaker-Dowling, P. and Youngner, J. S. Vaccinia-mediated rescue of encephalomyocarditis virus from the inhibitory effects of interferon. *Virology* 152:50–57 (1986).

Whitaker-Dowling, P. A., Wilcox, D. K., Widnell, C. C., and Youngner, J. S. Interferon-mediated inhibition of virus penetration. *Proc. Natl. Acad. Sci.* (USA) 80:1083–1086 (1983).

Wiebe, M. E. and Joklik, W. K. The mechanism of inhibition of reovirus replication by interferon. *Virology* 66:229–240 (1975).

Williams, B. R. G. and Kerr, I. M. Inhibition of protein synthesis by 2′-5′ linked adenine oligonucleotides in intact cells. *Nature* 276:88–89 (1978).

Wreschner, D. H., James, T. C., Silverman, R. H., and Kerr, I. M. Ribosomal RNA cleavage, nuclease activation and 2-5A (ppp $(A2'p)_nA$) in interferon-treated cells. *Nucl. Acids Res.* 9:1571–1581 (1981).

7 THE EFFECTS OF INTERFERONS ON CELL GROWTH AND DIVISION

Cell growth and division are influenced by IFNs (Paucker et al., 1962). There are examples of tumor cells whose replication is stimulated by IFNs, but this is exceptional, and the usual effect of all three IFN species is the inhibition of cellular replication. However, and contrary to IFN-α and IFN-β, there is no evidence suggesting a role for IFN-γ in the autocrine regulation of cell proliferation, and therefore, this chapter will mainly deal with IFN-α and IFN-β, whereas the effect of IFN-γ on cell proliferation will be considered in the chapter on the antitumor effects of IFN (Chapter 14).

In many respects IFNs α and β behave as the opposites of growth factors, and of mitogens in general, with the difference that growth factors are frequently specific for cells of a given type, whereas as far as we know IFNs exert very little cell specificity.

Understanding how the antiproliferative effects of IFNs come about will provide an insight into some aspects of normal cell growth regulation and into regulatory mechanisms in immunity and hematopoiesis. It will also explain some of the antitumor effects of IFNs, although it is clear that the antitumor activity of IFNs in vivo is complex and is not limited to a direct antiproliferative effect on tumor cells (Gresser et al., 1972; Uno et al., 1985).

Because of the ease of culture, most studies on cell division and IFN have been performed with continuous lines of fibroblasts and of tumor cells, but in addition to fibroblasts, IFNs also inhibit the replication of other nonmalignant cells, such as liver cells, mammary epithelial cells, vascular smooth muscle cells, and endothelial cells (Frayssinet et al., 1973; Balkwill et al., 1978; Heyns et al., 1985).

There is no reason for thinking that certain cell types cannot be influenced by the antiproliferative action of IFNs, although within a given cell type, at least when malignant cells are concerned, there are many cell lines that do not show an antiproliferative response to IFNs. However, only a small number of differentiated and undifferentiated normal cells, mainly bone-marrow stem cells and cells of the immune system, have been systematically examined for their sensitivity to the antiproliferative effects of IFNs. An investigation of the antiproliferative effects of IFNs on the more than 200 differentiated cell types that make up the adult organism has never been attempted, and would be hard to accomplish in view of the difficulty with which many differentiated cells are maintained and proliferate in vitro.

I. EFFECTS ON THE CELL CYCLE

The four successive phases of the cell cycle are the M phase, consisting of mitosis and cell division, followed by a period of variable synthetic activity called the G1 phase, the S phase, when new DNA synthesis starts, and the period between the completion of DNA synthesis and mitosis called the G2 phase. The G1 phase is usually the most variable in length, and generally the cell's growth rate is inversely proportional to the length of G1. Cells whose growth and replication has been arrested for a prolonged time and are in a quiescent state are usually said to be in G0 phase, rather than in G1, although it is debatable whether G0 is really distinct from G1, or only a prolonged G1 state. IFNs α and β can affect all phases of the cell cycle.

Fibroblasts in culture can be selectively arrested in G1 by depriving the medium of serum and growth factors; this procedure, furthermore, synchronizes the cells since they all become arrested in the same phase. When such synchronized normal mouse or human fibroblasts are stimulated from quiescence to growth, by serum or by more defined growth stimulators like epidermal growth factor, insulin and vasopressin, treatment with Mu IFN-α/β or Hu lymphoblastoid IFN causes a marked prolongation of the G1 phase, a reduced rate of entry into the S phase, and a lengthening of the S and the G2 phase (Sokawa et al., 1977; Balkwill and Taylor-Papadimitriou, 1978; Sreevalsan et al., 1979).

Exposure of human diploid fibroblasts to Hu IFN-β results in a prolongation of all phases of the cell cycle, with an increase of doubling time that can reach as much as three times the control value. The rates of protein, RNA and DNA synthesis, however, are only marginally affected in these cells, with the result that DNA and proteins accumulate, and that the cells become enlarged (Pfeffer et al., 1979). This is a clear indication that in IFN-treated diploid fibroblasts the primary event is inhibition of cell replication, but not of cell growth.

Microcinematographic analysis of Mu IFN-α/β-treated mouse mammary tumor cells reveals that also in tumor cells IFN treatment results in an extended intermitotic time and delays in the G0/G1 phase. In the continuous presence of IFN, the effect becomes more marked with each succeeding generation (Collyn d'Hooghe et al., 1977). In human melanoma cells that are treated with Hu leukocyte IFN the G0/G1 phase is significantly prolonged, with the result that the entry into the S phase is delayed (Creasey et al., 1981).

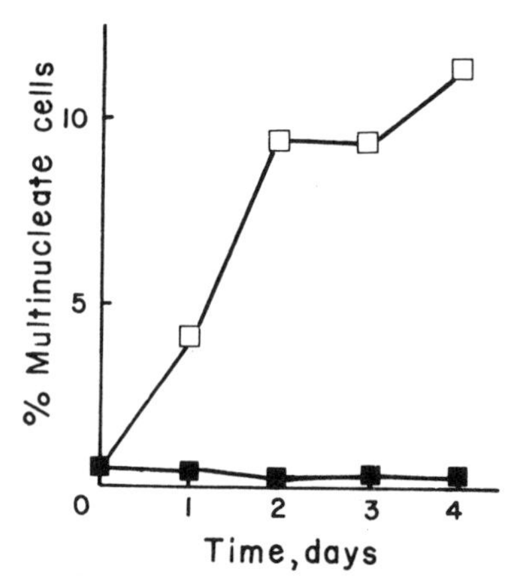

Figure 7.1 In some IFN-treated cells, mitosis is relatively less inhibited than cytokinesis, resulting in the formation of multinucleate cells. The figure shows the rate of formation of multinucleate cells of mouse Ehrlich ascites cells grown in culture in the presence (□) or absence (■) of 1,000 units/ml of Mu IFN-α/β. (Adapted from Panniers and Clemens, 1981.)

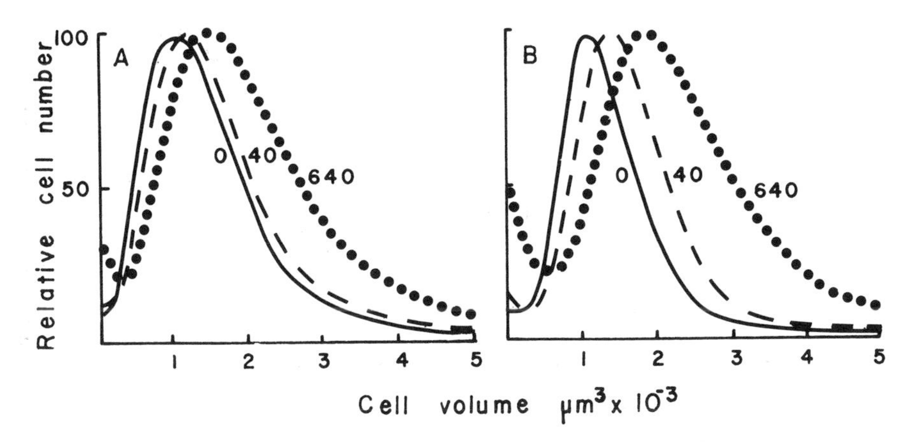

Figure 7.2 As a result of the prolongation of cell cycle in IFN-treated cells, without concomitant inhibition of macromolecular synthesis, cell size increases. The effect of two different concentrations of Hu IFN-β, 40 (- - - -) and 640 (• • • •) units/ml, on cell volume distribution after 3 days exposure of HeLa cells (left panel) or 4 days exposure of Daudi cells (right panel). (Adapted from Pfeffer and Tamm, 1983.)

Like in diploid fibroblasts, in Ehrlich ascites tumor cells Mu IFN-α/β has little effect on the rates of macromolecular synthesis but causes a prolongation of the cell cycle, and as a result, cell size increases. Moreover, mitosis is relatively less inhibited than cytokinesis (or cell separation), resulting in an augmentation of multinucleate cells (Fig. 7.1) (Panniers and Clemens, 1981). An increase in cell size also occurs in human tumor cells of different origins that are exposed to Hu lymphoblastoid IFN, Hu leukocyte IFN, or Hu IFN-β (Fig. 7.2) (Shibata and Taylor-Papadimitriou, 1981; Pfeffer and Tamm, 1983).

These examples make the point that prolongation of the cell cycle of both normal and tumor cells by IFN-α or IFN-β is a general phenomenon. The cumulative effects of this prolongation over several cell generations are cytostatic and can become cytocidal. For example, in the presence of Hu lymphoblastoid IFN, Daudi cells remain viable and continue to grow, with prolonged doubling times, for the first days after treatment. Subsequently, however, they cease to proliferate and start to die (Gewert et al., 1984). Similar observations have been made in other IFN-treated cell systems.

II. GROWTH FACTORS AND IFNs

A. Mutual Antagonism Between Growth Factors and IFNs

Growth factors are relatively cell-specific polypeptides acting through high affinity receptors at the cell surface to trigger cell growth and replication. Growth factors and IFNs mutually influence each other in an antagonistic way: Growth factors reverse the antimitotic effects of IFNs and IFNs block the growth-factor-induced entry of cells into the S phase.

Stimulation of quiescent cells into DNA synthesis either by growth factors such as platelet-derived growth factor (PDGF), epidermal growth factor (EGF), or fibroblast growth factor (FGF), or by other mitogenic stimuli, can be inhibited by IFN-α or IFN-β, even when up to three different mitogenic agents are combined. However, when cell growth is stimulated by a combination of four or more of these factors, the antiproliferative effect of IFN is antagonized, and cell growth is no longer inhibited, even by several thousand units of IFN (Fig. 7.3). Various combinations of growth factors are able to reverse the antiproliferative effect of IFN. Since the growth factors can be added to the cells after IFN-treatment, this excludes, among others, an effect on IFN binding. The antiviral state, however, is not reversed by the growth factors but remains established (Sreevalsan etal., 1980; Taylor-Papadimitriou et al., 1981).

In the presence of low amounts of serum, which normally already contains several mitogenic factors, PDGF, EGF, or fibroblast-derived growth factors by themselves are sufficient to overcome the antiproliferative effect of IFN-α/β (Oleszak and Inglot, 1980).

Conversely, IFNs can inhibit the mitogenic activity of growth factors. Rec Hu IFN-α2 inhibits PDGF-stimulated proliferation of bovine vascular smooth muscle cells and fibroblast growth-factor-induced replication of vascular endothelial cells (Heyns et al., 1985).

In quiescent mouse 3T3 fibroblasts, Mu IFN-α/β inhibits the serum-induced activation of the cell cycle and alters the pattern of secretion of

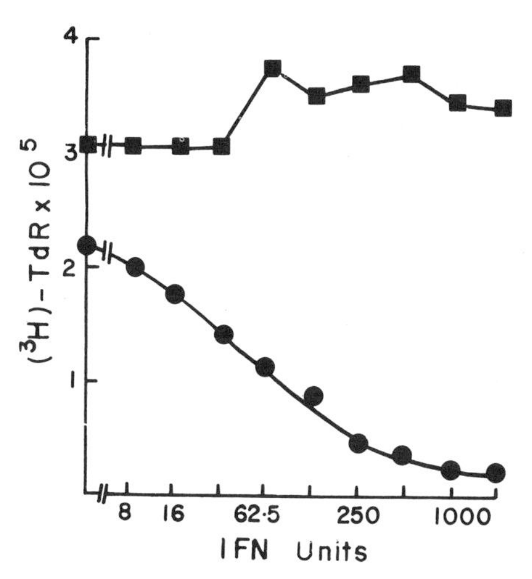

Figure 7.3 A demonstration of the increased effectiveness in reversing the antiproliferative effect of IFN when different mitogenic agents are combined. DNA synthesis, as indicated by thymidine uptake, is inhibited by Mu IFN-α/β in mouse 3T3 fibroblasts stimulated by two growth factors (●, EGF and insulin), but not when five growth factors (■, EGF, insulin, vasopressin, fibroblast-derived growth factor, and colchicine) are applied. (Adapted from Taylor-Papadimitriou et al., 1981.)

several PDGF-induced proteins. Whereas the secretion of some proteins is impaired, that of others is enhanced. The level of some PDGF-induced mRNAS, including c-*myc* mRNA, is actually slightly enhanced by IFN treatment (Tominaga and Lengyel, 1985; Lin et al., 1986). Manifestly, IFNs do not provoke a general inhibition of growth-factor-induced proteins, but they act selectively on some.

B. IFNs as Natural Down-Regulators of Growth Factor Activity

It is clear that, at least in cultured fibroblasts, IFN-α or IFN-β, on the one hand, and various growth factors and mitogenic agents, on the other hand, can act as mutual antagonists. Moreover, several growth factors induce the formation of IFN in the cells they stimulate, suggesting a natural role for IFNs in the down-regulation of growth factor activity.

The IFN-β gene is one of several late activated genes in PDGF-stimulated mouse 3T3 fibroblasts, which in view of the antiproliferative effects of IFNs, raises the possibility that the late activation of the IFN mechanism is part of the negative feedback regulation of the PDGF-induced cycle of cell replication (Zullo et al., 1985).

In addition to PDGF, several other growth factors and cytokines are IFN inducers. TNF-α stimulates the proliferation of human diploid fibroblasts, and addition of anti-Hu IFN-β serum enhances this mitogenic effect. This means that small amounts of IFN-β, probably induced by TNF, contribute to the natural down-regulation of TNF-α-induced cell growth. However, since Northern blot analysis in this particular system only shows the induction of Hu IFN-β2, and not of Hu IFN-β, mRNA, one cannot exclude that another feedback inhibitor in this system is Hu IFN-β2 (Kohase et al., 1986). In fact, the autoregulatory inhibition of cell division as a function of IFN-β2 is a working hypothesis currently receiving much attention (Zilberstein et al., 1986).

The macrophage growth factor CSF-1 causes proliferation and differentiation of bone-marrow stem cells into mononuclear phagocytes, and during this process induces the formation of Mu IFN-α/β (exactly which IFN species is induced has not been determined). This IFN acts as an antiproliferative feedback signal, because when it is neutralized by adding anti-Mu IFN-α/β serum, the CSF-1-induced proliferation of monocyte precursors is enhanced (Moore et al., 1984). This is in line with the observation that exogenous Mu IFN-α/β acts as an antagonist of bone-marrow-derived macrophage and granulocyte colony formation stimulated by CSF-1 (see Chapter 8) (Klimpel et al., 1982).

Hu IFN-β appears spontaneously in cultures of melanoma cells that are grown to high density. The addition of anti-Hu IFN-β antibodies to such cultures accelerates the entry into the S phase, when the cells are restimulated by the growth factors that are present in fresh serum (Creasey et al., 1983).

If IFNs are natural regulators of the cell cycle, we would expect to find evidence of their presence even in the absence of viruses or other infectious agents. Moreoever, since IFNs can only act from without, after binding to their specific plasma membrane receptors, they should be present extracellularly. Yet, usually "spontaneous" IFN production is not observed, either in cell cultures or in vivo. There are exceptions to this, and in some cell cultures, especially of tumor lines and of cells derived from the hemopoietic system, one can demonstrate constitutive IFN synthesis (see Chapter 3). Absence of evidence does not constitute evidence of absence, and the amounts of constitutive or growth factor-induced IFN synthesis may be so minute as to escape detection by routine antiviral assays.

III. HOW DO IFNs INHIBIT CELL PROLIFERATION?

Not much is known about the mechanism by which IFNs inhibit cell proliferation. An involvement of the 2-5A synthetase pathway (see Chapter 5) has been suggested because of the effect of 2-5A isoadenylate on cell replication and because of fluctuations of the 2-5A synthetase levels as a function of the cell cycle.

A. Evidence Supporting the Involvement of the 2-5A System

1. 2-5A Oligomers Inhibit Mitosis

There are several examples of the antimitogenic potential of 2-5A isoadenylate when taken up by cells. The 2-5A oligonucleotides can be introduced into cells by permeabilizing the cells with a hypertonic solution or by coprecipitation with calcium phosphate. Another procedure consists of removing the 5′-triphosphate from 2-5A oligonucleotide cores; such cores are more readily taken up by cells, and they are probably rephosphorylated once inside the cell, although this has not been directly shown to occur.

Addition of micromolar amounts of 2-5A to quiescent mouse fibroblasts, stimulated into growth by fresh serum, does not affect DNA synthesis per se, but does decrease the number of cells entering into the S phase, and therefore is comparable to the action of Mu IFN-α/β (Revel et al., 1982).

The mitogenic stimulation of mouse spleen lymphocytes by concanavalin A can be inhibited by adding 2-5A trimers, tetramers, or pentamers to the cell suspensions, for up to 24 hours after addition of con A (Kimchi et al., 1979; Hovanessian and Wood, 1980).

2. The levels of 2-5A Synthetase Activity Fluctuate with the Cell Cycle

A link between the 2-5A system and control of cell proliferation is, furthermore, suggested by the fluctuation of 2-5A synthetase activity as a function of the cell cycle. The concentration of enzyme activity decreases significantly in liver cells that regenerate after partial hepatectomy and goes up again upon completion of regeneration (Smekens-Etienne et al., 1983). Similarly, in mouse fibroblasts in culture the levels of 2-5A synthetase activity decrease in G2 phase, to go up again at the end of the S phase (Wells and Mallucci, 1985).

Friend erythroleukemic cells, when exposed to dimethylsulfoxide (DMSO), undergo a series of events that lead to a decreased rate of proliferation that is concomitant with terminal erythroid differentiation. In some Friend leukemia cell lines the levels of 2-5A synthetase activity, which are low during the fast growth period, increase five- to tenfold following entry into the stationary phase. Moreover, addition of anti-Mu IFN-α/β antibodies to the cultures abolishes the increase in enzyme activity, indicating that the enzyme increase is caused by the secretion of IFN-α/β by the cells. Measurable IFN-α/β can actually be isolated from the supernatants of the Friend cell cultures (Revel et al., 1982).

This suggests, but does not prove, a role for the 2-5A system in the IFN-mediated inhibition of cell proliferation, probably via ribonuclease L, which is the only known enzyme activated by the 2-5A pathway.

The levels of 2-5A-dependent ribonuclease L increase in 3T3 mouse fibroblasts following growth arrest, indicating that somehow the enzyme is either linked to or influenced by growth regulation. Endonuclease activity is minimal in actively dividing cells and goes up during growth inhibition, for example, as a result of confluency in cell cultures (Jacobsen et al., 1983; Krause et al., 1985).

3. The Correlation Between Activation of the 2-5A Pathway and the Antiproliferative Activity Is Not Perfect

It is also evident, however, that the activation of the 2-5A system alone cannot satisfactorily explain the antimitogenic effects of IFNs, and that other mechanisms must be involved. This is certainly not surprising in view of the complexity of IFN action in particular and of cellular control mechanisms in general.

For example, and in contrast to the preceding example of Friend cell differentiation, in some sublines of Friend leukemia cells DMSO-induced growth arrest and erythroid differentiation can occur without a concomitant rise in endogenous 2-5A synthetase activity (Mechti et al., 1984).

Some human fibroblasts can be completely resistant to the antiproliferative effects of Hu IFN-α or β, yet be fully inducible for 2-5A synthetase and endonuclease activity, resulting in the antiviral state (Vandenbussche et al., 1981; Verhaegen-Lewalle et al., 1982). Likewise, in murine embryonal carcinoma cells, Mu IFN-α/β enhances 2-5A synthetase without exerting any antiproliferative effect (Wood and Hovanessian, 1979). One can also isolate subclones of Daudi cells that are resistant to the antiproliferative effect of Hu IFN-α, yet remain fully inducible for 2-5A synthetase activity (Tovey et al., 1983).

4. Can These Apparently Contradictory Findings Be Reconciled?

It is clear that 2-5A under certain experimental conditions does have antiproliferative effects and that its concentration varies with the cell cycle. It is also evident; however, that the 2-5A system can be activated without concomitant inhibition of cell replication and, conversely, that there can be cell growth arrest in the absence of any measurable 2-5A activation.

The monitoring of 2-5A levels may not provide a precise reflection of the activation of the 2-5A pathway, since the levels of two enzymes of the pathway, the endonuclease that is activated by the oligomers, and the 2-5A phosphodiesterase that degrade them can vary greatly from one cell type to another, and are not necessarily directly correlated with 2-5A levels. It, furthermore, cannot be excluded that the 2-5A pathway is activated by other, as yet unknown, mechanisms and is not solely IFN-dependent. The 2-5A synthetases are present in a wide variety of cells and tissues, but the amount of enzyme activity can vary more than 1,000-fold and its levels are affected by agents other than IFNs. For example, withdrawal of estrogen results in an increase of 2-5A synthetase levels in chick oviducts, and administration of hydrocortisone or corticosteroid-stimulating hormone enhances 2-5A synthetase levels in mice and humans (Stark et al., 1979; Buffer-Janvresse et al., 1986).

It is generally accepted that in IFN-treated cells the 2-5A synthetase pathway is activated by dsRNA. This is usually the result of the presence of viral RNA, and it is not clear what activates 2-5A synthetase in the absence of a viral infection. Maybe double-stranded regions of cellular RNA, for example, of heterogeneous nuclear RNA, can fulfill this function, or other unknown activators are present in cells. There are, furthermore, several functional forms of 2-5A synthetase present in cells, and a large molecular weight form of the enzyme needs a thousand times less dsRNA to be activated (Ilson et al., 1986).

How could an involvement of the 2-5A pathway explain the antiproliferative effect of IFNs? The only function attributed to the 2-5A oligomers is the activation of the latent endoribonuclease L. This RNase cleaves preferably at the 3′ side of U(N) sequences of single-stranded RNA molecules and does not distinguish between RNA of viral or cellular origin. It can, therefore, degrade cellular mRNA and ribosomal RNA, which theoretically could explain the anticellular effect of IFNs.

However, the existence of a relationship between the antiproliferative effect and the inhibition of cellular protein synthesis by IFNs is the subject of some controversy. As we have seen previously, IFNs can exert their antiproliferative effects at concentrations that do not result in a significant inhibition of cellular macromolecular synthesis, and therefore do not inhibit as much cell growth as cell division. Possibly some cellular RNA molecules playing a key role in the onset of mitosis are preferentially cleaved by RNase L.

Some experiments in lymphoblastoid IFN-treated Daudi cells, however, do reveal quite a good correlation between the inhibition of amino acid incorporation and the inhibition of cell proliferation, which suggests that in these cells the determining factor in the inhibition of cell growth is the rate of protein synthesis (Clemens et al., 1984; McNurlan and Clemens, 1986). Daudi cells are among the most sensitive cells to the antiproliferative effects of IFNs, requiring IFN concentrations 10- to 100-fold less than many other cells, and therefore are well suited for an analysis of the antiproliferative effects. They are, however, malignant cells and thus may not necessarily provide a true picture of what goes on in normal cells.

B. Effects on the Utilization of Exogenous Thymidine

A large number of different cell types exhibit a slower rate of thymidine incorporation as a result of IFN treatment. This results from an effect on thymidine transport and uptake, rather than from a direct effect on DNA synthesis, and is partly due to a lowered thymidine kinase activity in IFN-treated cells (Tovey et al., 1975; Brouty-Boyé and Tovey, 1978; Gewert et al., 1981; Gewert et al., 1983; Jasny et al., 1985).

Cells of the human HeLa-S3 epidermoid carcinoma cell line are, like Daudi cells, rather sensitive to the antiproliferative effects of IFN. When these cells are stimulated from quiescence to growth, Hu IFN-β inhibits the transport of thymidine across the plasma membrane, resulting in reduced incorporation into DNA. The rate of transport of the nucleoside across the plasma membrane is reduced, probably reflecting IFN-induced changes in the plasma membrane-cytoskeletal complex. The intracellular uptake of thymidine, furthermore, is also reduced, as a result of inhibition of thymidine kinase activity (Pfeffer and Tamm, 1984).

Most of the effects of IFN on thymidine uptake and transport have been observed in malignant cells, and therefore it is important to compare these effects to what happens in normal cells. When normal mouse embryo fibroblasts in synchronized cultures are exposed to Mu IFN-α/β early in G1 or at the G1-S boudary, thymidine transport is not affected, but thymidine uptake is, and the S-phase-associated increase of thymidine kinase activity is also reduced. The duration of the S phase is not affected, contrary to G1 and G2, which are extended (Mallucci et al., 1983).

As a rule, no correlation is observed between the inhibition of thymidine uptake and the inhibition of cellular DNA synthesis. The various effects of IFN treatment on thymidine transport and uptake are probably not directly responsible for the antiproliferative effect, but rather, reflect IFN-induced changes in the plasma-membrane cytoskeletal complex.

C. Modulation of the Growth Factor Receptors

Modulation of growth factor receptors can influence cellular growth characteristics, and one way by which IFNs could conceivably act as antagonists of growth-promoting agents would be by down-regulating the specific plasma membrane receptors for these growth factors. This possibility is suggested by the down-regulation of low affinity insulin-binding sites on lymphoblastoid Daudi cells that have been exposed to Hu leukocyte IFN or Hu IFN-αConl. High affinity insulin-binding sites are not affected by the IFN treatment (Faltynek et al., 1984; Pfeffer et al., 1987). There is, however, no proof that the down-regulation of low affinity insulin receptors is actually involved in the antiproliferative effects of IFNs.

Madin-Darby bovine kidney (MDBK) cells are sensitive to the antiviral and antiproliferative activities of Hu IFN-α (not of Hu IFN-β or Hu IFN-γ), since they possess the specific high affinity receptors for Hu IFN-α. Treatment of MDBK cells with rec Hu IFN-α2 results in diminished binding of the epidermal growth factor (EGF). This is due to a decrease in the apparent number of specific cell-surface EGF receptors, accompanied by a reduction in receptor affinity as well (Fig. 7.4) (Zoon et al., 1986).

In murine peritoneal macrophages, Mu IFN-α/β down-regulates the number of receptors for the macrophage-granulocyte growth factor CSF-1, and also inhibits receptor-mediated uptake of this growth factor. It is not known whether IFN-mediated down-regulation of CSF-1 receptors also occurs on macrophage-granulocyte progenitor cells, in which case it could contribute to the inhibition of colony formation by IFN-α/β (see Chapter 8) (Chen et al., 1986).

D. Other IFN-Induced Changes Possibly Involved in the Antiproliferative Effects

Several other metabolic changes have been described in cells whose replication is decreased by IFNs, without really solving the problem of how IFNs influence cell proliferation. This is perhaps not surprising, since this question is intimately related to the problem of growth control in cells in general.

1. IFN Inhibits Induction of Enzymes Involved in the Synthesis of Polyamines

The polyamines putrescine, spermine, and spermidine play a role in cell growth and replication, and enzymes mediating polyamine biosynthesis are activated when cells go from quiescence to growth. Ornithine decarboxylase is a rate-limiting enzyme in the biosynthesis of polyamines. It catalyzes the decarboxylation of ornithine to produce one of the three major polyamines—putrescine. Putrescine then serves as a precursor for the synthesis of the two other major polyamines—spermine and spermidine. All three IFN species inhibit the induction of ornithine decarboxylase activity—and therefore the synthesis of putrescine—in human or murine fibroblast cultures stimulated into growth and replication; IFN-β is the most efficient of the three (Sreevalsan et al., 1980; Sekar et al., 1983).

To what extent the inhibition of ornithine decarboxylase synthesis is related to the effects of IFNs on the cell cycle is still an open question, with evidence pro and con. For example, in quiescent fibroblasts, that are stimu-

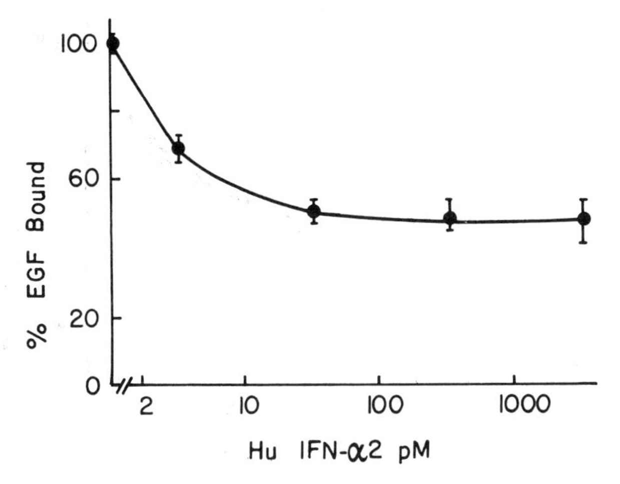

Figure 7.4 Down-regulation of the receptor for epidermal growth factor (EGF) on bovine MDBK cells by Hu IFN-α2, as demonstrated by decreased high affinity binding of radiolabeled EGF as a function of the concentration of IFN, indicated in picomoles, used to treat the cells. (Adapted from Zoon et al., 1986.)

lated into growth by serum or epidermal growth factor and insulin, induction of ornithine decarboxylase is inhibited if IFN is added to the cells for a short time and removed before the growth factors can act, but this inhibition does not result in decreased DNA synthesis and therefore does not seem causally related to the antimitogenic effect (Taylor-Papadimitriou et al., 1985).

On the other hand, some relationship is suggested by the observation that the inhibition by Mu IFN-α/β of mouse liver regeneration in partially hepatectomized mice can be reversed by the intraperitoneal administration of exogenous putrescine. Apparently, the administration of exogenous putrescine compensates for the decreased endogenous synthesis resulting from the IFN treatment (Nishiguchi et al., 1986).

2. *Effects on Nucleotide Metabolism*

A number of individual pathways of nucleotide metabolism is altered by rec Hu IFN-α2 in Daudi cells. Synthesis of ribonucleotides via both alternative pathways, purine biosynthesis de novo and purine and pyrimidine salvage pathways, is decreased. Similarly to the previously mentioned inhibition of thymidine kinase, adenosine kinase activity is also reduced. Nucleotide catabolism is, furthermore, markedly increased by the IFN treatment, resulting in a more than twice elevated excretion of products of ATP degradation. Synthesis of deoxyribonucleotides is also decreased in IFN-treated cells. The selective block of pathways of ribonucleotide and deoxyribonucleotide synthesis could limit the availability of nucleotides for many cellular processes, in addition to interfering with RNA and DNA synthesis (Barankiewicz et al., 1986).

IV. IS THE ANTIPROLIFERATIVE EFFECT RELATED TO THE ANTIVIRAL STATE?

It is a common observation that in many cells less IFN is required to induce the antiviral state than is needed to obtain the antiproliferative effect. Although this may be partly a result of methodology, since one usually measures different rounds of virus infection, resulting in amplification of the effect, differences in methodology do not provide a sufficient explanation to account for the different IFN concentrations that are needed to observe these two effects.

One can, furthermore, dissociate the induction of the antiviral state from the establishment of the antiproliferative effect. A short exposure of quiescent human or murine fibroblasts to IFN-α induces the antiviral state, but it has no antiproliferative effect when the cells are stimulated into growth. To obtain the antimitotic effect, IFN has to be present after growth factor stimulation during the G1 phase (Taylor-Papadimitriou et al., 1985).

If the IFN-induced metabolic changes that are responsible for the antiviral state were also responsible for the antiproliferative effects, one would expect to find a good correlation between both, but this is not the case. We have already seen that evidence exists that the 2-5A pathway might be involved in the antiproliferative action, but that activation of this pathway does not always result in an antiproliferative effect. In addition, normal diploid human fibroblasts of the MRC-5 line are very sensitive to the antiviral action of Hu IFN-β, and are extremely resistant to the antiproliferative effect of the same IFN, since concentrations that are 1,000-fold higher than those used to induce the antiviral state do not reduce the growth rate of these cells (Cook et al., 1983).

There are several examples of cellular mutants that have become resistant to the antiproliferative effects of IFNs, but in which the antiviral state can still be induced (Vandenbusshe et al., 1981; Leanderson et al., 1982). It can also occur, however, that mutants, selected for decreased sensitivity to the antiproliferative action, become at the same time less sensitive to the antiviral action. Since such mutants are not completely resistant to either IFN activity, the simultaneous selection for decreased sensitivity cannot be ascribed to a selection of receptor-less mutants. It could either mean selection of cells bearing receptors with less affinity, or, more likely, imply a common pathway for antiviral and antiproliferative action (Dron and Tovey, 1983).

The antiviral state is the result of a pleiotropic action, only partially characterized biochemically, which can influence virtually all stages of viral replication. The antiproliferative effects almost certainly result from an equally complex array of events, some of which probably overlap with the antiviral mechanism, like the 2-5A pathway, whereas others are seemingly unrelated. Just as one cannot hold a single mechanism responsible for the antiviral state, there may not be a single mechanism that explains the antiproliferative effect. This is evident from the many different IFN-induced changes in cells and from the fact that one can isolate mutants, for example, of Daudi cells, that have reverted from extreme sensitivity to the antiproliferative effect to intermediate sensitivity (Dron and Tovey, 1983). In the case of a single mechanism, one would expect to isolate totally resistant mutants, but not intermediates.

V. IS THE ANTIPROLIFERATIVE EFFECT RELATED TO THE POSSIBLE ROLE OF IFNs AS NATURAL CELL GROWTH REGULATORS?

As seen earlier in Section II, several growth factors induce the formation of IFN-α and IFN-β, and the addition of anti-IFN antibodies to growth-factor-stimulated cells in culture sometimes enhances cell proliferation. In view of the antiproliferative effects of IFNs, this is taken to mean that IFNs are part of the natural feedback mechanism of growth-factor-stimulated mitosis. There is, however, a discrepancy between the very small amounts of growth-factor-induced IFNs that one can isolate from such cultures and the rather high amounts of IFN-α or IFN-β that are sometimes required in culture to inhibit proliferation of normal cells such as murine 3T3 cells or human diploid fibroblasts. Frequently, concentrations from 100 to 10,000 units per milliliter have to be used to obtain measurable effects on cell proliferation. Therefore, one wonders how the relatively small amounts of IFN—at the most a few units per milliliter—isolated from growth-factor-stimulated or density-inhibited cells can affect mitosis. Yet, there is no doubt that they do so, since addition of anti-IFN antibodies to some cell cultures with no or hardly any measurable IFN in the supernatent does result in stimulation of cell proliferation. There are indications for the cell-to-cell transfer of IFN-induced antiproliferative activity, not due to the carryover of IFN itself, but of some other, undefined, IFN-induced substances. This transfer is mediated by a contact-dependent mechanism; it occurs only in high-density cell cultures and may help to explain the relative antiproliferative efficiency of very low amounts of growth-factor-induced IFNs (Lloyd et al., 1983).

A further explanation could be that fibroblasts, which are most frequently studied because of the ease of cultivation, are particularly resistant to the antiproliferative effects of IFNs, whereas some other cells are more sensitive. For example, the proliferation of undifferentiated erythroid precursor cells of mice is significantly inhibited by Mu IFN-α/β at concentrations as low as 5 units per milliliter (Gallien-Lartigue et al., 1980). This is discussed in more detail in the chapter on IFNs and hematopoiesis (Chapter 8).

It is also possible that we are not always working with the right IFNs when extrapolating from the antiproliferative effects of endogenous IFNs to the natural regulation of cell proliferation. For example, a spontaneous IFN secreted during growth and differentiation of Friend erythroleukemia cells, antigenically related to Mu IFN-β, exhibits a higher ratio of antigrowth to antiviral activity than a Sendai-virus-induced IFN (Friedman-Einat et al., 1982).

VI. TUMOR CELLS

A. Tumor Cells Are Frequently More Sensitive to the Antiproliferative Action of IFNs Than Are Untransformed Cells

Proliferation of many tumor cells, derived from either continuous lines or freshly explanted tumors, can be inhibited by IFN-α, β, or γ (Gresser and Tovey, 1978). The sensitivity to the antiproliferative effect varies widely between different tumors or cell lines (Paraf et al., 1983), and frequently tumor cells are more sensitive than normal fibroblasts. The already mentioned EBV transformed B-lymphoblastoid Daudi cells provide an example of high sensitivity, since a few units per milliliter of either Hu IFN-α or IFN-β are sufficient to induce a cytostatic and even a cytocidal effect.

Malignant transformation may render some cells more sensitive to the antimitotic effects of IFNs. For example, the cloning efficiency in a low serum medium of murine NIH/3T3 fibroblasts, is significantly impaired by Mu IFN-α/β after the cells have been transformed by murine sarcoma virus (Bakhanashvili et al., 1983). This correlation between malignancy and increased sensitivity to IFN, however, is far from absolute, and some malignant cells are quite resistant to the antiproliferative effects of IFNs. Paradoxically, proliferation of some freshly isolated tumor cells is actually enhanced by IFNs.

B. The Proliferation of Some Tumor Cells Is Stimulated by IFNs

The replication of cells from a small percentage of freshly collected human tumor samples, cultured as single cell suspensions in a clonogenic assay, is not inhibited, but on the contrary, is stimulated by Hu Leucocyte IFN or by rec Hu IFN-αA and αD (Ludwig et al., 1983). Hu leukocyte IFN also induces proliferation in cells from some patients with primary chronic lymphocytic leukemia (Robert et al., 1985). Thus, for some tumor cells IFNs seem to act as growth factors rather than as growth factor antagonists. This reminds us of the action of TNF-α, which usually causes cell killing, yet can act as a growth factor on human diploid fibroblasts (Kohase et al., 1986).

REFERENCES

Bakhanashvili, M., Wreschner, D. H., and Salzberg, S. Specific antigrowth effect of interferon on mouse cells transformed by murine sarcoma virus. *Cancer Res.* 43: 1289–1294 (1983).

Balkwill, F. and Taylor-Papadimitriou, J. Interferon affects both G1 and S+G2 in cells stimulated from quiescence to growth. *Nature* 274: 798–800 (1978).

Balkwill, F., Watling, D. and Taylor-Papadimitriou, J. Inhibition by lymphoblastoid interferon of growth of cells derived from the human breast. *Int. J. Cancer* 22: 258–265 (1978).

Barankiewicz, J., Kaplinsky, C., and Cohen, A. Modification of ribonucleotide and deoxyribonucleotide metabolism in interferon treated human B-lymphoblastoid cells. *J. Interferon Res.* 6: 717–727 (1986).

Brouty-Boye, D. and Tovey, M. G. Inhibition by interferon of thymidine uptake in chemostat cultures of L1210 cells. *Intervirology* 9: 243–252 (1978).

Buffet-Janvresse, C., Kuhn, J. M., Galabru, J., Svab, J., and Hovanessian, A. G. Interferon-induced enzymes in mice and in volunteers in response to glucocorticoids. *Ann. Inst. Pasteur/Virol.* 137E: 143–154 (1986).

Chen, B. D. M. Interferon-induced inhibition of receptor-mediated endocytosis of colony-stimulating factor (CSF-1) by murine peritoneal exsudate macrophages. *J Immunol.* 136: 174–180 (1986).

Clemens, M. J., McNurlan, M. A., Moore, G., and Tilleray, V. J. Regulation of protein synthesis in lymphoblastoid cells during inhibition of cell proliferation by human interferons. *FEBS Lett.* 171: 111–116 (1984).

Collyn d' Hooghe, M. C., Brouty-Boye, D., Malaise, E. P., and Gresser, I. Interferon and cell division: XII Prolongation by interferon of the intermitotic time of mouse mammary tumor cells in vitro. Microcinematographic analysis. *Exp. Cell. Res.* 105: 73–75 (1977).

Cook, A. W., Carter, W. A., Nidzgorski, F., and Akhtar, L. Human brain tumor-derived cell lines: Growth rate reduced by human fibroblast interferon. *Science* 219: 881–883 (1983).

Creasey, A. A., Bartholomew, J. C., and Merigan, T. C. The importance of Go in the site of action in the cell cycle. *Exp. Cell. Res.* 134: 155–160 (1981).

Creasey, A. A., Eppstein, D. A., Marsh, Y. V., Khan, Z., and Merigan, T. C. Growth regulation of melanoma cells by interferon and (2′-5′)oligoadenylate synthetase. *Mol. Cell. Biol.* 3: 780–786 (1983).

Dron, M. and Tovey, M. G. Isolation of Daudi cells with reduced sensitivity to interferon. I. Characterization. *J. Gen. Virol.* 64: 2641–2647 (1983).

Faltynek, C. R., McCandless, S., and Baglioni, C. Treatment of lymphoblastoid cells with interferon decreases insulin binding. *J. Cell. Physiol.* 121: 437–441 (1984).

Frayssinet, C., Gresser, I., Tovey, M., and Lindahl, P. Inhibitory effect of potent interferon preparations on the regeneration of mouse liver after partial hepatectomy. *Nature* 245: 146–147 (1973).

Friedman-Einat, M., Revel, M., and Kimchi, A. Initial characterization of a spontaneous interferon secreted during growth and differentiation of Friend erythroleukemia cells. *Mol. Cell. Biol.* 2: 1472–1480 (1982).

Gallien-Lartigue, O., Carrez, D., De Maeyer, E., and De Maeyer-Guignard, J. Strain dependence of the antiproliferative action of interferon on murine erythroid precursors. *Science* 209: 292–293 (1980).

Gewert, D. R., Moore, G., and Clemens, M. J. Inhibition of cell division by interferons. The relationship between changes in utilization of thymidine for DNA synthesis and control of proliferation in Daudi cells. *Biochem. J.* 214: 983–990 (1983).

Gewert, D. R., Moore, G., Tilleray, V. J., and Clemens, M. J. Inhibition of cell proliferation by interferons. 1. Effects on cell division and DNA synthesis in human lymphoblastoids (Daudi) cells. *Eur. J. Biochem.* 139: 619–625 (1984).

Gewert, D. R., Shah, S., and Clemens, M. J. Inhibition of cell division by interferons. Changes in the transport and intracellular metabolism of thymidine in human lymphoblastoid (Daudi) cells. *Eur. J. Biochem.* 116: 487–492 (1981).

Gresser, I., Maury, C., and Brouty-Boye, D. Mechanism of antitumour effect of interferon in mice. *Nature* 239: 167–168 (1972).

Gresser, I. and Tovey, M. Antitumour effects of interferon. *Biochim. Biophys. Acta* 516: 231–247 (1978).

Heyns, A. P., Eldor, Z., Vlodavsky, I., Kaiser, N., Fridman, R. and Panet, A. The antiproliferative effect of interferon and the mitogenic activity of growth factors are independent cell cycle events. Studies with vascular smooth muscle cells and endothelial cells. *Exp. Cell Res.* 161: 297–306 (1985).

Hovanessian, A. G. and Wood, J. N. Anticellular and antiviral effects of pp-pA(2′p5′)n. *Virology* 101: 81–90 (1980).

Ilson, D. H., Torrence, P. F., and Vilcek, J. Two molecular weight forms of human 2′,5′-oligoadenylate synthetase have different activation requirements. *J. Interferon Res.* 6: 5–12 (1986).

Jacobsen, H., Krause, D., Friedmann, R. M., and Silverman, R. H. Induction of ppp(A2′p)nA-dependent RNase in murine]LS-VgR cells during growth inhibition *Proc. Nat. Acad. Sci.* (USA.) 80: 4954–4958 (1983).

Jasny, B. R., Fried, J., and Tamm, I. The effects of treatment with human β-interferon on the stimulation of thymidine uptake and DNA synthesis by colchicine in human fibroblasts. *J. Interferon Res.* 5: 239–246 (1985).

Kimchi, A., Shure, H., and Revel, M. Regulation of lymphocyte mitogenesis by (2′-5′)-oligo-isoadenylates. *Nature* 282: 849–851 (1979).

Klimpel, G. R., Fleischmann, W. R., Jr., and Klimpel, K. D. Gamma interferon (IFN-γ) and IFN-α/β suppress murine myeloid colony formation (CFU-C): Magnitude of suppression is dependent upon level of colony-stimulating factor (CSF). *J. Immunol.* 129: 76–80 (1982).

Kohase, M., Henriksen-Destefano, D., May, L. T., Vilcek, J., and Sehgal, P. B. Induction of β2-interferon by tumor necrosis factor: A homeostatic mechanism in the control of cell proliferation. *Cell* 45: 659–666 (1986).

Krause, D., Panet, A., Arad, G., Dieffenbach, C. W., and Silverman, R. H. Independent regulation of ppp$(A2')_n$A-dependent RNase in NIH 3T3, clone 1 cells by growth arrest and interferon treatment. *J. Biol. Chem.* 260: 9501–9507 (1985).

Leanderson, T., Sundstrom, S., Martensson, I. L., Ny, T., and Lundgren, E. Interferon-specific effects on protein synthesis in P_3HR-1 cells. *EMBO J.* 1: 1505–1511 (1982).

Lin, S. L., Kikuchi, T., Pledger, W. J., and Tamm, I. Interferon inhibits the establishment of competence in Go/S-phase transition *Science* 233: 356–359 (1986).

Lloyd, R. E., Blalock, J. E., and Stanton, G. J. Cell-to-cell transfer of interferon-induced antiproliferative activity. *Science* 221: 953–955 (1983).

Ludwig, C. U., Durie, B. G. M., Salmon, S. E., and Moon, T. E. Tumor growth stimulation in vitro by interferon. *Eur. J. Cancer* 19: 1625–1632 (1983).

Mallucci, L., Rasbridge, S., and Wells, V. Cell cycle study on the effect of interferon on synchronized mouse embryo fibroblasts. *J. Interferon Res.* 3: 181–189 (1983).

McNurlan, M. A. and Clemens, M. J. Inhibition of cell proliferation by interferons. Relative contributions of changes in protein synthesis and breakdown to growth control of human lymphoblastoid cells. *Biochem. J.* 237: 871–876 (1986).

Mechti, N., Affabris, E., Romeo, G., Lebleu, B., and Rossi, G. B. Role of interferon and 2′,5′-oligoadenylate synthetase in erythroid differentiation of Friend leukemia cells. Studies with interferon-sensitive and -resistant variants. *J. Biol. Chem.* 259: 3261–3265 (1984).

Moore, R. N., Larsen, H. S., Horohov, D. W., and Rouse, B. T. Endogenous regulation of macrophage proliferative expansion by colony-stimulating factor-induced interferon. *Science* 223: 178–180 (1984).

Nishiguchi, S., Otani, S., Matsui-Yuasa, I., Morisawa, S., Monna, T., Kuroki, T., Kobayashi, K., and Yamamoto, S. Inhibition by interferon ($\alpha + \beta$) of mouse liver regeneration and its reversal by putrescine. *FEBS Lett.* 205: 61–65 (1986).

Oleszak, E. and Inglot, A. D. Platelet derived growth factor (PDGF) inhibits antiviral and anticellular action of interferon in synchronized mouse or human cells. *J. Interferon Res.* 1: 37–48 (1980).

Panniers, L. R. V. and Clemens, M. Inhibition of cell division by interferon: Changes in cell cycle characteristics and in morphology of Ehrlich ascites tumour cells in culture. *J. Cell. Sci.* 48: 259–279 (1981).

Paraf, A., Philips, N., Simonin, G., De Maeyer-Guignard, J., and De Maeyer, E. Differential cytostatic effect of interferon on murine BALB/c B and T-cell tumors. *J. Interferon Res.* 3: 253–260 (1983).

Paucker, K., Cantell, K., and Henle, W. Quantitative studies on viral interference in suspended L cells. III. Effect of interfering viruses and interferon on the growth rate of cells. *Virology* 17: 324–334 (1962).

Pfeffer, L. M., Donner, D. B., and Tamm, I. Interferon-α down-regulates insulin receptors in lymphoblastoid (Daudi) cells. *J. Biol. Chem.* 262: 3665–3670 (1987).

Pfeffer, L. M., Murphy, J. S., and Tamm, I. Interferon effects on the growth and division of human fibroblasts. *Exp. Cell Res.* 121: 111–120 (1979).

Pfeffer, L. M. and Tamm, I. Comparison of the effects of α and β interferons on the proliferation and volume of human tumor cells (HeLa-S3, Daudi, P_3HR-1). *J. Interferon Res.* 3: 395–408 (1983).

Pfeffer, L. M. and Tamm, I. Interferon inhibition of thymidine incorporation into DNA through effects on thymidine transport and uptake. *J. Cell. Physiol.* 121: 431–436 (1984).

Revel, M., Kimchi, A., Friedman, D., Wolf, D., Merlin, G., Panet, A., Rapoport, S., and Lapidot, Y. Cell-regulatory functions of interferon induced enzymes: Antimitogenic effect of (2-5) oligoA, growth-related variations in (2-5) oligo-A synthetase, and isolation of its mRNA. In: *The Interferon System: A Review to 1982 Texas Reports on Biology and Medicine*. S. Baron, Ed. 41: 452–458 (1982), University of Texas Medical Branch, Galveston, Texas.

Robert, K. H., Einhorn, S., Juliusson, G., Ostlund, L., and Biberfeld, P. Interferon induces proliferation and differentiation in primary chronic lymphocytic leukemia cells. *Clin. Exp. Immunol.* 62: 530–534 (1985).

Sekar, V., Atmar, V. J., Joshi, A. R., Krim, M., and Kuehn, G. D. Inhibition of ornithine decarboxylase in human fibroblast cells by type I and type II interferons. *Biochem. Biophys. Res. Commun.* 114: 950–954 (1983).

Shibata, H. and Taylor-Papadimitriou, J. Effects of human lymphoblastoid interferon on cultured breast cancer cells. *Int. J. Cancer 28:* 447–453 (1981).

Smekens-Etienne, M., Goldstein, J., Omms, H. A., and Dumont, J. E. Variation of the (2′-5′)oligo(adenylate) synthetase activity during rat-liver regeneration. *Eur. J. Biochem.* 130: 269–273 (1983).

Sokawa, Y., Watanabe, Y., Watanabe, Y., and Kawade, Y. Interferon suppresses the transition of quiescent 3T3 cells to a growing state. *Nature* 268: 236–238 (1977).

Sreevalsan, T., Rozengurt, E., Taylor-Papadimitriou, J., and Burchell, J. Differential effect of interferon on DNA synthesis, 2-deoxyglucose uptake and ornithine decarboxylase activity in 3T3 cells stimulated by polypeptide growth factors and tumor promotors. *J. Cell. Physiol.* 104: 1–9 (1980).

Sreevalsan, T., Taylor-Papadimitriou, J., and Rozengurt, E. Selective inhibition by interferon of serum-stimulated events in 3T3 cells. *Biochem. Biophys. Res. Commun.* 87: 679–684 (1979).

Stark, G. R., Dower, W. J., Schimke, R. T., Brown, R. E., and Kerr, I. M. 2-5A synthetase: Assay, distribution and variation with growth or hormone status. *Nature* 278: 471–473 (1979).

Taylor-Papadimitriou, J., Balkwill, F., Ebsworth, N., and Rozengurt, E. Antiviral and antiproliferative effects of interferons in quiescent fibroblasts are dissociable. *Virology* 147: 405–412 (1985).

Taylor-Papadimitriou, J., Shearer, M., and Rozengurt, E. Inhibitory effect of interferon on cellular DNA synthesis: Modulation by pure mitogenic factors. *J. Interferon Res.* 1: 401–410 (1981).

Tominaga, S. I. and Lengyel, P. β-interferon alters the pattern of proteins secreted from quiescent and platelet-derived growth factor-treated BALB/c-3T3 cells. *J. Biol. Chem.* 260: 1975–1978 (1985).

Tovey, M. G., Brouty-Boye, D., and Gresser, I. Early effect of interferon on mouse leukemia cells cultivated in a chemostat. *Proc. Natl. Acad. Sci.* (USA) 72: 2265–2269 (1975).

Tovey, M. G., Dron, M., Mogensen, K. E., Lebleu, B., Mechti, N., and Begon-Lours-Guymarho, J. Isolation of Daudi cells with reduced sensitivity to interferon. II. On the mechanisms of resistance. *J. Gen. Virol.* 64: 2649–2653 (1983).

Uno, K., Shimizu, S., Ido, M., Naito, K., Inaba, K., Oku, T., Kishida, T., and Muramatsu, S. Direct and indirect effects of interferon on in vivo murine tumor cell growth. *Cancer Res.* 45: 1320–1327 (1985).

Vandenbussche, P., Divizia, M., Verhaegen-Lewalle, M., Fuse, A., Kuwata, T., De Clercq, E., and Content, J. Enzymatic activities induced by interferon in human fibroblast cell lines differing in their sensitivity to the anticellular activity of interferon. *Virology* 111: 11–22 (1981).

Verhaegen-Lewalle, M., Kuwata, T., Zhang, Z. X., De Clercq, E., Cantell, K., and Content, J. 2-5A synthetase activity induced by interferon α,β and γ in human cell lines differing in their sensitivity to the anticellular and antiviral activities of these interferons. *Virology* 117: 425–434 (1982).

Wells, V. and Mallucci, L. Expression of the 2-5A system during the cell cycle. *Exp. Cell Res.* 159: 27–36 (1985).

Wood, J. N. and Hovanessian, A. G. Interferon enhances 2-5A synthetase in embryonal carcinoma cells. *Nature* 282: 74–76 (1979).

Zilberstein, A., Ruggieri, R., Korn, J., and Revel, M. Structure and expression of cDNA and genes for human interferon-β2, a distinct species inducible by growth-stimulatory cytokines. EMBO *J.* 5: 2529–2537 (1986).

Zoon, K. C., Karasaki, Y., Zur Nedden, D. L., Hu, R., and Arnheiter, H. Modulation of epidermal growth factor receptors by human α-interferon. *Proc. Natl. Acad. Sci.* (USA) 83: 8226–8230 (1986).

Zullo, J. N., Cochran, B. H., Huang, A. S., and Stiles, C. D. Platelet-derived growth factor and double-stranded ribonucleic acids stimulate expression of the same genes in 3T3 cells. *Cell 43:* 793–800 (1985).

8 INTERFERONS AND HEMATOPOIESIS

The hematopoietic system is linked to the IFN system in many different ways. One of the most important manifestations of this interrelationship are the numerous up- and down-regulatory effects exerted by IFNs on the cells of the immune system. In other chapters we describe how IFNs profoundly influence the functions of T and B lymphocytes, macrophages, and NK cells, which all originate from hematopoietic stem cells. Cells derived from the hematopoietic system are, furthermore, important producers of both IFN-γ and IFN-α and IFN-β, and frequently display low levels of "constitutive" IFN production. Finally, as we will discuss in the present chapter, IFNs affect the maturation pathways of hematopoietic cells by influencing stem cell proliferation and sometimes progeny differentiation. To appreciate the different systems in which the effects of IFNs on hematopoiesis have been measured, we present a short introduction on stem cell proliferation and some of the cytokines involved.

I. BLOOD CELL FORMATION RESULTS FROM CONSTANT RENEWAL BY PLURIPOTENT STEM CELLS

Blood cell formation, called hematopoiesis, is characterized by its complexity, since precursor cells of nine distinct lineages, each with multiple maturation stages, live together as a mixed population in the adult bone marrow. All lineages originate from a small population of replicating multipotential stem cells, which maintain their identity and assure hematopoiesis throughout life.

During embryonic development, hematopoietic stem cells are first found in the yolk sac; then in the liver and spleen; and finally, they colonize the bone marrow, where they stay throughout the remainder of postembryonic life, although the spleen remains an important secondary site of hematopoiesis. The precursor stem cells are unable to survive or proliferate unless specifically stimulated by a group of regulatory polypeptides, designated as colony stimulating factors (CSF) because they were first identified by their capacity to stimulate precursor cells to form colonies of progeny cells in culture. This explains why stem cells are represented by the symbol CFU, standing for "colony forming unit." Colony assays have revealed a variety of different precursors of the myeloid lineage, some of which are pluripotent, others oligopotent, giving rise to a few lineages only, and some unipotent, giving rise to one lineage. A common totipotent stem cell gives rise to a distinct set of progeny cells that undergo commitment to the different myeloid or lymphoid developmental lineages. The myeloid stem cells (CFU-GEMM) then give rise to five different lineages: The CFU-GM lineage leads to neutrophils via CFU-G and macrophages via CFU-M; the CFU-MEGA lineage gives rise to megakaryocytes from which platelets are derived; the BFU-E lineage, via the CFU-E, leads to erythrocytes; the CFU-EO lineage leads to eosinophils; and a fifth lineage leads to the formation of mast cells (Fig. 8.1). In addition to the myeloid stem cell lineage, there are special B lineage precursors leading to the formation of B cells and T lineage precursors leading to the various T cell subsets. We thus have accounted for all the different cells that make up the hematopoietic system, except for NK cells, whose lineage is not known (see Chapter 11).

II. LINEAGE-SPECIFIC GROWTH FACTORS ARE RESPONSIBLE FOR GROWTH AND DIFFERENTIATION OF HEMOPOIETIC STEM CELLS

A. The Granulocyte and Macrophage Colony Stimulating Factors

Specific glycoprotein growth factors, required for proliferation and differentiation of the various committed precursor cells, have been identified. Like other growth factors, it is possible that they all induce IFN-α/β, although this has only been formally demonstrated for CSF-1.

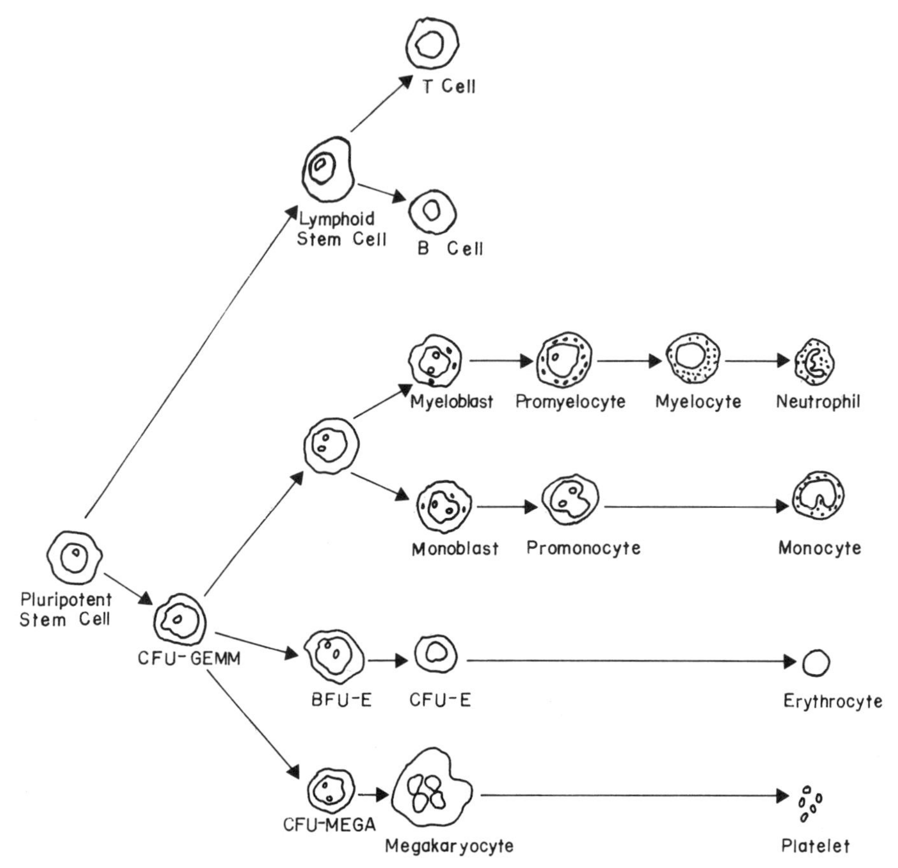

Figure 8.1 A schematic representation of the different pathways of human stem cell differentiation. (Adapted from Foon and Todd, 1986.)

1. Interleukin 3 or Multi-CSF

IL-3, also called Multi-CSF, stimulates the proliferation of a broad distribution of hematopoietic cell types: pure or mixed colonies of granulocytes and macrophages, eosinophil, mast cell, and erythroid colonies. Like many other growth factors, IL-3 stimulates the expression of the protooncogenes c-*myc* and c-*fos* (Sanderson et al., 1985; Begley et al., 1986; Conscience et al., 1986; Ishibashi and Burstein, 1986; Kanz et al., 1986; Koike et al., 1986; Metcalf, 1986; Suda et al., 1986).

The human IL-3 gene codes for a 152 amino acid polypeptide, comprising a 19-amino-acid-long signal peptide. The mature 133 amino acid protein has an expected molecular mass of 14.6 kDa. The gene coding for this protein consists of five exons separated by four introns (Yang et al., 1986). It is situated on human chromosome 5, at bands q23-31 (Le Beau et al., 1987).

The murine IL-3 gene has a structure that is very similar to that of the human gene, and is also composed of five exons and four introns. The predicted amino acid sequence is that of a 146 amino acid mature protein, containing four potential *N*-glycosylation sites, preceded by a 20 amino acid putative leader sequence. The nucleotide homology of the murine and human coding sequences is about 45%, the amino acid homology 28%. The gene is located on mouse chromosome 11, about 230 kb from the gene encoding GM-CSF (Fung et al., 1984; Yokota et al., 1984; Campbell et al., 1985; Miyatake et al., 1985a,b; Barlow et al., 1987; Ihle et al., 1987).

IL-3 is normally produced by antigen- or mitogen-stimulated T cells. Two types of T helper cells have been identified according to the cytokines produced; the first type produces IL-2, IL-3, GM-CSF, and IFN-γ, whereas the second type makes IL-3 and IL-4 (Kelso et al., 1986; Mosmann et al., 1986). IL-3, furthermore, is constitutively produced by transformed cell lines derived from the hemopoietic system, such as the myelomonocytic leukemia cell line WEHI-3B, and by a T-cell hybridoma (Schrader et al., 1980; Ihle et al., 1982).

Mice receiving rec Mu IL-3 exhibit marked rises in blood levels of eosinophils, neutrophils, and monocytes and their spleens show a 50% increase in weight, with elevated levels of maturing granulocytes, eosinophils, nucleated erythroid cells, megakaryocytes, and especially mast cells, which can increase up to a 100-fold. Moreover, in mice with depressed bone-marrow function as a result of sublethal total body irradiation, treatment with IL-3 markedly stimulates hematopoiesis in the spleen (Kindler et al., 1986).

Since IL-3 is not detectable in normal mouse serum, and since it remains unclear whether any IL-3 is synthesized in a normal spleen unless antigen stimulation occurs, it has been suggested that IL-3 is not an important regulator under steady-state conditions, but rather, functions as an effective regulator under conditions of T-lymphocyte activation (Metcalf et al., 1986).

2. Granulocyte-Macrophage Colony Stimulating Factor: GM-CSF

This glycoprotein has a more limited spectrum of activity than IL-3, since its stimulating effect is limited to granulocyte, macrophage, and eosinophil progenitor cells. Added to cultures of bone-marrow cells that are suspended in semisolid medium, it triggers proliferation and differentiation of single hematopoietic precursors and induces them to form colonies of mature granulocytes, macrophages and under certain conditions also eosinophils (Metcalf, 1986).

In vivo, continuous infusion of recombinant human GM-CSF in healthy monkeys causes high levels of circulating neutrophils, monocytes, lymphocytes, and eosinophils, which can be maintained for extended periods by continuous administration of this factor (Donahue et al., 1986). When administered to AIDS patients displaying leukopenia, recombinant human GM-CSF causes a significant increase in circulating neutrophils, monocytes, and eosinophils (Groopman et al., 1987). In addition to cell number, cell function is also affected by GM-CSF. GM-CSF influences the function of mature neutrophils by stimulating their cytotoxicity and accumulation at sites of inflammation (Lopez et al., 1983; Gasson et al., 1984). Moreover, human rec GM-CSF stimulates the cytotoxicity of human monocytes against various tumor cells (Grabstein et al., 1986).

The cDNA coding for human GM-CSF predicts a 144 amino acid polypeptide, consisting of a 17 amino acid leader sequence and a mature protein of 127 residues, with two *N*-glycosylation sites (Lee et al., 1985; Wong et al., 1985). The structural gene for human GM-CSF is about 2.5 kb pairs long and is composed of three introns and four exons. It is located on chromosome 5, region q23-q31 (Huebner et al., 1985; Miyatake et al., 1985a; Le Beau et al., 1986).

Murine GM-CSF, as predicted from the cDNA, consists of a polypeptide of 118 amino acids, containing two *N*-glycosylation sites. The coding sequence of murine GM-CSF displays 70% homology at the nucleotide level and 50% homology at the amino acid level with human GM-CSF. Like the human gene, the murine gene comprises four exons, encompassing 2.5 kb of genomic DNA. There is evidence that the upstream regulatory region of this gene contains two different promoters, resulting in mRNAs of different sizes. Somatic cell hybrid analysis and segregation in an interspecies *Mus musculus* × *Mus spretus* cross (see Chapter 2 for a more detailed description of such an interspecies cross) shows that the GM-CSF gene is closely linked to the IL-3 gene on murine chromosome 11 (Gough et al., 1984; Stanley et al., 1985; Barlow et al., 1987). The GM-CSF and IL-3 encoding genes thus provide an example of conserved linkage between humans and mice.

Although GM-CSF can be made by many different cells, including tumor cells, macrophages seem to provide an important source in vivo. Mouse peritoneal macrophages are induced to produce GM-CSF by exposure to LPS and also by adherence to fibronectin, conditions that mimic events occurring at sites of inflammation (Thorens et al., 1987). GM-CSF has been found in tight association with the stromal cells of the bone marrow. It is possible that GM-CSF is synthesized in small amounts by these cells, since hematopoietic progenitor cells proliferate in close association with stromal cells in vivo and in vitro (Collins and Dorshkind, 1987; Gordon et al., 1987).

3. Granulocyte Colony Stimulating Factor (G-CSF)

This 19 kDa protein stimulates the formation of granulocytic colonies from committed precursor cells and also initiates, but does not sustain, proliferation of erythroid and megakaryocytic precursors (Metcalf, 1986). In the presence of accessory cells, however, G-CSF has a broader spectrum of activity since it also supports the growth of early erythroid progenitors (BFU-E) and pluripotential progenitors (CFU-GEMM). Rec human G-CSF, furthermore, promotes the differentiation of myeloid leukemic cell lines and of fresh myeloid leukemic cells (Souza et al., 1986).

The cDNA encoding human G-CSF predicts a protein of 207 amino acids, consisting of a 30 amino acid leader sequence and a 177 amino acid mature protein; there is little homology with human GM-CSF or IL-3 (Nagata et al.,

1986). The gene encoding human G-CSF is situated on chromosome 17, at 17q11.2-q21 (Simmers et al., 1987). Contrary to the considerable species specificity that is usually exhibited by the various CSFs, human and murine G-CSF display cross-species activity. Human rec G-CSF stimulates granulocyte colony formation of human but also of murine cells, and when administered to mice, rec G-CSF stimulates granulopoiesis and causes splenomegaly (Tsuchiya et al., 1987). In cynomolgus monkeys with cyclophosphamide-induced myelosuppression, administration of human rec G-CSF considerably shortens the period of chemotherapy-induced bone-marrow aplasia (Welte et al., 1987).

4. Macrophage Colony Stimulating Factor (M-CSF or CSF-1)

CSF-1, also called M-CSF, stimulates proliferation of cells belonging to the mononuclear phagocytic lineage. In both humans and mice, CSF-1 is a dimer consisting of two equal subunits, linked by disulfide bonds. Murine CSF-1 is produced by a variety of cells and tissues, such as fibroblasts, embryonic yolk sac, and mouse uterus (Metcalf, 1986). During mouse fetal development, the major sources of CSF-1 are extraembryonic tissues, such as placenta and yolk sac (Azoulay et al., 1987). After postembryonic life, CSF-1, like GM-CSF, is produced by bone-marrow stromal cells, which thus provide the cellular environment that is necessary for the proliferation and differentiation of hematopoietic stem cells (Naparstek et al., 1986; Hunt et al., 1987).

One cDNA for human CSF-1 predicts a mature protein of 224 amino acids long with a putative leader peptide of 32 amino acids (Kawasaki et al., 1985). Comparison with a partial cDNA clone encoding 68 amino acids of the N-terminal region of mature murine CSF-1 reveals 80% nucleotide and 78% amino acid homology between the human and the murine protein in this particular region of the molecule (Rajavashisth et al., 1987). The human CSF-1 structural gene is made up of 10 exons and 9 introns, spanning 20 kb. Differential splicing of transcripts from this gene results in mRNAs that can encode either a 224 or a 522 amino acid CSF-1 molecule (Ladner et al., 1987). Like the gene for GM-CSF, the gene encoding human CSF-1 is on chromosome 5, region 5q33.1 (Pettenati et al., 1987).

The expression of the CSF-1 gene, in both humans and mice, is complex and multiple species of mRNA resulting from the differential processing of the primary transcript have been identified. These mRNAs encode substantially different forms of prepro-CSF-1 polypeptides, resulting in related but different forms of mature CSF-1 (Wong et al., 1987).

5. Macrophage-Granulocyte CSF Receptors

Colony stimulating factors are active at concentrations of $10^{-11} - 10^{-13}$ M, and the plasma membranes of the cells responding to the various growth and differentation factors of the hemopoietic system contain specific high affinity receptors for these cytokines. Different receptors of different molecular weights are involved in the binding of the various CSFs. The mean number of receptors per cell varies for each CSF, but it is generally rather low, consisting on the average of a few thousand high affinity binding sites per cell (Metcalf, 1985).

We will discuss in some detail the receptor for CSF-1, since interaction of murine CSF-1 with its receptor on the plasma membrane of macrophages results in the synthesis of Mu IFN-α/β (see subsequent discussion). The murine CSF-1 receptor is related, and possibly identical, to the c-*fms* protooncogene. V-*fms* is the transforming gene of the McDonough strain of feline sarcoma virus and belongs to the family of *src*-related oncogenes with tyrosine kinase activity. The feline c-*fms* gene product is a 165-kDa glycoprotein of the transmembrane tyrosine kinase type. Human c-*fms* and feline c-*fms* display extensive homology; the 972-amino-acid-long human protein has an extracellular domain, a membrane spanning region, and a cytoplasmic tyrosine kinase domain (Sherr et al., 1985; Coussens et al., 1986).

The murine CSF-1 receptor is immunologically related to the feline c-*fms* protooncogene and is active as a tyrosine-specific protein kinase. The binding of CSF-1 to the purified receptor results in tyrosine-specific autophosphorylation (Sacca et al., 1986).

In vivo, mature macrophages clear CSF-1 by receptor-mediated endocytosis and degrade the growth factor intracellularly. Since the CSF-1 receptor is selectively expressed on mononuclear phagocytes, a very specific feedback control system results, whereby mature macrophages negatively regulate macrophage production by lowering the CSF-1 concentration of the blood (Bartocci et al., 1987).

The FMS gene, encoding human c-*fms,* is located on the long arm of chromosome 5, band q33-q34, close to the structural genes for IL-3 GM-CSF and CSF-1 (Groffen et al., 1983; Le Beau et al., 1986, 1987; Pettenati et al., 1987). The FMS gene is expressed in human bone marrow and differentiated mononuclear cells, and the possibility has been raised that hemizygosity at this locus can lead to abnormalities in hematopoietic maturation (Nienhuis et al., 1985).

B. Stimulation of the Erythrocyte Lineage by Erythropoietin

Human erythropoietin is a glycoprotein with a molecular weight of 34–38 kDa, produced in liver and kidney by the fetus and in the kidney by the adult. Elevated circulating hormone levels trigger proliferation and differentiation of bone-marrow precursor cells, simulate hemoglobin synthesis in maturing erythroblasts, and accelerate the release of erythrocytes into the circulation. After pluripotent stem cells have been committed—for example, by IL-3—to the erythrocyte line, they become sensitive to erythropoietin, and in vitro form colonies giving rise to mature erythrocytes, via either CFU-E or BFU-E. The latter are probably derived from more primitive precursor cells than the CFU-E.

Human preerythropoietin consists of a 27 amino acid signal peptide followed by a 166 amino acid mature protein. The single copy structural gene has five exons and four introns and is located on the long arm of chromosome 7, region q11-q22 (Jacobs et al., 1985; Lin et al., 1985; Law et al., 1986; Powell et al., 1986).

Murine erythropoietin, like human erythropoietin, consists of a 166 amino acid mature protein, showing about 80% homology with the human product, and furthermore also containing three *N*-glycosylation sites. The gene, present as a single copy in the haploid genome, is made up of five exons and four introns. The first intron and much of the 5′ and 3′ flanking regions show a high degree of conservation between humans and mice (McDonald et al., 1986; Shoemaker and Mitsock, 1986).

III. IFNs AS DOWN-REGULATORS OF HEMATOPOIETIC CELL PROLIFERATION

The usual effect of IFN-α, β, or γ on the proliferation of CFUs of any lineage in inhibitory. The inhibition of stem cell development is frequently observed at the relatively low IFN concentration of a few units per milliliter. This suggests that IFNs may exert this effect under physiological conditions, in the microenvironment of the spleen or bone marrow, in which low amounts of IFN are frequently present. As is usually the case with IFNs, their effect is not exclusively down-regulatory, and Hu IFN-γ, which, as discussed subsequently, does inhibit CFU formation, on the other hand also stimulates monocytes to release GM-CSF and induces T cells to release GM-CSF and IL-3 (Piacibello et al., 1985, 1986).

A. Effects of IFNs on the Monocyte-Macrophage Lineage

Mu IFN-α/β can inhibit the proliferation of committed precursors leading to mononuclear phagocytes and appears as a natural feedback regulator of CSF-1-stimulated monocyte development.

Macrophages develop from committed stem cells in vitro when bone-marrow cells are cultured in the presence of CSF-1. When mouse bone marrow cells are thus cultured in liquid medium in the presence of CSF-1, cells of the granulocyte-monocyte lineage survive and proliferate at first, but after 4 or 5 days only the macrophages remain. It is, therefore, relatively easy to test the effect of IFNs on the generation of monocytes by adding the desired IFN concentrations to the culture medium. Under these conditions, Mu IFN-α/β inhibits the generation of monocytes in a dose-dependent way. The bone-marrow cells derived from some inbred mouse strains, for example, BALB/c, are more sensitive to this effect than those derived from other strains such as C57BL/6 (Fig. 8.2). This difference is due to the presence of different alleles at genes influencing the antiproliferative effect of IFN-α/β

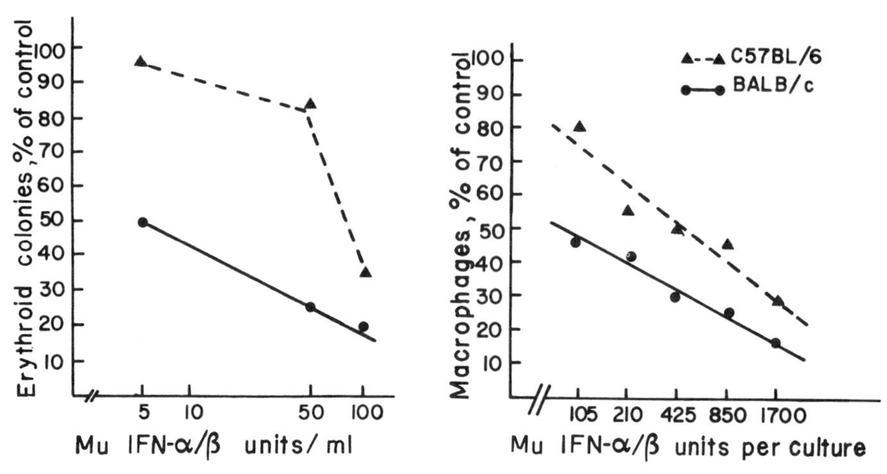

Figure 8.2 A demonstration that the antiproliferative effect of IFN is influenced by the genotype of the target cell. Hemopoietic stem cells from some murine genotypes, for example, BALB/c (●—●), are more sensitive to the antiproliferative effect of Mu IFN-α/β than are cells of other genotypes, for example, C57BL/6 (▲—▲). Left panel: Antiproliferative effect of different IFN concentrations on erythropoietin-stimulated CFU-E in vitro. Right panel: Antiproliferative effect of different IFN concentrations on CSF-1-stimulated macrophage development in vitro. (Adapted from Gallien-Lartigue et al., 1980, and from Dandoy et al., 1981.)

on the monocyte precursors (see Chapter 15) (Dandoy et al., 1981). Marked differences in sensitivity to the antiproliferative effects of Hu leukocyte and Hu IFN-β, ranging from extreme sensitivity to virtually total resistance, have also been observed with human granulocyte progenitor cells (CFU-C) that were obtained from different individuals (Oladipupo-Williams et al., 1981). Such individual variations in sensitivity explain why, as a result of IFN-therapy, the bone-marrow function of some patients becomes severely depressed, whereas that of others shows no apparent effect of the IFN treatment. In some patients, myelosuppression is the major dose-limiting factor in the course of IFN-α treatment, which has been specifically shown to inhibit maturation of bone-marrow progenitor cells, thus preventing the repopulation of the peripheral blood (Ernstoff et al., 1985).

The inhibitory effect of IFN is significantly influenced by the growth factor involved, since when GM-CSF is used as a growth factor instead of CSF-1, the in vitro proliferation of bone-marrow cells and their differentiation into macrophages and granulocytes is more resistant to inhibition by Mu IFN-α/β, and higher concentrations of IFN are required to obtain inhibition (Fig. 8.3) (Yamamoto-Yamaguchi et al., 1983).

Another way of analyzing the effects of IFNs on CFU-GM progeny is to follow cell proliferation in suspension cultures in a semisolid medium and to count the number of colonies. In such cultures, rec Hu IFN-γ and rec Hu IFN-αD significantly inhibit colony formation (Rigby et al., 1985). Mu IFN-γ

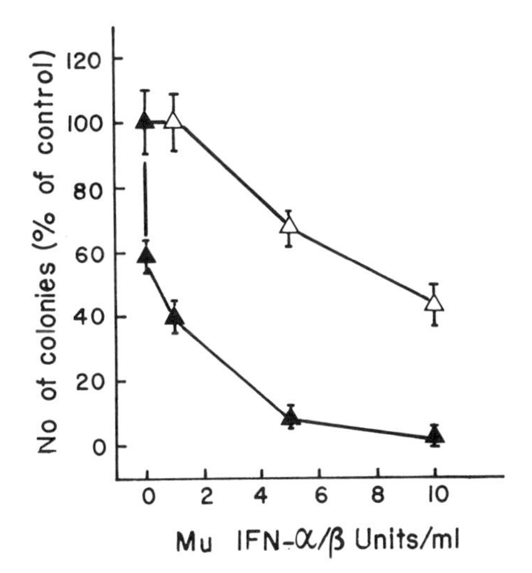

Figure 8.3 The antiproliferative effect of Mu IFN-α/β is influenced by the growth factor used to stimulate development, and GM-CSF stimulated macrophage and granulocyte colonies (△—△) are less inhibited by IFN than are CSF-1-stimulated colonies (▲—▲). (Adapted from Yamamoto-Yamaguchi et al., 1983.)

inhibits the formation of myeloid colonies derived from CFU-GM, and in doing so, it can act synergistically with IFN-α/β, with the result that the degree of suppression achieved by the mixture of the three IFNs is higher than the additive effects of each IFN separately. At high doses of IFN all colonies are suppressed, which means that both macrophage and granulocyte precursors are inhibited. The suppression of colony formation is dependent on the level of CSF-1 used to stimulate colony formation, and the antagonistic effect of all three IFNs on colony formation can be completely overcome by raising the amount of CSF-1 (Klimpel et al., 1982). Myelosuppression in vivo by Mu IFN-α, β, or γ can similarly be prevented by treatment of the mice with CSF-1 (Koren et al., 1986). This is in line with many other observations on the antagonism between growth factors and IFNs, as discussed in Chapter 7.

The antiproliferative effects exerted by IFNs in stem cell-derived cultures suggest a role for IFNs in the physiological control of progenitor cell proliferation under natural conditions. This is particularly evident in the case of the specific macrophage growth factor CSF-1.

Whereas CSF-1 acts as a proliferative signal for committed macrophage precursors, on more differentiated cells it acts as a stimulatory signal for the synthesis of effector molecules, such as prostaglandins E or IFN-α/β. For example, CSF-1 enhances LPS- or polyrIrC-induced Mu IFN-α/β production of murine macrophages, and human monocytes produce more than 20 times as much polyrIrC-induced IFN-α/β, if they are first exposed to CSF (Fleit and Rabinovitch, 1981; Warren and Ralph, 1986). Moreover, CSF-1 itself induces the synthesis of Mu IFN-α/β in murine bone-marrow cells, thereby down-regulating its own growth stimulating effect as well as inducing an antiviral state. Addition of anti-Mu IFN-α/β serum to such CSF-1-stimulated cultures results in an enhancement of the proliferative response, due to the inactivation of the CSF-1-induced IFN-α/β (Fig. 8.4) (Moore et al., 1984; Lee and Warren, 1987).

The molecular mechanism of the antiproliferative effects of IFN-α/β on CSF-1-stimulated monocytes is not known, reflecting our ignorance of the mechanism of the antiproliferative effects of IFNs in general. One possible mechanism is suggested by the observation that the specific CSF-1 receptor on murine macrophages is down-regulated after treatment with Mu IFN-α/β. Such down-regulation is one possible way of counteracting the stimulation of cell proliferation by CSF-1, and it strengthens the possibility that the induction of IFN-α/β by CSF-1 is part of the physiological feedback mechanisms regulating the activity of this growth factor (Chen, 1986).

B. Effects of IFNs on the Erythrocyte Lineage

IFNs can suppress the proliferation of erythroid progenitors (BFU-E) and colony forming units (CFU-E) in vitro. Murine IFN-α/β inhibits the development of murine early (BFU-E) and late (CFU-E) erythroid progenitors derived from fetal liver (Smith et al., 1977). In these assays the growth factor is erythropoietin, and contrary to what is observed with CSF-1 and GM-CSF, augmenting the concentration of erythropoietin has little or no effect on the antiproliferative action of IFN. The genes that affect the degree of susceptibility to the antiproliferative action of Mu IFN-α/β at some stage of monocyte precursor development also influence the antiproliferative action on the

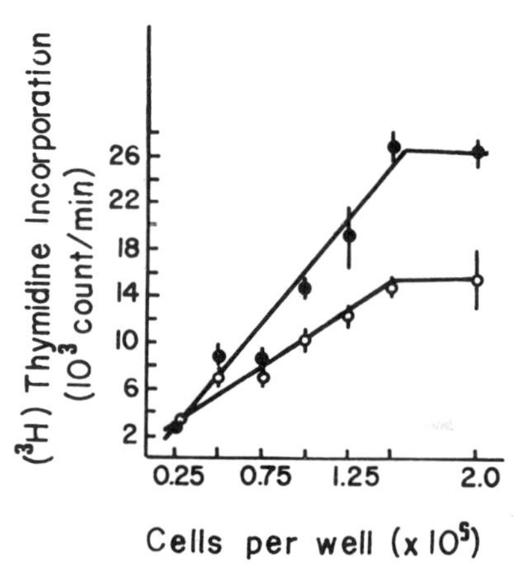

Figure 8.4 Evidence for down-regulation of cell proliferation by CSF-1-induced endogenous IFN. When the Mu IFN-α/β that appears in CSF-1-stimulated cultures of murine femoral bone-marrow cells is neutralized by antibody, cell proliferation is enhanced (●): with anti-IFN-α/β antibody; ○: without antibody. (Adapted from Moore et al., 1984).

late erythroid precursors (CFU-E) (Fig. 8.2). They do not, however, influence the antiproliferative effect on the very early BFU-E precursors, which are also more resistant to the antiproliferative effect than are the CFU-E. The genes influencing the degree of sensitivity to the antiproliferative effects of IFN are apparently expressed only at a certain stage of precursor development (Gallien-Lartigue et al., 1980, 1985).

The development of BFU-E and CFU-E is also inhibited by Hu IFN-γ, and as in the case of Mu IFN-α and β, the effect is not reversed by adding more erythropoietin. As observed in mouse cells with Mu IFN-α/β, CFU-E exhibit a greater sensitivity to the antiproliferative action of Mu IFN-γ than the earlier BFU-E. The possibility that at least part of the inhibition is due to an indirect effect, via accessory cells, is raised by the observation that inhibition by Hu IFN-γ of BFU-E and CFU-E is significantly decreased if the bone-marrow cell suspensions are first depleted of monocytes or T cells. One can also preexpose monocytes or T cells to Hu IFN-γ, then coculture these pretreated cells with untreated bone-marrow cells and obtain an inhibition of erythroid colony formation (Mamus et al., 1985). This would imply that at least part of the antiproliferative effects of IFN-γ on erythroid precursors does not simply occur as the result of the direct antimitotic effect of IFNs, but is more likely a consequence of the stimulation of cells that normally are involved in the homeostatic regulation of erythropoiesis. This point of view, though, is somewhat controversial, as there are also reports that depleting monocytes, T cells, and B cells from human bone-marrow cells does not affect the suppressive effects of either Hu IFN-α or Hu IFN-γ on CFU-E and BFU-E colony formation (Broxmeyer et al., 1983). It is possible, furthermore, to isolate CFU-E from colonies originating from BFU-E and to observe a direct inhibitory effect of Hu IFN-γ on the thus purified CFU-E (Raefsky et al., 1985). This shows that the inhibitory effect of IFN-γ on these progenitors is direct and does not require an accessory cell.

Generally speaking, it is impossible to recognize precursor cells other than by their progeny, so that there is no absolute way of judging the purity of stem cell preparations. Therefore, the possibility that some of the antiproliferative effects exerted on hemopoietic stem cells by IFNs take place via activation of accessory cells cannot be excluded. A role for NK cells in the normal control of stem cell proliferation has been postulated, and IFNs—especially of the α-and β-type are very efficient stimulators of NK cell activity (see Chapter 11).

C. IFN-γ and the Pathogenesis of Aplastic Anemia

A spontaneous suppressor activity of hematopoiesis has been observed with T cells from patients with aplastic anemia (Hoffman et al., 1977). In some of these patients there is an abnormally elevated level of activated CD8 positive T cells, expressing receptors for IL-2. Removal of T cells from the bone marrow of these patients results in enhanced erythropoietic colony formation by the marrow cells in vitro, whereas the addition of T cells from the same patients to bone-marrow cell cultures from normal individuals suppresses colony formation (Singer et al., 1979; Torok-Storb et al., 1980). The possibility that these suppressing effects are a result of IFN production by the T cells of these patients has been raised by the observation that lectin-stimulated peripheral blood mononuclear cells from aplastic anemia patients produce much higher levels of Hu IFN-γ, and for a longer period, than control cells. When put into culture, peripheral blood mononuclear cells of some patients with aplastic anemia, furthermore, spontaneously release IFN-γ, and also, but to a lesser extent, IFN-α. Some patients with aplastic anemia also have IFN in their circulation, with higher levels of IFN in the bone-marrow serum than in the peripheral blood serum (Zoumbos et al., 1985a,b).

Could the spontaneous production of IFN-γ and IFN-α possibly result in lowered stem cell proliferation and thus explain the anemia of these patients? What could be the primary cause of this enhanced IFN production? Spontaneous production of IFN has been observed in a variety of disease states, especially in autoimmune disease, without concomitant aplastic anemia (see Chapter 16). The origin of such spontaneous IFN production is not known. Bone-marrow cells from normal individuals, when put into culture, also display spontaneous IFN production, although usually of the IFN-α-type and in much lower quantity than what is observed in the case of aplastic anemia.

IFNs given by intravenous perfusion can cause depression of bone-marrow function, as observed in patients receiving IFN treatment. One such clinical case seems particularly relevant to the problem of IFN production and aplastic anemia. After a 10-day course of intramuscularly administered Hu leukocyte IFN, a patient with non-Hodgkin lymphoma developed aplastic anemia, characterized by near total disappearance of CFU-E, BFU-E, and CFU-GM bone-marrow progenitor cells. At the same time, an invasion of the bone marrow by CD8 positive T cells occurred. This is a demonstration that IFN can cause bone-marrow aplasia, most likely by activating marrow suppressor T cells, and hence that IFN could indeed be part of a pathogenic mechanism causing anemia (Mangan et al., 1985).

It is clear that not all patients suffer from this side effect of IFN treatment, which is indicative of genetically determined differences in stem cell susceptibility. Furthermore, in a number of cases of aplastic anemia there seems to be little evidence for the presence of IFN-γ, although this still leaves open the possibility of IFN-α (Torok-Storb et al., 1987).

REFERENCES

Azoulay, M., Webb, C. G., and Sachs, L. Control of hematopoietic cell growth regulators during mouse fetal development. *Mol. Cell. Biol.* 7: 3361–3364 (1987).

Barlow, D. P., Bucan, M., Lehrach, H., Hogan, B. L. M., and Gough, N. M. Close genetic and physical linkage between the murine haemopoietic growth factor genes GM-CSF and multi-CSF (IL3). *EMBO J.* 6: 617–623 (1987).

Bartocci, A., Mastrogiannis, D. S., Migliorati, G., Stockert, R. J., Wolkoff, A. W., and Stanley, E. R. Macrophages specifically regulate the concentration of their own growth factor in the circulation. *Proc. Natl. Acad. Sci.* (USA) 84: 6179–6183 (1987).

Begley, C. G., Lopez, A. F., Nicola, N. A., Warren, D. J., Vadas, M. A., Sanderson, C. J., and Metcalf, D. Purified colony-stimulating factors enhance the survival of human neutrophils and eosinophils in vitro: A rapid and sensitive microassay for colony-stimulating factors. *Blood* 68: 162–166 (1986).

Broxmeyer, H. E., Lu, L., Platzer, E., Feit, C., Juliano, L., and Rubin, B. Y. Comparative analysis of the influences of human γ, α and β interferons on human multipotential (CFU-GEMM), erythroid (BFU-E) and granulocyte-macrophage (CFU-GM) progenitor cells *J. Immunol.* 131: 1300–1305 (1983).

Campbell, H. D., Ymer, S., Fung, M. C., and Young, I. G. Cloning and nucleotide sequence of the murine interleukin-3 gene. *Eur. J. Biochem.* 150: 297–304 (1985).

Chen, B. D. M. Interferon-induced inhibition of receptor-mediated endocytosis of colony-stimulating factor (CSF-1) by murine peritoneal exudate macrophages. *J. Immunol.* 136: 174–180 (1986).

Collins, L. S. and Dorshkind, K. A stromal cell line from myeloid long-term bone marrow cultures can support myelopoiesis and B lymphopoiesis. *J. Immunol.* 138: 1082–1087 (1987).

Conscience, J. F., Verrier, B., and Martin, G. Interleukin-3-dependent expression of the c-*myc* and c-*fos* protooncogenes in hemopoietic cell lines. *EMBO J.* 5: 317–323 (1986).

Coussens, L., Van Beveren, C., Smith, D., Chen, E., Mitchell, R. L., Isacke, C. M., Verma, I. M., and Ullrich, A. Structural alteration of viral homologue of receptor proto-oncogene *fms* at carboxyl terminus. *Nature* 320: 277–280 (1986).

Dandoy, F., De Maeyer, E., and De Maeyer-Guignard, J. Antiproliferative action of interferon on murine bone-marrow derived macrophages is influenced by the genotype of the marrow donor. *J. Interferon Res.* 1: 263–270 (1981).

Donahue, R. E., Wang, E. A., Stone, D. K., Kamen, R., Wong, G. G., Sehgal, P. K., Nathan, D. G., and Clark, S. C. Stimulation of haematopoiesis in primates by continuous infusion of recombinant human GM-CSF. *Nature* 321: 872–875 (1986).

Ernstoff, M. S., Gallicchio, V., and Kirkwood, J. M. Analysis of granulocyte-macrophage progenitor cells in patients treated with recombinant interferon α-2. *Am. J. Med.* 79: 167–170 (1985).

Fleit, H. B. and Rabinovitch, M. Interferon induction in marrow-derived macrophages: Regulation by L cell conditioned medium. *J. Cell. Physiol.* 108: 347–352 (1981).

Foon, K. A. and Todd, III, R. F. Immunologic classification of leukemia and lymphoma. *Blood* 68: 1–31 (1986).

Fung, M. C., Hapel, A. J., Ymer, S., Cohen, D. R., Johnson, R. M., Campbell, H. D., and Young, I. G. Molecular cloning of cDNA for murine interleukin-3. *Nature* 307: 233–237 (1984).

Gallien-Lartigue, O., Carrez, D., De Maeyer, E., and De Maeyer-Guignard, J. Strain dependence of the antiproliferative action of interferon on murine erythroid precursors. *Science* 209: 292–293 (1980).

Gallien-Lartigue, O., De Maeyer-Guignard, J., Carrez, D., and De Maeyer, E. The antiproliferative effect of murine interferon α/β on early bone marrow-derived erythroid precursors (BFU/e). *J. Interferon Res.* 5: 347–354 (1985).

Gasson, J. C., Weisbart, R. H., Kaufman, S. E., Clark, S. C., Hewick, R. M., Wong, G. G., and Golde, D. W. Purified human granulocyte-macrophage colony stimulating factor: Direct action on neutrophils. *Science* 226: 1339–1342 (1984).

Gordon, M. Y., Riley, G. P., Watt, S. M. and Greaves, M. F. Compartmentalization of a haematopoietic growth factor (GM-CSF) by glycosaminoglycans in the bone marrow microenvironment. *Nature* 326: 403–405 (1987).

Gough, N. M., Gough, J., Metcalf, D., Kelso, A., Grail, D., Nicola, N. A., Burgess, A. W., and Dunn, A. R. Molecular cloning of cDNA encoding a murine haematopoietic growth regulator, granulocyte-macrophage colony stimulating factor. *Nature* 309: 763–767 (1984).

Grabstein, K. H., Urdal, D. L., Tushinski, R. J., Mochizuki, D. Y., Price, V. L., Cantrell, M. A., Gillis, S., and Conlon, P. J. Induction of macrophage tumoricidal activity by granulocyte-macrophage colony-stimulating factor. *Science* 232: 506–598 (1986).

Groffen, J., Heisterkamp, N., Spurr, N., Dana, S., Wasmuth, J. J., and Stephenson, J. R. Chromosomal location of the human c-*fms* oncogene. *Nucl. Acids Res.* 11: 6331–6339 (1983).

Groopman, J. E., Mitsuyasu, R. T., DeLeo, M. J., Oette, D. H., and Golde, D. W. Effect of recombinant human granulocyte-macrophage colony-stimulating factor on myelopoiesis in the acquired immuno-deficiency syndrome. *N. Eng. J. Med.* 317: 593–598 (1987).

Hoffman, R., Zanjani, E. D., Lutoon, J. D., Zalusky, R., and Wasserman, L. R. Suppression of erythroid colony formation by lymphocytes from patients with aplastic anemia. *N. Engl. J. Med.* 296: 10–13 (1977).

Huebner, K., Isobe, M., Croce, C. M., Golde, D. W., Kaufman, S. E., and Gasson, J. C. The human gene encoding GM-CSF is at 5q21-q32, the chromosome region deleted in the $5q^-$ anomaly. *Science* 230: 1282–1285 (1985).

Hunt, P., Robertson, D., Weiss, D., Rennick, D., Lee, F., and Witte, O. N. A single bone marrow-derived stromal cell type supports the in vitro growth of early lymphoid and myeloid cells. *Cell* 48: 997–1007 (1987).

Ihle, J. N., Keller, J., Henderson, L., Klein, F., and Palaszynski, E. Procedures for the purification of interleukin 3 to homogeneity. *J. Immunol.* 129: 2431–2436 (1982).

Ihle, J. N., Silver, J., and Kozak, C. A. Genetic mapping of the mouse interleukin 3 gene to chromosome 11. *J. Immunol.* 138: 3051–3054 (1987).

Ishibashi, T. and Burstein, S. A. Interleukin 3 promotes the differentiation of isolated single megakaryocytes. *Blood* 67: 1512–1514 (1986).

Jacobs, K., Shoemaker, C., Rudersdorf, R., Neill, S. D., Kaufman, R. J., Mufson, A., Seehra, J., Jones, S. S., Hewick, R., Fritsch, E. F., Kawakita, M., Shimizu, T., and Miyake, T. Isolation and characterization of genomic and cDNA clones of human erythropoietin. *Nature* 313: 806–810 (1985).

Kanz, L., Lohr, G. W., and Fauser, A. A. Lymphokine(s) from isolated T lymphocyte subpopulations support multilineage hematopoietic colony and megakaryocytic colony formation. *Blood* 68: 991–995 (1986).

Kawasaki, E. S., Ladner, M. B., Wang, A. M., Van Arsdell, J., Warren, M. K., Coyne, M. Y., Schweickart, V. L., Lee, M. T., Wilson, K. J., Boosman, A., Stanley, E. R., Ralph, P., and Mark, D. F. Molecular cloning of a complementary DNA encoding human macrophage-specific colony-stimulating factor (CSF-1). *Science* 230: 291–296 (1985).

Kelso, A., Metcalf, D., and Gough, N. M. Independent regulation of granulocyte-macrophage colony-stimulating factor and multilineage colony-stimulating factor production in T lymphocyte clones. *J. Immunol.* 136: 1718–1725 (1986).

Kindler, V., Thorens, B., De Kossodo, S., Allet, B., Eliason, J. F., Thatcher, D., Farber, N., and Vassalli, P. Stimulation of hematopoiesis in vivo by recombinant bacterial murine interleukin 3. *Proc. Natl. Acad. Sci.* (USA) 83: 1001–1005 (1986).

Klimpel, G. R., Fleischmann, W. R., Jr., and Klimpel, K. D. γ interferon (IFN-γ) and IFN-α/β suppress murine myeloid colony formation (CFU-C): Magnitude of suppression is dependent upon level of colony-stimulating factor (CSF). *J. Immunol.* 129: 76–80 (1982).

Koike, K., Stanley, E. R., Ihle, J. N., and Ogawa, M. Macrophage colony formation supported by purified CSF-1 and/or interleukin 3 in serum-free culture: Evidence for hierarchical difference in macrophage colony-forming cells. *Blood* 67: 859–864 (1986).

Koren, S., Klimpel, G. R., and Fleischmann, W. R., Jr. Treatment of mice with macrophage colony stimulating factor (CSF-1) prevents the in vivo myelosuppression induced by murine α, β, and γ interferons. *J. Biol. Resp. Modif.* 5: 481–489 (1986).

Ladner, M. B., Martin, G. A., Noble, J. A., Niloloff, D. M., Tal, R., Kawasaki, E. S., and White, T. J. Human CSF-1: Gene structure and alternative splicing of mRNA precursors. *EMBO J.* 6: 2693–2698 (1987).

Law, M. L., Cai, G. Y., Lin, F. K., Wei, Q., Huang, S. Z., Hartz, J. H., Morse, H., Lin, C. H., Jones, C., and Kao, F. T. Chromosomal assignment of the human erythropoietin gene and its DNA polymorphism. *Proc. Natl. Acad. Sci.* (USA) 83: 6920–6924 (1986).

Le Beau, M. M., Epstein, N. D., O'Brien, S. J., Nienhuis, A. W., Yang, Y. C., Clark, S. C., and Rowley, J. D. The interleukin 3 gene is located on human chromosome 5 and is deleted in myeloid leukemias with a deletion of 5q. *Proc. Natl. Acad. Sci.* (USA) 84: 5913–5917 (1987).

Le Beau, M. M., Westbrook, C. A., Diaz, M. O., Larson, R. A., Rowley, J. D., Gasson, J. C., Golde, D. W., and Sherr, C. J. Evidence for the involvement of GM-CSF and FMS in the deletion (5q) in myeloid disorders. *Science* 231: 984–987 (1986).

Lee, F., Yokota, T., Otsuka, T., Gemmell, L., Larson, N., Luh, J., Arai, K. I., and Rennick, D. Isolation of cDNA for a human granulocyte-macrophage colony-stimulating factor by functional expression in mammalian cells. *Proc. Natl. Acad. Sci.* (USA) 82: 4360–4364 (1985).

Lee, M. T. and Warren, M. K. CSF-1-induced resistance to viral infection in murine macrophages. *J. Immunol.* 138: 3019–3022 (1987).

Lin, F. K., Suggs, S., Lin, C. H., Browne, J. K., Smalling, R., Egrie, J. C., Chen, K. K., Fox, G. M., Martin, F., Stabinsky, Z., Badrawi, S. M., Lai, P. H., and Goldwasser, E. Cloning and expression of the human erythropoietin gene. *Proc. Natl. Acad. Sci.* (USA) 82: 7580–7584 (1985).

Lopez, A. F., Nicola, N. A., Burgess, A. W., Metcalf, D., Battye, F. L., Sewell, W. A., and Vadas, M. Activation of granulocyte cytotoxic function by purified mouse colony-stimulating factors. *J. Immunol.* 131: 2983–2988 (1983).

Mamus, S. W., Beck-Schroeder, S., and Zanjani, E. D. Suppression of normal human erythropoiesis by γ interferon in vitro. Role of monocytes and T lymphocytes. *J. Clin. Invest.* 75: 1496–1503 (1985).

Mangan, K. F., Zidar, B., Shadduck, R. K., Zeigler, Z., and Winkelstein, A. Interferon-induced aplasia: Evidence for T-cell mediated suppression of hematopoiesis and recovery after treatment with horse antihuman thymocyte globulin. *Am. J. Hematol.* 19: 401–413 (1985).

McDonald, J. D., Lin, F. K., and Goldwasser, E. Cloning, sequencing, and evolutionary analysis of the mouse erythropoietin gene. *Mol. Cell. Biol.* 6: 842–848 (1986).

Metcalf, D. The granulocyte-macrophage colony-stimulating factors. *Science* 229: 16–22 (1985).

Metcalf, D. The molecular biology and functions of the granulocyte-macrophage colony-stimulating factors. *Blood* 67: 257–267 (1986).

Metcalf, D., Begley, C. G., Johnson, G. R., Nicola, N. A., Lopez, A. F., and Williamson, D. J. Effects of purified bacterially synthesized murine multi-CSF (IL-3) on hematopoiesis in normal adult mice. *Blood* 68: 46–57 (1986).

Miyatake, S., Otsuka, T., Yokota, T., Lee, F., and Arai, K. Structure of the chromosomal gene for granulocyte-macrophage colony stimulating factor: Comparison of the mouse and human genes. *EMBO J.* 4: 2561–2568 (1985a).

Miyatake, S., Yokota, T., Lee, F., and Arai, K. I. Structure of the chromosomal gene for murine interleukin 3. *Proc. Natl. Acad. Sci.* (USA) 82: 316–320 (1985b).

Moore, R. N., Larsen, H. S., Horohov, D. W., and Rouse, B. T. Endogenous regulation of macrophage proliferative expansion by colony-stimulating factor-induced interferon. *Science* 223: 178–180 (1984).

Mosmann, T. R., Cherwinski, H., Bond, M. W., Giedlin, M. A., and Coffman, R. L. Two types of murine helper T cell clone. I. Definition according to profiles of lymphokine activities and secreted proteins. *J. Immunol.* 136: 2348–2357 (1986).

Nagata, S., Tsuchiya, M., Asano, S., Kaziro, Y., Yamazaki, T., Yamamoto, O., Hirata, Y., Kubota, N., Oheda, M., Nomura, H., and Ono, M. Molecular cloning and expression of cDNA for human granulocyte colony-stimulating factor. *Nature* 319: 415–418 (1986).

Naparstek, E., Donelly, T., Shadduck, R. K., Waheed, A., Wagner, K., Kase, K. R., and Greenberger, J. S. Persistent production of colony-stimulating factor (CSF-1) by cloned bone marrow stromal cell line D2XRII after X-irradiation. *J. Cell. Physiol.* 126: 407–413 (1986).

Nienhuis, A. W., Bunn, H. F., Turner, P. H., Gopal, T. V., Nash, W. G., O'Brien, S. J., and Sherr, C. J. Expression of the human c-*fms* proto-oncogene in hematopoietic cells and its deletion in the $5q^-$ syndrome. *Cell* 42: 421–428 (1985).

Oladipupo-Williams, C. K., Svet-Moldavskaya, I., Vilcek, J., Ohnuma, T., and Holland, J. F. Inhibitory effects of human leukocyte and fibroblast interferons on normal and chronic myelogenous leukemic granulocytic progenitor cells. *Oncology* 38: 356–360 (1981).

Pettenati, M. J., Le Beau, M. M., Lemons, R. S., Shima, E. A., Kawasaki, E. S., Larson, R. A., Sherr, C. J., Diaz, M. O., and Rowley, J. D. Assignment of CSF-1 to 5q33.1: Evidence for clustering of genes regulating hematopoiesis and for their involvement in the deletion of the long arm of chromosome 5 in myeloid disorders. *Proc. Natl. Acad. Sci.* (USA) 84: 2970–2974 (1987).

Piacibello, W., Lu, L., Wachter, M., Rubin, B., and Broxmeyer, H. E. Release of granulocyte-macrophage colony stimulating factors from major histocompatibility complex class II antigen-positive monocytes is enhanced by human γ interferon. *Blood* 66: 1343–1351 (1985).

Piacibello, W., Lu, L., Williams, D., Aglietta, M., Rubin, B. Y., Cooper, S., Wachter, M., Gavosto, F., and Broxmeyer, H. E. Human γ interferon enhances release from phytohemagglutinin-stimulated $T4^+$ lymphocytes of activities that stimulate colony formation by granulocyte-macrophage, erythroid, and multipotential progenitor cells. *Blood* 68: 1339–1347 (1986).

Powell, J. S., Berkner, K. L., Lebo, R. V., and Adamson, J. W. Human erythropoietin gene: High level expression in stably transfected mammalian cells and chromosome localization. *Proc. Natl. Acad. Sci.* (USA) 83: 6465–6469 (1986).

Raefsky, E. L., Platanias, L. C., Zoumbos, N. C., and Young, N. S. Studies of interferon as a regulator of hematopoietic cell proliferation. *J. Immunol.* 135: 2507–2512 (1985).

Rajavashisth, T. B., Eng, R., Shadduck, R. K., Waheed, A., Ben-Avram, C. M., Shively, J. E., and Lusis, A. J. Cloning and tissue-specific expression of mouse macrophage colony-stimulating factor mRNA. *Proc. Natl. Acad. Sci.* (USA) 84: 1157–1161 (1987).

Rigby, W. F. C., Ball, E. D., Guyre, P. M., and Fanger, M. W. The effects of recombinant-DNA-derived interferons on the growth of myeloid progenitor cells. *Blood* 65: 858–861 (1985).

Sacca, R., Stanley, E. R., Sherr, C. J., and Rettenmier, C. W. Specific binding of the mononuclear phagocyte colony-stimulating factor CSF-1 to the product of the v-*fms* oncogene. *Proc. Natl. Acad. Sci.* (USA) 83: 3331–3335 (1986).

Sanderson, C. J., Warren, D. J., and Strath, M. Identification of a lymphokine that stimulates eosinophil differentiation in vitro. Its relationship to interleukin 3, and functional properties of eosinophils produced in cultures. *J. Exp. Med.* 162: 60–74 (1985).

Schrader, J. W., Arnold, B., and Clark-Lewis, I. A ConA-stimulated T-cell hybridoma releases factors affecting haematopoietic colony-forming cells and B-cell antibody responses. *Nature* 283: 197–199 (1980).

Sherr, C. J., Rettenmier, C. W., Sacca, R., Roussel, M. F., Look, A. T., and Stanley, E. R. The c-*fms* proto-oncogene product is related to the receptor for the mononuclear phagocyte growth factor, CSF-1. *Cell* 41: 665–676 (1985).

Shoemaker, C. B. and Mitsock, L. D. Murine erythropoietin gene: Cloning, expression, and human gene homology. *Mol. Cell. Biol.* 6: 849–858 (1986).

Simmers, R. N., Webber, L. M., Shannon, M. F., Garson, O. M., Wong, G., Vadas, M. A., and Sutherland, G. R. Localization of the G-CSF gene on chromosome 17 proximal to the breakpoint in the t(15;17) in acute promyelocytic leukemia. *Blood* 70: 330–332 (1987).

Singer, J. W., Doney, K. C., and Thomas, E. D. Coculture studies of 16 untransfused patients with aplastic anemia. *Blood* 54: 180–185 (1979).

Smith, K. A., Fredrickson, T. N., Mobraaten, L. E., and De Maeyer, E. The interaction of erythropoietin with fetal liver cells. II. Inhibition of the erythropoietin effect by interferon. *Exp. Hematol.* 5: 333–340 (1977).

Souza, L. M., Boone, T. C., Gabrilove, J., Lai, P. H., Zsebo, K. M., Murdock, D. C., Chazin, V. R. Bruszewski, J., Lu, H., Chen, K. K., Barendt, J., Platzer, E., Moore, M. A. S., Mertelsmann, R., and Welte, K. Recombinant human granulocyte colony-stimulating factor: Effects on normal and leukemic myeloid cells. *Science* 232: 61–65 (1986).

Stanley, E., Metcalf, D., Sobieszczuk, P., Gough, N. M., and Dunn, A. R. The structure and expression of the murine gene encoding granulocyte-macrophage colony stimulating factor: Evidence for utilisation of alternative promoters. *EMBO J.* 4: 2569–2573 (1985).

Suda, J., Suda, T., Kubota, K., Ihle, J. N., Saito, M., and Miura, Y. Purified interleukin-3 and erythropoietin support the terminal differentiation of hemopoietic progenitors in serum-free culture. *Blood* 67: 1002–1006 (1986).

Thorens, B., Mermod, J. J., and Vassalli, P. Phagocytosis and inflammatory stimuli induce GM-CSF mRNA in macrophages through posttranscriptional regulation. *Cell* 48: 671–679 (1987).

Torok-Storb, B. J., Johnson, G. G., Bowden, R., and Storb, R. γ-interferon in aplastic anemia: Inability to detect significant levels in sera or demonstrate hematopoietic suppressing activity. *Blood* 69: 629–633 (1987).

Torok-Storb, B. J., Sieff, C., Storb, R., Adamson, J., and Thomas, E. D. In vitro tests for distinguishing possible immune-mediated aplastic anemia from transfusion-induced sensitization. *Blood* 55: 211–215 (1980).

Tsuchiya, M., Nomura, H., Asano, S., Kaziro, Y., and Nagata, S. Characterization of recombinant human granulocyte-colony-stimulating factor produced in mouse cells. *EMBO J.* 6: 611–616 (1987).

Warren, M. K. and Ralph, P. Macrophage growth factor CSF-1 stimulates human monocyte production of interferon, tumor necrosis factor, and colony stimulating activity. *J. Immunol.* 137: 2281–2285 (1986).

Welte, K., Bonilla, M. A., Gillio, A. P., Boone, T. C., Potter, G. K., Gabrilove, J. L., Moore, M. A. S., O'Reilly, R. J., and Souza, L. M. Recombinant human granulocyte colony-stimulating factor. Effects on hematopoiesis in normal and cyclophosphamide-treated primates. *J. Exp. Med.* 165: 941–948 (1987).

Wong, G. G., Temple, P. A., Leary, A. C., Witek-Giannotti, J. S., Yang, Y. C., Ciarletta, A. B., Chung, M., Murtha, P., Kriz, R., Kaufman, R. J., Ferenz, C. R., Sibley, B. S., Turner, K. J., Hewick, R. M., Clark, S. C., Yanai, N., Yokota, H., Yamada, M., Saito, M., Motoyoshi, K., and Takaku, F. Human CSF-1: Molecular cloning and expression of 4-kb cDNA encoding the human urinary protein. *Science* 235: 1504–1508 (1987).

Wong, G. G., Witek, J. S., Temple, P. A., Wilkens, K. M., Leary, A. C., Luxenberg, D. P., Jones, S. S. Brown, E. L., Kay, R. M., Orr, E. C., Shoemaker, C., Golde, D. W., Kaufman, R. J., Hewick, R. M., Wang, E. A., and Clark, S. C. Human GM-CSF: Molecular cloning of the complementary DNA and purification of the natural and recombinant proteins. *Science* 228: 810–815 (1985).

Yamamoto-Yamaguchi, Y., Tomida, M., and Hozumi, M. Effect of mouse interferon on growth and differentiation on mouse bone marrow cells stimulated by two different types of colony-stimulating factor. *Blood* 62: 597–601 (1983).

Yang, Y. C., Ciarletta, A. B., Temple, P. A., Chung, M. P., Kovacic, S., Witek-Giannotti, J. S., Leary, A. C., Kriz, R., Donahue, R. E., Wong, G. G., and Clark, S. C. Human IL-3 (multi-CSF): Identification by expression cloning of a novel hematopoietic growth factor related to murine IL-3. *Cell* 47: 3–10 (1986).

Yokota, T., Lee, F., Rennick, D., Hall, C., Arai, N., Mosmann, T., Nabel, G., Cantor, H., and Arai, K. Isolation and characterization of a mouse cDNA clone that expresses mast cell growth factor activity on monkey cells. *Proc. Natl. Acad. Sci.* (USA) 81: 1070–1074 (1984).

Zoumbos, N. C., Gascon, P., Djeu, J. Y., Trost, S. R., and Young, N. S. Circulating activated suppressor T lymphocytes in aplastic anemia. *N. Engl. J. Med.* 312: 257–265 (1985a).

Zoumbos, N. C., Gascon, P., Djeu, J. Y., and Young, N. S. Interferon is a mediator of hematopoietic suppression in aplastic anemia in vitro and possibly in vivo. *Proc. Natl. Acad. Sci.* (USA) 82: 188–192 (1985b).

9 MODULATION OF THE EXPRESSION OF THE MAJOR HISTOCOMPATIBILITY ANTIGENS

Modulation of the expression of cell-surface antigens of the major histocompatibility complex (MHC) is one of the essential effects by which all three IFN species influence the immune system.

The MHC encodes three families of genes, denoted class I, class II, and class III, and plays a primordial role in self-nonself recognition. The structural genes of the MHC are on chromosome 6 in humans and on chromosome 17 in the mouse. Class I antigens, normally expressed on most cells in the body, are involved in target recognition by CD8+ (frequently cytotoxic

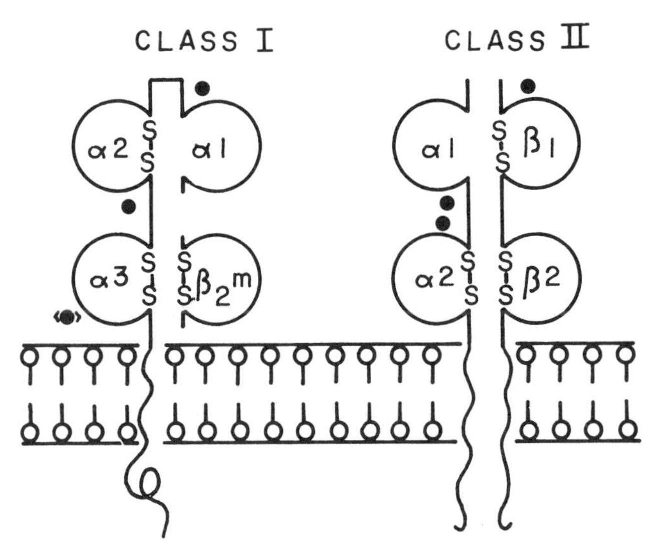

Figure 9.1 MHC Class I and Class II antigens. Class I: the 45kDa heavy chain that anchors the complex in the membrane is divided into three extracellular domains (α1,2, and 3), a hydrophobic transmembrane region, and a cytoplasmic region. The black dots represent glycosylation sites. The β2 microglobulin molecule is associated noncovalently. Class II: Class II antigens consist of the noncovalent association of the α and β chain.

and sometimes suppressor) T cells, whereas class II antigens, principally but not exclusively expressed on cells of the immune system, are involved in the restriction of CD4+ (frequently helper and sometimes suppressor) T cells (see Chapter 11). Class III genes do not code for cell-surface antigens but for components of the complement cascade. The highly polymorphic class I molecules are encoded by three distinct loci, H-2K, H-2D, and H-2L in the mouse and HLA-A, -B, and -C in humans. Each class I antigen, consisting of a 45 kDa glycoprotein, is, furthermore, associated noncovalently at the cell surface with a second 11.6 kDa polypeptide chain, β2 microglobulin, encoded by a gene that is not part of the MHC but is located on a different chromosome. Class II antigens, also highly polymorphic, include the Ia molecules that consist of the association of two polypeptide chains, the α and the β chain, encoded by the genes of the I-A and I-E region of the MHC cluster in the mouse and of the HLA-DP, DQ, and DR region in humans (Fig. 9.1). In addition, a third chain, called the invariant chain, is associated with the α and β subunits of class II antigens during their transport to the cell surface, but it is released from the dimer after transition of the Golgi complex. The expression of class II antigens is limited to certain cell types. It is constitutive on Langerhans and dendritic cells and on B cells and is usually transient but inducible on monocytes, macrophages, and some T cells. Class II antigens are important for cell-to-cell recognition and interaction of immune cells, and quantitative variation of these antigens at the cell surface plays a central role in immuno-regulation (Janeway et al., 1984; Kaufman et al., 1984).

The expression of both class I and class II antigens is modulated by IFNs.

I. EXPRESSION OF MHC ANTIGENS ON MACROPHAGES AND OTHER ACCESSORY CELLS

The first step leading to immunization is the uptake of antigen by accessory cells. The most important of these are macrophages, characterized by their phagocytic capacity and by the fact that they can "process" antigen for efficient presentation to other cells. Two additional cell types involved in antigen presentation and also derived from bone-marrow precursors are the dendritic cells of lymphoid organs and the Langerhans cells of the epithelial linings of the skin and intestine. Presentation of antigen to T cells can, under certain conditions, also become a function of several other cells, for example, B cells, vascular endothelial cells, and astrocytes.

The antigen-charged accessory cell presents the antigen to helper T lymphocytes (TH cells) and amplifying T lymphocytes (TA cells). The presentation of antigen by an accessory cells to a T cell can only proceed efficiently if both cells express identical class II antigens at their surface. This requirement has, therefore, been called "MHC" restriction, and also associated or dual recognition (Germain, 1981).

A. Modulation of Antigen Expression by IFN-γ

1. On Monocytes-Macrophages

Murine monocytes and peritoneal macrophages either do not express class II antigens constitutively or at a relatively low level (Unanue, 1981). Although in humans most monocytes do express class II antigens, this is also usually at a low level. Stimulation of the expression of class II antigens is, therefore, important to ensure efficient antigen presentation and immunization. This explains the considerable interest for many years in T-cell-derived soluble mediators that induce the enhanced expression of class II antigens on accessory cells. It has now become clear that one of the most powerful of these mediators is IFN-γ (Fig. 9.2). This has been demonstrated first by purifying and characterizing as IFN-γ the Ia-inducing activity released by mitogen-stimulated murine spleen cells (Steeg et al., 1982), and more directly by using recombinant IFN-γ (Basham and Merigan, 1983; Wong et al., 1983; Becker, 1984; Kelley et al., 1984; Tweardy et al., 1986).

A systematic study of the induction of class I (H-2K and H-2D) and class II (I-A^k) antigens in mice after intravenous administration of recombinant Mu IFN-γ shows a widespread, but selective increase of these antigens, not only on macrophages but in most tissues. In the lymphatic-hematopoietic system, bone-marrow cells show the most significant increase of both class I

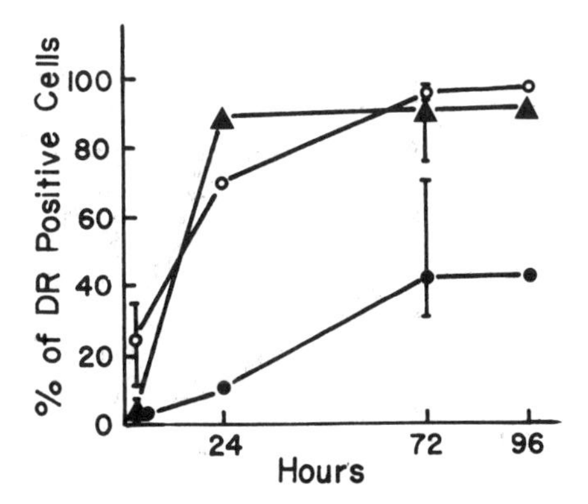

Figure 9.2 The kinetics of IFN-γ-enhanced expression of class II HLA-DR antigens on cells of three different human monocyte cell lines (○: THP-1 cells; ▲: U-937 cells; ●: HL-60 cells). (Adapted from Virelizier et al., 1984.)

and class II antigens. Of the nonlymphoid organs, increased class I antigen expression is mainly found on capillary endothelial cells, and on renal tubular cells and bronchiolar epithelial cells. There is, furthermore, a significant increase of class II antigen-bearing dendritic cells (Skoskiewicz et al., 1985; Momburg et al., 1986). Systemic administration of rec Mu IFN-γ to mice also results in an increased number of class II antigen-bearing dendritic cells in the lungs (Schneeberger et al., 1986).

Increased expression of class II antigens by IFN-γ-treated monocytes has functional implications since it results in an accelerated immune response. In vitro treatment with recombinant Hu IFN-γ of human monocytes, derived from individuals recently immunized with tetanus toxoid, results in an enhanced HLA-DR antigen expression on the monocytes followed by a significantly increased T-cell proliferation in the presence of the antigen-presenting monocytes. This stimulation is abolished by treatment with specific anti HLA-DR antibody, indicating that the stimulation of the T cells is indeed due to the enhanced expression of class II antigens (Becker, 1985).

2. *On Endothelial Cells*

A role for vascular endothelial cells as antigen-presenting cells has been suggested by their capacity, at least in culture, to present antigen to T cells in an MHC class II restricted manner. For example, human vascular endothelial cells that contain a small subpopulation of class II antigen-bearing cells can function as accessory cells to present mumps virus antigen to T cells in an HLA restricted manner (Nunez et al., 1983). The great majority of vascular endothelial cells, however, does not express class II antigens, but can be induced by IFN-γ to do so. Class I antigens, which are normally already expressed on endothelial cells, are simultaneously enhanced (Pober et al., 1983a,b).

As a result of induction by IFN-γ of HLA-DR, -DP, and -DQ class II antigens, endothelial vascular cells, furthermore, acquire an enhanced capacity to stimulate allogeneic T cells. Such a mechanism probably contributes to allograft rejection, since the vascular endothelial cells of the graft could serve to prime T cells of the host (Geppert and Lipsky, 1985). Moreover, recruitment of endothelial cells by IFN-γ to perform class II antigen-restricted functions and thereby stimulate T-cell activity has also implications for the pathogenesis of autoimmune disease. For example, in autoimmune experimental allergic encephalomyelitis, a T-cell-mediated neurologic disease usually provoked by immunization with myelin basic protein, the microvascular endothelial cells of the brain are characterized by a markedly enhanced class II antigen expression prior to inflammatory cell infiltration (Sobel et al., 1984). This suggests very strongly that induction and expression of class II antigens on endothelial cells is a prerequisite for starting the autoimmune reaction, and that IFN-γ, through a positive feedback loop, can act as a stimulator of autoimmunity. In line with this hypothesis is the observation that anti-class II MHC antibodies as well as anti-IFN-γ antibodies can decrease autoimmune reactions in mice. This is discussed in detail in Chapter 16.

3. On Astrocytes

Astrocytes belong to the different kinds of glial cells that provide metabolic and mechanical support for the neurones of the central nervous system. Astrocytes have the potential to act as antigen-presenting cells, and in vitro they are able to present myelin basic protein antigen to T cells in an MHC restricted manner (Fontana et al., 1984). Under normal conditions, however, brain cells, including astrocytes, do not express class II antigens; the signal for induction of expression is provided by T-cells, via the production of IFN-γ (Wong et al., 1984a,b; Fierz et al., 1985; Takiguchi and Frelinger, 1986). By regulating class II antigen expression and thereby stimulating the accessory function of astrocytes, IFN-γ participates in the control of the interactions between T cells and the central nervous system. This has implications for normal as well as pathological immune reactions, since malfunction of this system could result in aberrant immune reactions in the brain.

By inducing or augmenting the expression of class II antigens on many different accessory cells and thereby stimulating the interaction of these cells with T helper and T amplifying cells, IFN-γ promotes antibody formation and development of cytotoxic T cells. Since IFN-γ is mainly produced as a result of antigen recognition by sensitized T cells, there is an obvious requirement for a primary immunization event that takes place in the absence of IFN-γ. IFN-γ is an important amplifying signal for immune reactions, but it may not be involved in the very first events of immunization, unless one assumes the involvement of low levels of constitutive IFN-γ production (Palacios et al., 1983).

B. Modulation by IFN-α and IFN-β

1. IFN-α and IFN-β Also Sometimes Stimulate Class II Antigen Expression

IFN-α and IFN-β are often found to be less efficient stimulators of class II antigens on accessory cells, but there are, nonetheless, examples of enhanced expression due to these IFNs. When human peripheral blood monocytes are cultured in vitro they express progressively less HLA-DR antigens. After treatment with recombinant Hu IFN-γ, but also with recombinant Hu IFN-αA or Hu IFN-αD, the DR antigens are reexpressed (Sztein et al., 1984). It is important to realize that differential effects of individual IFN-α subspecies can occur: rec Hu IFN-α1 increases class II but not class I antigen expression on human monocytes, whereas rec Hu IFN-α2 increases class I but not class II (Rhodes et al., 1986). The existence of different effects depending on the IFN-α subspecies offers a possible explanation for several contradictory findings, since there can be different ratios of IFN-α subspecies in leukocyte and lymphoblastoid IFN preparations (see Appendix).

IFN-γ is undoubtedly one of the prime inducers of class II antigens on accessory cells, but IFN-β and some IFN-α species are also capable of doing so. However, under certain conditions, IFN-α and IFN-β can also exert the opposite effect, and down-regulate class II antigens.

2. Down-Regulation of IFN-γ-Stimulated Class II Antigen Expression on Macrophages by IFN-α and IFN-β

Mu IFN-α as well as Mu IFN-β are able to block the induction of class II antigens by IFN-γ in mouse peritoneal macrophages. This blocking occurs even when the macrophages are prestimulated for several hours with IFN-γ before being treated with either IFN-α or β. Furthermore, macrophages from newborn mice cannot be stimulated by IFN-γ to express Ia antigens unless in the presence of anti-IFN-β serum. This means that IFN-β, constitutively produced by these macrophages, acts as a natural antagonist for class II antigen induction by IFN-γ and down-regulates the expression of these antigens (Ling et al., 1985; Inaba et al., 1986). This is one of the very few and interesting examples of an antagonistic relationship between IFN-γ and IFN-α/β (see also Chapter 10). Usually these IFNs act synergistically, as in the case of their antiviral and anticellular effects (Fleischmann et al., 1979).

The possibility that the constitutive production of IFN-α/β by murine macrophages is one of the natural mechanisms that down-regulate the expression of class II antigens, and thereby the accessory function of these cells, is intriguing. It may explain why class II antigens are usually not expressed on murine peritoneal macrophages, since these cells have a low constitutive production of IFN-α/β, as do many other bone-marrow derived cells when put into culture (see Chapter 3).

The expression of class I antigens on macrophages and monocytes is enhanced by IFN-α and β. For example, Hu leukocyte IFN stimulates the expression of HLA-A, HLA-B, and β2 microglobulin on peripheral blood mononuclear cells (Fellous et al., 1979). This is in line with the stimulation of many other macrophage functions by these IFNs (see Chapter 10).

II. EXPRESSION OF MHC ANTIGENS ON T AND B CELLS

A. Modulation by IFN-γ

The induction of the expression of class II antigens by IFN gamma is not limited to accessory cells, but can also take place on cells of T and B lineage. For example, addition of Hu IFN-γ to cultures of pokeweed mitogen-stimulated T cells from human cord blood, or to adult T cells, significantly increases the expression of HLA-DR antigens (Miyawaki et al., 1984). Cells belonging to different murine and human B cell lines can also be stimulated by IFN-γ to enhanced the expression of class I and class II antigens (Kim et al., 1983; Wong et al., 1983; Capobianchi et al., 1985).

The maturation of bone-marrow-derived precursor T cells to immunocompetent virgin T cells takes place in the thymus, and the expression of class II antigens on the membranes of thymic accessory cells is a critical requirement in this differentiation step. Since Hu IFN-γ modulates the HLA class II antigen expression on cultured thymic epithelial cells (Fig. 9.3), it may possibly play a natural role in the mechanisms that assure the permanent expression of class II antigens on thymic accessory cells in vivo, with potentially important functional consequences in terms of education for self-recognition (Berrih et al., 1985).

B. Modulation by IFN-α and IFN-β

Although IFN-α and IFN-β can stimulate the expression of class II antigens on some cells of B and T lineage (Basham and Merigan, 1983; Kim et al., 1983; Virelizier et al., 1984; Capobianchi et al., 1985), their major effect is the induction of class I antigens. For example, the expression of HLA antigens on human B and T cells and on cultured thymic cells is stimulated by Hu leukocyte IFN (Heron et al., 1978; Diaz-Espada et al., 1986). These effects are not limited to in vitro systems, but can be obtained by systemic administration of IFN to animals, which means that they can take place as a result of IFN synthesis during a viral infection (Lindahl et al., 1976).

A natural role for Hu IFN-β in the turning on of class I HLA antigens during cell differentiation is suggested by experiments in which the appearance of these antigens during the differentiation of cells from a lymphoma line into macrophages can be inhibited by treatment with anti-IFN-β serum, but not with anti-IFN-α or anti-IFN-γ serum (Yarden et al., 1984). This

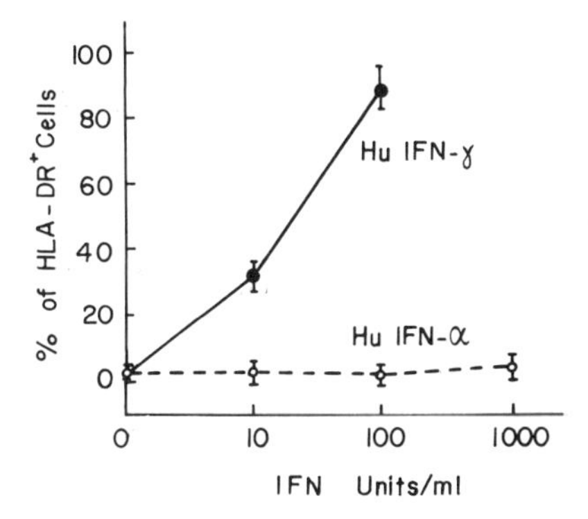

Figure 9.3 Stimulation by Hu IFN-γ (●—●) of HLA Class II antigens on thymic epithelial cells, and the lack of effect of Hu IFN-α2 (○—○). (Adapted from Berrih et al., 1985).

raises the possibility that endogenous IFN-β acts as a natural modulator of the antigen expression during the normal differentiation process of cells derived from the hematopoietic system.

Mouse embryonal carcinoma F9 cells are undifferentiated cells that exhibit a variety of properties that are characteristic of early embryos. Since they are undifferentiated, these cells do not normally express MHC class I antigens. They can, however, be induced into transcription and expression of MHC class I genes by Mu IFN-α/β and by Mu IFN-γ. The expression, however, is not an irreversible effect, as it disappears some time after IFN is withdrawn (Wan et al., 1987).

During embryonic development of the mouse, class I antigen expression in general begins at the midsomite stage (gestation day 10), but can be induced at an earlier stage by Mu IFN-α/β (Ozato et al., 1985). It is intriguing that the timing of class I antigen expression in the mouse embryo corresponds to the time when IFN-α/β inducibility appears (Barlow et al., 1984), but we can obviously not exclude that this is merely a correlation without functional implications.

III. EXPRESSION OF MHC ANTIGENS ON MAST CELLS

Mast cells, like basophils, are important mediators of immediate hypersensitivity responses of the anaphylactic type (see Chapter 12). In humans they are prominent in the lung and are one of the major sources of mediators in allergic asthma; they have receptors for IgE on their surface and participate in IgE-mediated hypersensitivity reactions. When lines of murine T-dependent mast cells or mast cell progenitors, which normally do not express class II antigens, are cultured in the presence of natural or recombinant Mu IFN-γ, class II antigens are induced (Wong et al 1982; Koch et al., 1984).

Whether or not this is followed by functional changes is not known; there is some evidence for an interaction of mast cells and T cells, which theoretically could be stimulated by the expression of class II antigens.

IV. EXPRESSION OF MHC ANTIGENS ON OTHER CELLS

A. On Normal Cells

Murine IFN-α/β enhances the expression of class I antigens on mouse fibroblasts (Vignaux and Gresser, 1978), and Mu IFN-γ, either in vitro or after systemic administration to mice, enhances the expression of class I antigens on practically every tissue in the body. In some tissues, for example the pancreas and small intestine, the increase of the H-2K antigen expression can be quite spectacular and reach values of up to 13- and 17-fold (Wong et al., 1984; Skoskiewicz et al., 1985).

As mentioned earlier, IFN-γ stimulates the expression of class II antigens on astrocytes, and thus is instrumental for the accessory role of these cells. Astrocytes can, furthermore, be induced to express class I antigens by Mu IFN-α/β. Since there is normally very low expression of MHC antigens in the brain, this enhanced class I antigen expression can favor interaction of virus-infected cells with cytotoxic T cells (Tedeschi et al., 1986).

Mouse pancreatic islet β cells are stimulated to a tenfold enhanced class I (H-2K) expression when cultured in vitro in the presence of rec Mu IFN-γ (Campbell et al., 1985). Insulin-dependent diabetes mellitus is believed to result from the autoimmune destruction of pancreatic islet β cells, which among others can be triggered by certain virus infections (see Chapter 16). Increased expression of class I antigens at the surface of these cells, by stimulating virus-infected target cell destruction by cytotoxic T cells, could therefore contribute to the pathogenesis of the disease. Like the enhanced expression of class II antigens on astrocytes in the brain, this is an example of a potential immunopathological role of IFN-γ.

Unlike most mature nucleated cells, but like other fetal cells residing near the maternal-fetal interface, human amnion cells either do not express or express very low levels of class I HLA antigens. This is possibly one of the reasons that the fetus is protected from maternal cytotoxic T cells. When human amnion cells are exposed in vitro to Hu IFN-γ, class I antigens are strongly expressed (Hunt and Wood, 1986). This shows that the normal lack of expression is not due to defective structural genes. One can ask which mechanism is responsible for the normal down-regulation of class I antigens in these cells. A special form of α-related Hu IFN is present in amniotic fluid (Lebon et al., 1982), and it would be worthwhile to examine its effects on MHC antigen modulation in fetal cells.

Like human amnion cells, murine trophoblast cells normally express only low levels of paternal class I MHC antigens. They do not, however, express class II antigens, and although treatment with Mu IFN-α/β or IFN-γ results in an enhanced class I expression, it does not induce class II expression (Zuckerman and Head, 1986).

The synthesis of class I HLA proteins can be stimulated in normal diploid human cells of the WISH cell line by all three Hu IFN species (Wallach, 1983).

This general property of all three IFN species to modulate expression of MHC antigens on normal cells has important implications as an antiviral mechanism, since the capability to mount a succesful antiviral cell-mediated immune response depends to a certain extent on the ability of the target cells to present the viral antigens in conjunction with class I antigens (Zinkernagel and Doherty, 1974). To produce molecules during a virus infection that can amplify the expression of class I antigens at the surface of infected cells appears as a good way to favor the elimination of such cells by cytotoxic T cells. For example, Mu IFN-α/β greatly enhances the susceptibility of vaccinia or lymphocytic choriomeningitis virus-infected mouse fibroblasts to lysis by cytotoxic T cells, and this correlates with an increased expression of MHC class I antigens (Bukowski and Welsh, 1985). Enhanced efficacy of cytotoxic T cells may be beneficial to the host in many instances, but one can also imagine situations when destruction of infected cells will do more harm than good, for example, in the nervous system. This raises the distinct possibility that IFNs are at times aggravating or possibly even initiating factors in autoimmune reactions (Chapter 16).

B. On Tumor Cells

We have seen that class I MHC antigens are found on all nucleated cells in the body, whereas the presence of class II antigens is generally limited to cells of the immune system and some endothelial cells and astrocytes. Epithelial cells, in the absence of any known immune function, can also express class II antigens (Wiman et al., 1978; Winchester et al., 1978). These antigens probably play a role in the interaction of tissue cells with cells of the immune system. Possibly as a consequence of malignant transformation, some nonlymphoid tumors, such as melanoma and colorectal cancer cells, also express class II antigens (Wilson et al., 1979; Daar et al., 1982). Some melanoma-derived cell lines continue to express class II antigens in vitro, but many others do not; they can be induced to expression by IFN-γ, as can normal melanocytes (Ameglio et al., 1983; Basham and Merigan, 1983; Dolei et al., 1983; Houghton et al., 1984). Hu IFN-$\alpha 2$ or Hu leukocyte IFN, even when used at much higher doses than Hu IFN-γ, do not induce the expression of class II antigens on melanoma cells. Hu leukocyte IFN, however, does stimulate the expression of HLA class I antigens (Basham et al., 1982; Liao et al., 1982; Basham and Merigan, 1983).

In addition to the expression, as well as the shedding of class II antigens, the shedding of specific melanoma antigens is also significantly increased by

IFN-γ (Gershon et al., 1985). Although theoretically such shedding could stimulate the host's immune response to premalignant and malignant melanocytes, there is no formal evidence that shedding of melanoma, or for that matter HLA-DR, antigens is in fact immunostimulatory in vivo (Herlyn et al., 1985). Recombinant Hu IFN-γ induces the expression and shedding of HLA class II antigens on melanoma cells that are resistant to its antiproliferative effect: This is a clear indication that the antiproliferative and antigen modulating effects of IFN-γ take place via a different metabolic pathway (Ziai et al., 1985).

The capacity of IFN-γ to stimulate the expression of class II antigens on tumor cells is not limited to melanoma cells, which often already express these antigens. Cells belonging to established lines derived from many different types of tumors, such as malignant glioma, breast cancer, bladder cancer, kidney cancer, colon cancer, cervix cancer, ovarian cancer, gastrointestinal cancer, and lung cancer, can be induced to express class II antigens when cultured with IFN-γ in vitro or when grown as solid tumors in IFN-γ-treated nude mice (Houghton et al., 1984; Schwartz et al., 1985; Takiguchi et al., 1985; Balkwill et al., 1987). This is, however, not a general property of all malignant cells, as some tumor cell lines are resistant to this effect. In addition, a study using malignant glioma-derived cell lines has shown that there is no correlation between inducibility for class I and class II antigens, and that some tumor cell lines can be induced by IFN-γ to express both, whereas others are only induced to express either class I or class II (Piguet et al., 1986). The same conclusion has been reached from an analysis of the effect of IFN-γ on three different neuroblastoma lines: Only class I but not class II HLA antigens were induced (Lampson and Fisher, 1984). Similarly, in cells of the human leukemic cell line K562, derived from a patient with chronic myelocytic leukemia, IFN-γ, as well as IFN-α and β, only induced class I HLA antigens but not class II (Sutherland et al., 1985).

Stimulation of the expression of class II antigens on some tumor cells is not a unique property of IFN-γ and has also been observed with IFN-α. In splenic hairy cells from patients with hairy cell leukemia, rec Hu IFN-α2C enhances the expression of class II antigens, with the functional result that the stimulatory capacity of the leukemic cells in a one-way mixed allogeneic lymphocyte reaction is increased (Baldini et al., 1986).

Is there any evidence that the enhanced expression of MHC antigens on tumor cells contributes to the antitumor effects of IFNs in vivo?

The most direct indication that class II antigen expression is important for the antigen-presenting ability of tumor cells is provided by the observation that melanoma cells expressing HLA-DR induce the in vitro proliferation of autologous T cells, and that this T-cell stimulation is abolished by the treatment of the melanoma cells with anti-HLA-DR antibodies. However, and for reasons unknown, DR positive melanoma cells from advanced disease lose the capacity to stimulate autologous T cells, in spite of the presence of significant amounts of class II antigens on the surface of these melanoma cells (Guerry et al., 1984).

C. Increased Expression of MHC Antigens on Tumor Cells Can Favor Rejection

Class I antigens are important for the interaction of cells bearing foreign antigens with effector T cells. Thus, cytotoxic T cells recognize cells that are infected with virus or altered by neoplastic transformation in association with class I antigens only (Zinkernagel and Doherty, 1979). Many tumor cells no longer express class I antigens and this is likely to promote escape from host surveillance and immune destruction and to lead to survival and further development of the tumor. When class I antigens are reexpressed on the tumor cells, their oncogenic potential diminishes.

There are indeed several examples of decreased tumorigenicity of malignant cells as a result of transfection with an MHC gene followed by antigen expression at the cell surface (Wallich et al., 1985). Transformation of murine cells with human adenovirus 12 results in greatly reduced levels of class I antigen expression, which favors survival of the tumor cells when injected into mice. However, when these cells are transfected with a functional class I gene and start expressing antigens at their surface, their oncogenic potential is significantly decreased (Fig. 9.4) (Tanaka et al., 1985). An AKR leukemia line, which normally does not express class I antigens and grows easily in AKR mice, becomes much less tumorigenic after transfection with a cosmid bearing the H2-K^k gene. Similarly, the metastatic properties of a murine fibrosarcoma line, which does not express H-2K antigens, are abro-

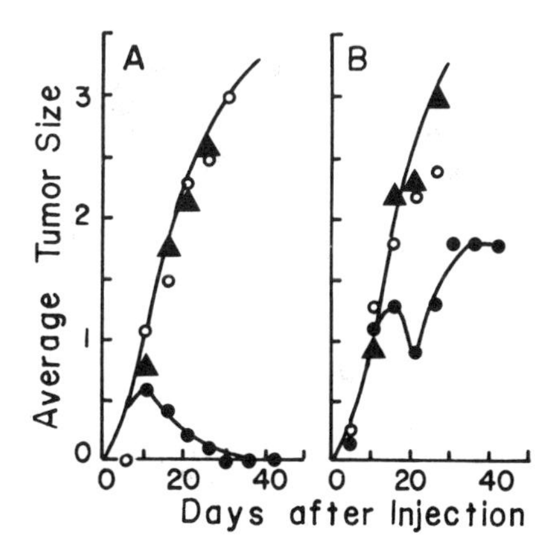

Figure 9.4 This experiment demonstrates that increased expression of MHC Class I antigens by tumor cells can favor rejection of the cells. Increased expression in this particular experiment was obtained by transfecting malignant cells with a functional MHC class I gene. An adenovirus 12-transformed cell line that expresses no detectable class I antigens is highly tumorigenic in BALB/c mice (▲). The same cells, transfected with an MHC class I gene, are much less tumorigenic when the cells are high antigen expressors (●), but retain their original oncogenic potential if they are low antigen expressors (○). The figure shows the average increase in tumor size after inoculation of 5×10^6 (A) or 1×10^7 (B) cells per animal. (Adapted from Tanaka et al., 1985.)

gated when these antigens are reexpressed after transfection with an H-2K gene containing plasmid (Hui et al., 1984; Wallich et al., 1985).

These are good reasons for believing that the increased expression of class I antigens is one of the ways by which IFNs inhibit tumor development. Moreover, the importance of an enhanced class I antigen expression for the antitumor effect of IFNs is directly shown by the observation that an increased expression of these antigens induced by Mu IFN-α/β on adenovirus transformed cells significantly reduces the tumorigenicity of these cells in immunocompetent hosts, an effect comparable to the higher mentioned abrogation of tumorigenicity after transfection with an MHC class I gene (Hayashi et al., 1985).

V. HOW DO IFNs ENHANCE THE EXPRESSION OF MHC ANTIGENS?

The MHC cell-surface antigens are, like most cellular proteins, in a state of continuous turnover, and except for IFNs, we have little or no information on what turns the synthesis of these membrane proteins on or off. Frequently, when cells are put into culture, the expression of MHC antigens decreases, suggesting that in vivo they are constantly reinduced, possibly by IFNs. There are several ways by which IFNs could theoretically augment antigen expression—increased transcription of the structural genes is one of them, and the use of cDNA probes to measure mRNA by Northern blotting shows that this is indeed a result of IFN treatment.

Runoff assays with nuclei from a human melanoma cell line that is exposed to either Hu IFN-β or Hu IFN-γ show that transcription of the HLA-A, -B and -C as well as of β2 microglobulin genes is significantly enhanced during at least 16 hours. Interestingly, the kinetics of the enhancement of transcription rates are slower with IFN-γ than with IFN-β (Rosa et al., 1986). The induction of the antiviral state is also slower in IFN-γ-treated cells than in IFN-α or IFN-β-treated cells, which suggests that activation of some genes occurs more rapidly with IFN-α and IFN-β than with IFN-γ.

Hu IFN-α, IFN-β, and to a much larger extent, IFN-γ enhance the levels of class II HLA mRNA, for both the α and the β chains as well as for the invariant chain, which occurs intracellularly together with the α and β chains but is not expressed at the cell surface. The enhancement of mRNA, however, is not always accompanied by an increased expression of the corresponding cell-surface antigen, indicating the existence of controls at the posttranscriptional level (Rosa et al., 1983a; Rahmsdorf et al., 1986). Murine cells can be transfected with cDNAs encoding various MHC molecules of human or porcine origin. One can turn on the transfected genes in these cells by IFN treatment, as witnessed by the onset of the synthesis of the corresponding mRNAs. This can only be done with IFNs derived from the same animal species as the transfected cell, and not by the IFN of the species from which the cDNA is derived: The IFN molecules do not act directly on the gene, but rather, act through the species-specific receptor at the cell surface (Rosa et al., 1983b; Yoshie et al., 1983; Satz and Singer, 1984; Yoshie et al., 1984). The events occurring after receptor binding are not known, but the molecular mechanisms that are set into motion must be very specific, since cells that are resistant to the other effects of IFNs can still be induced to express MHC antigens, and since antigen expression is selective and varies with IFN and cell type. The continuous presence of IFN seems to be required to keep the induced expression going, and when IFN is removed, the effect is reversed (Carrel et al., 1985).

By selecting IFN-α-treated cells from the leukemia cell line Molt 4 with the fluorescence-activated cell sorter for high and low HLA class I antigen induction, stable mutants can be obtained, displaying high or low inducibility of HLA antigens by IFN (Burrone et al., 1985). The study of such mutants may be rewarding for revealing the molecular mechanism of MHC antigen induction by IFNs.

The 5′ flanking regions of several IFN-inducible genes, including HLA class I and class II genes, contain a 30-nucleotide-long consensus sequence (Friedman and Stark, 1985). The promoter of several murine class I MHC genes contains a similar IFN response sequence (see Chapter 5, Table 5.4). The latter seems to be activable by IFN only when associated with a functional enhancer sequence, also present in the promoter region of murine class I genes, since the combination of these two sequences renders a heterologous promoter responsive to Mu IFN-α/β and to Mu IFN-γ (Israel et al., 1986; Vogel et al., 1986).

This aspect of gene derepression by IFNs is discussed in more detail in Chapter 5.

REFERENCES

Ameglio, F., Capobianchi, M. R., Dolei, A., and Tosi, R. Differential effects of gamma interferon on expression of HLA class II molecules controlled by the DR and DC loci. *Infect. Immun.* 42: 122–125 (1983).

Baldini, L., Cortelezzi, A., Polli, N., Neri, A., Nobili, L., Maiolo, A. T., Lambertenghi-Deliliers, G., and Polli, E. E. Human recombinant interferon-α-2C enhances the expression of class II HLA antigens on hairy cells. *Blood* 67: 458–464 (1986).

Balkwill, F. R., Stevens, M. H., Griffin, D. B., Thomas, J. A., and Bodmer, J. G. Interferon γ regulates HLA-D expression on solid tumors in vivo. *Eur. J. Cancer Clin. Oncol.* 23: 101–106 (1987).

Barlow, D. P., Randle, B. J., and Burke, D. C. Interferon synthesis in the early post-implantation mouse embryo. *Differentiation* 27: 229–235 (1984).

Basham, T. Y., Bourgeade, M. F., Creasey, A. A., and Merigan, T. C. Interferon increases HLA synthesis in Melanoma cells: interferon-resistant and -sensitive cell lines. *Proc. Natl. Acad. Sci.* (USA) 79: 3265–3269 (1982).

Basham, T. Y. and Merigan, T. C. Recombinant interferon-γ increases HLA-DR synthesis and expression. *J. Immunol.* 130: 1492–1494 (1983).

Becker, S. Interferons as modulators of human monocyte-macrophage differentiation. I. Interferon-γ increases HLA-DR expression and inhibits phagocytosis of zymosan. *J. Immunol.* 132: 1249–1254 (1984).

Becker, S. Interferon-γ accelerates immune proliferation via its effect on monocyte HLA-DR expression. *Cell. Immunol.* 91: 301–307 (1985).

Berrih, S., Arenzana-Seisdedos, F., Cohen, S., Devos, R., Charron, D., and Virelizier, J. L. Interferon-γ modulates HLA class II antigen expression on cultured human thymic epithelial cells. *J. Immunol.* 135: 1165–1171 (1985).

Bukowski, J. F. and Welsh, R. M. Interferon enhances the susceptibility of virus-infected fibroblasts to cytotoxic T cells. *J. Exp. Med.* 161: 257–262 (1985).

Burrone, O. R., Kefford, R. F., Gilmore, D., and Milstein, C. Stimulation of HLA-A,B,C by IFN-α. The derivation of Molt 4 variants and the differential expression of HLA-A,B,C subsets. *EMBO J.* 4: 2855–2860 (1985).

Campbell, I. L., Wong, G. H. W., Schrader, J. W., and Harrison, L. C. Interferon-γ enhances the expression of the major histocompatibility class I antigens on mouse pancreatic β cells. *Diabetes* 34: 1205–1209 (1985).

Capobianchi, M. R., Ameglio, F., Tosi, R., and Dolei, A. Differences in the expression and release of DR, BR and DQ molecules in human cells treated with recombinant interferon γ: Comparison to other interferons. *Human Immunol.* 13: 1–11 (1985).

Carrel, S., Schmidt-Kessen, A. and Giuffre, L. Recombinant interferon-γ can induce the expression of HLA-DR and DC on DR-negative melanoma cells and enhance the expression of HLA-ABC and tumor-associated antigens. *Eur. J. Immunol.* 15: 118–123 (1985).

Daar, A. S., Fuggle, S. V., Ting, A., and Fabre, J. W. Anomalous expression of HLA-DR antigens in human colorectal cancer cells. *J. Immunol.* 129: 447–449 (1982).

Diaz-Espada, Milstein, C., and Secher, D. Effect of purified human interferon-α on the expression of differentiation antigens and mitogen reactivity of cultured human thymic cells. *Cell. Immunol.* 100: 331–339 (1986).

Dolei, A., Capobianchi, M. R., and Ameglio, F. Human interferon-γ enhances the expression of class I and class II major histocompatibility complex products in neoplastic cells more effectively than interferon-α and interferon-β. *Infect. Immun.* 40: 172–176 (1983).

Fellous, M., Kamoun, M., Gresser, I., and Bono, R. Enhanced expression of HLA antigens and β2-microglobulin on interferon-treated human lymphoid cells. *Eur. J. Immunol.* 9: 446–449 (1979).

Fierz, W., Endler, B., Reske, K., Wekerle, H., and Fontana, A. Astrocytes as antigen-presenting cells. I. Induction of Ia antigen expression on astrocytes by T cells via immune interferon and its effect on antigen presentation. *J. Immunol.* 134: 3785–3793 (1985).

Fleischmann, W. R., Jr., Georgiades, J. A., Osborne, L. C., and Johnson, H. M. Potentiation of interferon activity by mixed preparations of fibroblast and immune interferon. *Infect. Immun.* 26: 248–253 (1979).

Fontana, A., Fierz, W., and Wekerle, H. Astrocytes present myelin basic protein to encephalitogenic T-cell lines. *Nature* 307: 273–276 (1984).

Friedman, R. L. and Stark, G. R. α-interferon-induced transcription of HLA and metallothionein genes containing homologous upstream sequences. *Nature* 314: 637–639 (1985).

Geppert, T. D. and Lipsky, P. E. Antigen presentation by interferon-γ-treated endothelial cells and fibroblasts: Differential ability to function as antigen-presenting cells despite comparable Ia expression. *J. Immunol.* 135: 3750–3762 (1985).

Germain, R. N. Accessory cell stimulation of T cell proliferation requires active antigen processing, Ia-restricted antigen presentation, and a separate nonspecific 2nd signal. *J. Immunol.* 127: 1964–1966 (1981).

Gershon, H. E., De Kuang, Y., Scala, G., and Oppenheim, J. J. Effects of recombinant interferon-γ on HLA-DR antigen shedding by human peripheral blood adherent mononuclear cells. *J. Leukocyte Biol.* 38: 279–291 (1985).

Guerry, D. P., Alexander, M. A., Herlyn, M. F., Zehngebot, L. M., Mitchell, K. F., Zmijewski, C. M., and Lusk, E. J. HLA-DR histocompatibility leukocyte antigens permit cultures human melanoma cells from early but not advanced disease to stimulate autologous lymphocytes. *J. Clin. Invest.* 73: 267–271 (1984).

Hayashi, H., Tanaka, K., Jay, F., Khoury, G., and Jay, G. Modulation of the tumorigenicity of human adenovirus-12-transformed cells by interferon. *Cell* 43: 263–267 (1985).

Herlyn, M., Guerry, D. and Koprowski, H. Recombinant γ-interferon induces changes in expression and shedding of antigens associated with normal human melanocytes, nevus cells, and primary and metastatic melanoma cells. *J. Immunol.* 134: 4226–4230 (1985).

Heron, I., Hokland, M., and Berg, K. Enhanced expression of β2-microglobulin and HLA antigens on human lymphoid cells by interferon. *Proc. Natl. Acad. Sci.* 75: 6215–6219 (1978).

Houghton, A. N., Thomson, T. M., Gross, D., Oettgen, H. F., and Old, L. J. Surface antigens of melanoma and melanocytes. Specificity of induction of Ia antigens by human γ-interferon. *J. Exp. Med.* 160: 255–269 (1984).

Hui, K., Grosveld, F., and Festenstein, H. Rejection of transplantable AKR leukaemia cells following MHC DNA-mediated cell transformation. *Nature* 311: 750–752 (1984).

Hunt, J. S. and Wood, G. W. Interferon-γ induces class I HLA and β2-microglobulin expression by human amnion cells. *J. Immunol.* 136: 364–367 (1986).

Inaba, K., Kitaura, M., Kato, T., Watanabe, Y., Kawade, Y., and Muramatsu, S. Contrasting effect of α/β- and γ-interferons on expression of macrophage Ia antigens. *J. Exp. Med.* 163: 1030–1035 (1986).

Israel, A., Kimura, A., Fournier, A., Fellous, M., and Kourilsky, P. Interferon response sequence potentiates activity of an enhancer in the promoter region of a mouse H-2 gene. *Nature* 322: 743–746 (1986).

Janeway, C. A., Jr., Bottomly, K., Babich, J., Conrad, P., Conzen, S., Jones, B., Kaye, J., Katz, M., McVay, L., Murphy, D. B., and Tite, J. Quantitative variation in Ia antigen expression plays a central role in immune regulation. *Immunol. Today* 5: 99–105 (1984).

Kaufman, J. F., Auffray, C., Korman, A. J., Shackelford, D. A., and Strominger, J. The class II molecules of the human and murine major histocompatibility complex. *Cell* 36: 1–13 (1984).

Kelley, V. E., Fiers, W., and Strom, T. B. Cloned human interferon-γ, but not interferon-β or γ, induces expression of HLA-DR determinants by fetal monocytes and myeloid leukemic cell lines. *J. Immunol.* 132: 240–245 (1984).

Kim, K. J., Chaout, G., Leiserson, W. M., King, J., and De Maeyer, E. Characterization of T-cell soluble factors modulating the expression of Ia and H-2 antigens on BALB/c B lymphoma cell lines. *Cell. Immunol.* 76: 253–267 (1983).

Koch, N., Wong, H. W., and Schrader, J. W. Ia antigens and associated invariant chain are induced simultaneously in lines of T-dependent mast cells by recombinant interferon-γ. *J. Immunol.* 132: 1361–1369 (1984).

Lampson, L. A. and Fisher, C. A. Weak HLA and β2-microglobulin expression of neuronal cell lines can be modulated by interferon. *Proc. Natl. Acad. Sci.* (USA) 81: 6476–6480 (1984).

Lebon, P., Girard, S., Thepot, F., and Chany, C. The presence of α-interferon in human amniotic fluid. *J. Gen. Virol.* 59: 393–396 (1982).

Liao, S. K., Kwong, P. C., Khosravi, M., and Dent, P. B. Enhanced expression of melanoma-associated antigens and β2-microglobulin on cultured human melanoma cells by interferon. *J. Natl. Cancer Inst.* 68: 19–25 (1982).

Lindahl, P., Gresser, I., Leary, P., and Tovey, M. Interferon treatment of mice: Enhanced expression of histocompatibility antigens on lymphoid cells. *Proc. Natl. Acad. Sci.* (USA) 73: 1284–1287 (1976).

Ling, P. D., Warren, M. K., and Vogen, S. N. Antagonistic effect of interferon-β on the interferon-γ-induced expression of Ia antigen in murine macrophages. *J. Immunol.* 135: 1857–1863 (1985).

Miyawaki, T., Seki, H., Taga, K., and Taniguchi, N. Interferon-γ can augment expression ability of HLA-DR antigens on pokeweed mitogen-stimulated human T lymphocytes. *Cell. Immunol.* 89: 300-309 (1984).

Momburg, F., Koch, N., Moller, P., Moldenhauer, G., and Hammerling, G. J. In vivo induction of H-2K/D antigens by recombinant interferon-γ. *Eur. J. Immunol.* 16: 551–557 (1986).

Nunez, G., Ball, E. J., and Stastny, P. Accessory cell function of human endothelial cells. I. A subpopulation of Ia positive cells is required for antigen presentation. *J. Immunol.* 131: 666–673 (1983).

Ozato, K., Wan, Y. J., and Orrison, B. M. Mouse major histocompatibility class I gene expression begins at midsomite stage and is inducible in earlier-stage embryos by interferon. *Proc. Natl. Acad. Sci.* (USA) 82: 2427–2431 (1985).

Palacios, R., Martinez-Maza, O., and De Ley, M. Production of human immune interferon (Hu IFN-γ) studied at the single cell level. Origin, evidence for spontaneous secretion and effect of cyclosporin A. *Eur. J. Immunol.* 13: 221–225 (1983).

Piguet, V., Carrel, S., Diserens, A. C., Mach, J. P., and De Tribolet, N. Heterogeneity of the induction of HLA-DR expression by human immune interferon on glioma cell lines and their clones. *J. Natl. Cancer Inst.* 76: 223–228 (1986).

Pober, J. S., Collins, T., Gimbrone, M. A., Jr., Cotran, R. S., Gitlin, J. D., Fiers, W., Clayberger, C., Krensky, A. M., Burakoff, S. J., and Reiss, C. S. Lymphocytes recognize human vascular endothelial and dermal fibroblast Ia antigens induced by recombinant immune interferon. *Nature* 305: 726–729 (1983a).

Pober, J. S., Gimbrone, M. A., Jr., Cotran, R. S., Reiss, C. S., Burakoff, S. J., Fiers, W., and Ault, K. A. Ia expression by vascular endothelium is inducible by activated T cells and by human γ interferon. *J. Exp. Med.* 157: 1339–1353 (1983b).

Rahmsdorf, H. J., Harth, N., Eades, A. M., Litfin, M., Steinmetz, M., Forni, L., and Herrlich, P. Interferon-γ, mitomycin C., and cycloheximide as regulatory agents of MHC class II-associated invariant chain expression. *J. Immunol.* 136: 2293–2299 (1986).

Rhodes, J., Ivanyi, J., and Cozens, P. Antigen presentation by human monocytes: Effects of modifying major histocompatibility complex class II antigen expression and interleukin 1 production by using recombinant interferons and corticosteroids. *Eur. J. Immunol.* 16: 370–375 (1986).

Rosa, F. M., Cochet, M. M., and Fellous, M. Interferon and major histocompatibility complex genes: A model to analyse eukaryotic gene regulation? In: *Interferon 7*. I. Gresser, Ed., Academic Press London, pp. 48–87 (1986).

Rosa, F. M., Hatat, D., Abadie, A., Wallach, D., Revel, M., and Fellous, M. Differential regulation of HLA-DR mRNAs and cell surface antigens by interferon. *EMBO J.* 2: 1585–1589 (1983a).

Rosa, F., Le Bouteiller, P. P., Abadie, A., Mishal, Z., Lemonnier, F. A., Bourrel, D., Lamotte, M., Kalil, J., Jordan, B., and Fellous, M. HLA class I genes integrated into murine cells are inducible by interferon. *Eur. J. Immunol.* 13: 495–499 (1983b).

Satz, L. and Singer, D. S. Effect of mouse interferon on the expression of a porcine major histocompatibility gene introduced into mouse L cells. *J. Immunol.* 132: 496–501 (1984).

Schneeberger, E. E., De Ferrari, M., Skoskiewicz, M. J., Russell, P. S., and Colvin, R. B. Induction of MHC-determined antigens in the lung by interferon-γ. *Lab. Invest.* 55: 138–144 (1986).

Schwartz, R., Momburg, F., Moldenhauer, G., Dorken, B., and Schirrmacher, V. Induction of HLA class II antigen expression on human carcinoma cell lines by interferon-γ. *Int. J. Cancer* 35: 245–250 (1985).

Sekaly, R. P., Tonnelle, C., Strubin, M., Mach, B., and Long, E. O. Cell surface expression of class II histocompatibility antigens occurs in the absence of the invariant chain. *J. Exp. Med.* 164: 1490–1504 (1986).

Skoskiewicz, M. J., Colvin, R. B., Schneeberger, E. E., and Russell, P. S. Widespread and selective induction on major histocompability complex-determined antigens in vivo by γ-interferon. *J. Exp. Med.* 162: 1645–1664 (1985).

Sobel, R. A., Blanchette, B. W., Bhan, A. K., and Colvin, R. B. The immunopathology of experimental allergic encephalomyelitis. II. Endothelial cell Ia increases prior to inflammatory cell infiltration. *J. Immunol.* 132: 2402–2407 (1984).

Steeg, P. S., Moore, R. N., Johnson, H. M., and Oppenheim, J. J. Regulation of murine macrophage Ia antigen expression by a lymphokine with immune interferon activity. *J. Exp. Med.* 156: 1780–1793 (1982).

Sutherland, J., Mannoni, P., Rosa, F., Huyat, D., Turner, A. R., and Fellous, M. Induction of the expression of HLA class I antigens on K562 by interferons and sodium butyrate. Hum. Immunol. 12: 65–73 (1985).

Sztein, M. B., Steeg, P. S., Johnson, H. M., and Oppenheim, J. J. Regulation of human peripheral blood monocyte DR antigen expression in vitro by lymphokines and recombinant interferons. *J. Clin. Invest.* 73: 556–565 (1984).

Takiguchi, M. and Frelinger, J. A. Induction of antigen presentation ability in purified cultures of astroglia by interferon-γ. *J. Mol. Cell. Immunol.* 2: 269–280 (1986).

Takiguchi, M., Ting, J. P.-Y., Buessow, S. C., Boyer, C., Gillespie, Y., and Frelinger, J. A. Response of glioma cells to interferon-γ: Increase in class II RNA, protein and mixed lymphocyte reaction-stimulating ability. *Eur. J. Immunol.* 15: 809–814 (1985).

Tanaka, K., Isselbacher, K. J., Khoury, G., and Jay, G. Reversal of oncogenesis by the expression of a major histocompatibility complex class I gene. *Science* 228: 26–30 (1985).

Tedeschi, B., Barrett, J. N., and Keane, R. W. Astrocytes produce interferon that enhances the expression of H-2 antigens on a subpopulation of brain cells. *J. Cell. Biol.* 102: 2244–2253 (1986).

Tweardy, D. J., Fujiwara, H., Scillian, J. J., and Ellner, J. J. Concurrent enhancement of monocyte immunoregulatory properties and effector functions by recombinant interferon-γ. *Cell. Immunol.* 100: 34–46 (1986).

Unanue, E. R. The regulatory role of macrophages in antigenic stimulation. Part 2. Symbiotic relationship between lymphocytes and macrophages. *Adv. Immunol.* 31: 1–136 (1981).

Vignaux, F. and Gresser, I. Enhanced expression of histocompatibility antigens on interferon-treated mouse embryonic fibroblasts. *Proc. Soc. Exp. Biol. Med.* 157: 456–460 (1978).

Virelizier, J. L., Perez, N., Arenzana-Seisdedos, F., and Devos, R. Pure interferon γ enhances class II antigens on human monocyte cell lines. *Eur. J. Immunol.* 14: 106–108 (1984).

Vogel, J., Kress, M., Khoury, G., and Jay, G. A transcriptional enhancer and an interferon-responsive sequence in major histocompatibility complex class I genes. *Mol. Cell. Biol.* 6: 3550–3554 (1986).

Wallach, D. The HLA proteins and a related protein of 28 kDa are preferentially induced by interferon-γ in human WISH cells. *Eur. J. Immunol.* 13: 794–798 (1983).

Wallich, R., Bulbuc, N., Hammerling, G. J., Katzav, S., Segal, S., and Feldman, M. Abrogation of metastatic properties of tumour cells by de novo expression of H-2K antigens following H-2 gene transfection. *Nature* 315: 301–305 (1985).

Wan, Y. J. Y., Orrison, B. M., Lieberman, R., Lazarovici, P., and Ozato, K. Induction of major histocompatibility class I antigens by interferons in undifferentiated F9 cells. *J. Cell. Physiol.* 130: 276–283 (1987).

Wilson, B. S., Indiveri, F., Pellegrino, M. A., and Ferrone, S. DR (Ia-like) antigens on human melanoma cells. *J. Exp. Med.* 149: 658–668 (1979).

Wiman, K., Curman, B., Forsum, U., Klareskog, L., Malmas-Tjernlund, U., Rask, L., Tragardh, L., and Peterson, P. A. Occurrence of Ia antigens on tissues of non-lymphoid origin. *Nature* 276: 711–713 (1978).

Winchester, R. J., Wang, C. Y., Gibofsky, A., Kunkel, H. G., Lloyd, K. O., and

Old, L. J. Expression of Ia-like antigens on cultured human malignant melanoma cell lines. *Proc. Natl. Acac. Sci.* (USA) 75: 6235–6239 (1978).

Wong, G. H. W., Bartlett, P. F., Clark-Lewis, I., Battye, F., and Schrader, J. W. Inducible expression of H-2 and Ia antigens on brain cells. *Nature* 310: 688–691 (1984).

Wong, G. H. W., Clark-Lewis, I., Harris, A. W., and Schrader, J. W. Effect of cloned interferon-γ on expression of H-2 and Ia antigens on cell lines of hemopoietic, lymphoid, epithelial, fibroblastic and neuronal origin. *Eur. J. Immunol.* 14: 52–56 (1984).

Wong, G. H. W., Clark-Lewis, I., McKimm-Breschkin, J. L., Harris, A. W., and Schrader, J. W. Interferon-γ induces enhanced expression of Ia and H-2 antigens on B lymphoid, macrophage, and myeloid cell lines. *J. Immunol.* 131: 788–793 (1983).

Wong, G. H. W., Clark-Lewis, I., McKimm-Breschkin, J. L., and Schrader, J. W. Interferon-γ-like molecule induces Ia antigens on cultured mast cell progenitors. *Proc. Natl. Acad. Sci.* (USA) 79: 6989–6993 (1982).

Yarden, A., Shure-Gottlieb, H., Chebath, J., Revel, M., and Kimchi, A. Autogenous production of interferon-β switches on HLA genes during differentiation of histiocytic lymphoma U937 cells. *EMBO J.* 3: 969–973 (1984).

Yoshie, O., Schmidt, H., Lengyel, P., Reddy, E. S. P., Morgan, W. R. and Weissman, S. M. Transcripts of human HLA gene fragments lacking the 5′-terminal region in transfected mouse cells. *Proc. Natl. Acad. Sci.* (USA) 81: 649–653 (1984).

Yoshie, O., Schmidt, H., Reddy, E. S. P., Weissman, S. M., and Lengyel, P. Mouse interferons enhance the accumulation of a human HLA RNA and protein in transfected mouse and hamster cells. *J. Biol. Chem.* 257: 13169–13172 (1983).

Ziai, M. R., Imberti, L., Tongson, A., and Ferrone, S. Differential modulation by recombinant immune interferon of the expression and shedding of HLA antigens and melanoma associated antigens by a melanoma cell line resistant to the antiproliferative activity of immune interferon. *Cancer Res.* 45: 5877–5882 (1985).

Zinkernagel, R. M., and Doherty, P. C. MHC-restricted cytotoxic T cells: studies on the biological role of polymorphic major transplantation antigens determining T-cell restriction-specificity, function, and responsiveness. In: *Advances in Immunology,* H. G. Kunkel and F. J. Dixon, Eds., Academic Press, New York, Vol. 27, pp. 51–177 (1979).

Zinkernagel, R. M. and Doherty, P. C. Restriction of in vitro T cell-mediated cytotoxicity in lymphocytic choriomeningitis within a syngeneic or semiallogeneic system. *Nature* 224: 701–702 (1974).

Zuckermann, F. A. and Head, J. R. Expression of MHC antigens on murine trophoblast and their modulation by interferon. *J. Immunol.* 137: 846–853 (1986).

10 MACROPHAGES AS INTERFERON PRODUCERS AND INTERFERONS AS MODULATORS OF MACROPHAGE ACTIVITY

Macrophages play a central role in nonantigen-specific as well as in antigen-specific immunity. Derived from myeloid stem cells of the bone marrow, they are ubiquitous in the body, both as circulating and as fixed cells. In the circulation they are present as monocytes, in lymphoid organs such as the spleen and lymph nodes as part of the mononuclear phagocyte system, in the liver as Kupffer cells, in the lungs as alveolar macrophages, and in many other tissues as "resident" cells. Macrophages are intimately related to the IFN system. Contrary to most of the other cells in the body, they can be induced not only by viruses but also by other infectious agents to make IFN-α/β, and they also cooperate with T cells in the production of IFN-γ. Moreover, some very important functions of macrophages are under the influence of IFN-α/β and IFN-γ.

I. A SHORT INTRODUCTION TO MACROPHAGES

A. The Major Functions of Macrophages

Mononuclear phagocytes play a central role as effector cells in inflammatory reactions and cell-mediated immune responses.

1. Phagocytosis, followed by the intracellular killing, of infectious agents such as viruses, bacteria, and parasites. This activity is partly due to the release of reactive metabolites of oxygen such as hydrogen peroxide and hydroxyl radicals, to the presence of many acid hydrolases in the phagolysosomes, and possibly to the depletion of essential metabolites.
2. Destruction of tumor cells. Since they can destroy tumor cells on a nonantigen-specific basis, the hypothesis that macrophages are an important component of the tumor surveillance mechanism is receiving considerable attention. The formation of toxic oxygen intermediates and perhaps other cytocidal agents is believed to be instrumental for this activity. In vivo, tumors are often infiltrated by macrophages, and adoptive transfer of macrophages under certain conditions can retard the growth of syngeneic tumors.
3. Production and secretion of many mediator substances such as complement components and arachidonic acid metabolites, neutral proteases like plasminogen activator, collagenase and elastase, growth factors, monokines and interleukins such as platelet-derived growth factor, tumor necrosis factor (TNF-α), myeloid precursor colony stimulating factor (G-CSF), and interleukin-1 and cytokines like IFN-α and β.
4. Antigen presentation to T cells in the context of MHC class II cell-surface antigens. The expression of class II MHC antigens on macrophages is required for their function as accessory cells for antigen presentation to $CD4^+$ T cells. This aspect of cell-surface changes induced by IFNs is treated in detail in Chapter 9.
5. Macrophages also play a role in wound healing through secretion of

angiogenic factors, mitogenic factors for fibroblasts like the platelet-derived growth factor, and in tissue renewal through the disposal of senescent cells, for example, erythrocytes. No attention has been paid to the possibility that this particular activity of macrophages may also be under the influence of IFNs.

B. Some Preliminary Remarks Relative to Macrophage Obtention

When interpreting the effects of IFNs on macrophage activity, one should be aware that monocyte/macrophage suspensions or cultures used in vitro may represent different populations, depending on their origin, mode of obtention, and culture conditions.

An important operational characteristic of macrophages is their capacity to adhere to glass or plastic, a property frequently used to isolate and separate them from other cells. The designation "glass- or plastic-adherent cells" is often considered to be synonymous of macrophages. One must bear in mind that cells isolated in this way can, nevertheless, still represent different functional subclasses, at various stages of differentiation and activation. This should be taken into account for the interpretation of many studies on IFN and macrophages. Moreover, cells of other types, including T and B cells, can show some degree of adherence in vitro. A staining method for macrophages, based on the detection of nonspecific esterase, does exist, but is not always used for identification. In addition to phagocytosis of antibody-coated erythrocytes or of latex particles, which represents another useful method for identifying macrophages in vitro, differentiation antigens and specific plasma membrane receptors provide markers to characterize cells of the mononuclear phagocyte system with the appropriate monoclonal antibodies.

Like many other organs, the peritoneal cavity contains resident macrophages, together with lymphocytes and granulocytes, and murine macrophages are frequently obtained by rinsing this cavity. The formation of an inflammatory exudate can be elicited by injecting an irritating substance such as thioglycollate, and as a result of this treatment the number of peritoneal macrophages increases considerably because of the recruitment of monocytes from the blood. The macrophages obtained from unstimulated or stimulated peritoneal cavities are referred to as resident or induced glass adherent peritoneal cells, respectively, and such cells have frequently been used for IFN studies in murine models.

Another way of obtaining murine macrophages is to put bone marrow cells into culture in the presence of a growth factor that is usually derived from supernatants of mouse L cell cultures. This growth factor, presently characterized as M-CSF or CSF-1 is capable of inducing IFN in macrophages (see Chapter 8). Putting bone-marrow cells into culture in the presence of CSF-1 favors the selective survival of the precursors of the granulocyte-monocyte lineage, and after 5 or 6 days almost pure cultures of murine macrophages are thus available (Neumann and Sorg, 1980).

Human macrophages for IFN studies have usually been obtained as circulating monocytes from peripheral blood. Monocytes and lymphocytes are first separated together from other blood cells by gradient centrifugation, and a second step allows isolating the monocytes by their capacity to adhere firmly to glass or plastic.

The cells prepared in these various ways are all referred to as macrophages or mononuclear phagocytes, and as monocytes when derived from peripheral blood. Manifestly, they may—and at times do—represent functionally different populations. For example, murine lung macrophages differ from peritoneal macrophages by their morphology, lysosomal enzyme activity, and phagocytic potential (Hearst et al., 1980), and thioglycollate-elicited peritoneal macrophages have surface antigens that are lacking on resident peritoneal macrophages (Ho and Springer, 1982).

In addition to freshly isolated cells, continuous lines of murine and human macrophages are available, and cells belonging to some of these lines have also been used for IFN studies.

II. MACROPHAGES AND INTERFERON PRODUCTION

A. Macrophages as Producers of IFN-α and IFN-β

1. Induction by Viruses

Like many other cells in the body, macrophages can produce IFN-α and IFN-β as a result of virus infection. Since macrophages are specialized in phagocytosis and digestion of antigens, including viruses, this means that at least some viruses can bypass this process and proceed to IFN induction. Rabbit macrophages originating from alveolar lavage or from the peritoneal cavity produce IFN when infected in vitro with myxoviruses (Smith and Wagner, 1967 a,b; Acton, 1973). Human peripheral blood monocytes, infected in vitro with influenza A virus, produce type I IFN (Roberts et al., 1979). Mouse peritoneal macrophages, induced in vitro with Newcastle disease virus, produce mainly IFN-β. Since macrophages derived from mice that are high IFN producers make more IFN-β in vitro than macrophages derived from mice belonging to low responder strains, genes influencing the levels of IFN production in vivo are manifestly expressed in macrophages (De Maeyer et al., 1979).

When the number of IFN-producing cells is measured in cultures of Newcastle disease virus-induced murine peritoneal macrophages by in situ hybridization, using Mu IFN-β cDNA as a probe to detect cells containing IFN-β mRNA, virtually all cells are found to be positive (Fig. 10.1). This suggests that they probably all are IFN-β producers, although it cannot be formally excluded that some cells contain IFN-β mRNA yet do not release IFN protein because of a block at the translational level. Newcastle disease virus-infected peritoneal macrophages also contain Mu IFN-α mRNA, as revealed by in situ hybridization, and yet they produce little or no IFN-α (De

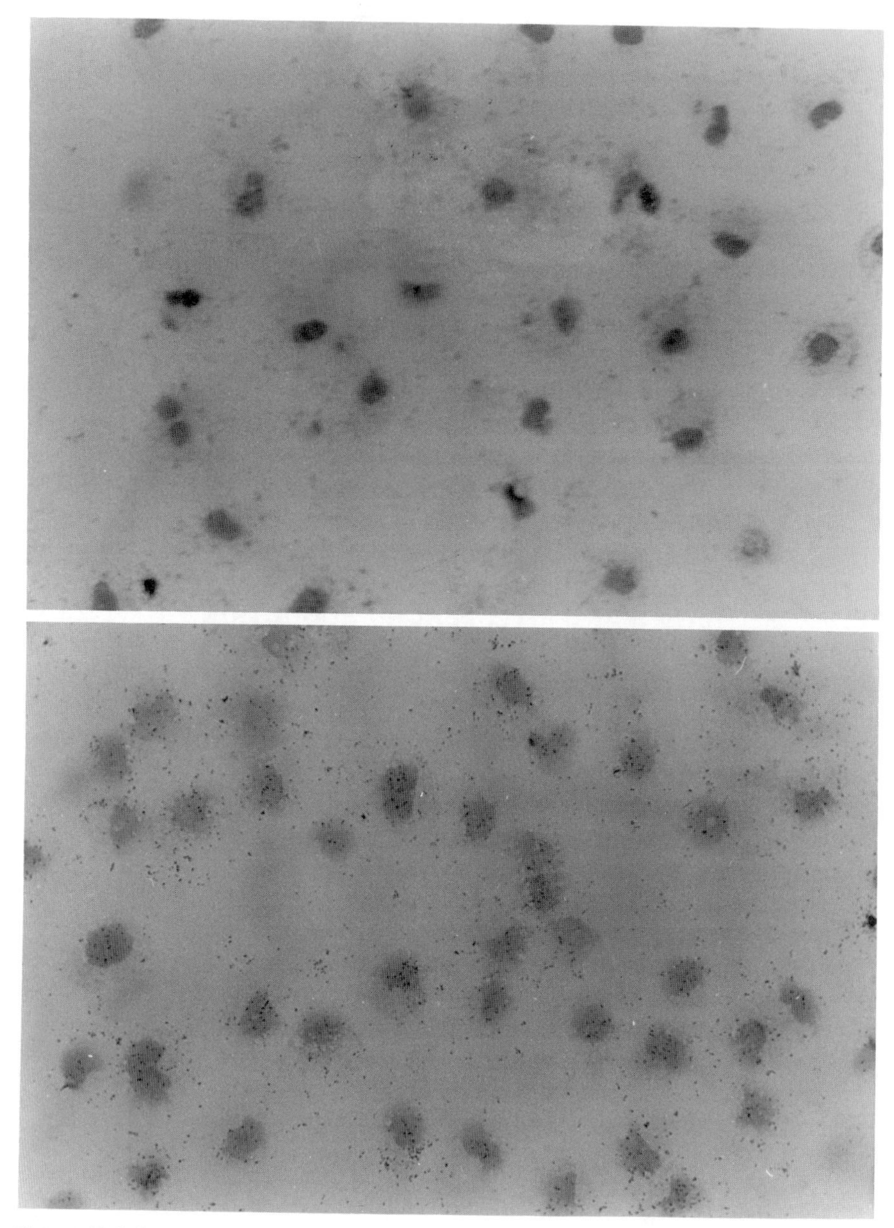

Figure 10.1 Detection by in situ hybridization of Mu IFN-β mRNA in murine peritoneal macrophages induced in vitro by Newcastle disease virus and probed with an ^{35}S-labeled Mu IFN-β cDNA. Upper pannel: Uninduced cells. Lower panel: Induced cells, fixed and hybridized 6 hours after induction. All cells of the induced culture are positive for the presence of IFN-β mRNA.

Maeyer and De Maeyer-Guignard, 1986). Apparently some posttranscriptional event prevents these cells from either synthesizing or secreting IFN-α. But when the macrophages originate from bone-marrow precursors in vitro rather than from the peritoneal cavity, a significant fraction of the IFN produced upon induction with Newcastle disease virus is IFN-α, as judged by its molecular weight of about 18 K (Brehm and Kirchner, 1986). This is a clear indication that the stage of differentiation of the macrophage has an influence on the type of IFN produced, which is in line with other observations that cell specific factors influence IFN production (see Chapter 3).

Using immunofluorescent staining with a monoclonal anti-IFN-α antibody for visualizing individual IFN-producing cells, one can show that human monocytes contain Hu IFN-α in their cytoplasm upon induction with Sendai virus (Saksela et al., 1984). Apparently, and contrary to what one observes in murine peritoneal macrophages, the major component of paramyxovirus-induced IFN in human monocytes is α and not β. The reason for this difference has never been systematically investigated, but it may again reflect the different origin and stage of differentiation of these macrophage populations—resident peritoneal macrophages, on the one hand, and circulating monocytes, on the other.

The induction of IFN-α and IFN-β by viruses in macrophages contributes to host resistance by more than one mechanism: It restricts virus replication in macrophages and surrounding cells through the direct antiviral action of IFNs, and as a result of the macrophage-activating activities of IFN-α and β, it stimulates other functions of macrophages that are relevant to host defense.

2. *Induction by dsRNA*

The synthetic double-stranded polyribonucleotide polyrI-rC induces Mu IFN-α/β in murine bone-marrow-derived macrophages in culture (Neumann, 1982). When inoculated intravenously into mice, polyrIrC induces the production of an IFN that is at least 60% of the β type and originates principally in macrophages (Jullien et al., 1974; Martinez et al., 1980; Fleit and Rabinovitch, 1981a; De Maeyer-Guignard et al., 1986). In vitro, human peripheral blood monocytes can be induced to make IFN-α by exposure to polyrI-rC (Stevenson et al., 1985). This is in contrast to what happens in human diploid fibroblasts, in which polyrI-rC induces IFN-β only. This points to the existence of cell-specific factors determining the type of IFN produced.

3. *Induction by Endotoxin in Murine Macrophages*

Endotoxin, the lipopolysaccharide moiety (LPS) derived from the cell walls of gram-negative bacteria, is an efficient IFN inducer in murine macrophages (Kono, 1967; Matisova et al., 1970; De Maeyer et al., 1971; Maehara and Ho, 1977; Fleit and Rabinovitch, 1981b; Neumann and Sorg, 1981). By their property to produce IFN-α/β in response to LPS, macrophages distinguish themselves from all other cells in the body. Unless special precautions are taken, low amounts of LPS are practically ubiquitous in vivo and in vitro, and this raises the possibility that LPS, through induction of IFN, is one of the natural stimulators of the antiviral state in macrophages.

When peritoneal macrophages from C57BL mice are put into culture, low levels of type I IFN are spontaneously released into the medium during the first days of the culture. After a few days this spontaneous production stops, but if the cultures are restimulated with endotoxin, IFN is again produced. Endotoxin is probably responsible for the low level of "spontaneous" IFN production by peritoneal macrophages when they are put into culture. This notion is reinforced by the observation that peritoneal macrophages from germ-free mice do not release spontaneous IFN in culture (Smith and Wagner, 1967a; De Maeyer et al., 1971; Havell and Spitalny, 1983).

Spontaneous IFN-α/β production also occurs in vivo and influences the antiviral resistance of macrophages. Macrophages derived from mice treated with anti-Mu IFN-α/β globulin are susceptible in vitro to infection with vesicular stomatitis virus and mouse encephalomyocarditis virus. In the absence of previous treatment of the donor animals with anti-IFN globulin, very few macrophages are permissive for these viruses (Belardelli et al., 1984). Whether or not the spontaneous IFN-α/β production in vivo that renders macrophages resistant to viral infection is actually coming from the macrophages themselves and linked to endotoxin action is an unresolved question, although this is suggested by the fact that macrophages from *Lps*^{d} mice (see subsequent discussion) are permissive for viral replication (Vogel and Fertsch, 1987). It certainly would seem to be the most likely explanation, in view of the ubiquitous nature of endotoxin. Another way of approaching this problem, instead of using *Lps*^{d} mice, would be to use germ-free mice, although the latter can still be in contact with endotoxin, which can contaminate sterilized water, food, or cage bedding. Undoubtedly, these low levels of endogenous IFN production in vivo are important since they provide a way for macrophages to be in a continuous antiviral state.

In the mouse, a gene for endotoxin responsiveness, the *Lps* locus, has been mapped to chromosome 4 (see Fig. 2.3, Chapter 4). Two allelic forms exist: *Lps*^{n} for normal responsiveness, and *Lps*^{d} for defective responsiveness (Sultzer, 1972; Watson et al., 1978). Defective mice are much more resistant to the immunological and toxic effects of LPS than mice having the allele for normal responsiveness. In addition, they do not produce IFN-α/β upon injection of LPS, whereas *Lps*^{n} do. Impaired IFN production is specifically observed with LPS, and induction with viruses remains normal in these animals (Ascher et al., 1981). Macrophages from defective mice, the prototype strain being C3H/HeJ, are defective in several functions, such as receptor-mediated phagocytosis (phagocytosis of antibody-coated particles via the Fc receptor) and tumoricidal capacity (Ruco and Meltzer, 1978; Vogel and Rosenstreich, 1979). The deficient receptor-mediated phagocytosis of *Lps*^{d} macrophages can be restored by treatment with either IFN-α, β, or γ (Vogel et al., 1983a,b). On the other hand, the normal receptor-mediated phagocytosis of *Lps*^{n} macrophages is decreased by the anti-IFN-α/β antibody (Vogel and Fertsch, 1984). This is in favor of the view that LPS responsive macrophages produce endogenous IFN that acts as an activator; LPS responsiveness seems to be the normal condition for macrophages in most inbred mouse strains. Whether this IFN is of the α or β type has not yet been determined, since the sera that have been used for characterization neutralize both types.

Since macrophages from mice with the defective allele at the *Lps* locus cannot be induced by LPS to make IFN-α/β, they have reduced Fc receptor-mediated phagocytic activity since they lack the stimulatory effect of IFN (Vogel and Fertsch, 1984). The view that endogenous IFN production by macrophages, probably triggered by small amounts of endotoxin, acts as an autostimulatory signal is an attractive one. The evidence that IFNs of all three species are macrophage-activating agents is indeed overwhelming, as we will see in section III of this chapter.

A locus corresponding to the *Lps* locus in mice has not been characterized in humans, and the role played by endotoxin in IFN induction in human macrophages in general has not been extensively investigated.

B. Macrophages as Accessory Cells for IFN-γ Production

IFN-γ production by T cells stimulated by bacterial and viral antigens or by mitogens in vitro is more efficient in the presence of macrophages (Epstein et al., 1971a,b, 1972; Valle et al., 1975). The antigen-presenting capacity of the macrophages may be involved in this effect, but the major reason is probably the release of T-cell-activating monokines by macrophages.

Indeed, the presence of monocytes is also required for optimal production of Hu IFN-γ by T cells induced with LPS or with IL-2. This is because monocytes are stimulated by LPS to produce IL-1, which acts synergistically with IL-2 to boost IFN-γ production by T cells (Blanchard et al., 1986; Le et al., 1986). The production of IL-1 itself is stimulated by IFN-γ, and to a lesser degree by IFN-α and β (Arenzana-Seisdedos et al., 1985). Apparently, the production of IL-1 and IFN-γ is interconnected by a positive feedback loop.

III. STIMULATION OF MACROPHAGE ACTIVITY BY INTERFERONS

Before they develop efficient phagocytic, bactericidal or tumoricidal activity, macrophages have to undergo a series of functional modifications, generally grouped under the term "activation." The importance of T-cell-derived lymphokines and of bacterial products for macrophage activation has been recognized for a long time (Cohn, 1978). It has now become clear that IFN α, IFN-β, and especially IFN-γ, play a role in macrophage activation.

Frequently, this activation results from effects on the expression and activity of several molecules at the macrophage surface.

Ingestion of foreign material by phagocytes is either nonspecific, meaning that it occurs without previous attachment of the ingested particle to a specific receptor, or it is receptor-mediated, requiring as a first step the attachment to a specific receptor. For phagocytosis, two kinds of cell-surface receptors are important. The first consists of Fc receptors, defined by their affinity for the constant regions of IgG antibodies. Any infectious agent, particulate antigen, or tumor cell coated with IgG antibody will bind to the Fc receptor and trigger phagocytosis. The second kind of receptors that are important for phagocytosis are the receptors for the C3 protein derivatives of the complement pathway. C3b and C3bi are cleavage products of complement component C3 that arise after the formation of a specific protease resulting from complement activation. They bind to antigen-antibody complexes, which then interact with the specific C3b and C3bi receptors on phagocytic cells.

A. Nonreceptor-Mediated Phagocytosis

Nonreceptor-mediated phagocytosis, also called nonspecific phagocytosis, is often measured by the uptake of carbon particles or latex beads. Murine IFN-α/β stimulates the uptake of carbon particles by resident peritoneal macrophages in vitro, and also when the IFN is administered systemically and the carbon particles injected into the peritoneal cavity (Huang et al., 1971; Donahoe and Huang, 1976). Treatment of human monocytes with Hu leukocyte IFN stimulates the uptake of latex beads: Both the number of phagocytic cells and the number of beads ingested per cell are enhanced (Imanishi et al., 1975). Of direct relevance to the role of macrophages in nonspecific immunity is the stimulation by Mu IFN-α/β of the uptake and killing of *E. Coli* and *Staphylococcus aureus* by peritoneal macrophages (Rollag and Degré, 1981; Imanishi et al., 1982).

B. Receptor-Mediated Phagocytosis

1. Fc Receptor-Mediated Phagocytosis

The binding of antibodies to effector cells via receptors to their constant regions (Fc receptors) is the first step of the pathway that leads to the uptake by mononuclear phagocytes of antibody-coated infectious agents or cells. The mouse macrophage has independent-binding sites for IgG2a, IgG2b, and IgG2b/IgG1 complexes. The Fc receptor proteins are members of the immunoglobulin supergene family, displaying homology with the major histocompatibility antigens (Ravetch et al., 1986).

Phagocytosis via Fc receptors is defective in macrophages from C3H/HeJ mice, hyporeactive to endotoxin as a result of the presence of the Lps^d allele. Treatment with Mu IFN-α/β enhances the expression of Fc receptors at the surface of these macrophages: This can be observed directly by electron microscopy of macrophages that are treated first with an IgG antibody that binds to the receptor, and subsequently with a ferritin-labeled secondary antibody that binds to the IgG (Vogel et al., 1983a). An increase in the number and density of Fc receptors could contribute to the stimulation of receptor-mediated phagocytosis by IFN-α/β, but does not seem to be a necessary condition. Indeed, a few hundred units of Mu IFN-α/β are sufficient to stimulate the uptake of IgG-coated sheep erthrocytes by resident murine peritoneal macrophages in culture, apparently without actually increasing the number of Fc receptors (Hamburg et al., 1980).

By controlling the levels of IFN-α/β induced by Newcastle disease virus in the mouse, the *If*-1 locus influences the efficacy of phagocytosis of IgG-coated sheep erythrocytes after IFN induction: Injection of NDV into mice of *If*-1^h and *If*-1^l genotype, enhances the phagocytic capacity of macrophages in high responders, whereas phagocytosis by macrophages of low responders is not boosted (see Chapter 15) (Manejias et al., 1978).

Much of the work on receptor-mediated phagocytosis with murine cells has been done with Mu IFN-α/β, which is a mixture of IFN-α and IFN-β. Essentially pure Mu IFN-α (but still a nondefined mixture of different IFN-α species), as well as essentially pure Mu-IFN β are capable of boosting the binding and phagocytosis of antibody-coated sheep erythrocytes by cells of a murine macrophagelike cell line, with very similar kinetics and dose dependence. Both IFN species induce a two- to fivefold increase in Fc receptors, as measured by the binding of specific antibodies. Enhancement of binding and phagocytosis of the antibody-coated erythrocytes is an active process that requires both RNA and protein synthesis, and the effect is abolished by the removal of IFN (Yoshie et al., 1982).

Although generally assumed to be the case, there is no formal proof that the increased uptake of antibody-coated particles or cells is a direct consequence of the increased number of Fc receptors. Quite the contrary, it has been shown that human monocytes, after treatment with Hu IFN-γ, display strongly enhanced affinity for antibody-coated erythrocytes, but with a decreased ability to phagocytose these erythrocytes. This suggests that the IFN-γ-induced high affinity Fc receptors may not efficiently promote phagocytosis (Wright et al., 1986). Since this is an effect of IFN-γ, one cannot directly extrapolate to what is observed with IFN-α and β, but it is clear that enhanced expression of Fc receptors and high affinity binding can be dissociated from enhanced phagocytosis.

2. Phagocytosis Mediated via the Receptor for the Third Complement Component

The receptor for the inactivated complement component C3bi corresponds to the cellular adhesion glycoprotein Mac-1 (Beller et al., 1983). Mac-1 is a surface glycoprotein expressed on monocytes, macrophages, granulocytes, and large granular lymphocytes. This molecule is involved in the adherence of monocytes and granulocytes, for example, to vascular endothelium, and in migration to inflammatory sites (Springer and Anderson, 1986). In humans, a disease has been recognized in which Mac-1 and LFA-1 (see subsequent discussion) are deficient on the surface of monocytes, granulocytes, and lymphocytes. The disease is inherited as an autosomal recessive mutation and is characterized by recurrent, life-threatening bacterial infections, a lack of pus formation and persistent granulocytosis (Springer and Anderson, 1986).

Mac-1 is composed of two subunits, α (Mr 170,000) and β (Mr 95,000), noncovalently associated in a dimer. It is structurally related to another leukocyte function—associated antigen, LFA-1, with which it shares the same β subunit. The α subunits of Mac-1 and LFA-1 have 33% amino acid homology in their aminoterminal region (Sastre et al., 1986) and, intriguingly, also share some homology with IFN-α in this region (Springer et al., 1985). The meaning of this homology is not understood; it could have a functional importance or point to a common evolutionary origin.

The expression of Mac-1 is under the direct influence of IFN-γ, which induces the transcription of Mac-1 α subunit mRNA when stimulating the murine premyelocytic cell line M1 into maturation (Sastre et al., 1986). In addition to stimulating its expression in undifferentiated cells, IFN-γ also influences the function of the C3 receptor in mature cells, by decreasing its affinity for the C3b protein.

In differentiated monocytes, the effect of IFN-γ on the Mac-1 receptor for the complement protein C3bi is the opposite of that on the Fc receptor: Affinity for ligand is reduced instead of enhanced. Receptors on human monocytes for C3b bind C3b-coated particles and generate signals that lead to the phagocytosis of these particles. In vitro, binding to the receptor does not automatically lead to engulfment and phagocytosis; the signal for phagocytosis has to be generated separately by a second signal, for example, after interaction of the monocyte with the tumor promoter phorbol myristate acetate (Wright and Silverstein, 1982) or with surfaces coated with fibronectin (Wright et al., 1983a,b). Fibronectins are large, asymmetric glycoproteins with multiple domains that can bind to various cellular components. These proteins are part of the extracellular matrix that provides a guiding substrate for cell movement, and thereby influence a wide variety of cellular properties such as adhesion, migration, differentation, and phagocytosis. Human monocytes as well as mouse peritoneal macrophages can synthesize and secrete fibronectin when cultured in vitro; the fibronectin is then organized at the macrophage surface in structures that mediate adhesion to the culture vessel. Human monocytes have cell-surface receptors for fibronectin (Bevilacqua et al., 1981). Under certain conditions, IFN-γ can stimulate fibronectin synthesis by macrophages, thereby increasing the phagocytic and tumoricidal potential of the cells (Cofano et al., 1984).

When human monocytes are exposed to Hu IFN-γ, a five- to tenfold decrease in the binding of C3b occurs after several hours, although the number of C3b receptors is not diminished. The capacity of IFN-γ-treated monocytes to bind C3b-coated erythrocytes is restored to normal levels if the cells are allowed to spread on surfaces coated with fibronectin. Apparently, under the influence of IFN-γ a reversible change in the nature of the C3b receptors occurs, which prevents them from interacting with ligand. Normal affinity is restored by spreading on fibronectin, but, in contrast to monocytes that are not exposed to IFN, in which spreading on fibronectin results in phagocytosis, the receptors remain incapable of generating signals for phagocytosis. For this to occur, IFN-γ-treated monocytes require exposure to both fibronectin and phorbol myristate acetate. Binding and signaling activities of the C3b receptor are manifestly regulated separately (Wright et al., 1986).

The effect of IFN-γ on the C3b receptor demonstrates that in some instances IFNs decrease receptor activity. A somewhat different example of this is provided by the down-regulation by Mu IFN-γ of the transferrin receptor on murine peritoneal macrophages. Transferrin, a plasma protein that binds iron and zinc, has been implicated in immuno-regulation. Treatment of murine macrophages with IFN-γ significantly decreases the number of active receptors without modifying their affinity (Hamilton et al., 1984). Withholding iron from invading microorganisms is part of the natural defenses (Weinberg, 1978), and down-regulation of the transferrin receptor on cells whose function is to be in contact with such organisms may be one way of doing so. IFNs clearly exert multiple effects on macrophage membrane composition.

C. Tumor Cell Killing

In contrast to cytotoxic T cells, but like natural killer cells, macrophages can destroy tumor cells without MHC restriction. Antibody-coated tumor cells bind to the Fc receptors on the macrophage surface, but how macrophages can recognize uncoated tumor cells is not known. To become activiated to kill tumor cells, macrophages need to traverse a series of steps (Ruco and Meltzer, 1978). The first of these primes macrophages to respond to a heterogeneous group of second signals, such as LPS or other bacterial products that trigger the actual cytolytic process (Meltzer, 1981).

1. Activation by IFN-γ

For many years considerable attention has been focused on lymphokines, released by activated T cells, that are capable of priming macrophages for nonspecific tumoricidal activity. This activity, appropriately called macrophage-activating factor or MAF, can be ascribed at least partly, if not entirely, to IFN-γ, for the following reasons. The macrophage priming activity ascribed to MAF copurifies with IFN-γ, is inactivated by highly specific antibodies to IFN-γ, and can be reproduced by recombinant Mu IFN-γ (Fig. 10.2) (Roberts and Vasil, 1982; Le et al., 1983; Nathan et al., 1983; Pace et

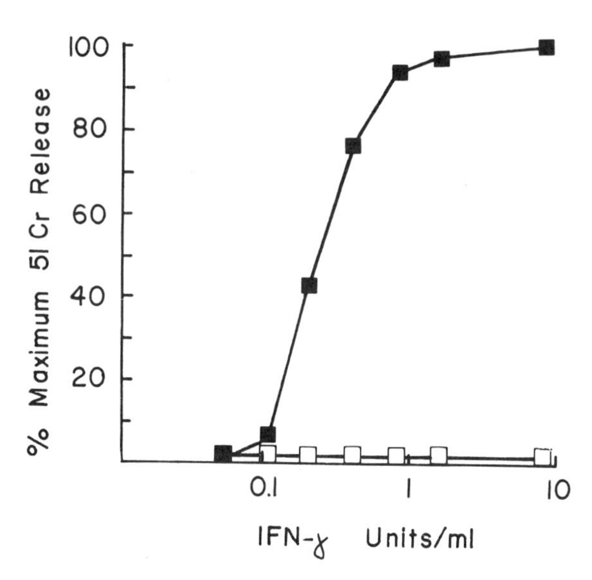

Figure 10.2 IFN-γ can prime macrophages to respond to a second signal that triggers tumor cell killing. The figure shows the priming activity of rec Mu IFN-γ on murine macrophages as a function of the treatment dose. The cytotoxic activity of mouse peritoneal macrophages, incubated with IFN-γ in the presence (■) or absence (□) of LPS, is represented by the release of radiolabeled chromium from the target cells, which in this particular case are mastocytoma cells. (Adapted from Pace et al., 1983.)

al., 1983; Rothermel et al., 1983; Schreiber et al., 1983; Schultz and Chirigos, 1983; Schultz and Kleinschmidt, 1983; Svedersky et al., 1984). Activation of cell killing by macrophages could be an important function for IFN-γ, since it is believed—but by no means proven—that tumor cell lysis by activated macrophages is part of natural host resistence to cancer. Although natural IFN-γ is glycosylated, unglycosylated recombinant Mu IFN-γ is just as efficient an activator of mouse macrophages; apparently the sugar moiety is not important for this function (Varesio et al., 1984).

Receptor engagement is a necessary first step for macrophage priming activity of IFN-γ. The number of receptors for IFN-γ on murine macrophages has been estimated to lie somewhere between 5,000 and 12,000 per cell, and the possibility has been raised that there are high and low affinity receptors (Finbloom et al., 1985; Aiyer et al., 1986; Celada et al., 1986). Furthermore there are indications that the receptor for Hu IFN-γ on peripheral blood monocytes, by its more stringent binding characteristics, differs from the IFN-γ receptor on fibroblasts (Orchansky et al., 1986). This is discussed in more detail in Chapter 4.

Although IFN-γ is undoubtedly a major macrophage priming agent, it shares this function with other lymphokines that also prime or activate macrophages (Andrew et al., 1984; Kleinerman et al., 1984; Grabstein et al., 1986). After priming by IFN-γ, a second signal is required to trigger the cytolytic process (Meltzer, 1981). In vitro, this second signal can be provided by LPS (Kleinerman et al., 1984) but other lymphokines, still to be characterized, are probably also involved in this process (Krammer et al., 1985).

IFN-γ and IFN-α or IFN-β have synergistic antiviral, anticellular and NK-cell stimulating activities, and a similar synergistic tumoricidal effect is obtained when macrophages are treated with IFN-α or IFN-β in addition to IFN-γ. However, even when the action of IFN-γ is boosted this way by IFN-α or IFN-β, LPS is still required as a second signal to trigger cytolysis (Pace et al., 1985). The possibility has been raised that IFN-γ acts via the Ca^{2+} dependent induction of protein kinase C activity, which is then further augmented by a diacylglycerol-like action of LPS in the cell membrane. Diacylglycerol is one of the intracellular messengers initiating the flow of information in the protein kinase C branch (Hamilton et al., 1985). Indeed, the induction of cytocidal activity by natural and rec Mu IFN-γ is blocked either by nonspecific inhibitors of protein kinase C, such as phenothiazines, or a Ca^{2+} chelator, provided the drugs are added together with IFN. Reciprocally, IFN-like priming of macrophages for tumoricidal activity can be obtained in the absence of IFN-γ with a combination of Ca^{2+} ionophores and protein kinase C activators, such as phorbol esters. Unlike IFN-γ, however, which at appropriate concentrations can trigger a full cytocidal response, these agents induce only priming, and a second signal, for instance LPS, is needed to trigger the cytotoxic reaction (Celada and Schreiber, 1986).

Activated monocytes or macrophages kill tumor cells by releasing reactive oxygen intermediates (see section E) and also by producing several cytotoxic molecules, one of which is TNF-α (Urban et al., 1986). IFN-γ induces the formation and release of TNF by macrophages, but IFN-α, although it also activates the cytolytic effect of macrophages, does not induce TNF formation (Philip and Epstein, 1986). However, IFN-α and IFN-β, like IFN-γ, enhance the expression of receptors for TNF on target cells, although IFN-α and IFN-β do so with lower efficacy than IFN-γ (Tsujimoto et al., 1986). The existence of different receptors and activation pathways for these IFNs explains why they can act synergistically when stimulating the tumoricidal effect of macrophages.

2. *Activation by IFN-α and IFN-β*

Most of the attention has been focused on macrophage activation by IFN-γ, mainly because as a lymphokine it appears as a natural activator of monocytes. However, IFN-α and IFN-β are also capable of boosting the tumoricidal activity of macrophages (Fig. 10.3). Phagocytosis of tumor cells by macrophages in the peritoneal cavity of mice is enhanced by treatment with IFN-α/β (Gresser and Bourali, 1970) and in vitro treatment of mouse peritoneal macrophages with Mu IFN-α or Mu IFN-β induces cytolytic activity against malignant target cells (Blasi et al., 1984). Furthermore, treatment of monocytes from human peripheral blood with Hu leukocyte IFN or Hu IFN-β, augments their cytolytic activity against malignant cells (Dean and Virelizier, 1983).

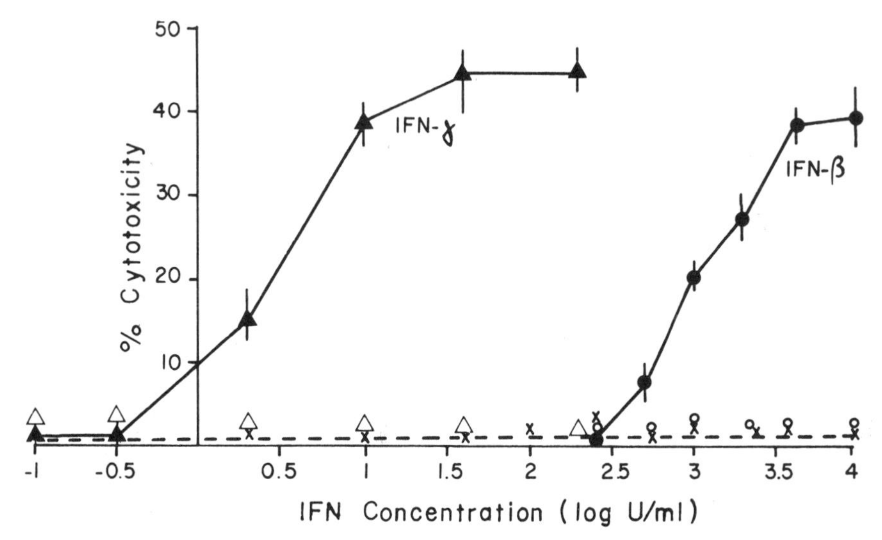

Figure 10.3 Both Mu IFN-γ and Mu IFN-β are capable of boosting the cytotoxic activity of mouse peritoneal macrophages against lymphoma cells. (Adapted from Varesio et al., 1984.)

D. Destruction of Parasites by IFN-Activated Macrophages

1. IFN-γ Stimulates Intracellular Parasite Killing in Leishmaniasis

Some parasites, like certain other infectious agents, instead of being killed once they are taken up by macrophages, survive and multiply inside the cells. This is the case for organisms of the Leishmania genus, obligate intracellular parasites, that live and multiply in cells of the mononuclear phagocyte system. Human Leishmaniasis is caused by at least 14 different species and subspecies of the genus Leishmania. Depending on the species, the clinical picture of the infection varies, as do the immune responses. The disease can range from localized cutaneous lesions to disseminated systemic infection. Systemic visceral infection results in the depression of cellular immunity but good antibody response, whereas cutaneous Leishmaniasis induces good cellular immunity and poor antibody response (Bray, 1974). Infections with *Leishmania major* and other cutaneous strains frequently result in a spontaneous cure with destruction of the intracellular parasite, whereas visceral disease, caused by *Leishmania donovani* and *Leishmania chagasi,* often proves fatal.

Since Leishmania disease results from intracellular infection of macrophages, it is important to understand the mechanisms that determine intracellular parasite killing or survival. Intracellular parasite killing is significantly activated when infected macrophages are exposed to immune lymphocytes or their lymphokines, and IFN-γ has revealed itself as one of the active components. Hu IFN-γ stimulates intracellular parasite killing in suspensions of human monocytes infected with *Leishmania donovani* but not in monocytes infected wtih *Leishmania major*. The effect can be either preventive or curative, since parasite killing is also enhanced when already infected monocytes are subsequently treated with IFN-γ. Hu IFN-α or Hu IFN-β has no effect in suspension cultures (Hoover et al., 1985). However, when monocytes are selected by adherence to the plastic of culture vessels, quite different results can be obtained. Under these conditions, both Hu IFN-γ and Hu IFN-β enhance parasite killing by the monocytes infected with *Leishmania major*: This effect can be preventive or curative (Passwell et al., 1986).

There is no contradiction between IFN effects that are observed when monocytes are selected and cultured under different conditions, since, as we have repeatedly stressed, monocytes comprise a heterogeneous population. Moreover, since the effects of IFN are influenced by the genotype of the cells (see Chapter 15), monocytes obtained from different individuals can react differently to IFN treatment.

Murine peritoneal macrophages infected with intracellular amastigotes of *Leishmania major* kill this parasite after treatment with IFN-γ (Fig. 10.4); this is no exclusive activity of IFN-γ, and other, not yet well-defined lymphokines derived from antigen- or mitogen-stimulated spleen cells can also induce killing (Nacy et al., 1985). Macrophage activation for parasite killing, like activation for other functions such as tumor cell lysis, can be initiated by

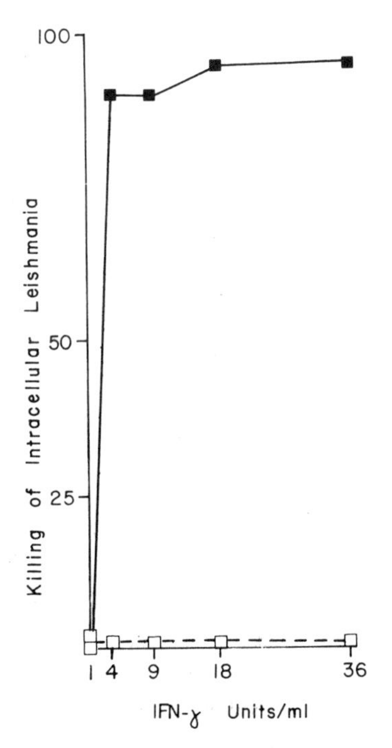

Figure 10.4 Addition of Mu IFN-γ to murine peritoneal macrophages infected with *Leishmania major* induces intracellular destruction of the parasites. IFN-γ: ■; Control: □. (Adapted from Nacy et al., 1985.)

several signals, of which IFN-γ is only one. We should not be surprised by the possibility of redundancy in the signals that regulate macrophage activation, since redundancy seems to be a common precaution of nature when immune defenses are involved. In addition, since macrophages do not constitute an homogeneous population, one can expect the existence of different activation signals (Nacy et al., 1985). Activation of macrophage oxidative metabolism is one of the major mechanisms that are responsible for enhanced killing by macrophages (see section E) and the IFN-γ-enhanced potential for intracellular killing in macrophages is accompanied by an increased capacity to release oxygen intermediates (Murray et al., 1985).

Several mouse models of Leishmania infection have been developed to study immunity to this protozoon. Infection of mice with *Leishmania major*, which causes the cutaneous form in humans, most closely resembles human visceral Leishmaniasis. Some mouse strains, like C57BL/6, are naturally resistant to this parasite and recover after infection, whereas other strains, like BALB/c, are susceptible and develop a lethal infection. When C57BL/6 and BALB/c mice are infected by footpad inoculation, local footpad lesions develop in both strains. After a few weeks the lesions regress and heal in C57BL/6 mice, but they continue to enlarge and become ulcerated in BALB/c mice. The ability of the resistant mice to recover from the infection correlates with the capacity of lymphocytes, taken from the local lymphnodes draining the lesions, to produce IFN-γ in response to Leishmania antigens. This capacity is lacking in lymphocytes from the susceptible BALB/c mice; these lymphocytes do, however, make IFN-γ upon other stimuli. The mechanism by which this Leishmania-antigen-specific unresponsiveness is established has not been elucidated; there are some indications that it could be due to antigen-specific T-suppressor cells (Howard et al., 1981).

How suppression is mediated by these T cells is not known, but at some stage it results in inhibition of antigen-specific IFN-γ induction. It is, however, possible to abrogate the specific suppression by irradiating (550 rads) susceptible BALB/c mice just before infection with Leishmania. Mice thus treated develop cellular responsiveness to Leishmania antigens in vitro, including the capacity to produce IFN-γ, and recover from their infection (Sadick et al., 1986). Active suppression of IFN-γ production on an antigen specific basis appears as part of the mechanism by which parasites can paralyze cellular immunity. An intriguing aspect of this particular mouse model is the fact that the susceptible mice belong to the BALB/c strain and the resistant mice to the C57BL/6 strains. The same pattern of susceptibility and resistance is encountered during several viral infections.

2. IFN-γ Restores the Capacity of Human Monocytes to Kill Malaria Parasites

Human malaria is caused by four major *Plasmodium* species, but in most parts of the world, the prevailing species is *Plasmodium falciparum*. The intraerythrocytic protozoon *Plasmodium falciparum* is lethally susceptible to killing by oxygen-dependent and oxygen-independent factors released by fresh human monocytes. Monocytes kept in cultures as adherent macrophages release only a fraction of the amount of hydrogen peroxide as compared to fresh monocytes, and as a result lose their killing potential. Treatment of such cells with IFN-γ restores the potential to secrete oxygen metabolites, and as a result, kill intraerythrocytic parasites (Ockenhouse et al., 1984).

3. *Other Parasites*

Mu IFN-γ can activate a macrophagelike cell line to destroy intracellular trypanosomes (*Trypanosoma cruzi*) and stimulates mouse peritoneal macrophages to kill the obligate intracellular protozoon *Toxoplasma gondii,* which normally replicates in mononuclear phagocytes. Administration of Mu IFN-γ to BALB/c mice stimulates their alveolar and peritoneal macrophages to inhibit or kill intracellular *Toxoplasma gondii.* The latter is not a specific effect on macrophages, since Hu IFN-γ also inhibits replication of *Toxoplasma gondii* in cultured fibroblasts. This is due to the induction by IFN-γ of indoleamine 2,3 dioxygenase, a tryptophan-degrading enzyme (see Chapter 5) (Pfefferkorn and Guyre, 1984; Plata et al., 1984; Sethi et al., 1985; Black et al., 1987).

E. Stimulation of the Secretion of Reactive Oxygen Intermediates: A Special Function of IFN-γ

Activated macrophages secrete chemically reactive, incompletely reduced metabolites of oxygen, including hydrogen peroxide, when confronted with appropriate triggering agents. Hydrogen peroxide is believed to be one of the secretory products of activated macrophages that can lyse tumor cells, and furthermore, it may cooperate with other macrophage-derived lytic effectors, such as cytolytic serine protease, in producing tumor cell injury (Adams and Nathan, 1983).

Production of reactive oxygen intermediates and secretion of hydrogen peroxide is closely correlated with the capacity of macrophages to kill intracellular parasites, including those that infect macrophages themselves (Murray 1981). A short exposure of a few hours to recombinant Hu IFN-γ leads to substantial activation of the hydrogen-peroxide-releasing capacity of macrophages derived from human peripheral blood. The activation of hydrogen peroxide can last several days, and is correlated with a stimulation of intracellular killing of *Toxoplasma gondii.* Mu IFN-γ has the same effect in mouse peritoneal macrophages. Stimulation of the secretion of reactive oxygen intermediates appears to be a special funtion of IFN-γ, but not of IFN-α, IFN-β, CSF-1, TNF, or IL-2 (Fig. 10.5) (Nathan et al., 1983; Murray et al., 1985; Nathan and Tsunawaki, 1986). Moreover, in human monocyte-derived macrophage cultures, rec Hu IFN-α2A and rec Hu IFN-β antagonize the hydrogen-peroxide-stimulating activity of Rec IFN-γ (Garotta et al., 1986).

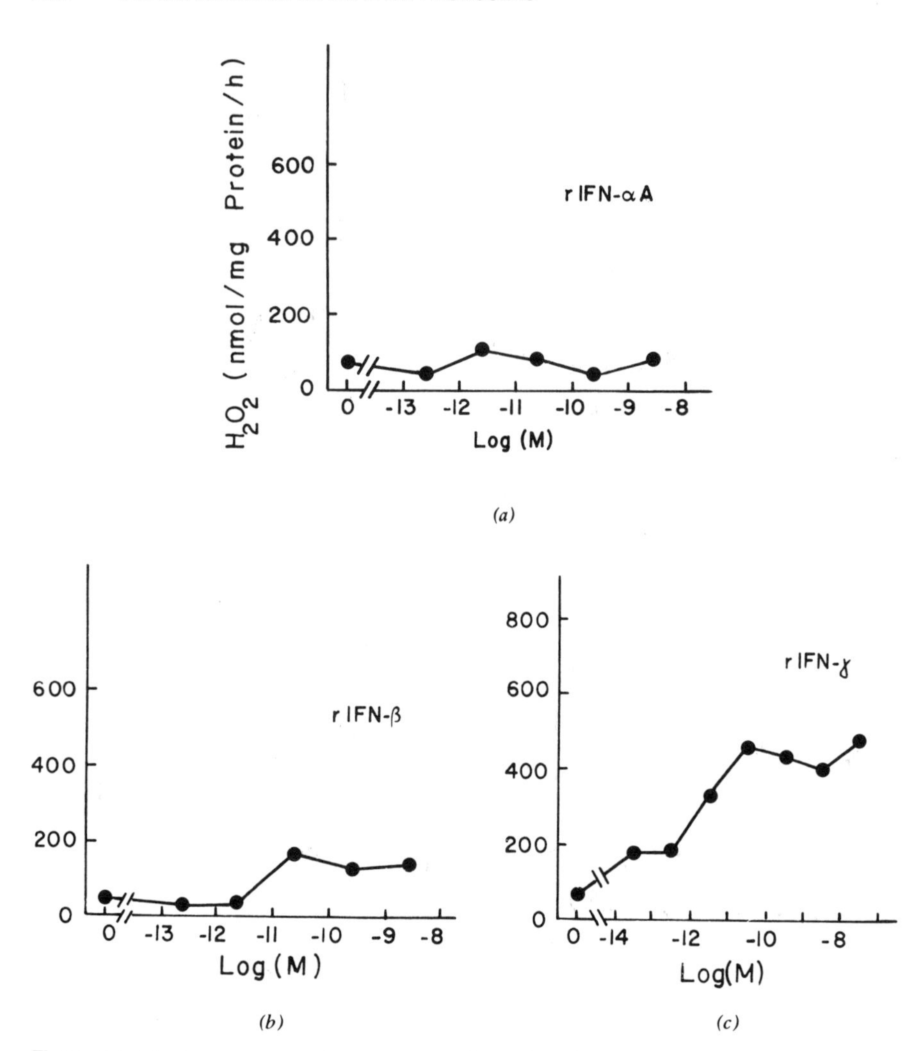

Figure 10.5 Hydrogen peroxide release by phorbol myristate acetate-challenged human mononuclear phagocytes after incubation for 3 days in rec Hu IFN-α A (panel *a*), rec Hu IFN-β (panel *b*) or rec Hu IFN-γ (panel *c*). Hydrogen peroxide release is only significantly boosted in the IFN-γ-treated cells. (Adapted from Nathan and Tsunawaki, 1986.)

Although antagonism between the different IFNs is very exceptional, this is not the only occurence, since Mu IFN-$\alpha\beta$ inhibits the Mu IFN-γ-induced enhanced expression of MHC class II antigens on peritoneal macrophages, and also inhibits the IFN-γ-induced down-regulation of mannosyl-fucosyl receptors on activated macrophages. The latter are lectinlike receptors that mediate endocytosis of glycoproteins or glycoconjugates with terminal mannose or fucose residues (Ling et al., 1985; Ezekowitz et al., 1986; and also see Chapter 9).

In humans, intravenous injection of Hu IFN-γ results in an augmentation of the hydrogen-peroxide-releasing capacity by monocytes that lasts for several days after the infusion of IFN. The actual metabolic changes induced by IFN-γ that give rise to the enhanced release of hydrogen peroxide have not been elucidated (Nathan et al., 1985).

REFERENCES

Acton, J. D. The lymphoreticular system and interferon production. *J. Reticuloendoth. Soc.* 14: 449–461 (1973).

Adams, D. O. and Nathan, C. F. Molecular mechanisms in tumor-cell killing by activated macrophages. *Immunol. Today* 4: 166–170 (1983).

Aiyer, R. A., Serrano, L. E., and Jones, P. P. Interferon-γ binds to high and low affinity receptor components on murine macrophages. *J. Immunol.* 136: 3329–3334 (1986).

Andrew, P. W., Rees, A. D. M., Scoging, A., Dobson, N., Matthews, R., Whittall, J. T., Coates, A. R. M., and Lowrie, D. B. Secretion of a macrophage-activating factor distinct from interferon-γ by human T cell clones. *Eur. J. Immunol.* 14: 962–964 (1984).

Arenzana-Seisdedos, F., Virelizier, J. L., and Fiers, W. Interferons as macrophage-activation factors. III. Preferential effects of interferon-γ on the interleukin 1 secretory potential of fresh or aged human monocytes. *J. Immunol.* 134: 2444–2448 (1985).

Ascher, O., Apte, R. N., and Pluznik, D. H. Generation of lipopolysaccharide-induced interferon in spleen cell cultures. I. Genetic analysis and cellular requirements. *Immunogenetics* 12: 117–127 (1981).

Belardelli, F., Vignaux, F., Proietti, E., and Gresser, I. Injection of mice with antibody to interferon renders peritoneal macrophages permissive for vesicular stomatitis virus and encephalomyocarditis virus. *Proc. Natl. Acad. Sci.* (USA) 81: 602–608 (1984).

Beller, D. I., Springer, T. A., and Schreiber, R. D. Anti-Mac-1 selectively inhibits the mouse and human type three complement receptor. *J. Exp. Med.* 156: 1000–1009 (1983).

Bevilacqua, M. P., Amrani, D., Mosesson, M. W., and Bianco, C. Receptors for cold-insoluble globulin (plasma fibronectin) on human monocytes. *J. Exp. Med.* 153: 42–60 (1981).

Black, C. M., Catterall, J. R., and Remington, J. S. In vivo and in vitro activation of alveolar macrophages by recombinant interferon-γ. *J. Immunol.* 138: 491–495 (1987).

Blanchard, D. K., Djeu, J. Y., Klein, T. W., Friedman, H., and Stewart II, W. E. Interferon-γ induction by lipopolysaccharide: Dependence on interleukin 2 and macrophages. *J. Immunol.* 136: 963–970 (1986).

Blasi, E., Herberman, R. B., and Varesio, L. Requirement for protein synthesis for induction of macrophage tumoricidal activity by IFN-α and IFN-β but not by IFN-γ. *J. Immunol.* 132: 3226–3228 (1984).

Bray, R. S. Leishmania. *Ann. Rev. Microbiol.* 28: 189–217 (1974).

Brehm, G. and Kirchner, H. Analysis of the interferons induced in mice in vivo and in macrophages in vitro by Newcastle disease virus and by polyinosinic-polycytidylic acid. *J. Interferon Res.* 6: 21–28 (1986).

Celada, A. and Schreiber, R. D. Role of protein kinase C and intracellular calcium mobilization in the induction of macrophage tumoricidal activity by interferon-γ. *J. Immunol.* 137: 2373–2379 (1986).

Cofano, F., Comoglio, P. M., Landolfo, S., and Tarone, G. Mouse immune interferon enhances fibronectin production of elicited macrophages. *J. Immunol.* 133: 3102–3106 (1984).

Cohn, Z. A. The activation of mononuclear phagocytes: Fact, fancy, and future. *J. Immunol.* 121: 813–816 (1978).

Dean, R. T. and Virelizier, J. L. Interferon as a macrophage activating factor. I. Enhancement of cytoxicity by fresh and matured human monocytes in the absence of other soluble signals. *Clin. Exp. Immunol.* 51: 501–510 (1983).

De Maeyer, E. and De Maeyer-Guignard, J. Interferon structural and regulatory genes in the mouse. In: *Interferons as Cell Growth Inhibitors and Antitumor Factors,* UCLA Symposia on Molecular and Cellular Biology New Series, R. M. Friedman, T. Merigan, and T. Sreevalsan, Eds., Alan R. Liss, New York, Vol. 50, pp. 435–445 (1986).

De Maeyer, E., Fauve, R. M., and De Maeyer-Guignard, J. Production d'interféron au niveau du macrophage. *Ann. Inst. Pasteur* 120: 438–446 (1971).

De Maeyer, E., Hoyez, M. C., De Maeyer-Guignard, J., and Bailey, D. W. Effect of mouse genotype on interferon production. III. Expression of *If*-1 by peritoneal macrophages in vitro. *Immunogenetics* 8: 257–263 (1979).

De Maeyer-Guignard, J., Dandoy, F., Bailey, D. W., and De Maeyer, E. Interferon structural genes do not participate in quantitative regulation of interferon production by *If* loci as shown in C57BL/6 mice that are congenic with BALB/c mice at the α interferon gene cluster. *J. Virol.* 58: 743–747 (1986).

Donahoe, R. M. and Huang, K. Y. Interferon preparations enhance phagocytosis in vivo. *Infect. Immun.* 13: 1250–1257 (1976).

Epstein, L. B., Cline, M. J., and Merigan, T. C. PPD-stimulated interferon: In vitro macrophage-lymphocyte interaction in the production of a cellular immunity. *Cell. Immunol.* 2: 602–613 (1971a).

Epstein, L. B., Cline, M. J., and Merigan, T. C. The interaction of human macrophages and lymphocytes in the phytohemagglutinin-stimulated production of interferon. *J. Clin. Invest.* 50: 744–753 (1971b).

Epstein, L. B., Stevens, D. A., and Merigan, T. C. Selective increase in lymphocyte interferon response to vaccinia antigen after revaccination. *Proc. Natl. Acad. Sci.* (USA) 69: 2632–2636 (1972).

Ezekowitz, R. A. B., Hill, M., and Gordon, S. Interferon α/β selectively antagonizes down-regulation of mannosyl-fucosyl receptors on activated macrophages by interferon γ. *Biochem. Biophys. Res. Commun.* 136: 737–744 (1986).

Finbloom, D. S., Hoover, D. L., and Wahl, L. M. The characteristics of binding of human recombinant interferon-γ to its receptor on human monocytes and human monocyte-like cell lines. *J. Immunol.* 135: 300–305 (1985).

Fleit, H. B. and Rabinovitch, M. Production of interferon by in vitro derived bone marrow macrophages. *Cell. Immunol.* 57: 495–504 (1981a).

Fleit, H. B. and Rabinovitch, M. Interferon in marrow-derived macrophages: Regulation by L cell conditioned medium. *J. Cell. Physiol.* 108: 347–352 (1981b).

Garotta, G., Talmadge, K. W., Pink, J. R. L., Dewald, B., and Aggiolini, M. Functional antagonism between type I and type II interferons on human macrophages. *Biochem. Biophys. Res. Commun.* 140: 948–954 (1986).

Grabstein, K. H., Urdal, D. L., Tushinski, R. J., Mochizuki, D. Y., Price, V. L., Cantrell, M. A., Gillis, S., and Conlon, P. J. Induction of macrophage tumoricidal activity by granulocyte-macrophage colony-stimulating factor. *Science* 232: 506–508 (1986).

Gresser, I. and Bourali, C. Antitumor effects of interferon preparations in mice. *J. Natl. Cancer Inst.* 45: 365–376 (1970).

Hamburg, S. I., Fleit, H. B., Unkeless, J. C., and Rabinovitch, M. Mononuclear phagocytes: Responders to and producers of interferon. *Ann. N.Y. Acad. Sci.* 350: 72–90 (1980).

Hamilton, T. A., Becton, D. L., Somers, S. D., Gray, P. W., and Adams, D. O. Interferon-γ modulates protein kinase C activity in murine peritoneal macrophages. *J. Biol. Chem.* 260: 1378–1381 (1985).

Hamilton, T. A., Gray, P. W., and Adams, D. O. Expression of the transferrin receptor on murine peritoneal macrophages is modulated by in vitro treatment with interferon-γ. *Cell. Immunol.* 89: 478–488 (1984).

Havell, E. A. and Spitalny, G. L. Endotoxin-induced interferon synthesis in macrophage cultures. *J. Reticuloendothel. Soc.* 33: 369–380 (1983).

Hearst, J. E., Warr, G. A., and Jakab, G. J. Characterization of murine lung and peritoneal macrophages. *J. Reticuloendothel. Soc.* 27: 443–454 (1980).

Ho, N. K. and Springer, T. Mac-2: A novel 32,000 Mr macrophage sub-population specific antigen defined by monoclonal antibodies. *J. Immunol.* 128: 1221–1225 (1982).

Hockmeyer, W. T., Walters, D., Gore, R. W., Williams, J. S., Fortier, A. H., and Nacy, C. A. Intracellular destruction of Leishmania donovani and Leishmania tropica amastigotes by activated macrophages: Dissociation of these microbicidal effector activities in vitro. *J. Immunol.* 132: 3120–3125 (1984).

Hoover, D. L., Nacy, C. A., and Meltzer, M. S. Human monocyte activation for cytotoxicity against intracellular Leishmania donovani amastigotes: Induction of microbicidal activity by interferon-γ. *Cell. Immunol.* 94: 500–511 (1985).

Howard, J. G., Hale, C., and Liew, F. Y. Immunological regulation of experimental cutaneous Leishmaniasis. IV. Prophylactic effect of sublethal irradiation as a result of abrogation of suppressor T cell generation in mice genetically susceptible to Leishmania tropica. *J. Exp. Med.* 153: 557–568 (1981).

Huang, K. Y., Donahoe, R. M., Gordon, F. B., and Dressler, H. R. Enhancement of phagocytosis by interferon-containing preparations. *Infect. Immun.* 4: 581–588 (1971).

Imanishi, J., Oishi, K., Suganuma, A., Yokota, Y., and Kishida, T. Enhancement of bactericidal activity of mouse peritoneal macrophages against *Staphylococcus aureus* by mouse interferon preparations. *Biken J.* 25: 71–77 (1982).

Imanishi, J., Yokota, Y., Kishida, T., Mukainaka, T., and Matsuo, A. Phagocytosis enhancing effect of human leukocyte interferon preparation on human peripheral monocytes in vitro. *Acta Virol.* 19: 52–58 (1975).

Jullien, P., De Maeyer-Guignard, J., and De Maeyer, E. Interferon synthesis in X-irradiated animals. V. Origin of mouse serum interferon induced by polyinosinic-polycytidylic acid and encephalomyocarditis virus. *Infect. Immun.* 10: 1023–1028 (1974).

Kleinerman, E. S., Zicht, R., Sarin, P. S., Gallo, R. C., and Fidler, I. J. Constitutive production and release of a lymphokine with macrophage-activating factor activity distinct from γ-interferon by a human T-cell leukemia virus-positive cell line. *Cancer Res.* 44: 4470–4475 (1984).

Kono, Y. Interferon-like inhibitor produced in bovine leukocyte cultures after inoculation with endotoxin. *Arch. Ges. Virusforsch.* 21: 276–281 (1967).

Krammer, P. H., Kubelka, C. F., Falk, W., and Ruppel, A. Priming and triggering of tumoricidal and schistosomulicidal macrophages by two sequential lymphokine signals: Interferon-γ and macrophage cytotoxicity inducing factor 2. *J. Immunol.* 135: 3258–3263 (1985).

Le, J., Lin, J. X., Henriksen-Destefano, D., and Vilcek, J. Bacterial lipopolysaccharide-induced interferon-γ production: Roles of interleukin 1 and interleukin 2. *J. Immunol.* 136: 4525–4530 (1986).

Le, J., Prensky, W., Yip, Y. K., Chang, Z., Hoffman, T., Stevenson, H. C., Balazs, I., Sadlick, J. R., and Vilcek, J. Activation of human monocyte cytotoxicity by natural and recombinant immune interferon. *J. Immunol.* 131: 2821–2826 (1983).

Ling, P. D., Warren, M. K., and Vogel, S. N. Antagonistic effect of interferon-β on the interferon-γ-induced expression of Ia antigen in murine macrophages. *J. Immunol.* 135: 1857–1863 (1985).

Maehara, N. and Ho, M. Cellular origin of interferon induced by bacterial lipopolysaccharide. *Infect. Immun.* 15: 78–83 (1977).

Manejias, R. E., Hamburg, S. I., and Rabinovitch, M. Serum interferon and phagocytic activity of macrophages in recombinant inbred mice inoculated with Newcastle disease virus. *Cell. Immunol.* 38: 209–213 (1978).

Martinez, D., Lynch, R. J., Meeker, J. B., and Field, A. K. Macrophage dependence of polyriboinosinic acid-polyribocytidylic acid-induced resistance to herpes simplex virus infection in mice. *Infect. Immun.* 28: 147–153 (1980).

Matisova, E., Butorova, E., Lackovic, V., and Borecky, L. Some properties of interferon released by mouse peritoneal leukocytes stimulated with endotoxin, mannan and Newcastle disease virus in vitro. *Acta Virol.* 14: 1–7 (1970).

Meltzer, M. S. Macrophage activation for tumor cytotoxicity: Characterization of priming and trigger signals during lymphokine activation. *J. Immunol.* 127: 179–183 (1981).

Murray, H. W. Susceptibility of Leishmania to oxygen intermediates and killing by normal macrophages. *J. Exp. Med.* 153: 1302–1315 (1981).

Murray, H. W., Spitalny, G. L., and Nathan, C. F. Activation of mouse peritoneal macrophages in vitro and in vivo by interferon-γ. *J. Immunol.* 134: 1619–1622 (1985).

Nacy, C. A., Fortier, A. H., Meltzer, M. S., Buchmeier, N. A., and Schreiber, R. D.

Macrophage activation to kill Leishmania major: Activation of macrophages for intracellular destruction of amastigotes can be induced by both recombinant interferon-γ and non-interferon lymphokines. *J. Immunol.* 135: 3505–3511 (1985).

Nathan, C. F., Horowitz, C. R., De La Harpe, J., Vadhan-Raj, S., Sherwin, S. A., Oettgen, H. F., and Krown, S. E. Administration of recombinant interferon γ to cancer patients enhances monocyte secretion of hydrogen peroxide. *Proc. Natl. Acad. Sci.* (USA) 82: 8686–8690 (1985).

Nathan, C. F., Murray, H. W., Wiebe, M. E., and Rubin, B. Y. Identification of interferon-γ as the lymphokine that activates human macrophage oxidative metabolism and antimicrobial activity. *J. Exp. Med.* 158: 670–689 (1983).

Nathan, C. F. and Tsunawaki, S. Secretion of toxin oxygen products by macrophages: Regulatory cytokines and their effects on the oxidase. In: *Biochemistry of Macrophages,* D. Evered, J. Nugent, and M. O'Connor, Eds., Pitman, London. Ciba Foundation Symposium 118, pp. 211–230 (1986).

Neumann, C. Mononuclear phagocytes as producers of interferon. In: *Lymphokines,* E. Pick, Ed., Vol. 7, Academic Press, New York, pp. 165–201 (1982).

Neumann, C. and Sorg, C. Sequential expression of functions during macrophage differentiation in murine bone marrow liquid cultures. *Eur. J. Immunol.* 10: 834–840 (1980).

Neumann, C. and Sorg, C. Heterogeneity of murine macrophages in response to interferon inducers. *Immunobiol.* 158: 320–329 (1981).

Ockenhouse, C. F., Schulman, S., and Shear, H. L. Induction of crisis forms in the human malaria parasite plasmodium falciparum by γ-interferon-activated, monocyte-derived macrophages. *J. Immunol.* 133: 1601–1608 (1984).

Orchansky, P., Rubinstein, M., and Fischer, D. G. The interferon-γ receptor in human monocytes is different from the one in nonhematopoietic cells. *J. Immunol.* 136: 169–173 (1986).

Pace, J. L., Russell, S. W., Leblanc, P. A., and Murasko, D. M. Comparative effects of various classes of mouse interferons on macrophage activation for tumor cell killing. *J. Immunol.* 134: 977–981 (1985).

Pace, J. L., Russell, S. W., Torres, B. A., Johnson, J. M., and Gray, P. W. Recombinant mouse γ interferon induces the priming step in macrophage activation for tumor cell killing. *J. Immunol.* 130: 2011–2013 (1983).

Passwell, J. H., Shor, R., and Shoham, J. The enhancing effect of interferon-β and -γ on the killing of Leishmania tropica major in human mononuclear phagocytes in vitro. *J. Immunol.* 136: 3062–3066 (1986).

Pfefferkorn, E. R. and Guyre, P. M. Inhibition of growth of toxoplasma gondii in cultured fibroblasts by human recombinant γ interferon. *Infect. Immun.* 44: 211–216 (1984).

Philip, R. and Epstein, L. B. Tumour necrosis factor as immunomodulator and mediator of monocyte cytotoxicity induced by itself, γ-interferon and interleukin-1. *Nature* 33: 86–89 (1986).

Plata, F., Wietzerbin, J., Garcia Pons, F., Falcoff, E., and Eisen, H. Synergistic protection by specific antibodies and interferon against infection by *Trypanosoma cruzi* in vitro. *Eur. J. Immunol.* 14: 930–935 (1984).

Ravetch, J. V., Luster, A. D., Weinshank, R., Kochan, J., Pavlovec, A., Portnoy,

D. A., Hulmes, J., Pan, Y. C. E., and Unkeless, J. C. Structural heterogeneity and functional domains of murine immunoglobulin G Fc receptors. *Science* 234: 718–725 (1986).

Roberts, N. J., Jr., Douglas, R. G., Jr., Simons, R. M., and Diamond, M. E. Virus-induced interferon production by human macrophages. *J. Immunol.* 123: 365–369 (1979).

Roberts, W. K. and Vasil, A. Evidence for the identity of murine γ interferon and macrophage activating factor. *J. Interferon Res.* 2: 519–532 (1982).

Rollag, H. and Degré, M. Effect of interferon preparations on the uptake on non-opsonized *E. coli* by mouse peritoneal macrophages. *Acta Path. Microbiol. Scand.* Sect. B 89: 153–159 (1981).

Rothermel, C. D., Rubin, B. Y., and Murray, H. W. γ-interferon is the factor in lymphokine that activates human macrophages to inhibit intracellular chlamydia psittaci replication. *J. Immunol.* 131: 2542–2544 (1983).

Ruco, L. P. and Meltzer, M. S. Macrophage activation for tumor cytotoxicity: Development of macrophage cytotoxic activity requires completion of a sequence of short-lived intermediary reactions. *J. Immunol.* 121: 2035–2042 (1978).

Sadick, M. D., Locksley, R. M., Tubbs, C., and Raff, H. V. Murine cutaneous Leishmaniasis: Resistance correlates with the capacity to generate interferon-γ in response to Leishmania antigens in vitro. *J. Immunol.* 136: 655–661 (1986).

Saksela, E., Virtanen, I., Hovi, T., Secher, D. S., and Cantell, K. Monocyte is the main producer of human leukocyte α interferons following Sendai virus induction. *Prog. Med. Virol.* 30: 78–86 (1984).

Sastre, L., Roman, J. M., Teplow, D. B., Dreyer, W. J., Gee, C. E., Larson, S., Roberts, T. M., and Springer, T. A. A partial genomic DNA clone for the α subunit of the mouse complement receptor type 3 and cellular adhesion molecule Mac-1. *Proc. Natl. Acad. Sci.* (USA) 83: 5644–5648 (1986).

Schreiber, R. D., Pace, J. L., Russell, S. W., Altman, A., and Katz, D. H. Macrophage-activating factor produced by a T cell hybridoma: Physiochemical and biosynthetic resemblance to γ-interferon. *J. Immunol.* 131: 826–832 (1983).

Schultz, R. M. and Chirigos, M. A. Selective neutralization by anti-interferon globulin of macrophage activation by L-cell interferon, *Brucella abortus* ether extract, *Salmonella typhimurium* lipopolysacchride, and polyanions. *Cell. Immunol.* 48: 52–58 (1979).

Schultz, R. M. and Kleinschmidt, N. J. Functional identity between murine γ interferon and macrophage activating factor. *Nature* 305: 239–240 (1983).

Sethi, K. K., Omata, Y., and Brandis, H. Contribution of immune interferon (IFN-γ) in lymphokine-induced anti-toxoplasma activity: Studies with recombinant murine IFN-γ. *Immunobiology* 170: 270–293 (1985).

Smith, T. J. and Wagner, R. R. Rabbit macrophages interferons. I. Conditions for biosynthesis by virus-infected and uninfected cells. *J. Exp. Med.* 125: 559–577 (1967a).

Smith, T. J. and Wagner, R. R. Rabbit macrophages interferons. II. Some physicochemical properties and estimations of molecular weights. *J. Exp. Med.* 125: 579–593 (1967b).

Springer, T. A. and Anderson, D. C. The importance of the Mac-1, LFA-1 glycoprotein family in monocyte and granulocyte adherence, chemotaxis, and migration into inflammatory sites: Insights from an experiment of nature. In: *Biochemistry of Macrophages,* D. Evered, J. Nugent, and M. O'Connor, Eds., Pitman, London, Ciba Foundation Symposium 118, pp. 102–126 (1986).

Springer, T. A., Teplow, D. B., and Dreyer, W. J. Sequence homology of the LFA-1 and Mac-1 leukocyte adhesion glycoproteins and unexpected relation to leukocyte interferon. *Nature* 314: 540–542 (1985).

Stevenson, H. C., Dekaban, G. A., Miller, P. J., Benyajati, C., and Pearson, M. L. Analysis of human blood monocyte activation at the level of gene expression. Expression of α interferon genes during activation of human monocytes by poly IC/LC and muramyl dipeptide. *J. Exp. Med.* 161: 503–513 (1985).

Sultzer, B. M. Genetic control of host responses to endotoxin. *Infect. Immun.* 5: 107–113 (1972).

Svedersky, L. P., Benton, C. V., Berger, W. H., Rinderknecht, E., Harkins, R. N., and Palladino, M. A. Biological and antigenic similarities of murine interferon-γ and macrophage-activating factor. *J. Exp. Med.* 159: 812–827 (1984).

Tsujimoto, M., Yip, Y. K., and Vilcek, J. Interferon-γ enhances expression of cellular receptors for tumor necrosis factor. *J. Immunol.* 136: 2441–2444 (1986).

Urban, J. L., Shepard, H. M., Rothstein, J. L., Sugarman, B. J., and Schreiber, H. Tumor necrosis factor: A potent effector molecule for tumor cell killing by activated macrophages. *Proc. Natl. Acad. Sci.* (USA) 83: 5233–5237 (1986).

Valle, M. J., Bobrove, A. M., Strober, S., and Merigan, T. C. Immune specific production of interferon by human T cells in combined macrophage-lymphocyte cultures in response to herpes simplex antigen. *J. Immunol.* 114: 435–441 (1975).

Varesio, L., Blasi, E., Thurman, G. B., Talmadge, J. E., Wiltrout, R. H., and Herberman, R. B. Potent activation of mouse macrophages by recombinant interferon-γ. *Cancer Res.* 44: 4465–4469 (1984).

Vogel, S. N., English, K. E., Fertsch, D., and Fultz, M. J. Differential modulation of macrophage membrane markers by interferon: Analysis of Fc and C3b receptors, MacI and Ia antigen expression. *J. Interferon Res.* 3: 153–160 (1983b).

Vogel, S. N. and Fertsch, D. Endogenous interferon production by endotoxin-responsive macrophages provides an autostimulatory differentiation signal. *Infect. Immun.* 45: 417–423 (1984).

Vogel, S. N. and Fertsch, D. Macrophages from endotoxin-hyporesponsive (Lps^d) C3H/HeJ mice are permissive for vesicular stomatitis virus because of reduced levels of endogenous interferon: Possible mechanism for natural resistance to virus infection. *J. Virol.* 61: 812–818 (1987).

Vogel, S. N., Finbloom, D. S., English, K. E., Rosenstreich, D. L., and Langreth, S. G. Interferon-induced enhancement of macrophage Fc receptor expression: β-interferon treatment of C3H/HeJ macrophages results in increased numbers and density of Fc receptors. *J. Immunol.* 130: 1210–1214 (1983a).

Vogel, S. N. and Rosenstreich, D. L. Defective Fc receptor-mediated phagocytosis in C3H/HeJ macrophages. I. Correction by lymphokine-induced stimulation. *J. Immunol.* 123: 2842–2850 (1979).

Watson, J., Kelly, K., Largen, M., and Taylor, B. A. Correction of defective macrophage differentiation in C3H/HeJ mice by an interferon-like molecule. *J. Immunol.* 120: 422–424 (1978).

Weinberg, E. D. Iron and infection. *Microbiol. Rev.* 42: 45–66 (1978).

Wright, S. D., Craigmyle, L. S., and Silverstein, S. C. Fibronectin and serum amyloid P component stimulate C3b- and C3bi-mediated phagocytosis in cultured human monocytes. *J. Exp. Med.* 158: 1338–1343 (1983a).

Wright, S. D., Detmers, P. A., Jong, M. T. C., and Meyer, B. C. Interferon-γ depresses binding of ligand by C3b and C3bi receptors on cultured human monocytes, an effect reversed by fibronectin. *J. Exp. Med.* 163: 1245–1259 (1986).

Wright, S. D., Rao, P. E., Van Voorhis, W. C., Craigmyle, L. S., Iida, K., Talle, M. A., Westberg, E. F., Goldstein, G., and Silverstein, S. C. Identification of the C3bi receptor of human monocytes and macrophages by using monoclonal antibodies. *Proc. Natl. Acad. Sci.* (USA) 80: 5699–5703 (1983b).

Wright, S. D. and Silverstein, S. C. Tumor-promoting phorbol esters stimulate C3b and C3b′ receptor-mediated phagocytosis in cultured human monocytes. *J. Exp. Med.* 156: 1149–1164 (1982).

Yoshie, O., Meliman, I. S., Broeze, R. J., Garcia-Blanco, M., and Lengyel, P. Interferon action: Effects of mouse and interferon on rosette formation, phagocytosis and surface-antigen expression of cells of the macrophage-type line RAW 309Cr.1. *Cell. Immunol.* 73: 128–140 (1982).

11 PRODUCTION OF IFN-γ BY T CELLS AND MODULATION OF T CELL, B CELL, AND NK CELL ACTIVITY BY INTERFERONS

By influencing the manifold functions of all lymphoid cells, IFNs can profoundly affect antigen-specific and nonspecific immunity.

T lymphocytes are the effector cells of cell-mediated immunity, and through their helper activity and the production of soluble mediators, known as lymphokines, they also contribute significantly to antibody formation. The activity of lymphokines, however, is much wider than merely providing help for antibody formation, and IFN-γ is one of the best-documented examples of a multifunctional lymphokine.

B lymphocytes are the precursors of antibody-producing cells, and their development and differentiation is influenced by many different cytokines, among which is IFN-γ. Although IFN-α and IFN-β can also influence B-cell development, a physiological role for these IFNs in B-cell development seems less obvious.

NK cells are nonantigen-specific cytolytic effector cells, and stimulation of NK-cell activity by IFNs may possibly contribute to the antitumor and also the antiviral effects of IFNs in vivo.

I. IFN-γ PRODUCTION BY T CELLS AND EFFECTS OF IFNs ON T-CELL FUNCTION

T lymphocytes represent 60 to 80% of peripheral blood lymphoid cells. They originate from the multipotential stem cells found in the liver during prenatal life and in the bone marrow after birth. The maturation of precursor T lymphocytes into immunocompetent virgin T cells, an antigen-independent step, takes place in the thymus. The thymic stage, characterized by rapid proliferation and death of a great number of cells, results in the selection of a repertoire of T cells with antigen-specific cell-surface receptors. From the thymus, immunocompetent T cells migrate to lymphoid organs, where, as a result of interaction with specific antigens, they are activated. This leads to production of effector molecules, among which is IFN-γ, and to expansion of the activated clone.

A. Intrathymic Differentiation of T Cells

In humans, the earliest identified T-lineage cells express the sheep erythrocyte receptor T11, now also called CD2 (stage I). Later, thymocytes express CD4, CD6, and CD8 cell-surface proteins (stage II). With further maturation, the CD6 antigen disappears and the thymocytes acquire the CD3/Ti receptor structure. When they leave the thymus, mature T cells all bear the sheep erythrocyte receptor CD2 (T11) and the CD3/Ti receptor complex, but they express either the CD4- or the CD8-surface protein (CD stands for "cluster of differentiation").

Murine immunocompetent thymocytes and peripheral T cells express the T-cell antigen-receptor complex together with either the Ly-2 antigen, analogous to the human CD8 protein, or the L3T4 antigen, corresponding to the human CD4 protein (Acuto and Reinherz, 1985; Ceredig et al., 1985).

1. The CD2 T-cell surface protein

CD2 (originally called T11) is a 50-kDa glycoprotein of the immunoglobulin supergene family that is expressed on all peripheral T cells, on NK cells, and on all but the most immature thymocytes. One of the earliest T-cell markers described, CD2 is also known as the sheep erythrocyte receptor, since it confers to T cells the ability to form rosettes with sheep red blood cells. In humans, the cDNA encoding the CD2 protein predicts a 336 amino acid mature polypeptide, and the structural gene is on the short arm of chromosome lp13 (Sewell et al., 1986; Sewell et al., 1987; Seed and Aruffo, 1987). The cDNA encoding the murine CD2 predicts a 322 amino acid polypeptide, after cleavage of the signal peptide, and the gene has been assigned to mouse chromosome 3 (Sewell et al., 1987). The amino acid homology between human and murine CD2 is about 50%. The CD2 surface antigen plays an important role in the activation of T lymphocytes.

2. The CD3/Ti Antigen Receptor Complex

a. The Ti Heterodimer. The great majority of human and murine T cells express a T-cell receptor consisting of a molecular complex comprising at least five polypeptides. Two of these are the disulfide linked Ti α and β chains, each having a molecular weight of about 45 kDa. The genes that encode these proteins have variable (V), diversity (D), joining (J), and constant (C) regions. In the formation of the mature message for the receptor proteins, genetic rearrangement occurs, and it is possible to distinguish germline gene configurations from mature rearranged genes. This results in a T-cell antigen-receptor repertoire displaying many similarities to the B-cell antibody repertoire (Acuto and Reinherz, 1985; Kronenberg et al., 1986).

In humans, the Ti α locus is on chromosome 14, region q11-12, and the Ti β locus is on chromosome 7, region q32-35. In mice, the Ti α locus is on chromosome 14, and the Ti β locus is on chromosome 6 (Caccia et al., 1984; Collins et al., 1984; Lee et al., 1984; Collins et al., 1985; Dembic et al., 1985; Isobe et al., 1985; Jones et al., 1985).

In a small subset of human and murine T cells, the α/β T-cell receptor heterodimer is replaced by either a γ/δ heterodimer or a γ chain homodimer associated at the cell surface with the CD3 complex. Such T cells are furthermore exceptional in that they display neither the CD4 nor the CD8 cell-surface antigen. The γ chain is organized in the same way as the α and β T-cell receptor chains, with variable, joining, and constant regions, but γ chain protein diversity seems to be more limited. The murine γ chain gene is situated on mouse chromosome 13 (Kranz et al., 1985; Kronenberg et al., 1986; Borst et al., 1987; Brenner et al., 1987; Maeda et al., 1987; Pelicci et al., 1987; Strauss et al., 1987).

b. The CD3 Complex. The heterodimeric Ti α/β chain is noncovalently associated on the T-cell surface with a complex of several membrane proteins, called the CD3 complex, because its presence was first detected by the affinity for monoclonal antibody OKT3. The CD3 complex consists of at least three invariable polypeptide chains, one 25 kDa chain called γ (not the same as the Ti γ chain!), and two 20 kDa chains called δ and ε (Weiss et al., 1986). The genes encoding the human CD3 γ, δ, and ε chains are clustered together on the long arm of chromosome 11, region q23, syntenic with the protooncogene c-*ets*-1. The murine CD3 δ and ε genes are on mouse chromosome 9 (Van den Elsen et al., 1985; Gold et al., 1987; Tunnacliffe et al., 1987).

After antigen recognition by the α and β chains of the Ti complex, signal transduction of the recognition event occurs via the CD3 complex (Yagüe et al., 1985; Marrack and Kappler, 1986; Samelson et al., 1986). Antibodies that are specific for CD3 structures can behave as T-cell specific agonists, and, among others, trigger IFN-γ production (see section C).

3. *The CD4 and CD8 T-Cell-Surface Proteins*

Some immature thymocytes express both the CD4 and CD8 cell surface proteins, but this would seem to represent an intermediate stage in thymocyte differentiation, since immunocompetent thymocytes and peripheral T cells express either CD4 (L3T4 in mice) or CD8 (Ly-2 or Lyt-2 in mice) cell-surface proteins, but not both. Both the CD4 (formerly T4) and the CD8 (formerly T8) molecules are membrane proteins with structures that bear homology to the variable regions of immunoglobulin light chains and belong to the immunoglobulin supergene family. (Littman et al., 1985; Maddon et al., 1985; Blue et al., 1987; Nakauchi et al., 1987).

B. Postthymic T-Cell Phenotype and Function

The CD4 glycoprotein is expressed on T cells that interact with targets bearing MHC class II antigens, whereas CD8-bearing T cells interact with targets bearing MHC class I antigens. As a rule, T cells recognize only antigens that are present on the surfaces of other cells in the context of the polymorphic cell-surface molecules encoded by the MHC, a phenomenon called "MHC restriction" (Meuer et al., 1982).

In general, the $CD4^+$ population of T lymphocytes contains helper cells, whereas the $CD8^+$ population contains the majority of cytotoxic and suppressor cells. However, $CD4^+$ cells can also function as cytotoxic or suppressor cells, showing that expression of CD4 or CD8 is more stringently associated with MHC class recognition that with the effector function and that phenotypically identical T cell can have distinct functions. The possibility has been raised that the CD4 or CD8 glycoproteins serve as ancillary-binding structures for an invariant portion of the MHC class I or class II antigens, functioning as stabilizing elements that facilitate cell-to-cell contact (Swain, 1983). Immunofluorescent monitoring of cell surface antigens shows that the specific binding of a helper T cell with an antigen-presenting cell results in the redistribution of the CD4 and T-cell receptor surface proteins, so that they become concentrated in the cell-to-cell contact region. The coclustering of CD4 and T-cell receptor proteins occurs only when the antigen-presenting cell presents the right antigen and expresses the appropriate MHC Class II cell surface molecules (Kupfer et al., 1987).

The binding of the CD4 protein to the MHC class II molecules of antigen-presenting cells is followed by T-cell activation and results in the triggering of IL-2 synthesis (Gay et al., 1987; Inaba and Steinman, 1987).

Functionally, T cells can be classified as follows:

1. Cytotoxic effector T cells (CTL), which are able to lyse target cells and are involved in contact sensitivity (Tc),

2. Immunoregulatory, lymphokine secreting T cells such as T helper cells (Th), involved in B-cell differentiation and proliferation; delayed hypersensitivity T cells (Td); amplifying T cells (Ta), involved in cytotoxic T-cell differentiation; and T suppressor (Ts) cells that dampen responses and may help prevent reaction to self-molecules.

The manifold activities of the various T-cell subsets are all intimately related to the IFN system. Interaction with antigen-presenting accessory cells and with target cells depends on the presence of MHC class I and class II molecules, whose expression is modulated by all three IFN species (see Chapter 9). T-suppressor cells can be up or down-regulated by IFNs and furthermore, T cells produce IFN-α and β, which also exert immunomodulating activities. Last, but no least, T cells are the major producers of IFN-γ.

C. IFN-γ Is a T-Cell Product

Contrary to IFN-α and IFN-β synthesis, which can take place in any cell, production of IFN-γ is a specialized function of T cells. IFN-γ is structurally unrelated to IFN-α and IFN-β, and acts through a different receptor (see Chapter 4). Its structure and genetics are discussed in Chapter 2.

All IFN-γ inducers activate T cells in either a polyclonal (mitogens or antibodies) or a clonally restricted manner (antigen-specific recognition) (Wheelock, 1965; Green et al., 1969).

1. Antigen Recognition Results in IFN-γ Production

If bacterial antigens like PPD (purified protein derivative of tubercle bacilli), diphtheria, or pertussis toxoid are added to human lymphocyte suspensions, IFN-γ only appears in the supernatants of cultures derived from individuals who have previously been immunized with the corresponding antigens: The production of IFN-γ clearly reflects the immune status of the lymphocyte donor (Green, 1969). Similarly, T lymphocytes from mice with an ongoing infection with Listeria monocytogenes develop the ability to produce IFN-γ upon stimulation by Listeria antigen (Havell et al., 1982). Viral antigens, for example, vaccinia, herpes simplex, or mumps virus, also induce IFN-γ in lymphocytes from immune donors only (Epstein et al., 1972; Rasmussen et al., 1974; Nakayama et al., 1983).

Cellular antigens also can stimulate IFN-γ production, as shown, for example, in one way mixed lymphocyte cultures of mice or humans (Virelizier et al., 1977a; Manger et al., 1981).

Almost from the very onset, T cells have been considered as the producers of IFN-γ in lymphocyte or mononuclear cell suspensions (Stobo et al., 1974; Valle et al., 1975; Archer et al., 1979; Epstein and Gupta, 1981; von Wussow et al., 1981; Landolfo et al., 1982a). Freshly isolated T cells are usually resting cells that enter the cell cycle upon stimulation, for example, with a mitogen, whereas T cells from long-term cultures are continuously cycling cells. The latter can be stimulated to produce IFN-γ in an accessory cell-independent way, whereas resting T cells need accessory cells.

a. Accessory Cells Stimulate IFN-γ Production by Resting T Cells. The presence of macrophages or other accessory cells, for example, dendritic cells, has a boosting effect on IFN-γ production by resting T cells (Fig. 11.1) (Epstein, 1976; Blanchard et al., 1986; Le et al., 1986). This is related to the antigen-presenting capacity as well as the IL-1 production by accessory cells. Apparently, the following series of events takes place. The interaction of T cells with antigen-presenting cells results in the production of IL-1. IL-1 then activates the T cells to produce IL-2, and IL-2, in addition to stimulating T-cell growth, also promotes IFN-γ production (Larsson et al., 1980; Marcucci et al., 1982; Palacios, 1984; Reem and Yeh, 1984; Vilcek et al., 1985). It can, therefore, be said that IL-1 and IL-2 act synergistically to boost IFN-γ production by T cells. Moreover, IFN-γ enhances IL-1 production by macrophages, which provides a positive feedback loop with IFN-γ enhancement of IL-1 production leading to IL-1-mediated enhancement of IL-2 production leading to IL-2-mediated enhancement of IFN-γ production (Newton, 1985). The various connections between these different effector molecules are discussed in Chapter 13.

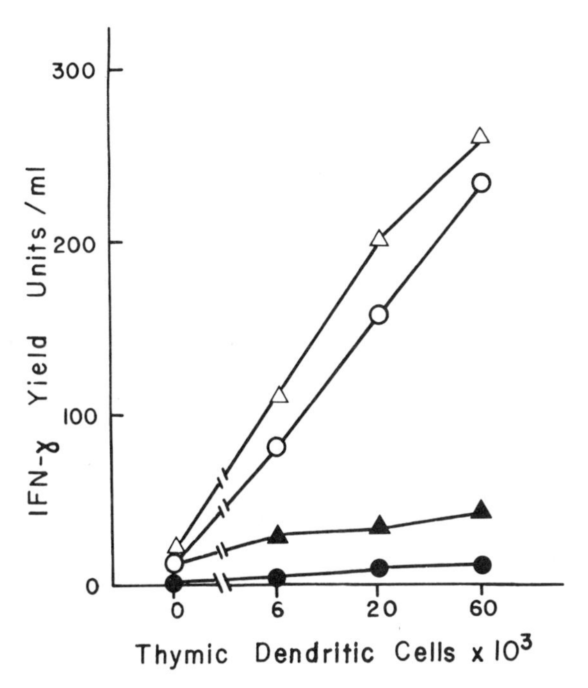

Figure 11.1 A demonstration of the accessory function of thymic dendritic cells for IFN-γ production by T cells, stimulated with either conA (△—△), PHA (○—○), or IL-2 (▲—▲). Unstimulated (●—●). (Adapted from Le et al., 1986.)

b. Production of IFN-γ by Cloned T-Cell Lines. The establishment of long-term cloned T-cell lines has made it possible to study the production of IFN-γ by T cells after mitogen stimulation or antigen recognition in the absence of accessory cells (Marcucci et al., 1981; Nathan et al., 1981; Le et al., 1982). The main line of evidence indicates that all subsets of cloned T cells can produce IFN-γ. An indication that all T cells have the potential for IFN-γ production is provided by the observation that binding of sheep erythrocytes to the ubiquitous CD2 sheep erythrocyte receptor on human T cells is capable of boosting mitogen-induced IFN-γ production (Wilkinson and Morris, 1984).

c. Activation of the CD3 Complex Induces IFN-γ synthesis. The OKT3 monoclonal antibody, specific for the CD3 complex, induces IFN-γ synthesis in human peripheral blood leukocytes, in T lymphocytes that are isolated from peripheral blood and in cells from the human Jurkat T cell line (von Wussow et al., 1981; Chang et al., 1982; Wiskocil et al,. 1985). This provides another indication that all T-cell subsets have at least the potential for making IFN-γ, although it is much more difficult to ascertain whether they all do so under physiological conditions.

2. IFN-γ Production by Different T-Cell Subsets

We will now briefly review the evidence for IFN-γ production by the different T-cell subsets.

a. Cells of the Cytotoxic-Suppressor Phenotype (Human CD8 or Murine Ly-2 Phenotype)

I. IFN-γ PRODUCTION BY T CELLS WITH CYTOLYTIC ACTIVITY. Cells from clonal lines representing cytotoxic T cells can be induced by mitogens to produce IFN-γ (Matsuyama et al., 1982), and specific antigen recognition also results in IFN-γ synthesis. T cells expressing the CD8 antigen, isolated from patients after an infection with influenza virus or immunization against rabies virus are stimulated to release IFN-γ when exposed to the corresponding viral antigens in vitro (Ennis and Meager 1981 Celis et al., 1986; Yamada et al., 1986). Cells from a line of cytotoxic T cells, derived from BALB/c mice immunized with influenza virus, react specifically against influenza virus-infected target cells, and release IFN-γ; like target cell lysis, IFN-γ production is MHC restricted (Fig. 11.2). Not all such cytotoxic Tc clones, however, secrete IFN-γ upon contact with their target cells, and cytotoxic activity can be exerted without concomitant secretion of IFN-γ (Morris et al., 1982; Taylor et al., 1985). Cytotoxic T cells also release IFN-γ in an antigen-specific way when cocultured with allogeneic target cells presenting a minor histocompatibility antigen (Klein et al., 1982).

The generation of cytolytic cells from their precursor cells can be inhibited by monoclonal antibodies to IFN-γ, suggesting that the production of

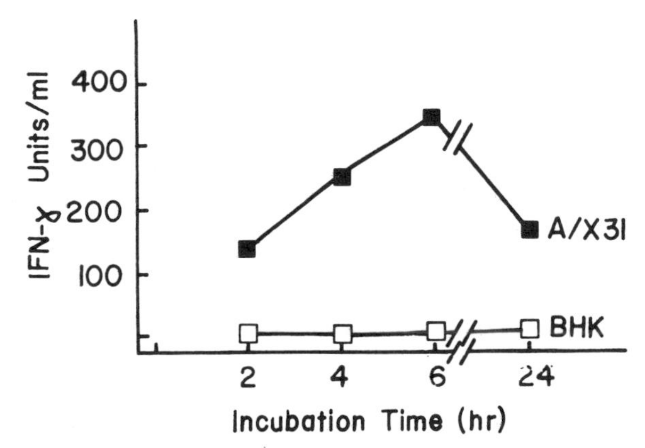

Figure 11.2 Cytotoxic T cells release IFN-γ when interacting with specific antigen-bearing target cells. Mastocytoma cells infected with influenza virus strain A/X31 (■—■) are incubated in the presence of cells from murine cytotoxic T-cell clones specifically sensitized against this virus, and as a result of this, IFN-γ is released into the culture medium. When cytotoxic cells from the same clone are incubated with noninfected mastocytoma cells, or with cells infected with an unrelated strain of virus (B/HK: □), there is no production of IFN-γ. (Adapted from Taylor et al., 1985.)

IFN-γ by cytolytic cells represents a case of autocrine growth regulation, in which the same cell synthesizes and responds to its factor (Simon et al., 1986a,b).

II. IFN-γ SECRETION BY T CELLS WITH SUPPRESSIVE ACTIVITY. Suppressor cells down-regulate antibody formation and cell-mediated immunity. Suppressor-cell activity is high in late embryonic life and early after birth, and declines subsequently. Certain types of antigen-specific suppressor cells, called veto cells, inhibit cytolytic T-cell activity against antigenic determinants that are part of the suppressor-cell membrane or attached to the suppressor-cell surface. The role of veto cells could be to inhibit autotoxic T cells, and thus prevent autoimmune tissue destruction (Miller, 1980; Rammensee et al., 1985). Although not all populations with suppressive activity exert their function in an antigen-specific way, antigen-specific suppressor cells are frequently generated in an MHC class II restricted way during an immune response (Hayglass et al., 1986).

In general, both in humans and in mice, IFN-γ synthesis has been demonstrated in T cells of the cytotoxic/suppressor phenotype, bearing either the CD8 or the Ly-2 antigen respectively. In the absence of concomitant functional tests, however, it is not always possible to distinguish between cytotoxic effector cells and suppressive cells (Guerne et al., 1984; Kees et al., 1984; Simon et al., 1984, 1985). In fact, what characterizes a suppressor T cell is currently a highly debated issue, and the phenotype characteristic for T suppressor cells may also be expressed by functionally distinct cells.

In humans, T cells of the CD8 phenotype that express the MHC class II HLA-DR antigen correspond to activated suppressor cells, These cells can synthesize Hu IFN-γ, and there is evidence that, through uncontrolled synthesis of IFN-γ, activated suppressor cells can be instrumental in the hematopoietic suppression that is typical of aplastic anemia (Zoumbos et al., 1985). The circulating mononuclear cells from a patient developing severe aplastic anemia during the course of non-A, non-B hepatitis almost entirely consisted of activated T-suppressor cells expressing the T8 and MHC class II antigens. When cultured in vitro, these cells released considerable amounts of Hu IFN-γ (Herrmann et al., 1986). The possible contribution of IFN-γ production by suppressor cells to the pathogenesis of certain cases of aplastic anemia is discussed in Chapter 8.

b. Cells of the T-Helper Phenotype (Human CD4 or Murine L3T4 Phenotype)

I. IFN-γ PRODUCTION BY T-HELPER CELLS. T-helper cells recognize processed antigen presented by accessory cells in conjunction with MHC class II determinants. The T-helper cell is stimulated by IL-1 to proliferate via IL-2 and induces other lymphoid cells to differentiate into effector cells. In many experimental systems the generation of optimal T-helper activity seems to require collaboration between different T-helper subsets, but most of these are presently ill characterized.

Mitogen or antigen-specific stimulation of murine T-helper cells as well as cells from T-helper cell lines or clones results in IFN-γ synthesis (Landolfo et al.,, 1982b; Conta et al., 1983; Hecht et al., 1983; Guerne et al., 1984). Human T helper cells expressing the CD4 cell-surface antigen can be induced to make IFN-γ when stimulated by mitogens (Abb et al., 1983; Kasahara et al., 1983b: Schober et al., 1984). Furthermore, addition of the monoclonal antibody OKT3, which activates the CD3 receptor, stimulates the synthesis of IFN-γ in T cells of the CD4 phenotype (Chang et al., 1982; Bhayani and Falcoff, 1985).

The foregoing examples only show that T cells of the helper phenotype have the potential for IFN-γ production, but do not tell us whether this actually takes place in vivo during the natural function of these cells. There are, however, indications from the clinic that T-helper cells do produce IFN-γ during infection. Patients with recurrent herpes labialis infection have IFN-γ-producing circulating T cells of the helper phenotype. Moreover, herpes simplex virus antigen-specific T-cell clones of the T helper type can be established from these patients, and the clones release IFN-γ when exposed in vitro to herpes virus antigen (Cunningham and Merigan, 1984; Cunningham et al., 1985). Similarly, rabies virus specific, MHC class II restricted T-helper cell clones, expressing the T4 antigen, can be isolated from rabies vaccine recipients (Celis et al., 1986).

II. IFN-γ PRODUCTION BY TD CELLS. Td cells are mediators of delayed hypersensitivity and express the CD4-surface protein; they are MHC class II restricted. Continuous T-cell lines, mediating hapten-specific delayed-type hypersensitivity to the contact sensitizing agent oxazolone, are induced to make IFN-γ only when encountering cells presenting the specific antigen, and not when encountering cells presenting another unrelated sensitizing hapten, such as picryl chloride (McKimm-Breshkin et al., 1982).

3. *Decreased IFN-γ Production in Newborns*

T cells from human neonates do not have a fully developed capacity for IFN-γ production, although their IL-2 production seems to be normal (Taylor and Bryson, 1985; Lewis et al., 1986; Wilson et al., 1986). The possibility has been raised that the decreased IFN-γ production is caused by specific suppressor cells, but there is presently no agreement as to the mechanism involved (Seki et al., 1986). The mitogen-activated suppressor cell present in cord blood is radiosensitive and expresses the CD4 surface antigen. The existence of $Ly\text{-}2^+$ suppressor cells of mitogen-induced IFN-γ production has been shown in the mouse (Torres et al., 1982).

In mice, antigen-specific memory T cells, both of helper and of cytotoxic/suppressor phenotype, display an increased capacity to produce IFN-γ as compared to virgin T cells. This indicates that attainment of maximal capacity to produce IFN-γ is a differentiation event linked to primary antigenic stimulation and offers another explanation for the lower IFN-γ production by T cells from neonates (Budd et al., 1987). The human fetus and neonate is unusually susceptible to infection with intracellular pathogens and viruses, and a low IFN-γ production could conceivably contribute to this enhanced susceptibility to infection.

D. T Cells Also Produce IFN-α and IFN-β

When myxoviruses and paramyxoviruses such as influenza, Newcastle disease, or Sendai virus are inoculated intravenously into mice, a significant fraction of the ensuing IFN-α and IFN-β production takes place in very radiosensitive, bone-marrow-derived cells (De Maeyer et al., 1970). These are characteristic properties of lymphocytes, and T lymphocytes have been specifically shown to be involved in IFN-α/β production. Cloned lines of murine cytotoxic T cells that produce IFN-α in an antigen-specific manner also produce IFN-α/β in a nonspecific way upon infection with viruses like Semliki Forest or Newcastle disease virus (Cooley et al., 1984; Pasternack et al., 1984). This shows that T cells that have acquired the specialized function of making IFN-γ do not lose their potential of being induced to make the other IFNs. We do not know whether the same T cells that have been sensitized against a viral antigen can simultaneously produce both IFN-γ and IFN-α/β when infected with this virus; only in situ hybridization experiments with the appropriate molecular probes will provide an answer to this question.

IFN-α/β production is usually induced by viruses, but exceptionally, antigen stimulation can also provide the trigger. Murine bone-marrow cells, cocultured with allogeneic spleen cells or syngeneic tumor cells produce significant levels of IFN-α/β. The presence of T cells of the Ly-2 phenotype is required, although it is not possible to decide whether they are the producers or prime other cells for production (Klimpel et al., 1982; Reyes et al., 1985). The production of IFN-α/β by alloantigen-stimulated mouse bone-marrow cells provides an intriguing parallel to the production of IFN-γ by alloantigen-stimulated spleen cells. Its physiological significance is not clear. It could be related to the activity of IFNs as B-cell differentiation factors, and in view of the high susceptibility of bone-marrow stem cells to the antiproliferative effects of IFNs (see Chapter 8), it may also be related to the regulation of hematopoiesis by T cells or NK cells.

E. Effects of IFNs on T-Cell function

1. *Cells of the Cytotoxic/Suppressor Phenotype (CD8, Ly-2)*

a. T Cells with Cytolytic Activity. Murine IFN-α/β enhances the specific cytotoxicity of sensitized lymphocytes against allogeneic tumor cells. This effect occurs very rapidly, after a few hours of treatment of the lymphocytes with IFN (Lindahl et al., 1972). Similarly, the cytotoxity of human lymphocytes in a mixed allogeneic lymphocyte reaction is augmented by treatment with Hu IFN-α or IFN-β (Heron et al., 1976; Zarling et al., 1978). Cytotoxic T cells react with specific antigen-bearing target cells in the context of class I MHC antigens, and both IFN-α/β and IFN-γ induce an expression of these cell-surface antigens (see Chapter 9). As a result of the enhanced expression of MHC class I antigens, the cytotoxic activity of T cells is boosted against target cells expressing viral antigens (Fig. 11.3). Thus, virus-induced IFN not only restricts the infection by inducing the antiviral state, but also by conditioning infected cells for destruction by cytotoxic T cells (Blackman and Morris, 1985; Bukowski and Welsh, 1985a, 1986).

The possibility has been raised that IFNs may act as a cytotoxic T-cell differentiation signal, since human T-cell clones, devoid of lytic activity, acquire antigen-specific cytolytic activity after the addition of IFN-α or IFN-γ (Chen et al., 1986b). Moreover, IFN-α induces a rearrangement of the TCRA locus, which codes for the α chain of the T-cell receptor. Whether there is a causal relationship of the IFN-induced rearrangement of the α-chain locus to the acquisition of cytotoxic potential is not established, since IFN-α can induce these rearrangements not only in precursor but also in mature cytotoxic T cells. In the latter, and in spite of the rearrangement, the

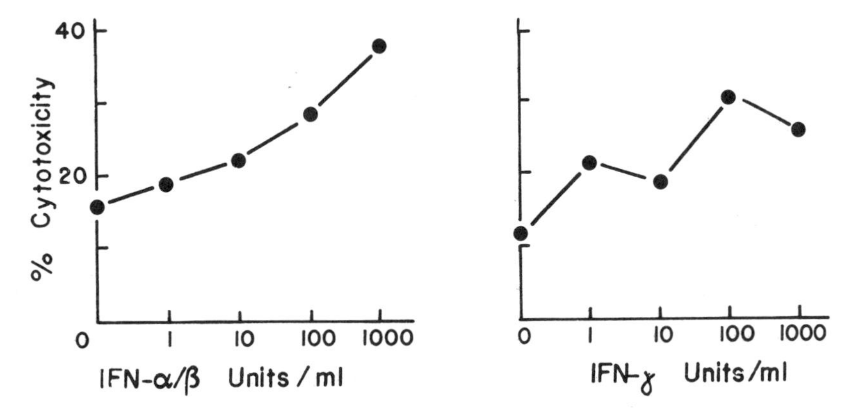

Figure 11.3 Treatment of target cells with IFN-α/β (left panel) or with IFN-γ (right panel) increases the cytotoxic effect of allogeneic cytolytic T cells. The target cells are C3H mouse fibroblasts, exposed to spleen cells from BALB/c mice sensitized to C3H cells. The figure shows increased cytotoxicity as a function of IFN treatment dose. (Adapted from Blackman and Morris, 1985.)

original antigen-specificity is retained, and therefore the functional meaning of the rearrangement is not clear (Chen et al., 1986a).

b. T Cells with Suppressive Activity. The effects of IFNs on T-suppressor cell generation and function are complex and can result in either boosting or down-regulation of suppression.

The principles involved in the generation of T-helper cells also apply to the induction of T-suppressor cells in that antigens must be presented by accessory cells in a genetically restricted way. MHC class II antigens are involved, but there is some discussion about the nature of other cell-surface molecules involved in the restriction and the chromosomal location of the genes coding for them (Minami et al., 1982; Aoki et al., 1983; Usui et al., 1984).

The stimulation by IFN-γ of the MHC class II (Ia antigens) expression on cultured mouse macrophages is rapidly followed by an increase in the ability of these macrophages to induce the generation of T suppressor cells. This shows that IFN-γ is capable of stimulating accessory cells to induce T-suppressor cells, but it does not prove that this induction is due to the increased expression of class II antigens, since the two phenomena could merely correlate without functional implications (Noma and Dorf, 1985).

The effect of IFN on the suppressor cell generation is not always stimulatory, and under certain conditions suppressor cell formation is decreased rather than increased. Allospecific T-suppressor cells are generated after stimulation with allogeneic cells, for example, in the mixed lymphocyte reaction in vitro. These suppressor cells are specific for the MHC antigens of the cells used for eliciting the response. The presence of low amounts of Hu leukocyte IFN in such mixed lymphocyte cultures results in a marked decrease in suppressor-cell activity. This effect is not the result of destruction by the IFN of presuppressor cells, but is due to inhibition of their differentiation into active T-suppressor cells (Fradelizi and Gresser, 1982).

As seen earlier, IFN-γ is produced as the result of alloantigen stimulation of T cells in the mixed lymphocyte reaction. IFN-γ in this system down-regulates the appearance of suppressor cells, since, when it is neutralized by monoclonal antibody, suppressor cells are quickly generated and the induction of T cells cytotoxic to alloantigens is inhibited. Similarly, anti-IFN-γ antibodies, when administered to mice, block the rejection of allogenic tumor cells. Presumably in vivo, as in vitro, the generation of cytotoxic T cells is inhibited by the suppressor cells that appear as a result of the inactivation of IFN-γ (Landolfo et al., 1985). The inhibition by Mu IFN-α/β of T-suppressor cells in delayed hypersensitivity is discussed in Chapter 12.

The fact that IFNs sometimes stimulate and at other times inhibit the generation and activity of T suppressor cells, depending on the experimental system, should not surprise us. The contradiction between these different observations is only apparent: IFNs do not act alone, but are influenced by a number of other cytokines, some of which have not even been characterized. Furthermore, the timing of IFN production or administration, IFN and antigen dosage, as well as genetic factors, all influence the outcome of the reaction. IFNs are only one link in a complex chain of interactions among these different factors, and we are far from having an integrated view of the whole.

2. *Cells of the Helper Phenotype (CD4, L3T4)*

IFNs can profoundly affect the T-helper cell function, and they do so in several ways. As seen earlier, T-helper cells interact with antigen-presenting accessory cells in an MHC class II antigen-restricted manner. One of the major activities of IFN-γ is to induce an expression of these antigens. The enhancement of the T-helper cell function most likely does not result from an enhanced expression of class II antigens on accessory cells only, but also results from a direct effect of IFN on the T-helper cell itself (Frasca et al., 1985).

II. EFFECTS OF INTERFERONS ON ANTIBODY FORMATION AND B-CELL DIFFERENTIATION

Like macrophages and T cells, B cells originate from the multipotential stem cells (CFUs) of the bone marrow, and before birth, of the fetal liver. The earliest elements of the B lineage carry no cell-surface immunoglobulins; they mature in the bone marrow to B cells with IgM on their surface. From the bone marrow they travel to lymph nodes, spleen, and other organs. Later in development, a new class of B cells, expressing both IgM and IgD, appear on lymphoid organs. These cells represent the predominant class of B cells, but cells expressing other immunoglobulins on their surface, for example, IgE, also develop. After antigen stimulation, in the presence of T-helper cells, B cells differentiate into mature, antibody-producing plasma cells. This is a highly complex process, involving rearrangement of the immunoglobulin genes, which is influenced by up and down-regulating cytokines, known as B-cell growth and differentiation factors. Results obtained with T-cell clones indicate that T-helper cells can induce production of all classes of immunoglobulin. In addition to immunoglobulins, several other cell-surface markers, for example, Fc receptors and the receptors for the activated complement molecule C3b, are found on plasma cells and their precursors.

Resting B cells can be induced to proliferate by anti-IgM antibody and growth-promoting factors. Differentiation of the cells to antibody production requires the addition of other cytokines. Several cytokines influencing and regulating growth and differentiation of B cells have been identified in humans and in mice. In humans, the most important cytokines influencing B-cell differentiation that have been characterized are IL-1, IL-2, IL-4 (BCGF-1), and BSF-2 (26 K protein or Hu IFN-β2). In the mouse, the best characterized are IL-1, IL-2, IL-4 (previously called BSF-1 or BCGF-1), IL-5 (BCGF-2), and BCDF-μ (Falkoff et al., 1982; Muraguchi et al., 1984; Hirano et al., 1985; Kehrl et al., 1985; Mishra et al., 1986). In both humans and mice there are indications that IFN-γ as well as IFN-α and IFN-β can also influence B-cell differentiation and function.

A. Effect of IFNs on Antibody Formation In Vivo

Without a doubt, IFNs can influence antibody formation in vivo, but it should be realized that in vivo or in mixed cell suspensions an effect of IFN on antibody formation does not necessarily imply a direct action of IFNs on B cells. In vivo, antibody formation can be indirectly influenced by IFNs, via effects on antigen-presenting accessory cells and on helper and T-suppressor cells. For example, hyperactive B cells from lupus patients can be down-regulated by IFN-γ-activated monocytes, and the number of B cells making antibodies to hapten-horse erythrocyte conjugates can be increased by stimulating T-helper activity with IFN-γ (Frasca et al., 1985; Golding et al., 1986).

1. Effects of IFN-α and IFN-β on Antibody Formation In Vivo

Antibody formation in mice can be modulated by Mu IFN-α/β. Small amounts of IFN, a few hundred units, when administered at the time of primary immunization with sheep erythrocytes (SRBC) slightly stimulate the number of antibody-forming cells as measured in a Jerne plaque assay 48 hours after immunization. Higher doses of IFN-α/β, however, significantly decrease the number of antibody-secreting cells (Braun and Levy, 1972). The effects of IFN-α/β on the generation of antibody-forming cells in vivo are influenced by the timing of IFN administration. When IFN is administered before immunization with SRBC, the number of antibody-forming cells is severely inhibited. Both IgM and IgG antibodies are decreased, as is the number of memory cells that are generated as a result of the primary immunization. When IFN is administered several days after immunization, an increase in antibody titers usually results (Chester et al., 1973; Brodeur and Merigan, 1974; Strannegard et al., 1978). This influence of timing is comparable to the effect of IFN-α/β on the afferent pathway of DTH, in that inhibition is obtained when IFNs act before sensitization and enhancement when IFNs act after sensitization (see Chapter 12). However, an enhancing effect on the number of antibody-forming cells when IFN is given before immunization has also been observed (Vignaux et al., 1980). Although the reasons for these different results are not clear, it is quite possible that they can be attributed to different compositions of the IFN-α/β mixtures that have been used for these experiments, since, as will be discussed in section C, different IFN-α subtypes can exert differential effects on B-cell function.

The in vivo effects of IFNs on antibody formation in humans have not been studied in detail, mainly because IFNs are usually administered to patients in which immune mechanisms are severely disturbed. Prolonged treatment with rec Hu IFN-α2 of patients with chronic type B hepatitis, who have relatively normal immunological functions, results in suppression of mitogen-induced immunoglobulin production (Peters et al., 1986b).

2. *Effects of IFN-γ on Antibody Formatin In Vivo*

In mice sensitized to *Mycobacterium bovis,* a correlation exists between the induction of IFN-γ with old tuberculin and the suppression of antibody formation to SRBC. The greatest reduction occurs when IFN-γ is induced 24 hours before the erythrocytes are given (Sonnenfeld et al., 1977). Since many other lymphokines are induced by this treatment, such an observation is only suggestive for an effect of IFN-γ, unless one can demonstrate that the effect is abrogated by specific anti-IFN-γ antibodies. This has not been the case for many in vivo experiments in which a role for endogenously produced IFN-γ has been implied, since specific monoclonal antibodies have only become available relatively recently.

Recombinant Mu IFN-γ, produced by either vertebrate cells or *E. coli,* influences the generation of antibody-forming cells in mice immunized with hen egg lysozyme or with dinitrophenol-conjugated bovine γ globulin. The number of antibody-forming cells, as well as the serum titer of circulating antibodies, is significantly stimulated if Mu IFN-γ is given to the mice at the same time as the antigen. This timing is important, and it indicates that an early step in the immune response is influenced. The concentration of IFN also plays a role, since giving more than an optimal amount does not result in any stimulation (Nakamura et al., 1984). The importance of these two parameters, timing and concentration, is the leitmotiv for immunomodulation by IFN-γ, as it is for IFNs in general, and variations in either one or both between experiments explain many apparently contradictory results. This in vivo effect of IFN-γ on antibody formation is quite comparable to the effect obtained with IFN-α/β.

B. Effects of IFNs on Immunization In Vitro

1. Effects of IFN-α and IFN-β

IFN-α and IFN-β also influence antibody formation in vitro. This can be shown by using techniques for in vitro immunization and generation of antibody-forming cells (Mishell and Dutton, 1967; Gisler and Ducor, 1972). Addition of IFN-α/β to such in vitro systems prior to, at the same time, or very shortly after the antigen, inhibits the generation of antibody-forming cells. If, however, IFN is added after the antigen, at a later stage of the culture, the number of antibody-forming cells can actually be boosted, not unlike some effects observed in vivo (Gisler et al., 1974; Johnson et al., 1975; Booth et al., 1976a,b). In a human system, however, exactly the reverse has been observed: Enhancement of the number of plaque-forming cells when Hu leukocyte IFN is added to the cultures prior to antigen, and suppression when IFN is added together with the antigen (Parker et al., 1981). This again emphasizes the importance of the system used in examining the effects of interferons.

In these in vitro systems, Mu IFN-α/β directly acts on B cells: If the major cellular components of the cultures—B cells, T cells, and macrophages—are treated separately with IFN, and then recombined with untreated complementary cells, the number of antibody-forming cells is decreased only in the cultures in which the B cells have been pretreated with IFN (Gisler et al., 1974). This interpretation is obviously directly dependent on the degree of purity of the different cell populations used for such experiments. It is, therefore, important that other experimental approaches also point to a direct effect of Mu IFN-α/β on B cells. For example, when the effect of Mu IFN-α/β on the development of clones of antibody-forming cells is studied, one observes a reduction of the number and the size of these clones, suggesting that IFN inhibits clonal initiation or activation (Booth et al., 1976a,b).

2. Effects of IFN-γ

When Mu IFN-γ is added to mouse spleen cell cultures prior to the antigen, a severe suppression of the number of antibody-forming cells results. The opposite effect is obtained when Mu IFN-γ is added 48 hours after the antigen: The number of antibody-forming cells is enhanced (Virelizier et al., 1977b; Sonnenfeld et al., 1978). This effect again is not different from what one observes with Mu IFN-α/β, except that, if the activity of these IFNs is expressed as antiviral units, less IFN-γ is required. This suggests that IFN-γ is a more potent immunomodulator on the basis of antiviral titer. When concentrations are expressed on a molar basis, one can make more valid comparisons of activities, and it appears that, depending on the activity measured, IFN-α/β can at times be a more efficient immunomodulator than IFN-γ.

C. Effects of IFNs on Proliferation and Differentiation of B Cells In Vitro

1. IFN-α and IFN-β

Several experimental approaches have shown that Hu IFN-α of different origins and composition influences immunoglobulin secretion by B cells.

Pokeweed mitogen induces IgG synthesis in peripheral blood lymphocytes on a nonantigen specific basis, and thus can be used to study immunoglobulin synthesis by undifferentiated B cells. This activity of pokeweed mitogen appears to result from the fact that it interacts with T-helper cells to trigger helper factor production. When human peripheral blood cell suspensions are pretreated with Hu leukocyte IFN before stimulation with pokeweed mitogen, IgG synthesis by B cells is significantly enhanced. It is essential that IFN acts before the mitogen is added, otherwise there is no stimulation and sometimes inhibition. The degree of stimulation varies from donor to donor, indicating the existence of individual differences in the capacity to respond to this particular effect of Hu IFN-α (Härfast et al., 1981). From studies in mice of different inbred strains, we know that genotype influences IFN action, and therefore it is not surprising to find individual differences in response to IFN action also in humans.

Human leukocyte IFN preparations are mixtures of different IFN-α subtypes, and some IFN-α species have different degrees of activity on B-cell function. Human consensus α IFN (Hu IFN-αCon1) is a consensus form of many subtypes, generated by recombinant DNA technology, and incorporates the most frequently observed amino acid residues at each position. It is nearly identical to the IFN-α consensus sequence presented in Table 2.1 of Chapter 2 (Alton et al., 1983). When Hu IFN-αCon1 is compared to human leukocyte IFN derived from Sendai virus-stimulated buffy coats, significant differences in the stimulation of immunoglobulin synthesis are observed in unfractionated peripheral blood lymphocytes, depleted of monocytes. Both IFN preparations stimulate the synthesis of IgG and IgM, but IFN-αCon1, at comparable dosage, stimulates Ig synthesis several times more than does leukocyte IFN (Fig. 11.4). When one compares the effect of three different recombinant IFN-α species under the same conditions, IFN-αF, at the optimal concentration, is about ten times more efficient in stimulating IgM synthesis than is IFN-αCon1, and about twice as efficient as IFN-αCon2 (αCon2 is a variant of αCon1 in which amino acid positions 14 and 16 are different). When purified B cells are used instead of total lymphocytes, IFN-αCon1 also increases the Ig production by B cells, but the effect is more pronounced in the presence of T-helper cells (Neubauer et al., 1985).

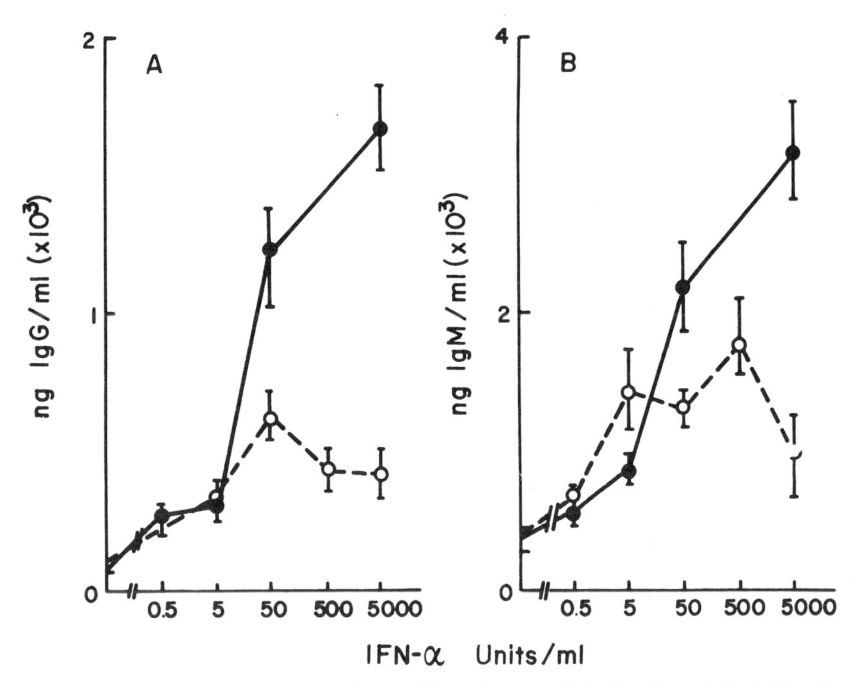

Figure 11.4 A comparison of the effect of Hu leukocyte IFN (○—○) and Hu-IFN-α Con1 (●—●) on stimulation of IgG (panel A) and IgM (panel B) production by peripheral blood leukocytes depleted of monocytes. (Adapted from Neubauer et al., 1985.)

We have seen that Hu IFN-αF is about ten times more efficient in stimulating Ig synthesis than is Hu IFN-αCon1. The existence of a tenfold difference in activity between two different subspecies of Hu IFN-α clearly shows that it is impossible to obtain consistent results and standardize experiments with human leukocyte IFN preparations of different origins and different compositions. The consensus IFN-αCon1 is not a naturally occurring IFN species, but it has been engineered in the laboratory, which theoretically could explain the difference with the natural IFN-αF. However, when natural species of IFN-α are compared, marked differences in the action on B cells are also observed, and some preparations of Hu lymphoblastoid IFN, for example, can inhibit polyclonal immunoglobulin synthesis rather than stimulate it (Fleisher et al., 1982). Two different fractions of Hu leukocyte IFN can be obtained by affinity chromatography on different monoclonal antibodies. One fraction significantly enhances B-cell proliferation induced by *Staphylococcus aureus* Cowan 1 bacterium; the other fraction has exactly the opposite effect and inhibits B-cell proliferation (Harada et al., 1983). These fractions still consist of several IFN-α subspecies each, and, as with so many other effects of IFNs, the immunomodulating effects of every subspecies have to be analyzed separately.

Furthermore, within each IFN-α subspecies there is a striking effect of IFN dosage. Low concentrations of a few hundred units of rec Hu IFN-α2 enhance pokeweed mitogen-stimulated Ig synthesis by peripheral blood mononuclear cells, but high concentrations of 10^5 units suppress Ig production (Fig. 11.5). Both enhancement and suppression occur in the absence of effects on B-cell proliferation, which means that IFN-α can directly affect Ig synthesis, without influencing the number of B cells (Peters et al., 1986a).

Hu IFN-β, natural or recombinant, stimulates immunoglobulin synthesis by B cells as efficiently as IFN-α, and more efficiently than natural or recombinant Hu IFN-γ, (Siegel et al., 1986).

These observations raise the possibility that IFNs of the α and β type are natural differentiation factors of B cells. But where would the IFN come from in the absence of viral infection? It could originate from antigenic stimulation of bone-marrow and spleen cells, which is known to induce IFN-α and IFN-β synthesis without requiring previous immunization (Ito and Hosaka, 1983; Smith et al., 1983; Reyes al., 1985). The IFN produced this way could conceivably act as a signal for B cells and amplify the antibody response.

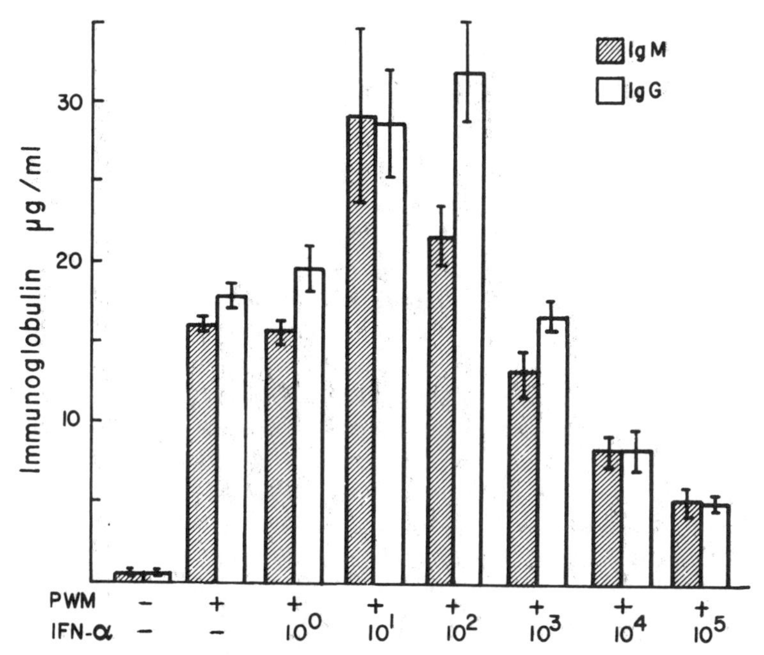

Figure 11.5 Effect of different concentrations of Hu IFN-α2 on pokeweed mitogen-stimulated IgG and IgM synthesis by peripheral blood mononuclear cells. Low concentrations, up to 100 units, stimulate synthesis, and high concentrations (10^4 or 10^5 units) are inhibitory. The bars represent the average value of seven individuals. (Adapted from Peters et al., 1986a.)

2. *IFN-γ*

Since many B-cell activating factors are prepared from mitogen-stimulated T cells or from allogeneic-mixed lymphocyte reactions, there is an excellent possibility that some of their activity is due to IFN-γ. This is borne out by experiments using pure or recombinant material. Hu IFN-γ promotes proliferation of anti-μ activated human B cells, as does IL 2 (Romagnani et al., 1986). In cultures of human B cells, IFN-γ can act synergistically with IL 2 to enhance immunoglobulin light chain synthesis (Bich-Thuy et al., 1986). This raises the possibility that IFN-γ is one of the natural B-cell differentiation factors, but it does not tell us whether IFN-γ is actually a component of the lymphokines that are normally required for B-cell stimulation, since one cannot exclude that IFN-γ could substitute for some of these factors, without being involved under natural conditions. Experiments using specific anti-IFN-γ antibodies show that this is not the case, since the addition of such antibodies to activated T-cell supernatants abrogates the capacity of these supernatants to stimulate B cells into production of antibody-forming cells as measured by plaque-forming cells in a Jerne assay (Sidman et al., 1984; Brunswick and Lake, 1985). This suggests very strongly that IFN-γ is a natural B-cell-maturing lymphokine. Interleukin 2 (IL-2), initially thought to be only a growth factor for T cells, also influences B-cell differentiation since it can stimulate activated B cells to proliferate and secrete immunoglobulins. The general conclusion from these experiments points to IL-2 as one of the growth factors for B cells and IFN-γ as one of the differentiation factors (Ralph et al., 1984; Mingari et al., 1984; Nakagawa et al., 1985).

The many different lymphokines involved in B-cell differentiation have not yet all been properly identified and characterized, and moreover, some observations raise the possibility that there are different lymphokine requirements for the induction of different Ig secreting B cells. Recombinant IFN-γ in conjunction with recombinant IL-2 is sufficient to support the induction of an antidextran IgA response in splenic B-cell cultures from BALB/c mice, but it does not suffice to support the generation of IgM producing B cells in response to that same antigen. The induction of a substantial IgM response to dextran requires a combination of IFN-γ and other, as yet uncharacterized lymphokines, distinct from IL-2. Although it is, therefore, possible that B cells require different lymphokines for different differentiation pathways, it cannot be excluded that the different requirements observed in vitro merely reflect different stages of differentiation of these cells, and that B cells making antidextran IgM antibodies require additional factors because they are at an earlier stage of differentiation (Murray et al., 1985). What seems well established is the capacity of IFN-γ to act as a B-cell differentiation factor and stimulate immunoglobulin synthesis.

Interestingly, such stimulation of immunoglobulin synthesis is isotype specific. In resting murine B cells, stimulated into IgG synthesis by LPS, rec Mu IFN-γ selectively enhances the synthesis of IgG2a and at the same time causes a near complete suppression of IgG3, IgG1, and IgG2b synthesis

(Snapper and Paul, 1987). It is revealing for the function of IFN-γ that the IgG2A isotype is the most efficient antibody for the induction of antibody-dependent macrophage and killer cell cytotoxicity for tumor cells.

There are other indications that the effect of IFN-γ is not always stimulatory, and that under certain conditions, IFN-γ inhibits Ig synthesis. Several T-cell clones with different characteristics and cell-surface markers can be maintained in cell culture. One of these is a T-helper cell clone, which, when induced by the mitogen concanavalin A, produces supernatants that stimulate LPS-treated mouse B cells to synthesize IgG, IgA, and especially IgE (Nabel et al., 1981). IgE is normally one of the least abundant Ig isotypes; clinical allergy is caused by an exaggerated IgE response. The enhancement of IgE and also of IgG1 synthesis by the supernatants of the cloned T-helper cells is completely inhibited by IFN-γ; IgA synthesis is considerably less affected, whereas IgM synthesis is moderately enhanced at some concentrations of IFN-γ, and moderately inhibited at increasing doses of IFN-γ (Coffman and Carty, 1986).

IFN-γ can, furthermore, block the stimulating effect of IL-4 on resting B cells, and consequently, inhibit the anti-IgM-induced blast formation of these cells. This is discussed in detail in the next section.

D. IL-4 and Its Down-Regulation by IFN-γ

IL-4 is a T-helper cell-derived lymphokine, initially called B-cell stimulating factor 1 (BSF-1) or B-cell growth factor (BCGF-1). In spite of its initial name, IL-4 does not limit its activity to B cells, since in addition to being instrumental in B-cell activation and proliferation, it can also induce and maintain a state of activation in mast cells and T cells. IFN-γ, as an effective suppressor of most IL-4 activities, is a natural down-regulator of IL-4 activity.

1. Structure and Genetics

The cDNA for human IL-4 predicts a polypeptide of 153 amino acids, including a putative signal peptide. The predicted amino acid sequence for murine IL-4, also based on the cDNA, reveals an amino acid sequence of 140 residues, including a putative signal peptide of 20 amino acids (Table 11.1). The deduced molecular weight of the mature protein is about 15 kDa and the difference with the natural product, which has a molecular weight of about 20 kDa, is the result of glycosylation. The protein contains three glycosylation sites. Human and murine IL-4 share about 40% homology at the amino acid level and 70% at the DNA level. The mouse chromosomal gene occurs as a single copy in the haploid genome and contains four exons and three introns, extending about 6 kb in length. Extending over 200 bp, the nucleotide sequence upstream of the TATA box shows homology with the corresponding region of murine IL-2. This sequence is probably involved in the regulated, inducible expression of these genes during the activation of T cells

TABLE 11.1 Alignment of the Human and Murine IL-4-Deduced Amino Acid Sequences

```
                10         20          30         40          50
Human:  MGLTSQLLPP LFFLLACAGN FVHGHKCD-IT LQEIIKTLNS LT-EQKTLCTE
        ---  --    - - - -      --  --    - ---  --   - -  - ---
Murine: MGLNPQLVVI LLFFLECTRS HIHG--CDKNH LREIIGILNE VTGE-GTPCTE
                10         20          29         39          49

                60         70         80         90        100
Human:  LTVTDIFAAS KNTTEKETFC RAATVLRQFY SHHEKDTRCL GATAQQFHRH
          -     -  -----    - --  --- --   - - - --
Murine: MDVPNVLTAT KNTTESELVC RASKVLRIFY LKHGK-TPCL KKNSSVLMEL
                59         69         79         88         98

                  110        120        130           140
Human:  KQ-LIR-FLKR-L DRNLWGLAGL NSCPVKEANQ S--T-LENFLERL
         - - - - - - -                --    -  - -  --- -
Murine: -QRLFRAF--RCL DS-------S ISCTM---NE SKSTSLKDFLESL
                  108        111        118           131

               150 153
Human:  KTIMREKYSK CSS
        - --   --
Murine: KSIMQMDYS
              140
```

Source: Based on Lee et al, 1986; Noma et al., 1986; Yokota et al., 1986.

The putative signal peptide is included in the sequence.

The one-letter amino acid code is explained in Chapter 2, Table 2.5.

(Ohara et al., 1985; Grabstein et al., 1986; Lee et al., 1986; Noma et al., 1986; Yokata et al., 1986; Ohara et al., 1987; Otsuka et al., 1987).

2. The Receptor for IL-4

High affinity cell-surface receptors for Mu IL-4 are found on B cells and T cells, and on several other cells of hemopoietic lineage, including mast cells and macrophages as well as undifferentiated hemopoietic cell lines. It is revealing for the function of IL-4 that these receptors are already present on resting B and T cells, although receptor expression on such cells is low, since there are less than 100 binding sites per cell. The number of receptors increases substantially upon activation of B cells by LPS or anti-IgM and of T cells by the polyclonal activator concanavalin A. In view of the inhibitory effect of IFN-γ on IL-4 activity, it is important to note that Mu IFN-γ fails to block binding of IL-4 to mouse spleen cells. Apparently, the antagonistic action of IFN-γ is not the result of competition for the same receptor or to receptor down-regulation (Ohara and Paul, 1987; Park et al., 1987).

3. Production of IL-4

Studies with mouse T-helper-cell clones have revealed the existence of at least two T-helper-cell types with distinct patterns of lymphokine production. The first type, referred to as T_H1 cells, produces IL-2, IL-3, GM-CSF, and IFN-γ, in response to antigen-presenting cells or to concanavalin A. The second type, T_H2 cells, produces IL-4 and can also make IL-3. The secretion of IL-4 by these T cells is antigen-specific and MHC-restricted (Fernandez-Botran et al., 1986b; Mosmann et al., 1986a).

4. Activities of IL-4

a. Stimulation of B Cells. Stimulation of B-cell growth and proliferation is an important activity of IL-4. For this, IL-4 synergizes with B-cell stimuli such as LPS and anti-IgM antibody, to cause polyclonal B-cell proliferation. Preculture of resting B cells with IL-4 speeds the entry into the S phase upon subsequent stimulation of the B cells with the polyclonal activators LPS or anti-IgM antibody (Howard et al., 1982; Rabin et al., 1985, 1986a).

Together with B-cell growth and proliferation, the synthesis of some immunoglobulins by LPS-stimulated B cells can be enhanced by IL-4. These immunoglobulins are IgG1 and IgE; the synthesis of the latter can be enhanced over 100-fold. This is not an indiscriminate effect on all immunoglobulin species, but is isotype specific since the synthesis of IgM, IgG2a, IgG2b, and IgA is not enhanced but rather is inhibited by IL-4. (Vitteta et al., 1985; Coffman and Carty, 1986; Coffman et al., 1986, Snapper and Paul, 1987).

An additional effect of IL-4 is the stimulation of the MHC class II antigen expression on resting B cells, whereas IFN-γ does not seem to have this effect on resting B cells. This is one of the rare instances in which IFN-γ does not induce or enhance class II antigen expression (Noelle et al., 1984).

b. Effects on T Cells and Mast Cells. Like IL-2, which was initially characterized as a T-cell growth factor and later shown to affect also B cells, IL-4, discovered as a B-cell stimulatory agent, can also affect T-cell function.

In mice, IL-4 can act as a costimulator, in combination with concanavalin A or phorbol myristate acetate, to induce proliferation of immature L3T4$^-$ Ly-2$^-$ thymocytes. Since immature thymocytes are capable of producing IL-4, it is possible that IL-4 has a natural role as growth factor in T-cell ontogeny (Zlotnik et al., 1987).

IL-4 can also act as a growth factor for mature T cells. Murine IL-4, together with phorbol myristate acetate, stimulates resting T cells, derived from mesenteric lymph nodes, to enter the S phase of the cell cycle and to proliferate. Under certain conditions IL-4 promotes the proliferation of clones or lines of antigen-specific T-helper cells, but not the proliferation of alloreactive or cytotoxic T cells, whereas under different experimental conditions IL-4 stimulates the growth of both helper CD4$^+$ and cytotoxic CD8$^+$ cell clones. Since MHC-restricted stimulation of T_H2 helper cells by antigen presenting B cells or accessory cells results in the synthesis of IL-4, we have an example of an antigen-induced autocrine pathway for T-helper-cell activity. In addition to its effect on T cells, IL-4 cooperates with IL-3 to stimulate proliferation of mast cell lines (Fernandez-Botran et al., 1986a,b; Mosmann et al., 1986b; Yokota et al., 1986; Hu-Li et al., 1987; Spits et al., 1987).

5. *IFN-γ Inhibits B-Cell Stimulation by IL-4*

The main activity of IL-4 is the stimulation of cell growth, and, like in many other growth factor-target cell systems, the expression of c-*myc* in B cells is stimulated by IL-4 (Lacy et al., 1986). IFN-γ, which like all IFNs can exert antiproliferative activity, inhibits the IL-4-induced cell proliferative response, which is in line with the antagonistic effect of IFNs on growth factors in general. Rec Mu IFN-γ is a potent inhibitor of the action of IL-4 on resting B cells, blocking entry into the S phase and induction of MHC class II molecules. IFN-γ is, like IL-4, produced by T cells upon MHC-restricted antigen stimulation, and therefore appears as a natural down-regulator of the autocrine IL-4-induced pathway of T-helper cells, in that it blocks the response of all small resting B cells to this growth factor (Noelle et al., 1984; Mond et al., 1985; Rabin et al., 1986b).

E. B-Cell Differentiation Factor (BCDF or BSF-2) and IFN-β2 Are One and the Same Cytokine

Stimulation of Ig synthesis is the hallmark of the human B-cell differentiation factor BCDF, also called BSF-2, which induces the final differentiation of B cells into high-rate Ig-secreting cells (Hirano et al., 1985). Natural BSF-2 is usually produced either by activated T cells or constitutively by human T-cell leukemia virus (HTLV-1) transformed T-cell lines. The structure of the natural B-cell differentiation factor prepared this way is completely identical with that of a protein that has been independently described as Hu IFN-β2 or 26K protein (Content et al., 1985; Haegeman et al., 1986; Hirano et al., 1986; Revel et al., 1986; Zilberstein et al., 1986a).

The gene for this protein has been assigned to chromosome 7 (May et al., 1986; Sehgal et al., 1986; Zilberstein et al., 1986b). It contains at least four introns and codes for a 212 amino acid polypeptide; the mature protein probably consists of 184 amino acids (Table 11.2). The amino acid sequence homology between Hu IFN-β and Hu IFN-β2 is very low and the designation of this protein as an IFN is controversial. Its specific antiviral activity is less than 10^2 antiviral units per milligram, which is at least 10^6 times lower than the specific antiviral activity of Hu IFN-β (Poupart et al., 1987; Van Damme et al., 1987). Hu IFN-β2 or BSF-2 is, furthermore, identical to the independently described and characterized B-cell hybridoma/plasmacytoma growth factor (Van Damme et al., 1987).

The synthesis of Hu IFN-β2 is inducible in many different cells, including diploid fibroblasts, by cytokines such as IL-1 and TNF-α (see Chapter 13), as well as by polyrIrC in a superinduction schedule. There is a low level of constitutive production in monocytes.

F. Conclusions

IFNs can affect B-cell differentiation in different ways; whether the effect is up- or down-regulation seems to depend on the Ig isotype and on the lym-

TABLE 11.2 The Deduced Amino Acid Sequence of BSF-2 or Hu IFN-β2

10	20	30	40	50
MNSFSTSAFG	PVAFSLGLLL	VLPAAFPAPV	PPGEDSKDVA	APHRQPLTSS
60	70	80	90	100
ERIDKQIRYI	LDGISALRKE	TCNKSNMCES	SKEALAENNL	NLPKMAEKDG
110	120	130	140	150
CFQSGFNEET	CLVKIITGLL	EFEVYLEYLQ	NRFESSEEQA	RAVQMSTKVL
160	170	180	190	200
IQFLQKKAKN	LDAITTPDPT	TNASLLTKLQ	AQNQWLQDMT	THLILRSFKE
210	212			
FLQSSLRALR	QM			

Source: Hirano et al., 1986; May et al., 1986; Zilberstein et al., 1986b.
The 27 aa putative signal sequence is included.
The one-letter amino acid code is explained in Chapter 2, Table 2.5.

phokines present in the supernatants from the activated T cells. Another determining factor in this interaction is the state of activation of the B cells derived from either mice or humans. This variable is not easily controlled under experimental conditions. B-cell suspensions are invariably and of necessity mixtures of cells in different states of activation and differentiation. One must, furthermore, bear in mind that most B-cell differentiation experiments are performed in vitro, with the mediators under study added to the culture medium. B-cell differentiation in vivo takes place in the special microenvironment of bone marrow and lymphoid tissues, where the molecules providing the specific signals are in all likelihood delivered by a direct transfer from T-helper cells to B cells. In vitro, one can only reproduce approximations of the natural conditions, and we can bring out potential activities of different mediator molecules, but not always in the right context.

Since many experimental findings are based on the addition of exogenous IFNs, caution has to be exerted when extrapolating from the in vitro results to a possible physiological role in vivo.

III. THE EFFECT OF INTERFERONS ON NATURAL KILLER CELLS

Besides MHC-restricted, antigen-specific cytotoxic T cells, active against allografts, tumor cells, and virus-infected cells, some cells are capable of mounting a cytotoxic response against malignant and virus-infected cells—and even against normal cells—without the need for prior immunization and without MHC restriction. Because of their spontaneous cytotoxicity and lack of antigen specificity, these cells have been called natural killer (NK) cells. Some cytotoxic T cells are also non-MHC-restricted, but unlike cytotoxic T cells, NK cells are not thymus-dependent, since they are present in nude and in neonatally thymectomized mice (Saksela, 1981; Herberman 1984). The origin of NK cells is still a matter of debate. They are not derived from the macrophage and B-cell lineage and are usually believed not to be of the T-cell lineage either, since, contrary to non-MHC-restricted T cells, they do not express the CD3 T-cell antigen receptor complex, and their T-cell receptor germline genes are not rearranged (Lanier et al., 1986; Trinchieri, 1986). Indeed, although some cloned murine cell lines that mediate NK-like killing have been shown to display gene rearrangement and transcription of the variable chains of the T-cell receptor (Yanagi et al., 1985; Ikuta et al., 1986), NK cells that are directly isolated from mice, after in vivo stimulation by an IFN-α/β inducer, do not express any of the T-cell receptor variable chains (Biron et al., 1987).

However, since some NK cells express characteristic T-cell-surface antigens like CD2, CD8, and CD11, and since the growth of a subset of NK cells can be stimulated by IL-2, the possibility has been raised that NK cells are of T-cell lineage but diverge from them before the early stage of thymocyte development, when all T cells undergo antigen receptor γ gene rearrangement (Kaplan, 1986). It is also possible that, rather than representing prethymic T cells, NK cells develop along a separate pathway altogether.

Morphologically, NK cells have been identified as large granular lymphocytes (LGL) that, unlike mature T cells, have a high cytoplasmic-nuclear ratio with azurophilic granules in their cytoplasm (de Landazuri et al., 1981; Timonen et al., 1981). Human peripheral blood contains about 5% LGL, of which up to 80% can function as NK cells (Timonen et al., 1982). The human NK-cell population is in all likelihood heterogeneous, since NK lysis of tumor target cells and lysis of herpes-virus-infected cells are independently regulated in patients with primary immunodeficiencies (Messina et al., 1986). The heterogeneity of the NK-cell population is, furthermore, suggested by the different cell surface and target cell specificities displayed by clones of large granular lymphocytes (Ortaldo and Herberman, 1984).

The cytoplasmic granules of NK cells contain cytolytic proteins involved in target cell destruction by a mechanism of cell lysis that is also present in cytotoxic T cells. Purified cytoplasmic granules of large granular lymphocytes contain potent cytolysins, and granule exocytosis appears to be instrumental in delivering lethal damage to target cells. During interaction with the target cell, the proteins present in the granules polymerize into tubular complexes, called perforins. These perforin tubules are then inserted into the target cell membrane and function as transmembrane channels causing leakage of water, ions, and macromolecules across the plasma membrane, resulting in lysis of the cell (Henkart et al., 1984; Podack and Konigsberg, 1984; Podack et al., 1985; Henkart et al., 1986; Liu et al., 1986). Interestingly, this pore-forming protein, found in cytolytic T cells and NK cells, is antigenically related to some components of the complement cascade, and perforin (also called cytolysin) and the C9 component display structural, immunological and functional similarities (Young et al., 1986; Zalman et al., 1986).

NK-cell activity is usually measured by an in vitro test, in which the lysis of target cells—generally belonging to one of several continuous lines of tumor cells—is quantified by the release of radioactive chromium. Since NK cells are not MHC-restricted, target cells can be syngeneic, allogeneic, and even xenogeneic, quite unlike the highly specific requirements for dual recognition in the case of antigen-specific cytotoxic T cells. Because NK cells can be activated without previous sensitization, they are believed to be, like macrophages, in the first line of defense against tumor cells and infectious agents. The boosting by IFNs of NK activity has, therefore, received considerable attention, since it seemed to offer an explanation for some of the antitumor effects of IFNs; the clinical data, however, do not indicate a good correlation between tumor regression and NK stimulation (see section D). Activation of NK activity by IFNs, furthermore, contributes to host defense during infection with some viruses (Welsh, 1981; Fitzgerald et al., 1982; Chmielarczyk et al., 1983; Kirchner et al., 1983; Munoz et al., 1983; Bukowski and Welsh, 1985b).

Cells of the hematopoietic system, especially bone-marrow cells, can be "normal" target cells for NK activity, and for this reason, a role of NK cells in regulating hematopoiesis and bone-marrow activity has been proposed (Kiessling et al., 1977). This proposal is based on the observation that parental bone-marrow-graft rejection in F1 hybrids is apparently mediated by NK cells, which implies that NK cells can recognize syngeneic hematopoietic progenitors. According to the laws of transplantation, applying to individuals of highly inbred strains, grafts from either parent to F1 hybrids succeed, since F1 hybrids do not recognize the parental cells as foreign. If the transplanted parental cells are lymphocytes, however, they can recognize and react against those F1 antigens that come from the other parent, thus causing a graft-versus-host disease.

The fact that, in contradiction to the laws of transplantation, parental bone-marrow stem cells, as opposed to skin grafts, for example, are sometimes rejected by F1 hybrids indicates the existence of a special, probably NK cell-mediated mechanism that recognizes and controls, the proliferation of hematopoietic stem cells. Indeed, in vitro, large granular lymphocytes having NK cell characteristics can suppress the development of autologous hematopoietic colonies such as erythrocyte colonies (CFU-E) and the more primitive CFU-S bone-marrow stem cell colonies (Holmberg et al., 1984). Some of this suppressive effect is due to the synthesis of TNF-α by NK cells (Degliantoni et al., 1985). Obviously, once it is firmly established that NK cells play a role in the regulation of hematopoiesis, it will be necessary to determine the specific features of hematopoietic cells that trigger NK-cell activity.

NK-cell activity is greatly influenced by IFNs. It can be up-regulated or down-regulated, either directly through effects of IFNs on the NK cells themselves, or indirectly through effects on the target cells.

A. Stimulation of NK-cell Activity by IFNs

1. Stimulation by IFN-α and IFN-β

In mice, intraperitoneal administration of IFN-α/β leads to a marked increase of NK-cell activity as soon as 3 hours afterward (Gidlund et al., 1978). In vitro, NK-cell activity from mouse spleen cells or from human peripheral blood leukocytes is boosted by Mu IFN-α/β or by Hu IFN-α or IFN-β, and there is usually an optimal IFN dosage, above which NK activity instead of being enhanced, is often decreased (Kuribayashi et al., 1981; Herberman et al., 1982; Brunda and Rosenbaum, 1984). Hyporesponsiveness to augmentation of NK-cell activity also occurs in vivo, in mice receiving multiple inoculations of Mu IFN-α/β and in patients treated with recombinant Hu IFN-α or lymphoblastoid IFN derived from Namalwa cells (Edwards et al., 1985; Talmadge et al., 1985). How NK-cell activity is boosted by moderate amounts of IFN, and how hyporesponsiveness is induced by larger doses or by prolonged treatment is not well understood. Down-regulation of growth and differentiation of NK-cell precursors by prolonged treatment with Mu IFN-α/β seems to occur indirectly, via a T cell-mediated suppressive activity. For example, nonadherent cells, probably T lymphocytes, which inhibit NK cell activity, can be isolated from mice after several days of treatment with IFN-α/β (Riccardi et al., 1983, 1986). Suppressor cells of NK activity are naturally present in human umbilical cord blood, and have also been found in patients with malignant disease. An interesting paradox is provided by the fact that in vitro, suppressor cells of NK activity are readily inactivated by Hu leukocyte IFN or rec Hu IFN-α, whereas in vivo, prolonged treatment with IFN seems to favor the generation of these suppressor cells (Mantovani et al., 1980; Tarkkanen and Saksela, 1982; Nair et al., 1986). This paradox is more apparent than real, and reflects the isolation of an in vitro system. Without any doubt, the regulation of NK-cell activity in vivo involves much more than just IFN, as NK cells can be influenced, among others, by IL-1, IL-2, thymosin, and the neurohormone β-endorphin (Dempsey et al., 1982; Favalli et al., 1985; Mandler et al., 1986; Papa et al., 1986).

IFN-mediated enhancement of NK-cell activity seems to involve a recruitment of cells from a noncytolytic "pre-NK" pool, and to a lesser extent an augmentation of the activity of already mature NK cells (Trinchieri et al., 1978; Herberman et al., 1979; Heron et al., 1979; Saksela et al., 1979; Moore and Potter, 1980; Targan and Dorey, 1980). Exposure to IFN increases the number of killing cells among pre-NK cells that are capable of binding but not of lysing target cells, and it also increases the number of binding cells.

IFN-mediated boosting of NK-cell activity in mouse spleens is due both to an increase in the number of large granular lymphocytes, as well as to an increased activity of already existing LGL (Santoni et al., 1985). Multiple mechanisms are at play, since there is also evidence that IFNs not only stimulate conversion of inactive, target binding, large granular lymphocytes into active cytolytic cells but also increase the degree of recycling of NK cells, with the result that more target cells are lysed per NK cell present (Bloom et al., 1982; Timonen et al., 1982).

2. *Stimulation by IFN-γ*

NK-cell activity is also boosted by IFN-γ, either natural or recombinant (Weigent et al., 1983; Platsoucas, 1986). Human neonatal NK cells, however, are for some reason relatively resistant to boosting by IFN-γ, although they can be boosted by Hu IFN-α, or by IL-2. Whether this means that there are different stages of NK-cell maturation, or distinct populations with different activation requirements, has not been determined (Ueno et al., 1985; Oh et al., 1986).

B. IFNs Protect Some Cells Against Lysis by NK Cells

In the absence of an antigen-specific recognition mechanism, there is no answer to the major question: What are the structural features that cause a cell to become a target for NK activity? In vitro, NK cells act not only against tumor cells and virus-infected cells, but also against a wide variety of normal cells as well. Binding of NK cells to target cells is not necessarily followed by destruction of the latter, which shows that there are at least two levels of recognition, the first one resulting in binding and the second in triggering lysis (Collins et al., 1981). When normal murine fibroblasts are transformed to express the cellular Harvey *ras* (c-Ha-*ras*) p21 oncogene protein, their sensitivity to lysis by NK cells increases, indicating that the expression of some proteins by the transformed cell is recognized by NK cells (Trimble et al., 1986). Why NK cells act selectively in vivo against cancer cells and virus-infected cells, without destroying normal cells, except maybe hematopoietic stem cells, is not understood.

In vitro, IFN treatment of normal cells confers protection against killing by NK cells. When pretreated with Hu leukocyte IFN, Hu IFN-β or Hu IFN-γ, human foreskin fibroblasts and thymocytes become almost completely resistant to NK-cell activity, whereas IFNs do not confer this protective effect to cells of many tumor cell lines (Trinchieri et al., 1981; Hansson et al., 1980). This has been envisaged as a mechanism through which NK cells could selectively kill tumor cells without harming normal cells (Trinchieri et al., 1978; Djeu et al., 1980), but this view is not really tenable, since some leukemic cell lines, and other tumor-derived cells as well, are also protected against NK activity by preexposure to IFN (Fig. 11.6) (Moore et al., 1980; Welsh et al., 1981). Target cell protection by IFNs also operates in

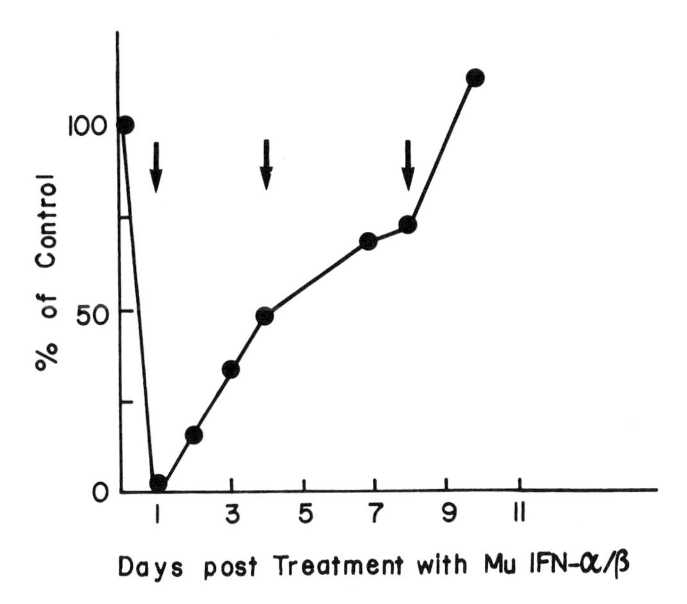

Figure 11.6 Decrease of target cell sensitivity to NK cell-mediated lysis after treatment with Mu IFN-α/β, and its regeneration. Mouse L-929 cells were treated with 1,000 units of IFN for 20 hours, and reincubated without IFN. Regeneration of sensitivity to NK cells, contained in mouse spleen cell suspensions, is a gradual process, requiring several days. (Adapted from Welsh et al., 1981.)

vivo: Cells of several murine lymphoma lines can be protected against NK activity in vitro by pretreatment with Mu IFN-α/β; when inoculated into mice, these IFN-treated lymphoma cells have become more resistant to NK cell-mediated tumor rejection (Greenberg et al., 1984). The lack of selectivity between normal and transformed cells clearly shows that the protective action conferred by IFN on target cells is not the overall mechanism that enables NK cells to distinguish between transformed and untransformed cells.

IFNs of all three species can protect cells against NK activity as shown, for example, in HeLa cells, representing a human tumor cell line, which can be protected by either Hu IFN-α, β, or γ (Wallach, 1983). Protection of cells by IFN against NK killing is an active process since it requires protein synthesis, but there is no indication as to the mechanism leading to this effect. Although it is logical to assume that IFN-induced cell-surface changes are involved, evidence for this is lacking.

C. NK Cells as IFN Producers

IFN-α and IFN-β production has been demonstrated in all nucleated cells in the body that have been properly examined, and therefore it is not a surprise that NK cells can also be induced to make these IFNs. For example, murine

NK cells infected with Sendai virus produce Mu IFN-α/β (Handa et al., 1983). More interesting, NK cells can be stimulated into making IFN-γ, a property shared with mature T cells only; this provides an argument in favor of the thesis that NK cells are of the T-cell lineage. Cloned lines of murine NK cells, or freshly isolated human large granular lymphocytes, cultivated in the presence of IL-2, produce Mu IFN-γ without any other stimulus. (Handa et al., 1983; Young and Ortaldo, 1987). The cell killing capacity of human and murine NK cells can be boosted by IL-2; the enhancement of activity is mainly due to a direct effect of IL-2 on NK function and probably not to the IL-2-induced IFN-γ production, since monoclonal antibodies to IFN-γ have little effect on the stimulation of NK activity by IL-2 (Sayers et al., 1986).

The production of IFN-γ by NK cells offers an alternative pathway for macrophage activation in the case of T-cell deficiency. An interesting demonstration of this is provided by mice with the severe combined immunodeficiency (scid) mutation, which have no detectable T or B cell activity. Infection of such lymphocyte-deficient mice with *Listeria monocytogenes* nevertheless triggers IFN-γ production, resulting in activation of macrophages and partial control of the bacterial infection (Bancroft et al., 1987).

Since to some extent IFNs regulate NK-cell activity, IFN production by these cells provides an example of autocrine stimulation. However, and like practically all other experiments involving IFN production, the studies on IFN production by large granular lymphocytes have been performed by integrating the production of a large number of cells, with no attempt made to identify individual IFN-producing cells. Large granular lymphocytes appear as a heterogeneous population when their cell surface antigens are analyzed with monoclonal antibodies, and therefore we can not be totally sure that the same large granular lymphocytes that make IFN are also responsible for the cytotoxic effect.

NK cells can be stimulated into IFN production by viruses, bacteria, mitogens, and cells. When suspensions of large granular lymphocytes are infected with either influenza or herpes simplex virus type 1, they produce mainly IFN-α. But they also produce IFN-γ if the cells are derived from individuals that have previously been exposed to these viruses as ascertained by the presence of antibodies (Djeu et al., 1982). The latter observation, taken at face value, does indicate a potential for antigen recognition by at least some large granular lymphocytes, without necessarily implying that these are the same lymphocytes that have NK capacity. Human large granular lymphocytes, when treated with lectins such as concanavalin A or phytohemagglutinin, produce several lymphokines including IL-2, colony stimulating factor, and IFN-γ. In addition to IFN-γ, concanavalin A also induces IFN-β (Kasahara et al., 1983a). Bacteria such as the tuberculosis bacillus of the strain Calmette-Guérin (BCG) and *Corynebacterium parvum* induce IFN-α in LGL isolated from peripheral blood, as do mycoplasma-contaminated tumor cells, but not tumor cells that are free of mycoplasma (Djeu,

1983). Fractions of human peripheral blood, highly enriched for large granular lymphocytes (about 75% of the cells) produce spontaneously IFN-γ and, interestingly in view of its production during autoimmune diseases, produce acid labile IFN-α (Fisher and Rubinstein, 1983).

IFN production by NK cells, triggered by a variety of inducers, suggests the existence of a positive feedback loop, whereby upon contact with transformed or infected cells, NK-cell activity is upregulated by the IFN produced as a result of the first contact. Since, on the other hand, an excess of IFN down-regulates NK activity (see section B) and can induce resistance of target cells to the cytolytic activity, the conditions for a complete autoregulatory mechanism would seem to be fulfilled. This is, however, by no means proven, and for the time being can best be considered as an attractive working hypothesis, keeping in mind that other cytokines, such as IL-2, are also involved in the regulation of the NK system (Kuribayashi et al., 1981; Brunda et al., 1986; Sayers et al., 1986; Van de Griend et al., 1986).

Several studies in animal models lend support to the view that NK-cell activity is intimately linked to the IFN system. Treatment of athymic nude mice with anti Mu IFN-α/β globulin favors the development of persistently virus-infected HeLa or BHK-cell tumors, which normally do not grow in these mice. One possible explanation would be that the stimulation of tumor growth in these animals is caused by the decreased NK-cell activity resulting from the treatment with anti-IFN globulin (Reid et al., 1981). One can specifically deplete mice of NK cells by treating them with antibodies directed against the NK cell-surface antigen asialo GM1. In such animals Mu IFN-α/β still inhibits the replication of Moloney Sarcoma virus-transformed cells when optimal conditions of timing and IFN dosage are used. Under suboptimal conditions, however—with greater tumor load or shorter duration of IFN treatment—treatment with anti-asialo-GM1, antibody partially abrogates the anti-tumor effect of IFN (Fresa and Murasko, 1986). This is highly suggestive of at least some contribution of NK cells to IFN-mediated inhibition of tumor growth.

D. Stimulation of NK Activity in IFN-Treated Patients

NK-cell activity is usually measured with a limited number of tumor-cell lines as target cells, but to ascertain the clinical relevance of enhanced NK-cell activity, it would be desirable to use the patient's tumor cells as a target. This is very difficult to do in practice, as one does not always have access to the tumor, and, moreover, tumor cells from most surgically removed cancers are hard to maintain in culture. There is no lack of evidence that IFN treatment of cancer patients with Hu IFN-α of different origins and with Hu IFN-β can boost the NK-cell activity of some individuals, at least when the standard allogeneic cells are used as targets for the test (Lucero et al., 1981; Pape et al., 1981; Lotzova et al., 1982; Maluish et al., 1983).

Frequently, however, with continuing IFN treatment, the first rise in NK activity is followed by a return to pretreatment levels, and NK cells from such patients can no longer be boosted in vitro by IFN (Golub et al., 1982a,b). In addition, when the level of NK activity is already relatively high before IFN is given, a depression of NK activity sometimes occurs right at the onset of IFN treatment (Lotzova et al., 1983; Spina et al., 1983). Generation of suppressor cells by IFN has been proposed as one of the mechanisms to explain the decreased NK activity in IFN-treated patients, but the importance of this phenomenon in the downregulation of NK activity in patients treated with IFN remains to be determined (Uchida et al., 1984).

The specificity of the target cells poses another problem for the interpretation of results when one follows NK activity in IFN-treated patients. Although a boosting of NK-cell activity is readily observed when the allogeneic tumor cells serve as targets, there is frequently no evidence for an increased NK activity against the patient's own tumor cells (Vanky et al., 1980; Vanky and Klein, 1982). This probably explains the lack of correlation between enhanced NK-cell activity, as measured on the usual allogeneic target cell lines, and the clinical remission of cancer patients after IFN treatment (Golub et al., 1982; Borden, 1983; Lotzova et al., 1983).

NK cells, because they can be recruited on short notice without necessitating previous immunization, may play a role in tumor surveillance and in the defense against some virus infections. But all the evidence is not yet in to decide how important their contribution is. The activation of NK cells by IFNs is an established fact; the importance of this activation in some of the remissions obtained after IFN treatment in cancer patients is still an open question.

References

Abb, J., Abb, H., and Deinhardt, F. Characterization of human interferon-γ-producing leukocytes with monoclonal antileukocyte antibodies. *Med. Microbiol. Immunol.* 171: 215–223 (1983).

Acuto, O. and Reinherz, E. L. The human T-cell receptor. Structure and function. *N. Engl. J. Med.* 312: 1100–1111 (1985).

Alton, K., Stabinsky, Y., Richards, R., Ferguson, B., Goldstein, L., Altrock, B., Miller, L., and Stebbing, N. Production, characterization and biological effects of recombinant DNA derived human IFN-α and IFN-γ analogs. In: *The Biology of the Interferon System 1983,* E. De Maeyer and H. Schellekens, Eds, Elsevier, Amsterdam (1983).

Aoki, I., Minami, M., and Dorf, M. E. A mechanism responsible for the induction of H-2 restricted second order suppressor T cells. *J. Exp. Med.* 157: 1726–1735 (1983).

Archer, D. L., Smith, B. G., Ulrich, J. T., and Johnson, H. M. Immune interferon induction by T-cell mitogens involves different T-cell subpopulations. *Cell. Immunol.* 48: 420–426 (1979).

Bancroft, G. J., Schreiber, R. D., Bosma, G. C., Bosma, M. J., and Unanue, E. R. A T Cell-independent mechanism of macrophage activation by interferon-γ. *J. Immunol.* 139: 1104–1107 (1987).

Bhayani, H. and Falcoff, R. T-cell surface antigens defined by monoclonal antibodies, involved in the induction of human interferon-γ and interleukin 2. *Cell. Immunol.* 536–546 (1985).

Bich-Thuy, L. T., Queen, C., and Fauci, A. S. Interferon-γ induces light chain synthesis in interleukin 2 stimulated human B cells. *Eur. J. Immunol.* 16: 547–550 (1986).

Biron, C. A., Van den Elsen, P., Tutt, M. M., Medveczky, P., Kumar, V., and Terhorst, C. Murine natural killer cells stimulated in vivo do not express the T cell receptor α, β, γ, T3δ or T3ε genes. *J. Immunol.* 139: 1704–1710 (1987).

Blackman, M. J. and Morris, A. G. The effect of interferon treatment of targets on susceptibility to cytotoxic T-lymphocyte killing: Augmentation of allogeneic killing and virus-specific killing relative to viral antigen expression. *Immunology* 56: 451–457 (1985).

Blanchard, D. K., Djeu, J. Y., Klein, T. W., Friedman, H., and Stewart II, W. E. Interferon-γ induction by lipopolysaccharide: dependence on interleukin 2 and macrophages. *J. Immunol.* 136: 963–970 (1986).

Bloom, B. R., Schneck, J., Rager-Zisman, B., Quan, P. C., Neighbour, A., Minato, N., Reid, L., and Rosen, O. Interactions between interferon and cells of the immune system. In: *Interferons, UCLA Symposia on Molecular and Cellular Biology,* T. C. Merigan and R. M. Friedman, Eds., Vol. XXV, Academic Press, New York, pp. 269–278 (1982).

Blue, M. L., Daley, J. F., Levine, H., Craig, K. A., and Schlossman, S. F. Identification and isolation of a T4^{+}T8^{+} cell with high T3 expression in human thymus: A possible late intermediate in thymocyte differentiation. *J. Immunol.* 139: 1065–1069 (1987).

Booth, R. J., Booth, J. M., and Marbrook, J. Immune conservation: A possible consequence of the mechanism of interferon-induced antibody suppression. *Eur. J. Immunol.* 6: 769–772 (1976a).

Booth, R. J., Rastrick, J. M., Bellamy, A. R., and Marbrook, J. Modulating effects of interferon preparations on an antibody response in vitro. *Austr. J. Exp. Biol. Med. Sci.* 54: 11–25 (1976b).

Borden, E. C. Interferons and cancer: How the promise is being kept. In: *Interferon 5,* I. Gresser, Ed., Academic Press, London, pp. 43–83 (1983).

Borst, J., Van de Griend, R. J., Van Oostveen, J. W., Ang, S. L., Melief, C. J., Seidman, J. G., and Bolhuis, R. L. H. A T-cell receptor γ/CD3 complex found on cloned functional lymphocytes. *Nature* 325: 683–688 (1987).

Braun, W. and Levy, H. B. Interferon preparations as modifiers of immune responses. *Proc. Soc. Exp. Biol. Med.* 141: 769–773 (1972).

Brenner, M. B., McLean, J., Scheft, H., Riberdy, J., Ang, S. L., Seidman, J. G., Devlin, P., and Krangel, M. S. Two forms of the T-cell receptor γ protein found on peripheral blood cytotoxic T lymphocytes. *Nature* 325: 689–694 (1987).

Brodeur, B. R. and Merigan, T. C. Suppressive effect of interferon on the humoral immune response to sheep red blood cells in mice. *J. Immunol.* 113: 1319–1325 (1974).

Brunda, M. J. and Rosenbaum, D. Modulation of murine natural killer cell activity in vitro and in vivo by recombinant human interferons. *Cancer Res.* 44: 597–601 (1984).

Brunda, M. J., Tarnowski, D., and Davatelis, V. Interaction of recombinant interferons with recombinant interleukin-2: Differential effects on natural killer cell activity and interleukin-2-activated killer cells. *Int. J. Cancer* 37: 787–793 (1986).

Brunswick, M. and Lake, P. Obligatory role of γ interferon in T cell-replacing factor-dependent antigen-specific murine B cell responses. *J. Exp. Med.* 161: 953–971 (1985).

Budd, R. C., Cerottini, J. C., and MacDonald, H. R. Selectively increased production of interferon-γ by subsets of Lyt-2^+ and L3T4^+ T cells identified by expression of Pgp-1. *J. Immunol.* 138: 3583–3586 (1987).

Bukowski, J. F. and Welsh, R. M. Interferon enhances the susceptibility of virus-infected fibroblasts to cytotoxic T cells. *J. Exp. Med.* 161: 257–262 (1985a).

Bukowski, J. F. and Welsh, R. M. Inability of interferon to protect virus-infected cells against lysis by natural killer (NK) cells correlates with NK cell-mediated antiviral effects in vivo. *J. Immunol.* 135: 3537–3541 (1985b).

Bukowski, J. F. and Welsh, R. M. Enhanced susceptibility to cytotoxic T lymphocytes of target cells isolated from virus-infected or interferon-treated mice. *J. Virol.* 59: 735–739 (1986).

Caccia, N., Kronenberg, M., Saxe, D., Haars, R., Bruns, G. A. P., Goverman, J., Malissen, M., Willard, H., Yoshikai, Y., Simon, M., Hood, L., and Mak, T. W. The T cell receptor β chain genes are located on chromosome 6 in mice and chromosome 7 in humans. *Cell* 37: 1091–1099 (1984).

Celis, E., Miller, R. W., Wiktor, T. J., Dietzschold, B., and Koprowski, H. Isolation and characterization of human T cell lines and clones reactive to rabies virus: Antigen specificity and production of interferon-γ. *J. Immunol.* 136: 692–697 (1986).

Ceredig, R., Lowenthal, J. W., Nabholz, M., and MacDonald, H. R. Expression of interleukin-2 receptors as a differentiation marker on intrathymic stem cells. *Nature* 314: 98–100 (1985).

Chang, T. W., Testa, D., Kung, P. C., Perry, L., Dreskin, H. J., and Goldstein, G. Cellular origin and interactions involved in γ-interferon production induced by OKT3 monoclonal antibody. *J. Immunol.* 128: 585–589 (1982).

Chen, L. K., Mathieu-Mahul, D., Back, F. H., Dausset, J., Bensussan, A., and Sasportes, M. Recombinant interferon-α can induce rearrangement of T-cell antigen receptor α-chain genes and maturation to cytotoxicity in T-lymphocyte clones in vitro. *Proc. Natl. Acad. Sci.* (USA) 4887–4889 (1986a).

Chen, L., Tourvieille, B., Burns, G. F., Mathieu-Mahul, D., Sasportes, M., and Bensussan, A. Interferon: A cytotoxic T lymphocyte differentiation signal. *Eur. J. Immunol.* 16: 767–770 (1986b).

Chester, T. J., Paucker, K., and Merigan, T. C. Suppression of mouse antibody producing spleen cells by various interferon preparations. *Nature* 246: 92–94 (1973).

Chmielarczyk, W., Engler, H., Brucher, J., and Kirchner, H. Herpes simplex virus-induced interferon production and activation of natural killer in SM/J mice. Relation to antiviral resistance. *Antiv. Res.* 3: 325–333 (1983).

Coffman, R. L. and Carty, J. A T cell activity that enhances polyclonal IgE production and its inhibition by interferon-γ. *J. Immunol.* 136: 949–954 (1986).

Coffman, R. L., Ohara, J., Bond, M. W., Carty, J., Zlotnik, A., and Paul, W. E. B cell stimulatory factor-1 enhances the IgE response of lipopolysaccaride-activated B cells. *J. Immunol.* 136: 4538–4541 (1986).

Collins, J. L., Patek, P. Q., and Cohn, M. Tumorigenicity and lysis by natural killers. *J. Exp. Med.* 153: 89–106 (1981).

Collins, M. K. L., Goodfellow, P. N., Dunne, M. J., Spurr, N. K., Solomon, E., and Owen, M. J. A human T-cell antigen receptor β chain gene maps to chromosome 7. *EMBO J.* 3: 2347–2349 (1984).

Collins, M. K. L., Goodfellow, P. N., Spurr, N. K., Solomon, E., Tanigawa, G., Tonegawa, S., and Owen, M. J. The human T-cell receptor α-chain gene maps to chromosome 14. *Nature* 314: 273–274 (1985).

Conta, B. S., Powell, M. B., and Ruddle, N. H. Production of lymphotoxin, IFN-γ and IFN-α,β by murine T cell lines and clones. *J. Immunol.* 130: 2231–2235 (1983).

Content, J., De Wit, L., Poupart, P., Opdenakker, G., Van Damme, J., and Billiau, A. Induction of a 26 kDa-protein mRNA in human cells treated with an interleukin-1-related, leukocyte-derived factor. *Eur. J. Biochem.* 152: 253–257 (1985).

Cooley, M. A., Blackman, M. J., and Morris, A. G. Production of type I (α/β) interferon after virus infection of cloned, allo-antigen-sensitized mouse T lymphocytes. *Eur. J. Immunol.* 14: 376–379 (1984).

Cunningham, A. L. and Merigan, T. C. Leu-3$^+$ T cells produce γ-interferon in patients with recurrent herpes labialis. *J. Immunol.* 132: 197–202 (1984).

Cunningham, A. L., Nelson, P. A., Fathman, C. G., and Merigan, T. C. Interferon γ production by herpes simplex virus antigen-specific T cell clones from patients with recurrent herpes labialis. *J. Gen. Virol.* 66: 249–258 (1985).

Degliantoni, G., Murphy, M., Kobayashi, M., Francis, M. K., Perussia, B., and Trinchieri, G. Natural killer (NK) cell-derived hematopoietic colony-inhibiting activity and NK cytotoxic factor. Relationship with tumor necrosis factor and synergism with immune interferon. *J. Exp. Med.* 162: 1512–1530 (1985).

De Maeyer, E., De Maeyer-Guignard, J., and Jullien, P. Circulating interferon production in the mouse. Origin and nature of cells involved and influence of animal genotype. *J. Gen. Virol.* 56: 43s–56s (1970).

Dembic, Z., Bannwarth, W., Taylor, B. A., and Steinmetz, M. The gene encoding the T-cell receptor α-chain maps close to the Np-2 locus on mouse chromosome 14. *Nature* 314: 271–273 (1985).

Dempsey, R. A., Dinarello, C. A., Mier, J. W., Rosenwasser, L. J., Allegretta, M., Brown, T. E., and Parkinson, D. R. The differential effects of human leukocytic pyrogen-lymphocyte-activating factor, T cell growth factor, and interferon on human natural killer activity. *J. Immunol.* 129: 2504–2510 (1982).

Djeu, J. Y. Production of interferon by natural killer cells. *Clin. Immunol. Allergy* 3: 561–568 (1983).

Djeu, J. Y., Huang, K. Y., and Herberman, R. B. Augmentation of mouse natural killer activity and induction of interferon by tumor cells in vivo. *J. Exp. Med.* 151: 781–789 (1980).

Djeu, J. Y., Stocks, N., Zoon, K., Stanton, G. J., Timonen, T., and Herberman, R. B. Positive self-regulation of cytotoxicity in human natural killer cells by production of interferon upon exposure to influenza and herpes viruses. *J. Exp. Med.* 156: 1222–1234 (1982).

Edwards, B. S., Merritt, J. A., Fuhlbridge, R. C., and Borden, E. C. Low doses of interferon α result in more effective clinical natural killer cell activation. *J. Clin. Invest.* 75: 1908–1913 (1985).

Ennis, F. A. and Meager, A. Immune interferon produced to high levels by antigenic stimulation of human lymphocytes with influenza virus. *J. Exp. Med.* 154: 1279–1289 (1981).

Epstein, L. B. The ability of macrophages to augment in vitro mitogen- and antigen-stimulated production of interferon and other mediators of cellular immunity by lymphocytes. In: *Immunobiology of the Macrophage,* D. S. Nelson, Ed., Academic Press, New York, pp. 201–234 (1976).

Epstein, L. B. and Gupta, S. Human T-lymphocyte subset production of immune (γ) interferon. *J. Clin. Immunol.* 1: 186–194 (1981).

Epstein, L. B., Stevens, D. A., and Merigan, T. C. Selective increase in lymphocyte interferon response to Vaccinia antigen after revaccination. *Proc. Natl. Acad. Sci.* (USA) 69: 2632–2636 (1972).

Falkoff, R. J. M., Zhu, L. P., and Fauci, A. S. Separate signals for human B cell proliferation and differentiation in response to *Staphylococcus aureus:* Evidence for a two-signal model of B cell activation. *J. Immunol.* 129: 97–102 (1982).

Favalli, C., Jezzi, T., Mastino, A., Rinaldi-Garaci, C., Riccardi, C., and Garaci, E. Modulation of natural killer activity by thymosin α1 and interferon. *Cancer Immunol. Immunother.* 20: 189–192 (1985).

Fernandez-Botran, R., Krammer, P. H., Diamanstein, T., Uhr, J. W., and Vitetta, E. S. B-cell-stimulatory factor 1 (BSF-1) promotes growth of helper T cell lines. *J. Exp. Med.* 164: 580–593 (1986a).

Fernandez-Botran, R., Sanders, V. M., Oliver, K. G., Chen, Y. W., Krammer, P. H., Uhr, J. W., and Vitetta, E. S. Interleukin 4 mediates autocrine growth of helper T cells after antigenic stimulation. *Proc. Natl. Acad. Sci.* (USA) 83: 9689–9693 (1986b).

Fischer, D. G. and Rubinstein, M. Spontaneous production of interferon-γ and acid labile interferon-α by subpopulations of human mononuclear cells. *Cell. Immunol.* 81: 426–434 (1983).

Fitzgerald, P. A., Von Wussow, P., Lopez, C. Role of interferon in natural kill of HSV-1-infected fibroblasts. *J. Immunol.* 129: 819–823 (1982).

Fleischer, T. A., Attallah, A. M., Tosato, G., Blease, R. M., and Greene, W. C. Interferon-mediated inhibition of human polyclonal immunoglobulin synthesis. *J. Immunol.* 129: 1099–1103 (1982).

Fradelizi, D. and Gresser, I. Interferon inhibits the generation of allospecific suppressor T lymphocytes. *J. Exp. Med.* 155: 1610–1622 (1982).

Frasca, D., Adorini, L., Landolfo, S., and Doria, G. Enhancing effect of interferon-γ on helper T cell activity and IL2 production. *J. Immunol.* 134: 3907–3911 (1985).

Fresa, K. L. and Murasko, D. M. Role of natural killer cells in the mechanism of the antitumor effect of interferon on Moloney sarcoma virus-transformed cells. *Cancer Res.* 46: 81–88 (1986).

Gay, D., Maddon, P., Sekaly, R., Talle, M. A., Godfrey, M., Long, E., Goldstein,

G., Chess, L., Axel, R., Kappler, J., and Marrack, P. Functional interaction between human T-cell protein CD4 and the major histocompatibility complex HLA-DR antigen. *Nature* 328: 626–629 (1987).

Gidlund, M., Orn, A., Wigzell, H., Senik, A., and Gresser, I. Enhanced NK cell activity in mice injected with interferon and interferon inducers. *Nature* 273: 759–761 (1978).

Gisler, R. H. and Dukor, P. A three cell mosaic culture: In vitro immune response by combination of pure B- and T-cells with peritoneal macrophage. *Cell. Immunol.* 4: 341–350 (1972).

Gisler, R. H., Lindahl, P., and Gresser, I. Effects of interferon on antibody synthesis in vitro. *J. Immunol.* 113: 438–444 (1974).

Gold, D. P., Van Dongen, J. J. M., Morton, C. C., Bruns, G. A. P., Van Den Elsen, P., Geurts van Kessel, A. H. M., and Terhorst, C. The gene encoding the ε subunit of the T3/T-cell receptor complex maps to chromosome 11 in humans and to chromosome 9 in mice. *Proc. Natl. Acad. Sci.* (USA) 84: 1664–1668 (1987).

Golding, B., Tsokos, G. C., Fleisher, T., Muchmore, A. V., and Blaese, R. M. The role of nonactivated and interferon-γ activated monocytes in regulating normal and SLE patient B cell responses to TNP-brucella abortus. *J. Immunol.* 137: 103–107 (1986).

Golub, S. H., d'Amore, P., and Rainey, M. Systemic administration of human leukocyte interferon to melanoma patient. II. Cellular events associated with changes in natural killer cytotoxicity. *J. Nat. Cancer Inst.* 68: 711–717 (1982a).

Golub, S. H., Dorey, F., Hara, D., Morton, D. L., and Burk, M. W. Systemic administration of human leukocyte interferon to melanoma patients. I. Effects on natural killer function and cell populations. *J. Nat. Cancer Inst.* 68: 703–710 (1982b).

Grabstein, K., Eisenman, J., Mochizuki, D., Shanebeck, K., Conlon, P., Hopp, T., March, C., and Gillis, S. Purification to homogeneity of B cell stimulating factor. A molecule that stimulates proliferation of multiple lymphokine-dependent cell lines. *J. Exp. Med.* 163: 1405–1414 (1986).

Green, J. A., Cooperband, S. R., and Kibrick, S. Immune specific induction of interferon production in cultures of human blood lymphocytes. *Science* 164: 1415–1417 (1969).

Greenberg, A. H., Miller, V., Jablonski, T., and Pohajdak, B. Suppression of NK-mediated natural resistance by interferon treatment of murine lymphomas. *J. Immunol.* 132: 2129–2134 (1984).

Guerne, P. A., Piguet, P. F., and Vassalli, P. Production of interleukin 2, interleukin 3, and interferon by mouse T lymphocyte clones of Lyt-2^+ and -2^- phenotype. *J. Immunol.* 132: 1869–1871 (1984).

Haegeman, G., Content, J., Volckaert, G., Derynck, R., Tavernier, J., and Fiers, W. Structural analysis of the sequence coding for an inducible 26-kDa protein in human fibroblasts. *Eur. J. Biochem.* 159: 625–632 (1986).

Handa, K., Suzuki, R., Matsui, H., Shimizu, Y., and Kumagai, K. Natural killer (NK) cells as a responder to interleukin 2 (IL 2). II. IL2-induced interferon γ production. *J. Immunol.* 130: 988–992 (1983).

Hansson, M., Kiessling, R., Andersson, B., and Welsh, R. M. Effect of interferon and interferon inducers on the NK sensitivity of normal mouse thymocytes. *J. Immunol.* 125: 2225–2231 (1980).

Harada, H., Shioiri-Nakano, K., Mayumi, M., and Kawai, T. Distinction of two subtypes of human leukocyte interferon (IFN-α) on B cell activation. B cell proliferation by two subtypes of IFN-α. *J. Immunol.* 131: 238–243 (1983).

Harfast, B., Huddlestone, J. R., Casali, P., Merigan, T. C., and Oldstone, M. B. A. Interferon acts directly on human B lymphocytes to modulate immunoglobulin synthesis. *J. Immunol.* 127: 2146–2150 (1981).

Havell, E. A., Spitalny, G. L., and Patel, P. J. Enhanced production of murine interferon γ by T cells generated in response to bacterial infection. *J. Exp. Med.* 156: 112–127 (1982).

Hayglass, K. T., Benacerraf, B., and Sy, M. S. The influence of B-cell idiotypes on the repertoire of suppressor T cells. *Immunol. Today* 7: 179–183 (1986).

Hecht, T. T., Longo, D. L., and Matis, L. A. The relationship between immune interferon production and proliferation in antigen specific, MHC restricted T cell lines and clones. *J. Immunol.* 131: 1049–1055 (1983).

Henkart, P. A., Millard, P. J., Reynolds, C. W., and Henkart, M. P. Cytolytic activity of purified cytoplasmic granules from cytotoxic rat large granular lymphocyte tumors. *J. Exp. Med.* 160: 75–93 (1984).

Henkart, P. A., Yue, C. C., Yang, J., and Rosenberg, S. A. Cytolytic and biochemical properties of cytoplasmic granules of murine lymphokine-activated killer cells. *J. Immunol.* 137: 2611–2617 (1986).

Herberman, R. B. Interferon and cytotoxic effector cells. In: *Interferon 2: Interferons and the Immune System,* J. Vilcek and E. De Maeyer, Eds. Elsevier Science Publishers Amsterdam. pp. 61–84 (1984).

Herberman, R. B., Ortaldo, J. R., and Bonnard, G. D. Augmentation by interferon of human natural and antibody-dependent cell-mediated cytotoxicity. *Nature* 277: 221–223 (1979).

Herberman, R. B., Ortaldo, J. R., Mantovani, A., Hobbs, D. S., Kung, H. F., and Pestka, S. Effect of human recombinant interferon on cytotoxic activity of natural killer (NK) cells and monocytes. *Cell. Immunol.* 67: 160–167 (1982).

Heron, I., Berg, K., and Cantell, K. Regulatory effect of interferon on T cells in vitro. *J. Immunol.* 117: 1370–1373 (1976).

Heron, I., Hokland, M., Moller-Larsen, A., and Berg, K. The effect of interferon on lymphocyte-mediated effector cell functions: Selective enhancement of natural killer cells. *Cell. Immunol.* 183–187 (1979).

Herrmann, F., Griffin, J. D., Meuer, S. G., and Meyer ZumBuschenfelde, K. H. Establishment of an interleukin 2-dependent T cell line derived from a patient with severe aplastic anemia, which inhibits in vitro hematopoiesis. *J. Immunol.* 136: 1629–1634 (1986).

Hirano, T., Taga, T., Nakano, N., Yasukawa, K., Kashiwamura, S., Shimizu, K., Nakajima, K., Pyun, K. H., and Kishimoto, T. Purification to homogeneity and characterization of human B-cell differentiation factor (BCDF or BSFp-2). *Proc. Natl. Acad. Sci.* (USA) 82: 5490–5494 (1985).

Hirano, T., Yasukawa, K., Harada, H., Taga, T., Watanabe, Y., Matsuda, T., Kashiwamura, S. I., Nakajima, K., Koyama, K., Iwamatsu, A., Tsunasawa, S., Sakiyama, F., Matsui, H., Takahara, Y., Taniguchi, T., and Kishimoto, T. Complementary DNA for a novel interleukin (BSF-2) that induces B lymphocytes to produce immunoglobulin. *Nature* 324: 73–76 (1986).

Holmberg, L. A., Miller, B. A., and Ault, K. A. The effect of natural killer cells on the development of syngeneic hematopoietic progenitors. *J. Immunol.* 133: 2933–2939 (1984).

Howard, M., Farrar, J., Hilfiker, M., Johnson, B., Takatsu, K., Hamaoka, T., and Paul, W. E. Identification of a T cell-derived B cell growth factor distinct from interleukin 2. *J. Exp. Med.* 155: 914–923 (1982).

Hu-Li, J., Shevach, E. M., Mizuguchi, J., Ohara, J., Mosmann, T., and Paul, W. E. B cell stimulatory factor 1 (interleukin 4) is a potent costimulant of normal resting T lymphocytes. *J. Exp. Med.* 165: 157–172 (1987).

Ikuta, K., Hattori, M., Wake, K., Kano, S., Honjo, T., Yodoi, J., and Minato, N. Expression and rearrangement of the α, β, and γ chain genes of the T cell receptor in cloned murine large granular lymphocyte lines. No correlation with the cytotoxic spectrum. *J. Exp. Med.* 164: 428–442 (1986).

Inaba, K. and Steinman, R. M. Monoclonal antibodies to LFA-1 and to CD4 inhibit the mixed leukocyte reaction after the antigen-dependent clustering of dendritic cells and T lymphocytes. *J. Exp. Med.* 165: 1403–1417 (1987).

Isobe, M., Erikson, J., Emanuel, B. S., Nowell, P. C., and Croce, C. M. Location of gene for β subunit of human T-cell receptor at band 7q35, a region prone to rearrangements in T cells. *Science* 228: 580–582 (1985).

Ito, Y. and Hosaka, Y. Component(s) of Sendai virus that can induce interferon in mouse spleen cells. *Infect. Immun.* 39: 1019–1023 (1983).

Jacobson, S., Richert, J. R., Biddison, W. E., Satinsky, A., Hartzman, R. J., and McFarland, H. F. Measles virus-specific T4 human cytotoxic T cell clones are restricted by class II HLA antigens. *J. Immunol.* 133: 754–757 (1984).

Johnson, H. M., Smith, B. G., and Baron, S. Inhibition of the primary in vitro antibody response by interferon preparations. *J. Immunol.* 114: 403–409 (1975).

Jones, C., Morse, H. G., Kao, F. T., Carbone, A., and Palmer, E. Human T-cell receptor α-chain genes: Location on chromosome 14. *Science* 228: 83–85 (1985).

Kaplan, J. NK cell lineage and target specificity: A unifying concept. *Immunol. Today* 7: 10–13 (1986).

Kasahara, T., Djeu, J. Y., Dougherty, S. F., and Oppenheim, J. J. Capacity of human large granular lymphocytes (LGL) to produce multiple lymphokines: Interleukin 2, interferon, and colony stimulating factor. *J. Immunol.* 131: 2379–2385 (1983a).

Kasahara, T., Hooks, J. J., Dougherty, S. F., and Oppenheim, J. J. Interleukin-2 mediated immune interferon (IFN-γ) production by human T cells and T cell subsets. *J. Immunol.* 130: 1784–1789 (1983b).

Kees, U., Kaltmann, B., Marcucci, F., Hultner, L., Staber, F., and Krammer, P. H. Frequency and activity of immune interferon (IFN-γ)-and colony-stimulating factor-producing human peripheral blood T lymphocytes. *Eur. J. Immunol.* 14: 368–373 (1984).

Kehrl, J. H., Muraguchi, A., Goldsmith, P. K., and Fauci, A. D. The direct effects of interleukin 1, interleukin 2, interferon-α, interferon-γ, B-cell growth factor, and a B-cell differentiation factor on resting and activated human B cells. *Cell. Immunol.* 96: 38–48 (1985).

Kiessling, R., Hochman, P. S., Haller, O., Shearer, G. M., Wigzell, H., and Cudcowicz, G. Evidence for a similar or common mechanism for NK cell activity and resistance to hematopoietic grafts. *Eur. J. Immunol.* 7: 655–663 (1977).

Kirchner, H, Engler, H., Schroder, C. H., Zawatzky, R., and Storch, E. Herpes simplex virus type 1-induced interferon production and activation of natural killer cells in mice. *J. Gen. Virol.* 64: 437–441 (1983).

Klein, J. R., Raulet, D. H., Pasternack, M. S., and Bevan, M. J. Cytotoxic T lymphocytes produce immune interferon in response to antigen or mitogen. *J. Exp. Med.* 155: 1198–1203 (1982).

Klimpel, G. R., Fleischmann, W. R., Jr., Baron, S., and Klimpel, K. D. Mouse bone marrow cells produce a different interferon (IFN) than do spleen cells in response to alloantigens. *J. Immunol.* 129: 946–949 (1982).

Kranz, D. M., Saito, H., Disteck, C. M., Swisshelm, K., Pravtcheva, D., Ruddle, F. H., Eisen, H. N., and Tonegawa, S. Chromosomal locations of the murine T-cell receptor alpha-chain gene and the T-cell gamma gene. *Science* 243: 941–945 (1985).

Kronenberg, M., Siu, G., Hood, L. E., and Shastri, N. The molecular genetics of the T-cell antigen receptor and T-cell antigen recognition *Ann. Rev. Immunol.* 4: 529–591 (1986).

Kupfer, A., Singer, S. J., Janeway, C. A., Jr., and Swain, S. L. Coclustering of CD4 (L3T4) molecule with the T-cell receptor is induced by specific direct interaction of helper T cells and antigen-presenting cells. *Proc. Natl. Acad. Sci. USA* 84: 5888–5892 (1987).

Kuribayashi, K., Gillis, S., Kern, D. E., and Henney, C. S. Murine NK cell cultures: Effects of interleukin-2 and interferon on cell growth and cytotoxic reactivity. *J. Immunol.* 126: 2321–2327 (1981).

Lacy, J., Sarkar, S. N., and Summers, W. C. Induction of c-myc expression in human B lymphocytes by B-cell growth factor and anti-immunoglobulin. *Proc. Natl. Acad. Sci.* (USA) 83: 1458–1462 (1986).

de Landazuri, M. O., Lopez-Botet, M., Timonen, T., Ortaldo, J. R., and Herberman, R. B. Human large granular lymphocytes: Spontaneous and interferon-boosted NK activity against adherent and nonadherent tumor cell lines. *J. Immunol.* 127: 1380–1383 (1981).

Landolfo, S., Arnold, B., and Suzan, M. Immune (γ) interferon production by murine T cell lymphomas. *J. Immunol.* 128: 2807–2809 (1982a).

Landolfo, S., Cofano, F., Giovarelli, M., Prat, M., Cavallo, G., and Forni, G. Inhibition of interferon γ may suppress allograft reactivity by T lymphocytes in vitro and in vivo. *Science* 229: 176–179 (1985).

Landolfo, S., Kirchner, H., and Simon, M. M. Production of immune interferon is regulated by more than one T cell subset: Lyt1,2,3 and Qat-5 phenotypes of murine T lymphocytes involved in IFN-γ production in primary and secondary mixed lymphocyte reaction. *Eur. J. Immunol.* 12: 295–299 (1982b).

Lanier, L. L., Benike, C. J., Phillips, J. H., and Engleman, E. G. Recombinant interleukin 2 enhanced natural killer cell-mediated cytotoxicity in human lymphocyte subpopulations expressing the Leu7 and Leu11 antigens. *J. Immunol.* 134: 794–801 (1985).

Lanier, L. L., Le, A. M., Cwirla, S., Federspiel, N., and Phillips, J. H. Antigenic, functional, and molecular genetic studies of human natural killer cells and cytotoxic T lymphocytes not restricted by the major histocompatibility complex. *Fed. Proceed.* 45: 2823–2828 (1986).

Larsson, E. L., Iscove, N. N., and Coutinho, A. Two distinct factors are required for induction of T-cell growth. *Nature* 283: 664–666 (1980).

Le, J., Vilcek, C., Saxinger, C., and Prensky, W. Human T cell hybridomas secreting immune interferon. *Proc. Natl. Acad. Sci.* (USA) 79: 7857–7861 (1982).

Le, J., Yao, J. S., Knowles II, D. M., and Vilcek, J. Accessory function of thymic and tonsillar dendritic cells in interferon γ production by T lymphocytes. *Lymphokine Res.* 5: 205–213 (1986).

Lee, F., Yokota, T., Otsuka, T., Meyerson, P., Villaret, D., Coffman, R., Mosmann, T., Rennick, D., Roehm, N., Smith, C., Zlotnik, A., and Arai, K. I. Isolation and characterization of a mouse interleukin cDNA clone that expresses B-cell stimulatory factor 1 activities and T-cell- and mast-cell-stimulating activities. *Proc. Natl. Acad. Sci.* (USA) 83: 2061–2065 (1986).

Lee, N. E., D'Eustachio, P., Pravtcheva, D., Ruddle, F. H., Hedrick, S. M., and Davis, M. M. Murine T cell receptor β chain is encoded on chromosome 6. *J. Exp. Med.* 160: 905–913 (1984).

Lewis, D. B., Larsen, A., and Wilson, C. B. Reduced interferon-γ mRNA levels in human neonates. Evidence for an intrinsic T cell deficiency independent of other genes involved in T cell activation. *J. Exp. Med.* 163: 1018–1023 (1986).

Lindahl, P., Leary, P., and Gresser, I. Enhancement by interferon of the specific toxicity of sensitized lymphocytes. *Proc. Natl. Acad. Sci.* (USA) 69: 721–725 (1972).

Littman, D. R., Thomas, Y., Maddon, P. J., Chess, L., and Axel, R. The isolation and sequence of the gene encoding T8: A molecule defining functional classes of T lymphocytes. *Cell* 40: 237–246 (1985).

Liu, C. C., Perussia, B., Cohn, Z. A., and Young, J. D.-E. Identification and characterization of a pore-forming protein of human peripheral blood natural killer cells. *J. Exp. Med.* 164: 2061–2076 (1986).

Lotzova, E., Savary, C. A., Gutterman, J. U., and Hersh, E. M. Modulation of natural killer cell-mediated cytotoxicity by partially purified and cloned interferon-α. *Cancer Res.* 42: 2480–2488 (1982).

Lotzova, E., Savary, C. A., Quesada, J. R., Gutterman, J., and Hersh, E. M. Analysis of natural killer cell cytotoxicity of cancer patients treated with recombinant interferon. *J. Nat. Cancer Inst.* 71: 903–910 (1983).

Lucero, M. A., Fridman, W. H., Provost, M. A., Billardon, C., Pouillart, P., Dumont, J., and Falcoff, E. Effect of various interferons on the spontaneous cytotoxicity exerted by lymphocytes from normal and tumor-bearing patients. *Cancer Res.* 41: 294–299 (1981).

Maddon, P. J., Littman, D. R., Godfrey, M., Maddon, D. E., Chess, L., and Axel, R. The isolation and nucleotide sequence of a cDNA encoding the T cell surface protein T4: A new member of the immunoglobulin gene family. *Cell* 42: 93–104 (1985).

Maeda, K., Nakanishi, N., Rogers, B. L., Haser, W. G., Shitara, K., Yoshida, H., Takagaki, Y., Augustin, A. A., and Tonegawa, S. Expression of the T-cell receptor γ-chain gene products on the surface of peripheral T cells and T-cell blasts generated by allogeneic mixed lymphocyte reaction. *Proc. Natl. Acad. Sci. USA* 84: 6536–6540 (1987).

Maluish, A. E., Ortaldo, J. R., Conlon, J. C., Sherwin, S. A., Leavitt, R., Strong, D.

M., Weirnik, P., Oldham, R. K., and Herberman, R. B. Depression of natural killer cytotoxicity after in vivo administration of recombinant leukocyte interferon. *J. Immunol.* 131: 503–507 (1983).

Mandler, R. N., Biddison, W. E., Mandler, R., and Serrate, S. A. β-endorphin augments the cytolytic activity and interferon production of NK cells. *J. Immunol.* 136: 934–939 (1986).

Manger, B., Kalden, J. R., Zawatzky, R., and Kirchner, H. Interferon production in the human mixed lymphocyte culture. *Transplantation* 32: 149–152 (1981).

Mantovani, A., Allevena, P., Sessa, C., Bolis, G., and Mangioni, C. Natural killer activity of lymphoid cells isolated from human ascitic ovarian tumors. *Int. J. Cancer* 25: 573–582 (1980).

Marcucci, F., Nowak, M., Krammer, P., and Kirchner, H. Production of high titres of interferon-γ by cells derived from short-term cultures of murine spleen leukocytes in T-cell growth factor-conditioned medium. *J. Gen. Virol.* 60: 195–198 (1982).

Marcucci, F., Waller, M., Kirchner, H., and Krammer, P. Production of immune interferon by murine T-cell clones from long-term cultures. *Nature* 291: 79–81 (1981).

Marrack, P. and Kappler, J. The antigen-specific, major histocompatibility complex-restricted receptor on T cells. *Adv. Immunol.* 38: 1–30 (1986).

Matsuyama, M., Sugamura, K., Kawade, Y., and Hinuma, Y. Production of immune interferon by human cytotoxic T cell clones. *J. Immunol.* 129: 450–451 (1982).

May, L. T., Helfgott, D. C., and Sehgal, P. B. Anti-β-interferon antibodies inhibit the increased expression of HLA-B7 mRNA in tumor necrosis factor-treated human fibroblasts: Structural studies of the β2 interferon involved. *Proc. Natl. Acad. Sci.* (USA) 83: 8957–8961 (1986).

McKimm-Breschkin, J. L., Mottram, P. L., Thomas, W. R., and Miller, J. F. A. P. Antigen-specific production of immune interferon by T cell lines. *J. Exp. Med.* 155: 1204–1209 (1982).

Messina, C., Kirkpatrick, D., Fitzgerald, P. A., O'Reilly, R. J., Siegal, F. P., Cunningham-Rundles, C., Blaese, M., Oleske, J., Pahwa, S., and Lopez, C. Natural killer cell function and interferon generation in patients with primary immunodeficiencies. *Clin. Immunol. Immunopathol.* 39: 394–404 (1986).

Meuer, S. C., Schlossman, S. F., and Reinherz, E. Clonal analysis of human cytotoxic T lymphocytes T4 and T8 effector T cells recognize products of different major histocompatibility complex regions. *Proc. Natl. Acad. Sci.* (USA) 79: 4395–4399 (1982).

Miller, R. G. An immunological suppressor cell inactivating cytotoxic T-lymphocyte precursor cells recognizing it. *Nature* 287: 544–546 (1980).

Minami, M., Honji, N., and Dorf, M. E. Mechanism responsible for the induction of I-J restrictions on Ts3 suppressor cells. *J. Exp. Med. 156: 1502–1515 (1982).*

Mingari, M. C., Gerosa, F., Carra, G., Accolla, R. S., Moretta, A., Zubler, R. H., Waldmann, T. A., and Moretta, L. Human interleukin-2 promotes proliferation of activated B cells via surface receptors similar to those of activated T cells. *Nature* 312: 641–643 (1984).

Mishell, I. and Dutton, R. W. Immunization of dissociated spleen cultures from normal mice. *J. Exp. Med.* 126: 423–442 (1967).

Mishra, G. C., Berton, M. T., Oliver, K. G., Krammer, P. H., Uhr, J. W., and Vitetta, E. S. A monoclonal anti-mouse LFA-1α antibody mimics the biological effects of B cell stimulatory factor-1 (BSF-1). *J. Immunol.* 137: 1590–1598 (1986).

Mond, J. J., Finkelman, F. D., Sarma, C., Ohara, J., and Serrate, S. Recombinant interferon-γ inhibits the B cell proliferative response stimulated by soluble but not by sepharose-bound anti-immunoglobulin antibody. *J. Immunol.* 135: 2513–2517 (1985).

Moore, M. and Potter, M. R. Enhancement of human natural cell-mediated cytotoxicity by interferon. *Br. J. Cancer* 41: 378–387 (1980).

Moore, M., White, W. J., and Potter, M. R. Modulation of target cell susceptibility to human natural killer cells by interferon. *Int. J. Cancer* 25: 565–572 (1980).

Morris, A. G., Lin, Y. L., and Askonas, B. A. Immune interferon release when a cloned cytotoxic T-cell line meets its correct influenza-infected target cell. *Nature* 295: 150–152 (1982).

Mosmann, T. R., Bond, M. W., Coffman, R. L., Ohara, J., and Paul, W. E. T-cell and mast cell lines respond to B-cell stimulatory factor 1. *Proc. Natl. Acad. Sci.* (USA) 83: 5654–5658 (1986b).

Mosmann, T. R., Cherwinski, H., Bond, M. W., Giedlin, M. A., and Coffman, R. L. Two types of murine helper T cell clone. I. Definition according to profiles of lymphokine activities and secreted proteins. *J. Immunol.* 136: 2348–2357 (1986a).

Munoz, A., Carrasco, L., and Fresno, M. Enhancement of susceptibility of HSV-1-infected cells to natural killer lysis by interferon. *J. Immunol.* 131: 783–787 (1983).

Muraguchi, A., Kehrl, J. H., Butler, J. L., and Fauci, A. S. Sequential requirements for cell cycle progression of resting human B cells after activation by anti-Ig. *J. Immunol.* 132: 176–180 (1984).

Murray, P. D. and Kagnoff, M. F. Differential effect of interferon-γ and interleukin-2 on the induction of IgA and IgM antidextran responses. *Cell. Immunol.* 95: 437–442 (1985).

Nabel, G., Greenberger, J. S., Sakakeeny, M. A., and Cantor, H. Multiple biologic activities of a cloned inducer T-cell population. *Proc. Natl. Acad. Sci.* (USA) 78: 1157–1161 (1981).

Nair, M. P. N., Cilik, J. M., and Schwartz, S. A. Histamine-induced suppressor factor inhibition of NK cells. Reversal with interferon and interleukin 2. *J. Immunol.* 136: 2456–2462 (1986).

Nakagawa, T., Hirano, T., Nakagawa, N., Yoshizaki, K., and Kishimoto, T. Effect of recombinant IL2 and γ-IFN on proliferation and differentiation of human B cells. *J. Immunol.* 134: 959–966 (1985).

Nakamura, M., Manser, T., Pearson, G. D. N., Daley, M. J., and Gefter, M. L. Effect of IFN-γ on the immune response in vivo and on gene expression in vitro. *Nature* 307: 381–382 (1984).

Nakauchi, H., Tagawa, M., Nolan, G. P., and Herzenberg, L. A. Isolation and characterization of the gene for the murine T cell differentiation antigen and immunoglobulin-related molecule, Lyt-2. *Nucl. Acids Res.* 15: 4337–4347 (1987).

Nakayama, T. Immune-specific production of γ interferon in human lymphocyte cultures in response to mumps virus. *Infect. Immun.* 40: 486–492 (1983).

Nathan, I., Groopman, J. E., Quan, S. G., Bersch, N., and Golde, D. W. Immune (γ) interferon produced by a human T-lymphoblast cell line. *Nature* 292: 842–844 (1981).

Neubauer, R. H., Goldstein, L., Rabin, H., and Stebbing, N. Stimulation of in vitro immunoglobulin production by interferon-α. *J. Immunol.* 134: 299–304 (1985).

Newton, R. C. Effect of interferon on the induction of human monocyte secretion of interleukin-1 activity. *Immunology* 56: 441–449 (1985).

Noelle, R., Krammer, P. H., Ohara, J., Uhr, J. W., and Vitetta, E. S. Increased expression of Ia antigens on resting B cells: An additional role for B-cell growth factor. *Proc. Natl. Acad. Sci.* (USA) 81: 6149–6153 (1984).

Noma, T. and Dorf, M. E. Modulation of suppressor T cell induction with γ-interferon. *J. Immunol.* 135: 3655–3660 (1985).

Noma, Y., Sideras, P., Naito, T., Bergstedt-Lindquist, S., Azuma, Y., and Honjo, T. Cloning of cDNA encoding the murine IgG1 induction factor by a novel strategy using SP6 promoter. *Nature* 319: 640–646 (1986).

Oh, S. H., Gonik, B., Greenberg, S. B., and Kohl, S. Enhancement of human neonatal natural killer cytotoxicity to herpes simplex virus with use of recombinant human interferons: Lack of neonatal response to γ interferon. *J. Infect. Dis.* 153: 791–793 (1986).

Ohara, J., Coligan, J. E., Zoon, K., Maloy, W. L., and Paul, W. E. High-efficiency purification and chemical characterization of B cell stimulatory factor-1/interleukin 4. *J. Immunol.* 139: 1127–1134 (1987).

Ohara, J., Lahet, S., Inman, J., and Paul, W. E. Partial purification of murine B cell stimulatory factor (BSF)-1. *J. Immunol.* 135: 2518–2523 (1985).

Ohara, J. and Paul, W. E. Receptors for B-cell stimulatory factor-1 expressed on cells of haemopoietic lineage. *Nature* 325: 537–540 (1987).

Ortaldo, J. R. and Herberman, R. B. Heterogeneity of natural killer cells. *Ann. Rev. Immunol.* 2: 359–394 (1984).

Otsuka, T., Villaret, D., Yokota, T., Takebe, Y., Lee, F., Arai, N., and Arai, K. I. Structural analysis of the mosue chromosomal gene encoding interleukin 4 which expresses B cell, T cell and mast cell stimulating activities. *Nucl. Acids Res.* 15: 333–344 (1987).

Palacios, R. Production of lymphokines by circulating human T lymphocytes that express lack of receptors for interleukin 2. *J. Immunol.* 132: 1833–1836 (1984).

Papa, M. Z., Vetto, J. T., Ettinghausen, S. E., Mule, J. J., and Rosenberg, S. A. Effect of corticosteroid on the antitumor activity of lymphokine-activated killer cells and interleukin 2 in mice. *Cancer Res.* 46: 5618–5623 (1986).

Pape, G. R., Hadam, M. R., Eisenburg, J., and Riethmuller, G. Kinetics of natural cytotoxicity in patients treated with human fibroblast interferon. *Cancer Immunol. Immunother.* 11: 1–6 (1981).

Park, L. S., Friend, D., Grabstein, K., and Urdal, D. L. Characterization of the high-affinity cell-surface receptor for murine B-cell-stimulating factor 1. *Proc. Natl. Acad. Sci.* (USA) 84: 1669–1673 (1987).

Parker, M. A., Mandel, A. D., Wallace, J. H., and Sonnenfeld, G. Modulation of the human in vitro antibody response by human leukocyte interferon preparations. *Cell. Immunol.* 58: 464–469 (1981).

Pasternack, M. S., Bevan, M. J., and Klein, J. R. Release of discrete interferons by cytotoxic T lymphocytes in response to immune and nonimmune stimuli. *J. Immunol.* 133: 277–280 (1984).

Pelicci, P. G., Subar, M., Weiss, A., Dalla-Favera, R., Littman, D. R. Molecular diversity of the human T-γ constant region genes. *Science* 237: 1051–1055 (1987).

Peters, M., Ambrus, J. L., Zheleznyak, A., Walling, D., and Hoffnagle, J. H. Effect of interferon-α on immunoglobulin synthesis by human B cells. *J. Immunol.* 137: 3153–3157 (1986a).

Peters, M., Walling, D. M., Kelly, K., Davis, G. L., Waggoner, J. G., and Hoffnagle, J. H. Immunologic effects of interferon-α in man: Treatment with human recombinant interferon-α suppresses in vitro immunoglobulin production in patients with chronic type B hepatitis. *J. Immunol.* 137: 3147–3152 (1986b).

Platsoucas, C. D. Regulation of natural killer cytotoxicity by *E. coli*-derived human interferon γ. *Scand. J. Immunol.* 24: 93–108 (1986).

Podack, E. R. and Konigsberg, P. J. Cytolytic T cell granules. Isolation, structural, biochemical, and functional characterization. *J. Exp. Med.* 160: 695–710 (1984).

Podack, E. R., Young, J. D. E., and Cohn, Z. A. Isolation and biochemical and functional characterization of perforin 1 from cytolytic T-cell granules. *Proc. Natl. Acad. Sci.* (USA) 82: 8629–8633 (1985).

Poupart, P., Vandenabeele, P., Cayphas, S., Van Snick, J., Haegeman, G., Kruys, V., Fiers, W., and Content, J. B cell growth modulating and differentiating activity of recombinant human 26-kd protein (BSF-2, HuIFN-β2, HPGF). *EMBO J.* 6: 1219–1240 (1987).

Rabin, E. M., Mond, J. J., Ohara, J., and Paul, W. E. B cell stimulatory factor 1 (BSF-1) prepares resting B cells to enter S phase in response to anti-IgM and lipopolysaccharide. *J. Exp. Med.* 164: 517–531 (1986a).

Rabin, E. M., Mond, J. J., Ohara, J., and Paul, W. E. Interferon-γ inhibits the action of B cell stimulatory factor (BSF)-1 on resting B cells. *J. Immunol.* 137: 1573–1576 (1986b).

Rabin, E. M., Ohara, J., and Paul, W. E. B-cell stimulatory factor 1 activates resting B cells. *Proc. Natl. Acad. Sci.* (USA) 82: 2935–2939 (1985).

Ralph, P., Jeong, G., Welte, K., Mertelsmann, R., Rabin, H., Henderson, L. E., Souza, L. M., Boone, T. C., and Robb, R. J. Stimulation of immunoglobulin secretion in human B lymphocytes as a direct effect of high concentrations of IL2. *J. Immunol.* 133: 2442–2445 (1984).

Rammensee, H. G., Bevan, M. J., and Fink, P. J. Antigen specific suppression of T-cell responses—The veto concept. *Immunol. Today* 6: 41–43 (1985).

Rasmussen, L. E., Jordan, G. W., Stevens, D. A., and Merigan, T. C. Lymphocyte interferon production and transformation after herpes simplex infections in humans. *J. Immunol.* 112: 728–735 (1974).

Reem, G. H. and Yeh, N. H. Regulation by interleukin 2 of interleukin 2 receptors and γ-interferon synthesis by human thymocytes: Augmentation of interleukin 2 receptors by interleukin 2. *J. Immunol.* 134: 953–958 (1985).

Reid, L. M., Minato, N., Gresser, I., Holland, J., Kadish, A., and Bloom, B. R. Influence of anti-mouse interferon serum on the growth and metastasis of tumor cells persistently infected with virus and of human prostatic tumors in athymic nude mice. *Proc. Natl. Acad. Sci.* (USA) 78: 1171–1175 (1981).

Revel, M., Ruggieri, R., and Zilberstein, A. Expression of human IFN-β2 genes in rodent cells. In: *The Biology of the Interferon System 1985,* W. E. Stewart II and H. Schellekens, Eds., Elsevier Science Publishers, Amsterdam, pp. 207–216 (1986).

Reyes, V. E., Klimpel, K. D., and Klimpel, G. R. T cell regulation of interferon-α/β (IFN-α/β) production by alloantigen-stimulated bone marrow cells. *J. Immunol.* 134: 3137–3141 (1985).

Riccardi, C., Giampietri, A., Migliorati, G., Cannarile, L., D'Adamio, L., and Herberman, R. B. Generation of mouse natural killer (NK) cell activity: Effect of interleukin-2 (IL-2) and interferon (IFN) on the in vivo development of natural killer cells from bone marrow (MB) progenitor cells. *Int. J. Cancer* 38: 553–562 (1986).

Riccardi, C., Vose, B. M., and Herberman, R. B. Modulation of IL 2-dependent growth of mouse NK cells by interferon and T lymphocytes. *J. Immunol.* 130: 228–232 (1983).

Romagnani, S., Giudizi, M. G., Biagiotti, R., Almerigogna, F., Mingari, C., Maggi, E., Liang, C. M., and Moretta, L. B cell growth factor activity of interferon-γ. Recombinant human interferon-γ promotes proliferation of anti-μ-activated human B lymphocytes. *J. Immunol.* 136: 3513–3516 (1986).

Saksela, E. Interferon and natural killer cells. In: *Interferon 3,* I. Gresser, Ed., Academic Press, New York, pp. 45–63 (1981).

Saksela, E., Timonen, T., and Cantell, K. Human natural killer cell activity is augmented by interferon via recruitment of "preNK" cells. *Scand. J. Immunol.* 10: 257–266 (1979).

Samelson, L. E., Patel, M. D., Wessman, A. M., Harford, J. B., and Klausner, R. D. Antigen activation of murine T cells induces tyrosine phosphorylation of a polypeptide associated with the T cell antigen receptor. *Cell* 46: 1083–1090 (1986).

Santoni, A., Piccoli, M., Ortaldo, J. R., Mason, L., Wiltrout, R. H., and Herberman, R. B. Changes in number and density of large granular lymphocytes upon in vivo augmentation of mouse natural killer activity. *J. Immunol.* 134: 2799–2810 (1985).

Sayers, T. J., Mason, A. T., and Ortaldo, J. R. Regulation of human NK cell activity by interferon-γ: Lack of a role in interleukin 2-mediated augmentation. *J. Immunol.* 136: 2176–2180 (1986).

Schober, I., Braun, R., Reiser, H., Munk, K., Leroux, M., and Kirchner, H. Ia-positive T lymphocytes are the producer cells of interferon γ. *Exp. Cell Res.* 152: 348–356 (1984).

Seed, B. and Aruffo, A. Molecular cloning of the CD2 antigen, the T-cell erythrocyte receptor, by a rapid immunoselection procedure. *Proc. Natl. Acad. Sci. USA* 84: 3365–3369 (1987).

Sehgal, P. B., Zilberstein, A., Ruggieri, R. M., May, L. T., Ferguson-Smith, A., Slate, D. L., Revel, M., and Ruddle, F. H. Human chromosome 7 carries the β2 interferon gene. *Proc. Natl. Acad. Sci.* (USA) 83: 5219–5222 (1986).

Seki, H., Taga, K., Matsuda, A., Uwadana, N., Hasui, M., Miyawaki, T., and Taniguchi, N. Phenotypic and functional characteristics of active suppressor cells against IFN-γ production in PHA-stimulated cord blood lymphocytes. *J. Immunol.* 137: 3158–3161 (1986).

Sewell, W. A., Brown, M. H., Dunne, J., Owen, M. J., and Crumpton, M. J. Molecular cloning of the human T-lymphocyte surface CD2 (T11) antigen. *Proc. Natl. Acad. Sci. USA* 83: 8718–8722 (1986).

Sewell, W. A., Brown, M. H., Owen, M. J., Fink, P. J., Kozak, C. A., and Crumpton, M. J. The murine homologue of the T lymphocyte CD2 antigen: Molecular cloning, chromosome assignment and cell surface expression. *Eur. J. Immunol.* 17: 1015–1020 (1987).

Sidman, C. L., Marshall, J. D., Shultz, L. D., Gray, P. W., and Johnson, H. M. γ-interferon is one of several direct B cell-maturing lymphokines. *Nature* 309: 801–803 (1984).

Siegel, D. S., Le, J., and Vilcek, J. Modulation of lymphocyte proliferation and immunoglobulin synthesis by interferon-γ and "type I" interferons. *Cell. Immunol.* 101: 380–390 (1986).

Simon, M. M., Hochgeschwender, U., Brugger, U., and Landolfo, S. Monoclonal antibodies to interferon-γ inhibit interleukin 2-dependent induction of growth and maturation in lectin/antigen reactive cytolytic T lymphocyte precursors. *J. Immunol.* 136: 2755–2762 (1986a).

Simon, M. M., Landolfo, S., Diamantstein, T., and Hochgeschwender, U. Antigen- and lectin-sensitized murine cytolytic T lymphocyte-precursors require both interleukin 2 and endogenously produced immune (γ) interferon for their growth and differentiation into effector cells. *Curr. Topics Microbiol. Immunol.* 126: 173–185 (1986b).

Simon, M. M., Moll, H., Prester, M., Nerz, G., and Eichmann, K. Immunoregulation by mouse T-cell clones. I. Suppression and amplification of cytotoxic responses by cloned H-Y-specific cytolytic T lymphocytes. *Cell. Immunol.* 86: 206–221 (1984).

Simon, M. M., Nerz, G., Prester, M., and Moll, H. Immunoregulation by mouse T cell clones. III. Cloned H-Y-specific cytotoxic T cells secrete a soluble mediator(s) that inhibits cytotoxic responses by acting on both Lyt-2^- and L3T4^- lymphocytes. *Eur. J. Immunol.* 15: 773–783 (1985).

Smith, E. M., Hughes, T. K., and Blalock, J. E. Transformed and normal cell surface glycoproteins that induce interferon production by nonsensitized lymphocytes. *Infect. Immun.* 39: 220–224 (1983).

Snapper, C. M. and Paul, W. E. Interferon-γ and B cell stimulatory factor-1 reciprocally regulate Ig isotype production. *Science* 236: 944–947 (1987).

Sonnenfeld, G., Mandel, A. D., and Merigan, T. C. The immunosuppressive effect of type II mouse interferon preparations on antibody production. *Cell. Immunol.* 34: 193–206 (1977).

Sonnenfeld, G., Mandel, A. D., and Merigan, T. C. Time and dosage dependence of immunoenhancement by murine type II interferon preparations. *Cell. Immunol.* 40: 285–293 (1978).

Spina, C. A., Fahey, J. L., Durkos-Smith, D., Dorey, F., and Sarna, G. Suppression of natural killer cell cytotoxicity in the peripheral blood of patients receiving interferon therapy. *J. Biol. Resp. Modif.* 2: 458–469 (1983).

Spits, H., Yssel, H., Takebe, Y., Arai, N., Yokota, T., Lee, F., Arai, K. I., Banchereau, J., and De Vries, J. E. Recombinant interleukin 4 promotes the growth of human T cells. *J. Immunol.* 139: 1142–1147 (1987).

Stobo, J., Green, I., Jackson, L., and Baron, S. Identification of a subpopulation of mouse lymphoid cells required for interferon production after stimulation with mitogens. *J. Immunol. 112: 1589–1593 (1974).*

Strannegard, O., Larsson, I., Lundgren, E., Miorner, H., and Persson, H. Modulation of immune responses in newborn and adult mice by interferon. *Infect. Immun.* 20: 334–339 (1978).

Strauss, W. M., Quertermous, T., and Seidman, J. G. Measuring the human T cell receptor γ-chain locus. *Science* 237: 1217–1219 (1987).

Swain, S. L. T cell subsets and the recognition of MHC class. *Immunol. Rev.* 74: 129–142 (1983).

Talmadge, J. E., Herberman, R. B., Chirigos, M. A., Maluish, A. E., Schneider, M. A., Adams, J. S., Philips, H., Thurman, G. B., Varesio, L., Long, C., Oldham, R. K., and Wiltrout, R. H. Hyporesponsiveness to augmentation of murine natural killer cell activity in different anatomical compartments by multiple injections of various immunomodulators including recombinant interferons and interleukin 2. *J. Immunol.* 135: 2483–2491 (1985).

Targan, S. and Dorey, F. Interferon activation of "prespontaneous killer" (PRE-SK) cells and alteration in kinetics of lysis of both "PRE-SK" and active SK cells. *J. Immunol.* 124: 2157–2161 (1980).

Tarkkanen, J. and Saksela, E. Umbilical-cord-blood-derived suppressor cells of the human natural killer cell activity are inhibited by interferon. *Scand. J. Immunol.* 15: 149–157 (1982).

Taylor, P. M., Wraith, D. C., and Askonas, B. A. Control of immune interferon release by cytotoxic T-cell clones specific for influenza. *Immunology* 54: 607–614 (1985).

Taylor, S. and Bryson, Y. J. Impaired production of γ-interferon by newborn cells in vitro is due to a functionally immature macrophage. *J. Immunol.* 134: 1493–1502 (1985).

Timonen, T., Ortaldo, J. R., and Herberman, R. B. Characteristics of human large granular lymphocytes and relationship to natural killer and K cells. *J. Exp. Med.* 153: 569–582 (1981).

Timonen, T., Ortaldo, J. R., and Herberman, R. B. Analysis by a single cell cytotoxicity assay of natural killer (NK) cell frequencies among human large granular lymphocytes and of the effects of interferon on their activity. *J. Immunol.* 128: 2514–2521 (1982).

Torres, B. A., Yamamoto, J. K., and Johnson, H. M. Cellular regulation of γ interferon production: Lyt phenotype of the suppressor cell. *Infect. Immun.* 35: 770–776 (1982).

Trimble, W. S., Johnson, P. W., Hozumi, N., and Roder, J. C. Inducible cellular transformation by a metallothionein-ras hybrid oncogene leads to NK cell susceptibility. *Nature* 321: 782–784 (1986).

Trinchieri, G. Surface phenotype of natural killer cells and macrophages. *Fed. Proceed.* 45: 2821–2822 (1986).

Trinchieri, G., Granato, D., and Perussia, B. Interferon-induced resistance of fibroblasts to cytolysis mediated by natural killer cells: Specificity and mechanism. *J. Immunol.* 126: 335–340 (1981).

Trinchieri, G., Santoli, D., Dee, R. R., and Knowles, B. B. Antiviral activity induced by culturing lymphocytes with tumor-derived or virus-transformed cells. Identification of the antiviral activity as interferon and characterization of the human effector lymphocyte subpopulation. *J. Exp. Med.* 147: 1299–1313 (1978).

Tunnacliffe, A., Buluwela, L., and Rabbits, T. H. Physical linkage of three CD3 genes on human chromosome 11. *EMBO J.* 6: 2953–2957 (1987).

Uchida, A., Yanagawa, E., Kokoschka, E. M., Micksche, M., and Koren, H. S. In vitro modulation of human natural killer cell activity by interferon: Generation of adherent suppressor cells. *Br. J. Cancer* 50: 483–492 (1984).

Ueno, Y., Miyawaki, T., Seki, H., Matsuda, A., Taga, K., Sato, H., and Taniguchi, N. Differential effects of recombinant human interferon-γ and interleukin 2 on natural killer cell activity of peripheral blood in early human development. *J. Immunol.* 135: 180–184 (1985).

Usui, M., Aoki, I., Sunshine, G. H., and Dorf, M. E. A role for macrophages in suppressor cell induction. *J. Immunol.* 132: 1728–1734 (1984).

Valle, M. J., Bobrove, A. M., Strober, S., and Merigan, T. C. Immune specific production of interferon by human T cells in combined macrophage-lymphocyte cultures in response to herpes simplex antigen. *J. Immunol.* 114: 435–441 (1975).

Van Damme, J., Opdenakker, G., Simpson, R. J., Rubira, M. R., Cayphas, S., Vink, A., Billiau, A., and Van Snick, J. Identification of the human 26-kD protein, interferon β2 (IFN-β2) as a B cell hybridoma/plasmacytoma growth factor induced by interleukin 1 and tumor necrosis factor. *J. Exp. Med.* 165: 914–919 (1987).

Van de Griend, R. J., Ronteltap, C. P. M., Gravekamp, C., Monnikendam, D., and Bolhuis, R. L. H. Interferon-β and recombinant IL2 can both enhance, but by different pathways, the nonspecific cytolytic potential of $T3^-$ natural killer cell-derived clones rather than that of $T3^+$ clones. *J. Immunol.* 136: 1700–1707 (1986).

Van den Elsen, P., Bruns, G., Gehard, D. S., Pravtcheva, D., Jones, C., Housman, D., Ruddle, F. A., Orkin, S., and Terhorst, C. Assignment of the gene coding for the T3-ω subunit of the T3-T-cell receptor complex to the long arm of human chromosome 11 and to mouse chromosome 9. *Proc. Natl. Acad. Sci.* (USA) 82: 2920–2924 (1985).

Vanky, F. T., Argov, S. A., Einhorn, S. A., and Klein, E. Role of alloantigens in natural killing. Allogeneic but not autologous tumor biopsy cells are sensitive for interferon-induced cytotoxicity of human blood lymphocytes. *J. Exp. Med.* 151: 1151–1165 (1980).

Vanky, F. T. and Klein, E. Alloreactive cytotoxicity of interferon-triggered human lymphocytes detected with tumor biopsy targets. *Immunogenetics* 15: 31–39 (1982).

Vignaux, F., Gresser, I., and Fridman, W. H. Effect of virus-induced interferon on the antibody response of suckling and adult mice. *Eur. J. Immunol.* 10: 767–772 (1980).

Vilcek, J., Henriksen-Destefano, D., Siegel, D., Klion, A., Robb, R. J., and Le, J. Regulation of IFN-γ induction in human peripheral blood cells by exogenous and endogenously produced interleukin 2. *J. Immunol.* 135: 1851–1856 (1985).

Virelizier, J. L., Allison, A. C., and De Maeyer, E. Production by mixed lymphocyte cultures of a type II interferon able to protect macrophages against virus infections. *Infect. Immun.* 17: 282–285 (1977a).

Virelizier, J. L., Chan, E. L., and Allison, A. C. Immunosuppressive effects of lymphocyte (type II) and leucocyte (type I) interferon on primary antibody responses in vivo and in vitro. *Clin. Exp. Immunol.* 30: 299–304 (1977b).

Vitetta, E. S., Ohara, J., Myers, C. D., Layton, J. E., Krammer, P. H., and Paul, W. E. Serological, biochemical, and functional identity of B cell-stimulatory factor 1 and B cell differentiation factor for IgGl. *J. Exp. Med.* 162: 1726–1731 (1985).

von Wussow, P., Platsoucas, C. D., Wiranowska-Stewart, M., and Stewart II, W. E. Human γ interferon production by leukocytes induced with monoclonal antibodies recognizing T cells. *J. Immunol.* 127: 1197–1200 (1981).

Wallach, D. Interferon-induced resistance to the killing by NK cells: A preferential effect of IFN-γ. *Cell Immunol.* 75: 390–395 (1983).

Weigent, D. A., Stanton, G. J., and Johnson, H. M. Recombinant gamma interferon enhances natural killer cell activity similar to natural gamma interferon. *Biochem. Biophys. Res. Comm.* 111: 525–529 (1983).

Weiss, A., Imboden, J., Hardy, K., Manger, B., Terhorst, C., and Stobo, J. The role of the T3/antigen receptor complex in T-cell activation. *Ann. Rev. Immunol.* W. E. Paul, C. G. Fathman, and H. Metzger, Eds., 4: 593–619 (1986).

Welsh, R. M. Do natural killer cells play a role in virus infections? *Antiv. Res.* 1: 5–12 (1981).

Welsh, R. M., Karre, K., Hansson, M., Kunkel, L. A., and Kiessling, R. W. Interferon-mediated protection of normal and tumor target cells against lysis by mouse natural killer cells. *J. Immunol.* 126: 219–225 (1981).

Wheelock, E. F. Interferon-like virus-inhibitor induced in human leukocytes by photohemagglutinin. *Science* 149: 310–311 (1965).

Wilkinson, M. and Morris, A. G. Role of the E receptor in interferon-γ expression: Sheep erythrocytes augment interferon-γ production by human lymphocytes. *Cell. Immunol.* 86: 109–117 (1984).

Wilson, C. B., Westall, J., Johnston, L., Lewis, D. B., Dower, S. K., and Alpert, A. R. Decreased production of interferon-γ by human neonatal cells. Intrinsic and regulatory deficiencies. *J. Clin. Invest.* 77: 860–867 (1986).

Wiskocil, R., Weiss, A., Imboden, J., Kamin-Lewis, R., and Stobo, J. Activation of a human T cell line: A two-stimulus requirement in the pretranslational events involved in the coordinate expression of interleukin 2 and γ-interferon genes. *J. Immunol.* 134: 1599–1603 (1985).

Yagüe, J., White, J., Coleclough, C., Kappler, J., Palmer, E., and Marrack, P. The T cell receptor: The α and β chains define idiotype, and antigen and MHC specificity. *Cell* 42: 81–87 (1985).

Yamada, Y. K., Meager, A., Yamada, A., and Ennis, F. A. Human interferon-α and γ production by lymphocytes during the generation of influenza virus-specific cytotoxic T lymphocytes. *J. Gen. Virol.* 67: 2325–2334 (1986).

Yanagi, Y., Caccia, N., Kronenberg, M., Chin, B., Roder, J., Rohel, D., Kiyohara, T., Lauzon, R., Toyonaga, B., Rosenthal, K., Dennert, G., Acha-Orbea, H., Hengartner, H., Hood, L., and Mak, T. W. Gene rearrangement in cells with natural killer activity and expression of the β-chain of the T-cell antigen receptor. *Nature* 314: 631–633 (1985).

Yokota, T., Otsuka, T., Mosmann, T., Banchereau, J., Defrance, T., Blanchard, D., De Vries, J. E., Lee, F., and Arai, K. I. Isolation and characterization of a human interleukin cDNA clone, homologous to mouse B-cell stimulatory factor 1, that expresses B-cell- and T-cell-stimulating activities. *Proc. Natl. Acad. Sci.* (USA) 83: 5894–5898 (1986).

Young, J. D. E., Liu, C. C., Leong, L. G., and Cohn, Z. A. The pore-forming protein (perforin) of cytolytic T lymphocytes is immunologically related to the components of membrane attack complex of complement through cystein-rich domains. *J. Exp. Med.* 164: 2077–2082 (1986).

Young, H. A. and Ortaldo, J. R. One-signal requirement for interferon-gamma production by human large granular lymphocytes. *J. Immunol.* 139: 724–727 (1987).

Zalman, L. S., Brothers, M. A., Chiu, F. J., and Muller-Eberhard, H. J. Mechanism of cytotoxicity of human large granular lymphocytes: Relationship of the cytotoxic lymphocytes-protein to the ninth component (C9) of human complement. *Proc. Natl. Acad. Sci.* (USA) 83: 5262–5266 (1986).

Zarling, J. M., Sosman, J., Eskra, L., Borden, E. C., Horoszewicz, J. S., and Carter, W. A. Enhancement of T cell cytotoxic responses by purified human fibroblast interferon. *J. Immunol.* 121: 2002–2004 (1978).

Zilberstein, A., Nissim, A., Ruggieri, R., Shulman, L., Chen, L., and Revel, M. The mode of action of human interferon-β2. In: *The Biology of the Interferon System 1985,* W. E. Stewart II and H. Schellekens, Eds., Elsevier Science Publishers, Amsterdam, pp. 119–124 (1986a).

Zilberstein, A., Ruggieri, R., Korn, J. H., and Revel, M. Structure and expression of cDNA and genes for human interferon-β2, a distinct species inducible by growth-stimulatory cytokines. *EMBO J.* 5: 2529–2537 (1986b).

Zlotnik, A., Ransom, J., Frank, G., Fischer, M., and Howard, M. Interleukin 4 is a growth factor for activated thymocytes: Possible role in T-cell ontogeny. *Proc. Natl. Acad. Sci.* (USA) 84: 3856–3860 (1987).

Zoumbos, N. C., Gascon, P., Djeu, J. Y., Trost, S. R., and Young, N. S. Circulating activated suppressor T lymphocytes in aplastic anemia. *N. Engl. J. Med.* 312: 257–265 (1985).

12 THE EFFECTS OF INTERFERONS ON IMMEDIATE AND DELAYED HYPERSENSITIVITY

Hypersensitivity is a special manifestation of immunity, frequently associated with pathology. Hypersensitive reactions can be either immediate, starting within minutes to hours after contact with allergen, or delayed, requiring from 24 to 48 hours to appear after antigen contact. Immediate and delayed hypersensitivity are mediated by two different mechanisms: immediate sensitivity is either complement- or IgE-mediated, and mobilizes basophils and mast cells, whereas delayed hypersensitivity is mainly a T-cell dependent reaction, which may also mobilize basophils, as well as many other inflammatory cells. Both types of hypersensitivity are influenced by IFNs.

I. IMMEDIATE HYPERSENSITIVITY

A special form of immediate hypersensitivity is mediated by the immunoglobulin IgE. Exposure to some antigens, in this particular case called allergens, stimulates plasma cells to produce allergen-specific antibodies of the

IgE class. These antibodies then interact with basophils and mast cells, inflammatory cells that produce histamine and store it in their basophilic granules. Basophils and mast cells have high affinity receptors for IgE, and when stimulated by antigen-IgE complexes, release histamine and other mediators of anaphylaxis causing vasodilatation, edema, and smooth muscle contraction. A classical example of immediate hypersensitivity is bronchial asthma, which can arise either from contact with allergen or be precipitated by upper respiratory tract infections, including viral infections.

A. Human Leukocyte IFN Enhances Histamine Release by Basophils

As an in vitro correlate of immediate hypersensitivity, one can measure histamine release by basophils and mast cells, present in peripheral leukocyte suspensions. When human leukocyte cultures, derived from allergic donors, are stimulated with the corresponding allergen or with anti-IgE antibody, histamine is released. When these cultures are in addition infected with viruses like HSV-1 or Influenza A, the release of histamine is significantly enhanced. This enhancement is the result of the IFN production induced by these viruses, as it can also be obtained by adding Hu leukocyte IFN to the allergen-stimulated cell suspensions, in the absence of any viral infection (Fig. 12.1) (Ida et al., 1977). Furthermore, leukocyte IFN enhances immediate hypersensitivity reactions by acting as a chemotactic agent for basophils, thus favoring their accumulation at sites of viral infection (Lett-Brown et al., 1981). By attracting mediator-releasing cells and by stimulating the activity of these cells, IFNs are potential contributors to the pathology caused by virus infections in allergic individuals, for example, the exacerbation of asthma in young children during upper respiratory tract infections

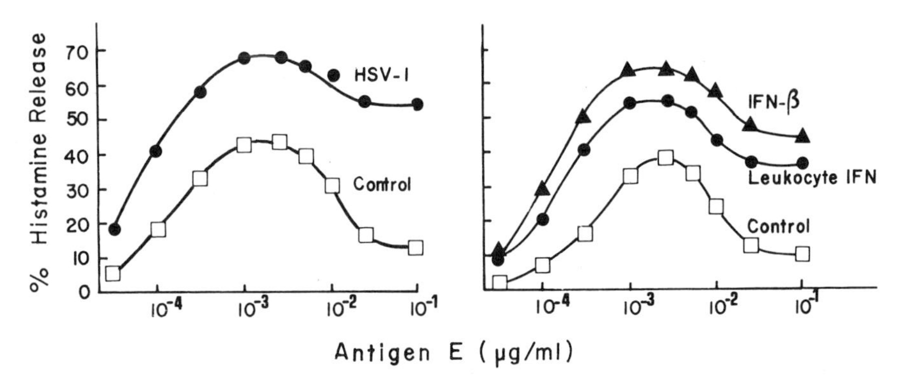

Figure 12.1 Enhancement of histamine release by human leukocytes infected with herpes simplex virus (left panel) or treated with Hu IFN-β or Hu leukocyte IFN (right panel). After 24 hours incubation, the leukocytes are challenged with ragweed antigen E and histamine release is determined. (Adapted from Ida et al., 1977.)

(McIntosh et al., 1973). This is one of the examples of exacerbated disease resulting from immuno-stimulation by IFN-α/β.

B. IFN-γ Also Enhances Histamine Release and Induces Class II Antigens on Mast Cells

Like IFN-α/β, IFN-γ enhances IgE-mediated histamine release by peripheral blood leukocytes (Hooks et al., 1980). In addition, IFN-γ enhances the MHC class II antigen expression at the surface of mast cells, and therefore could favor the interaction of these cells with T and B cells. This remains a theoretical possibility, however, and has not been directly demonstrated (Wong et al., 1982).

II. THE EFFECTS OF IFNs ON DELAYED HYPERSENSITIVITY

Delayed hypersensitivity (DH) is one of the many expressions of cell-mediated immunity, of which the skin reaction to tuberculin is probably the most widely known. DH develops following infection or immunization with a great variety of infectious agents, like viruses, bacteria, and parasites, but it can also be directed against proteins or haptens. In the usual course of events, after the first sensitizing contact with antigen or hapten, a minimum of 4 to 5 days are required to obtain an optimal DH reaction upon renewed contact with the antigen. During this period, corresponding to the afferent limb of the reaction, helper and effector T cells, as well as suppressor cells are generated. The helper cells, called amplifying T lymphocytes or T_A cells, are instrumental in the generation of the two major species of effector cells, cytotoxic T lymphocytes or T_C cells and the special DH effector T lymphocytes, T_D cells. T_C cells are cytotoxic for antigen-bearing target cells, whereas T_D cells are instrumental in initiating an inflammatory response at the site of entry of the antigen, and it is the size and extent of this inflammatory response that usually serves to measure the degree of DTH. During the afferent limb of the DH reaction, antigen-specific T-suppressor lymphocytes, T_S cells, are also generated. An antigen dosage that is optimal for antibody production also appears to be optimal for the induction of DH T_S cells (Mackaness et al., 1974). As discussed in section A, T-suppressor cells are extremely sensitive to inhibition by IFNs.

The effector or efferent limb of the DH reaction is triggered by a renewed antigen challenge of a sensitized individual. The hallmark of the reaction is the time it takes to reach its peak: To qualify as delayed, the reaction should peak at 24 to 48 hours. It consists of perivascular infiltration by lymphocytes and macrophages, followed by secondary infiltration of the lesion by a number of other cells such as eosinophils, monocytes, and sometimes mast cells (Turk, 1980). This stage of the reaction is also influenced by IFNs (Fig. 12.2).

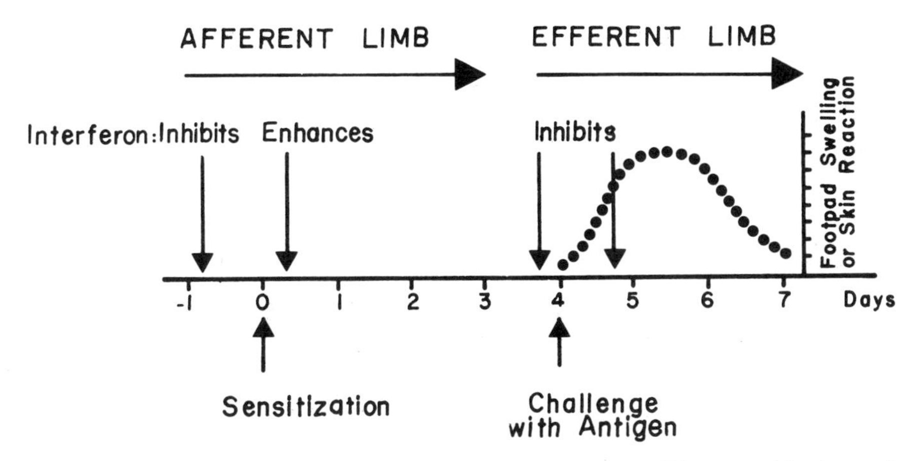

Figure 12.2 A schematic overview of the effects of exogenous IFN on sensitization and expression of delayed hypersensitivity in the mouse. (Adapted from De Maeyer and De Maeyer-Guignard, 1980b.)

Infection with many different viruses can lead to the development of DH, and animal models have been described for herpes simplex virus, lymphocytic choriomeningitis virus, Semliki forest virus, and different myxo- and paramyxoviruses (Hotchin, 1962; Lausch et al., 1966; Feinstone et al., 1969; Wetherbee, 1973; Kraaijeveld et al., 1979). Since virus infections are accompanied by the synthesis of IFN-α and IFN-β, the various effects of these IFNs on DH are relevant to the establishment of cell-mediated immunity to viruses.

A. Effects of IFN-α/β on the Afferent Limb of DH

IFNs can either inhibit or, on the contrary, stimulate the afferent limb of DH. Studies in mice show that the outcome of IFN action on sensitization is determined by the timing of IFN administration and by the antigen dosage used for sensitization. When Mu IFN-α/β is induced or administered systemically, before immunization, sensitization is inhibited and the inflammatory reaction upon subsequent challenge with antigen is either significantly reduced or completely absent. This occurs with a variety of antigens used for sensitization, such as sheep erythrocytes, Newcastle disease virus, and the hapten picryl chloride, which is a contact allergen (De Maeyer-Guignard et al., 1975; De Maeyer, 1976; Kraaijeveld et al., 1986). Under these conditions, IFN-α/β acts as an inhibitor of cell-mediated immunity by preventing sensitization. How this down-regulation takes place is a matter of conjecture. A reasonable hypothesis would be to invoke the antimitotic activity of IFNs, resulting in an inhibition of the expansion of T_H and T_A cells, but there is no proof that this is the actual mechanism.

When administered shortly after the antigen instead of before, IFNs have the opposite effect: They enhance DH. The enhancement depends to a certain extent on the amount of antigen used to immunize. For example, when mice are sensitized with an optimal dose of sheep red blood cells (SRBC), DH is not significantly altered by administration of IFN-α/β, within 24 hours after the antigen. However, when a suboptimal amount of antigen is used, followed within a few hours by the administration of IFN-α/β, one observes a significant enhancement of DH (De Maeyer and De Maeyer-Guignard, 1980a,b; Bartocci et al., 1982; Ron et al., 1984).

A similar stimulatory effect develops in mice sensitized with the contact allergen dinitrofluorobenzene (DNFB) and treated with IFN-α/β soon after sensitization: DH is enhanced upon subsequent challenge with DNFB. In this experimental model the stimulatory effect of IFN, when given after the hapten, is due to the specific inhibition of either the generation or the expansion of T_S cells. This selective blockade of Ts cells results from a direct effect of IFN-α/β on the Ts subpopulation and not through an effect on antigen-presenting macrophages (Knop et al., 1982; Knop et al., 1987). The T_D effector cells, on the contrary, are much more resistant to inhibition by IFN (Knop et al., 1984). T_S cells have been found to be particularly sensitive to inhibition by IFNs, not only in DH but also in other immune reactions. Hu leukocyte IFN, for example, inhibits the generation of allogeneic T_S cells, and it also inhibits the activity of umbilical-cord-derived cells that suppress NK-cell activity (Fradelizi and Gresser, 1982; Tarkkanen and Saksela, 1982).

B. Febrile Temperature and the Effect of IFN-α/β on Suppressor Cells

There is a longstanding interest in fever as part of the defense mechanism during acute infections. Yet, the influence of body temperature is not always taken into consideration during in vivo experiments, especially in small animals like mice, in which temperature is not so easily measured. To neglect the effects of temperature during IFN experiments in vivo is probably unwise, especially in view of the observation that pure Hu IFN-α is pyrogenic in patients (Scott et al., 1981; Ackerman et al., 1984). This pyrogenicity is due to a direct action of IFN on the thermoregulatory center of the brain, by inducing prostaglandin E2 in or near the anterior hypothalamus. The pyrogenic action of IFN-α is comparable to that of TNF-α and of IL-1, formerly called endogenous pyrogen (Chapter 13) (Dinarello et al., 1984, 1986). Several observations indicate that fever may have a beneficial influence on IFN action.

Like Hu leukocyte IFN in humans, Mu IFN-α/β induces fever in mice. This pyrogenic action of IFN creates the optimal thermal environment for its antisuppressive activity in DH. In mice sensitized with a suboptimal dose of SRBC, Mu IFN-α/β, as discussed earlier, can enhance DH to levels obtained with an optimal dose of antigen, provided the IFN is administered a few hours after immunization. Under these conditions, IFN treatment is accompanied by an augmentation of body temperature of about 0.4°C. This raise in temperature results from the administration of 10,000 IFN-α/β units, which is a small amount of IFN compared to the total production that can occur during some virus infections of the mouse. When the fever induced by IFN in these animals is prevented by the antipyretic drug acetaminophen, the antisuppressive effect of IFN is abrogated and there is no enhancement of DH. Thus, the lowering of the body temperature abrogates immuno-stimulation by IFN (Fig. 12.3). The incidence of temperature on the interaction of IFN and suppressor cells is also observed in vitro. At the temperature of 39.3°C, Mu IFN-α/β completely blocks the generation of T_S cells in splenic cells exposed to SRBC, whereas it has no effect on suppressor cell generaton at 37°C (Ron et al., 1984).

The optimal inhibition of suppressor cell formation at febrile temperatures

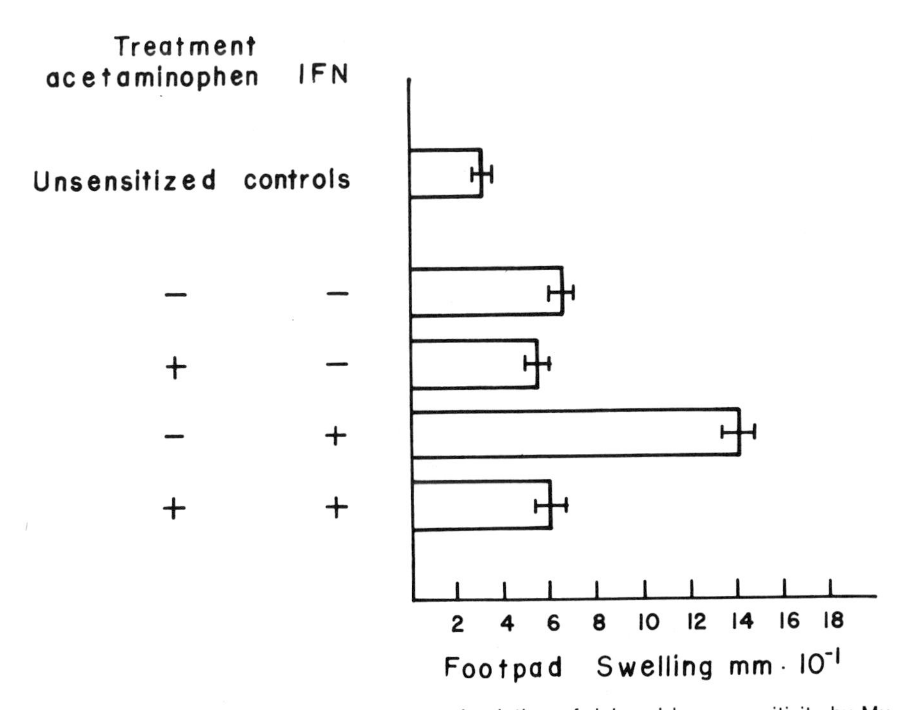

Figure 12.3 Inhibition of fever abrogates stimulation of delayed hypersensitivity by Mu IFN-α/β. Two hours after sensitization with a suboptimal dose of sheep erythrocytes, some animals received 10,000 units of Mu IFN-α/β. As a result of this treatment, footpad swelling was significantly enhanced upon challenge with sheep red cells 4 days later (next to last group). However, when IFN-treated mice received the antipyretic drug acetaminophen, IFN did not block the generation of suppressor cells, and thus, footpad swelling in mice sensitized to sheep red blood cells was not enhanced by the IFN treatment (last group). (Adapted from Ron et al., 1984.)

is probably a reflection of the increased efficacy of IFN action at such temperatures in general. In vitro, human and murine IFNs of different origins exert enhanced antiproliferative and antiviral activities when they are allowed to act on cells at 38 to 39°C, instead of at 37°C (Fig. 12.4), and an enhanced antiproliferative effect is probably the reason for the blocked expansion of T-suppressor cell clones (Heron and Berg, 1978; Yerushalmi et al., 1982; Hirai et al., 1984; Fleischmann et al., 1986).

The enhanced inhibition of suppressor cells during the febrile state links the intrinsic pyrogenic effect of IFNs to some of their immunoregulatory actions, and suggests one possible mechanism for the often observed beneficial effect of fever or artificial hyperthermia during virus infection, especially since IFN production itself is not adversely affected by moderate hyperthermia (Roberts, 1986). We can formulate the hypothesis that, favored by the higher body temperature, IFNs, through their antisuppressive effect, stimulate DH, and thus augment cell-mediated immunity directed against the IFN-inducing virus. If this mechanism is operative during viral infection, one would expect that occasionally the action of antipyretic drugs might be to delay or impede the establishment of immunity to viruses.

There is one well-known clinical entity, Reye's syndrome, in which the administration of antipyretic drugs during a viral infection has been linked to a sometimes fatal outcome of the infection. Reye's syndrome is a major cause of death after viral illness—frequently influenza or varicella—in children. The pathogenesis of this condition is unknown, but ingestion of salicylates appears to be a contributary factor (Mortimer and Lepow, 1962). The possibility that Reye's syndrome is related to an impairment of cell-mediated immunity as a result of the antipyretic activity of the drugs administered is

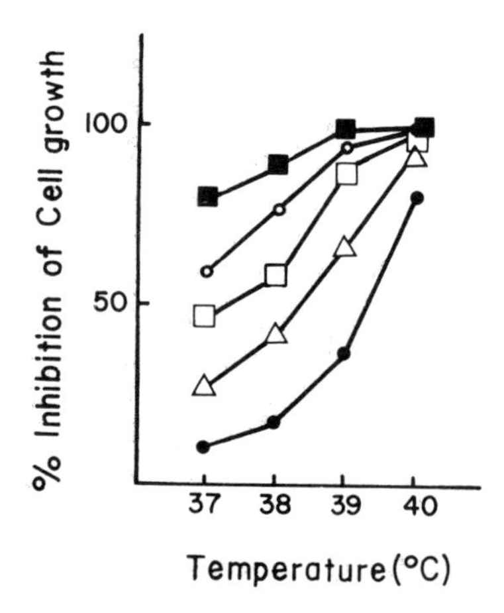

Figure 12.4 Potentiation by febrile temperatures of the antiproliferative effect of five different concentrations of Hu leukocyte IFN on lymphoblastoid Daudi cells. ●: 1 unit/ml; △: 2 units/ml; □:3 units/ml; ○: 5 units/ml; ■: 10 units/ml. (Adapted from Heron and Berg, 1978.)

raised by the higher mentioned findings that fever stimulates cell-mediated immunity by inhibition of suppressor cells due to IFN-α/β production in response to a viral infection (Ron et al., 1984).

Lending extra weight to the possibility that antipyretic drugs may favor virus infection is the fact that under certain conditions the direct antiviral activity of IFNs is also more pronounced at elevated temperatures; this suggests that abrogation of fever might favor virus replication by impeding an efficient establishment of the antiviral state by interferon in addition to decreasing cell-mediated immunity (Heron and Berg, 1978). Antipyretic drugs like aspirin and indomethacin can also decrease IFN production in tissue culture (Sekellick and Marcus, 1985). When infected with the Newcastle disease virus, lymphocytes obtained from patients during the acute phase of influenza B-associated Reye's syndrome produce significantly less IFN than lymphocytes obtained from the same patients during convalescence (Rozee et al., 1982). The combination of decreased IFN production and decreased IFN action would certainly be sufficient cause for enhanced pathogenicity during a viral infection.

C. Effects of IFN-α/β on the Efferent Limb of DH

The effect of IFNs on the efferent pathway is inhibitory. To block the expression of DH, in SRBC or hapten-sensitized mice, IFNs can be administered systemically and do not have to be injected into the site of antigen challenge. Adequate amounts of IFN completely block the inflammatory reaction that characteristically occurs at the site of antigen challenge, and histopathological examination shows the absence of the typical cellular infiltration. A renewed antigen challenge of the same animals a few days later, this time without previous IFN-α/β treatment, results in a normal DH reaction. This shows that IFN given to sensitized individuals only inhibits the expression of the sensitized state, without affecting the overall immune status (De Maeyer et al., 1975b; Kraaijeveld et al., 1986).

It has been a common observation that during or after a viral infection, some manifestations of cell-mediated immunity can be decreased. The classical observation in this respect was made in 1908 by the Viennese doctor Clemens von Pirquet, who described a decreased reactivity to the tuberculin skin test as a result of infection with the measles virus (von Pirquet, 1908). The occurrence of anergy as a result of measles has since been repeatedly confirmed, and the observation has been extended to other virus infections, such as varicella, influenza, and rubella. Moreover, anergy has also been observed after vaccination with a number of viruses, for example, polio, yellow fever, measles, and mumps (Brody and McAlister, 1964; Brody et al., 1964; Starr and Berkovich, 1964; Lamb, 1969; Kupers et al., 1970; Hall and Kantor, 1972). Since IFN-α/β production takes place as a result of a viral infection or vaccination, it can undoubtedly contribute to the anergy that may occur during or after a virus infection. Experimental evidence for this

IFN effect has been provided by an animal model, in which the effect of endogenous IFN-α/β production on the expression of DH has been examined. This model consists of two strains of mice of nearly identical genetical background that differ mainly by the amount of IFN-α/β produced as a result of infection with the Newcastle disease virus. One mouse strain has the low producer allele and the other strain the high producer allele at the *If*-1 locus, which controls the amount of IFN-α/β induced by NDV. The same amount of NDV is less immunosuppressive in mice carrying the low IFN producer allele than in mice carrying the high IFN producer allele which produce ten times as much IFN-α/β (De Maeyer et al., 1975a). This shows unambiguously that the extent of anergy resulting from virus infection can be related to the amount of IFN produced during this infection.

Another example is provided by LDH virus, a togavirus that causes an inapparent and persistent infection in mice. This virus is often found as a contaminant of transplanted murine tumors, and has in several instances been found to account for immunosuppressive or anti-inflammatory effects of these tumors or their extracts. These effects are for a large part, if not entirely, due to IFN-α/β induced by the virus, and can be abrogated by the treatment of infected mice with anti-Mu IFN-α/β antibodies (Heremans et al., 1987a).

Dysfunction of the immune system caused by IFN-α/β is an epiphenomenon of many virus infections. Viruses obviously can also disturb immune functions by mechanisms other than IFN production, such as, by directly infecting lymphocytes or macrophages and thereby derange the functions of these cells (Woodruff and Woodruff, 1975). The viruses causing AIDS offer the most dramatic illustration of this point (see Chapter 16).

D. A Natural Role for IFN-α/β in the Establishment of DH to a Virus: The *If*-1 Locus as an Immune Response Gene

Since IFNs α and β influence the afferent and efferent limbs of DH and are produced during virus infection, one can ask to what extent IFN-α/β influences the establishment of DH to the IFN-inducing virus. Because of the many factors involved in the establishment of cell-mediated immunity to viral antigens, it is not easy to single out the role of IFN.

It has, however, been possible to do so in one system, due to the existence of the previously mentioned *If*-1 locus, which influences the levels of the Newcastle disease virus-induced IFN-α/β production and which we will now discuss in some more detail (see also Chapter 15). Mice of the C57BL/6 By strain are *If*-1^{h}; they have the high producer allele and produce on the average ten times as much IFN-α/β as mice of the congenic B6.C-H-28c strains (there are three such strains), which are *If*-1^{l}, and have the low producer allele. B6.C-H-28c mice are otherwise genetically identical to C57BL/6 By mice, except for a small region of approximately 5 to 6 cM at the distal end of chromosome 3, on which *If*-1 is located (Chapter 15) (De Maeyer et al., 1975a; Mobraaten et al., 1984). When immunized under proper conditions with the Newcastle disease virus, mice of both strains develop DH to the virus. The reaction, however, is much more pronounced and of longer duration in the high IFN-α/β-producing C57BL/6 mice than in the low IFN-α/β-producing mice of the congenic strains (Fig. 12.5). In the latter, DH to NDV can be enhanced by compensating for the low IFN production with additional IFN-α/β, given intravenously a few hours after the virus has been administered. In C57BL/6 mice, on the other hand, DTH can be down-regulated by neutralizing the high IFN production with specific anti-IFN-α/β globulins (De Maeyer and De Maeyer-Guignard, 1983). The fact that high IFN production results in more pronounced DTH than low IFN production can be ascribed to the enhancing effect of IFN-α/β on the afferent loop of DTH, since IFN synthesis starts almost immediately and reaches peak levels a few hours after the virus has been inoculated, which corresponds to the optimal timing for up-regulation of sensitization. In this respect, the *If*-1 locus functions as an immune response locus, regulating the duration and intensity of DTH to NDV by influencing the levels of IFN-α/β

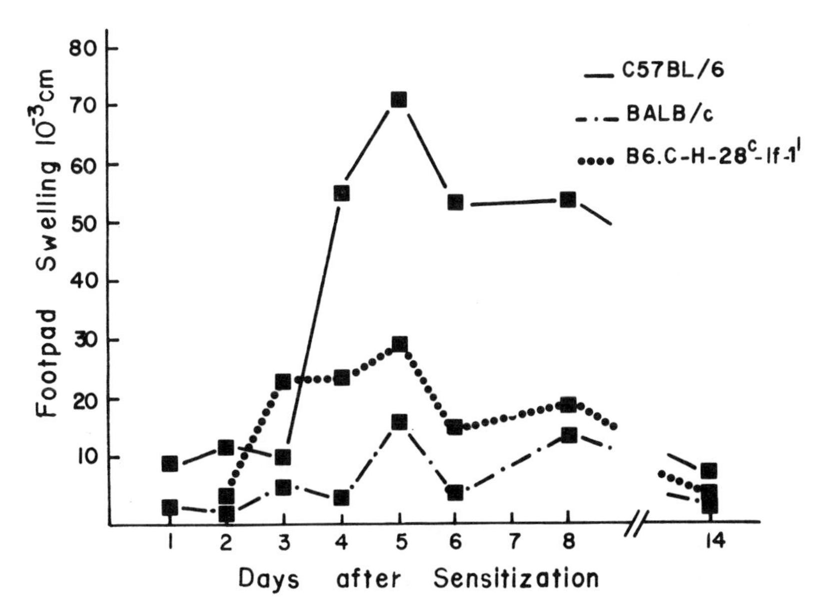

Figure 12.5 The intensity of delayed hypersensitivity to the Newcastle disease virus is influenced by the levels of Mu IFN-α/β that are induced by the virus at the time of sensitization. The reaction is much more pronounced in high IFN producer C57BL/6 mice than in the low IFN producer BALB/c mice, and the difference is due to the allele at the *If*-1 locus, since C57BL/6 mice congenic with BALB/c at the *If*-1 locus (B6.C-H-28^c-*If*-1^l) have, like BALB/c mice low-intensity-delayed hypersensitivity. The figure shows the 48-hour footpad swelling reaction in mice challenged on different days after sensitization. (Adapted from De Maeyer and De Maeyer-Guignard, 1983.)

production. Studies in F2 and backross generations show that several other genes are involved in this regulation (De Maeyer and De Maeyer-Guignard, 1980b).

E. Effects of IFN-γ

The possible role of IFN-γ in the establishment and expression of DH has received less attention. Like Mu IFN-α/β, Mu IFN-γ preferentially inhibits the T-suppressor circuit of contact allergy, and T cells of mice sensitized to the contact-sensitizing agent oxazolone produce IFN-γ upon a challenge with the hapten (McKimm-Breschkin et al., 1982; Taborski et al., 1986). Similarly to IFN-α/β, Mu IFN-γ can inhibit the local inflammatory reaction and the swelling developing after injection of LPS into the footpads of mice (Heremans et al., 1987b).

Undoubtedly, in view of its mode of induction, IFN-γ is produced during most, if not all, DH reactions that are triggered by a rechallenge with antigen, and it is one of the many lymphokines instrumental in the reaction.

III. CONCLUSION

The effects of IFN-α/β on the afferent and efferent pathways of DH reflect the complexity of the interaction of IFNs and the immune system, since either immuno-suppression or immuno-enhancement is obtained, depending on the timing of IFN action. Most studies concerning the effects of IFN on DH have been done with unfractionated murine IFN, which is a mixture of Mu IFN-β and an undetermined number of Mu IFN-α species. It will be interesting to determine separately the effects of the different IFN-α subtypes and of IFN-β.

REFERENCES

Ackerman, S. K., Hochstein, H. D., Zoon, K., Browne, W., Rivera, E. and Elisberg, B. Interferon fever: Absence of human leukocytic pyrogen response to recombinant α-interferon. *J. Leukocyte Biol.* 36: 17–25 (1984).

Bartocci, A., Read, E. L., Welker, R. D., Schlick, E., Papademetriou, V., and Chirigos, M. A. Enhancing activity of various immunoaugmenting agents on the delayed-type hypersensitivity response in mice. *Cancer Res.* 43: 3514–3518 (1982).

Brody, J. A. and McAlister, R. Depression of tuberculin sensitivity following measles vaccination. *Am. Rev. Resp. Dis.* 90: 607–611 (1964).

Brody, J. A., Overfield, T., and Hammes, L. M. Depression of the tuberculin reaction by viral vaccines. *N. Engl. J. Med.* 271: 1294–1296 (1964).

De Maeyer, E. Interferon and delayed-type hypersensitivity to a viral antigen. *J. Infect. Dis.* 133: A63–A65 (1976).

De Maeyer, E. and De Maeyer-Guignard, J. Host genotype influences immunomodulation by interferon. *Nature* 284: 173–175 (1980a).

De Maeyer, E. and De Maeyer-Guignard, J. Immunoregulatory action of type I interferon in the mouse. *Ann. N.Y. Acad. Sci.* 350: 1–11 (1980b).

De Maeyer, E. and De Maeyer-Guignard, J. Delayed hypersensitivity to Newcastle disease virus in high and low interferon producing mice. *J. Immunol.* 130: 2392–2396 (1983).

De Maeyer, E., De Maeyer-Guignard, J., and Bailey, D. W. Effect of mouse genotype on interferon production. Lines congenic at the If-1 locus. *Immunogenetics* 1: 439–445 (1975a)

De Maeyer-Guignard, J., Cachard, A., and De Maeyer, E. Delayed-type hypersensitivity to sheep red blood cells: Inhibition of sensitization by interferon. *Science* 190: 574–576 (1975).

De Maeyer, E., De Maeyer-Guignard, J., and Vandeputte, M. Inhibition by interferon of delayed-type hypersensitivity in the mouse. *Proc. Natl. Acad. Sci.* (USA) 72: 1753–1757 (1975b).

Dinarello, C. A., Bernheim, H. A., Duff, G. W., Le, H. V., Nagabhushan, T. L., Hamilton, N. C., and Coceani, F. Mechanisms of fever induced by recombinant human interferon. *J. Clin. Invest.* 74: 906–913 (1984).

Dinarello, C. A., Cannon, J. G., Wolff, S. M., Bernheim, H. A., Beutler, B., Cerami, A., Figari, I. S., Palladino, M. A., Jr., and O'Connor, J. V. Tumor necrosis factor (cachectin) is an endogenous pyrogen and induces production of interleukin 1. *J. Exp. Med.* 163: 1433–1450 (1986).

Feinstone, S. M., Beachey, E. H., and Rytel, M. W. Induction of delayed hypersensitivity to influenza and mumps viruses in mice. *J. Immunol.* 103: 844–849 (1969).

Fleischmann, W. R., Jr., Fleischmann, C. M., and Gindhart, T. D. Effect of hyperthermia on combination interferon treatment: Enhancement of the antiproliferative activity against murine B-16 melanoma. *Cancer Res.* 46: 1722–1726 (1986).

Fradelizi, D. and Gresser, I. Interferon inhibits the generation of allospecific suppressor T lymphocytes. *J. Exp. Med.* 155: 1610–1622 (1982).

Hall, C. B. and Kantor, F. S. (1972). Depression of established delayed hypersensitivity by mumps virus. *J. Immunol.* 108: 81–85 (1972).

Heremans, H., Billiau, A., Coutelier, J. P., and De Somer, P. The inhibition of endotoxin-induced local inflammation by LDH virus or LDH virus-infected tumors is mediated by interferon. *Proc. Soc. Exp. Biol. Med.* 185: 6–15 (1987a).

Heremans, H., Dijkmans, R., Sobis, H., Vanderkerckhove, F., and Billiau, A. Regulation by interferons of the local inflammatory response to bacterial lipopolysaccharide. *J. Immunol.* 138: 4175–4179 (1987b).

Heron, I. and Berg, K. The actions of interferon are potentiated at elevated temperature. *Nature* 274: 508–510 (1978).

Hirai, N., Hill, N. O., and Osther, K. Temperature influences on different human α interferon activities. *J. Interferon Res.* 4: 507–516 (1984).

Hooks, J. J., Moutsopoulos, H. M., and Notkins, A. L. The role of interferon in immediate hypersensitivity and autoimmune diseases. In: *Regulatory Functions of Interferons,* J. Vilcek, I. Gresser, and T. C. Merigan, Eds., *Ann. N.Y. Acad. Sci.* 350: 21–32 (1980).

Hotchin, J. The footpad reaction of mice to lymphocytic choriomeningitis virus. *Virology* 17: 214–216 (1962).

Ida, S., Hooks, J. J., Siraganian, R. P., and Notkins, A. L. Enhancement of IgE-mediated histamine release from human basophils by viruses: Role of interferon. *J. Exp. Med.* 145: 892–904 (1977).

Knop, J., Stremmer, R., Neumann, C., De Maeyer, E., and Macher, E. Interferon inhibits the suppressor T cell response of delayed-type hypersensitivity. *Nature* 296: 775–776 (1982).

Knop, J., Stremmer, R., Taborski, U., Freitag, W., De Maeyer-Guignard, J., and Macher, E. Inhibition of the T suppressor circuit of delayed-type hypersensitivity by interferon. *J. Immunol.* 133: 2412–2416 (1984).

Knop, J., Taborski, B., and De Maeyer-Guignard, J. Selective inhibition of the generation of T suppressor cells of contact sensitivity in vitro by interferon. *J. Immunol.* 138: 3684–3687 (1987).

Kraaijeveld, C. A., Harmsen, M., and Khader Boutahar-Trouw, B. Delayed-type hypersensitivity against Semliki forest virus in mice. *Infect. Immun.* 23: 219–223 (1979).

Kraaijeveld, C. A., Kamphuis, W., Benaissa-Trouw, B. J., Harmsen, M., and Snippe, H. Modulation of adjuvant-enhanced delayed-type hypersensitivity by the interferon inducers poly I : C and Newcastle disease virus. *Int. Archs Allergy Appl. Immun.* 79: 86–89 (1986).

Kupers, T. A., Petrich, J. M., Holloway, A. W., and St. Geme, J. W., Jr. Depression of tuberculin delayed hypersensitivity by live attenuated mumps virus. *J. Pediatr.* 76: 716–721 (1970).

Lamb, G. A. Effect of HPV-80 rubella vaccine on the tuberculin reaction. *Am. J. Dis. Child.* 118: 261 (1969).

Lausch, R. N., Swyers, J. S., and Kaufman, H. E. Delayed hypersensitivity to herpes simplex virus in the guinea pig. *J. Immunol.* 96: 981–987 (1966).

Lett-Brown, M. A., Aelvoet, M., Hooks, J. J., Georgiades, J. A., Thueson, D. O., and Grant, J. A. Enhancement of basophil chemotaxis in vitro by virus-induced interferon. *J. Clin. Invest.* 67: 547–552 (1981).

Mackaness, G. B., Lagrange, P. H., Miller, T. E., and Ishibashi, T. Feedback inhibition of specifically sensitized lymphocytes. *J. Exp. Med.* 139: 543–559 (1974).

McIntosh, K., Ellis, E. F., Hoffman, L. S., Lybass, T. G., Eller, J. J., and Fulginiti, V. A. The association of viral and bacterial respiratory infections with exacerbations of wheezing in young asthmatic children. *J. Pediatr.* 82: 578–590 (1973).

McKimm-Breschkin, J. L., Mottram, P. L., Thomas, W. R., and Miller, J. F. A. P. Antigen-specific production of immune interferon by T cell lines. *J. Exp. Med.* 155: 1204–1209 (1982).

Mobraaten, L. E., Bunker, H. P., De Maeyer-Guignard, J., De Maeyer, E., and Bailey, D. W. Location of histocompatibility and interferon loci on chromosome 3 of the mouse. *J. Hered.* 75: 233–234 (1984).

Mortimer, E. A., Jr. and Lepow, M. L. Varicella with hypoglycemia possibly due to salicylates. *Am. J. Dis. Child.* 130: 583–590 (1962).

Roberts, N. J., Jr. Differential effects of hyperthermia on human leukocyte production of interferon-α and interferon-γ. *Proc. Soc. Exp. Biol. Med.* 183: 42–47 (1986).

Ron, Y., Dougherty, J. P., Duff, G. W., and Gershon, R. K. The effect of febrile temperatures on biologic actions of interferons: Abrogation of suppression of delayed-type hypersensitivity and antibody production. *J. Immunol.* 133: 2037–2042 (1984).

Rozee, K., Lee, S. H. S., Crocker, J. F. S., Digout, S., and Arcinue, E. Is a compromised interferon response an etiologic factor in Reye's syndrome? *Can. Med. Assoc. J.* 126: 798–802 (1982).

Scott, G. M., Secher, D. S., Flowers, D., Bate, J., Cantell, K., and Tyrrell, D. A. J. Toxicity of interferon. *Br. Med. J.* 282: 1345–1348 (1981).

Sekellick, M. J. and Marcus, P. I. Interferon induction by viruses. XIV. Development of interferon inducibility and its inhibition in chick embryo cells "aged" in vitro. *J. Interferon Res.* 5: 651–667 (1985).

Starr, S. and Berkovich, S. Effects of measles, γ-globulin-modified measles and vaccine measles on the tuberculin test. *N. Engl. J. Med.* 270: 386–391 (1964).

Taborski, U., Freitag, W., Heremans, H., and Knop, J. Inhibitory effects of interferon-γ on the T suppressor cell circuit in contact sensitivity. *Immunobiology* 171: 329–338 (1986).

Tarkkanen, J. and Saksela, E. Umbilical-cord-blood-derived suppressor cells of the human natural killer cell activity are inhibited by interferon. *Scand. J. Immunol.* 15: 149–157 (1982).

Turk, J. L. Delayed Hypersensitivity. *Research Monographs in Immunology. Vol. 1,* Elsevier/North-Holland Biomedical Press, Amsterdam (1980).

Von Pirquet, C. Das Verhalten der kutanen Tuberkulin-Reaktion während der Masern. *Dtsch. Med. Wochenschr.* 34: 1297–1300 (1908).

Wetherbee, R. E. Induction of systemic delayed hypersensitivity during experimental viral infection of the respiratory tract with a myxovirus or paramyxovirus. *J. Immunol.* 111: 157–163 (1973).

Wong, G. H. W., Clark-Lewis, I., McKimm-Breschkin, J. L., and Schrader, J. W. Interferon-γ-like molecule induces Ia antigens on cultured mast cell progenitors. *Proc. Natl. Acad. Sci.* (USA) 79: 6989–6993 (1982).

Woodruff, J. F. and Woodruff, J. J. The effect of viral infections on the function of the immune system. In: *Viral Immunology and Immunopathology,* A. L. Notkins, Ed., Academic Press, New York, pp. 393–418 (1975).

Yerushalmi, A., Tovey, M. G., and Gresser, I. Antitumor effect of combined interferon and hyperthermia in mice. *Proc. Soc. Exp. Biol. Med.* 169: 413–415 (1982).

13 INTERACTION OF INTERFERONS, TUMOR NECROSIS FACTOR, INTERLEUKIN-1, AND INTERLEUKIN-2 AS PART OF THE CYTOKINE NETWORK

In the vertebrate organism, cells, tissues, and organs are in constant communication with one another. Two important ways of communication are the nervous system, on the one hand, and hormones—effector molecules made by highly specialized cells—on the other hand. A third way of intercellular communication is provided by a network of polypeptides whose production is usually not restricted to a given type of cells and that act as intercellular messengers. These cytokines are frequently involved in intimate cell-to-cell contact and communication but, like hormones, also act at a distance. They can exert an autocrine effect on the cells that produce them or influence other cells in a paracrine way. When released into the circulation, they can have profound systemic effects.

Many cytokines interact in various ways, either synergistically or as antagonists. They can reciprocally up- or down-regulate their production and in some cases also receptor expression. This results in a highly complex series of pathways that has been called the cytokine network, a tangled web of interactions and of positive and negative feedback loops branching off into many directions. Many of these interactions are being defined in *in vitro* systems, in which the components have been removed from their natural environment in the organism, and we have to exert due caution when extrapolating to physiological conditions.

Several cytokines were originally named according to their first known biological effect. Since it is now clear that they exert pleiotropic effects, there is a tendency to abandon descriptive names and replace them by a nomenclature that employs the term "interleukin" followed by a number. The interactions of IFNs and several cytokines, including IL-3 and IL-4, are discussed in various other chapters, as they relate to the subject matter of the chapter. Here we deal with cytokines that have pleiotropic functions of such magnitude and diversity that, like IFNs, they are not easily classified into any particular category of activities.

I. TUMOR NECROSIS FACTOR (TNF-α) AND LYMPHOTOXIN (LT OR TNF-β)

A. TNF-α and TNF-β Are Related Polypeptides with Pleiotropic Activity

1. Genetics and Structure

Tumor necrosis factor (TNF-α) and lymphotoxin (TNF-β) are two structurally and functionally related polypeptides with similar pleiotropic biological activities. The genes for TNF-α and TNF-β probably arose by duplication since they share structural features and are arranged in tandem near the major histocompatibility complex, between HLA-DR and HLA-A, on the short arm of chromosome 6 in humans and on chromosome 17 in the mouse, 70 kb proximal to the D gene (Nedwin et al., 1985a; Nedospasov et al., 1986; Spies et al., 1986; Müller et al., 1987).

In humans, the single copy genes of TNF-α and TNF-β are about 3 kb pairs long and are interrupted by three introns. Only the last exons of each of these two genes, coding for about 80% of the secreted proteins, share over 50% homology at the nucleotide level (Nedwin et al., 1985a). Like Hu IFN-α and -β, human TNF-α and β share about 30% homology at the amino acid level (Table 13.1) (Gray et al., 1984; Pennica et al., 1984a; Aggarwal et al., 1985a). The 157-amino-acid-long, mature sequence of human TNF-α, with a monomeric molecular weight of about 17 kDa, shares about 80% homology with murine TNF-α (Table 13.2) (Fransen et al., 1985; Marmenout et al., 1985). Human TNF-β is a glycosylated protein of 171 amino acids with a monomeric molecular weight of 25 kDa, and murine TNF-β contains 169 amino acids (Aggarwal et al., 1984, 1985a; Gray et al., 1987). Both mouse and human TNF genes contain in their 3′ untranslated region a 33-nucleotide-long conserved sequence, composed entirely of A and T residues, which in the corresponding mRNA consists of several AUUUA or UUAUUUAU sequence motifs. This sequence, which is uncommon among mammalian mRNAs, is also found in the 3′ untranslated region of other transiently expressed genes, encoding, for example, IFNs and other cytokines like IL 1 and GM CSF. The possibility has been raised that this

TABLE 13.1 Alignment of the Human TNF-α and TNF-β Deduced Amino Acid Sequences

			1 10	20	30	40
α:			VRSSSRTPSD	KPVAHVVANP	QAEGQLQWLN	RRANALLANG
β:	LPGVGLT	PSAAQTARQH	PKMHLAHSTL	KPAAHLIGDP	SKQNSLLWRA	NTDRAFLQDG
	1 7	17	27	37	47	57

	50	60	70	80	90	100
α:	VELRDNQLVV	PSEGLYLIYS	QVLFKGQGCP	STHVLLTHTI	SRIAVSYQTK	VNLLSAIKSP
β:	FSLSNNSLLV	PTSGIYFVYS	QVVFSGKAYS	PQATSSPLYL	AHEVQLFSSQ	YPFHVPLLSS
	67	77	87	97	107	117

	110	120	130	140	150	157
α:	CQRETPEGAE	AKPWYEPIYL	GGVFQLEKGD	RLSAEINRPD	YLDFAESGQV	YFGIIAL
β:	QKMVYPGLQE	--PWLHSMYH	GAAFQLTQGD	QLSTHTDGIP	HLVLSPST-V	FFGAFAL
	127	135	145	155	164	171

References: Gray et al., 1984; Pennica et al., 1984; Aggarwal et al., 1985b; Nedwin et al., 1985.

The one-letter amino acid code is explained in Chapter 2, Table 2.5.

region is a common feature and a special regulatory element of proteins that are mobilized during the inflammatory response, since it appears to confer instability upon an mRNA molecule and may be involved in the control of gene expression at the transcriptional level (Caput et al., 1986; Shaw and Kamen, 1986).

The TNF-α and TNF-β genes are closely linked and encode two partially homologous proteins with rather similar biological functions. Nevertheless, the expression of these genes is independently regulated, and one can establish conditions for producing each cytokine separately (Cuturi et al., 1987).

Like IFNs α and β, TNF-α and TNF-β share a common high affinity receptor at the cell surface (Aggarwal et al., 1985a).

TABLE 13.2 Comparison of the Human and Murine TNF-α Signal Peptide and Mature Protein Deduced Amino Acid Sequences

-76
MSTESMIRDVELAEEALPKKTGGPQGSRRCLFLSLFSFLIVAGATTLFCLLHFGVIGPQR_

MSTESMIRDVELAEEALPQKMGGFQNSRRCLCLSLFSFLLVAGATTLFCLLNFGVIGPQRD

	1 10	20	30	40	
EEFPRDLSLISPLAQA__	VRSSSRTPSD	KPVAHVVANP	QAEGQLQWLN	RRANALLANG	
EKFPNGLPLISSMAQTLT	LRSSSQNSSD	KPVAHVVANH	QVEEQLEWLS	QRANALLANG	

50	60	70	80	90	100
VELRDNQLVV	PSEGLYLIYS	QVLFKGQGCP	STHVLLTHTI	SRIAVSYQTK	VNLLSAIKSP
MDLKDNQLVV	PADGLYLVYS	QVLFKGQGCP	DY_VLLTHTV	SRFAISYQEK	VNLLSAVKSP

110	120	130	140	150	157
CQRETPEGAE	AKPWYEPIYL	GGVFQLEKGD	RLSAEINRPD	YLDFAESGQV	YFGIIAL
CPKDTPEGAE	LKPWYEPIYL	GGVFQLEKGD	QLSAEVNLPK	YLDFAESGQV	YFGVIAL

References: Fransen et al., 1985; Marmenout et al., 1985.

The one-letter amino acid code is explained Chapter 2, Table 2.5.

2. Properties of TNF-α

TNF-α, mainly produced by macrophages in response to endotoxin, has been known under different names, reflecting its pleiotropic activity. It was originally described as a monokine causing hemorrhagic necrosis of mouse tumors (Carswell et al., 1975), and furthermore is identical to cachectin, the name given to a protein that is responsible for inducing shock and cachexia (Beutler and Cerami, 1986; Tracey et al., 1986; Beutler and Cerami, 1987). Cachexia accompanies a variety of diseases, including cancer and infection, and is characterized by general ill health and malnutrition. Some of the metabolic derangements responsible for this wasting are the result of the activity of TNF-α. In the final stage of the disease, cachexia is accompanied by lipaemia; the serum triglyceride levels are elevated and there is an associated inhibition of lipoprotein lipase biosynthesis and expression. Furthermore, TNF-α is a potent pyrogen, causing fever through a direct effect on the hypothalamic thermoregulatory center by releasing prostaglandin E2. Enhanced production of prostaglandin E2 as well as of collagenase, also stimulated by TNF-α, may play an important role in the pathogenesis of inflammatory disease of joints. In this respect osteoclast-mediated bone resorption, stimulated by TNF-α (and also by TNF-β), may also contribute to this disease (Bertolini et al., 1986; Thomson et al., 1987). Finally, TNF-α stimulates the production of procoagulant activity by vascular endothelial cells, and thus is involved in the clotting abnormalities associated with endotoxaemia (Beutler and Cerami, 1986).

In addition to being produced by macrophages, TFN-α is also made by human large granular lymphocytes (NK cells), and although it does not appear as one of the major mediators of cytotoxicity of the latter, since tumor cells resistant to TNF-α can still be killed by NK cells, there is evidence that at times it is instrumental in cell killing (Degliantoni et al., 1985; Peters et al., 1986; Urban et al., 1986). Furthermore, TNF-α is one of the major mediators of cytotoxicity of natural cytotoxic cells, a type of effector cells that are different from NK cells and related to mast cells (Stutman et al., 1978; Jadus et al., 1986; Ortaldo et al., 1986).

Release of large amounts of TNF-α during malaria infection may represent an important effector mechanism for originating cerebral malaria. Cerebral malaria is a complication of *Plasmodium falciparum* infection, which develops in a minority of malaria patients. In a mouse model for this disease, high levels of TNF-α appear in the serum of mice belonging to susceptible strains, and treatment of the infected animals with anti-TNF-α antibodies prevents the neurological syndrome (Grau et al., 1987).

Surprisingly, in addition to these multiple catabolic and cytotoxic effects, TNF-α can stimulate cell proliferation and even induce differentiation under certain conditions. The replication of human diploid cells of different origins, such as colon, skin, or lung, is enhanced in vitro by the presence of low amounts of TNF-α. Stimulation of cell proliferation occurs at concentrations ranging from 10^{-10} to 10^{-13} M, which is equal to or lower than concentrations required to produce cytotoxicity in highly sensitive malignant cells. The addition of antibodies directed against Hu IFN-β to human fibroblasts whose growth is stimulated by TNF-α amplifies the effect of TNF-α in low serum medium (Fig. 13.1) (Sugarman et al., 1985; Vilcek et al., 1986). Stimulation of growth factor activity by anti-IFN-β antibodies is not unique for TNF, and has been observed with several other growth factors that provoke the synthesis of IFN in a negative feedback loop (see Chapters 7 and 8). The increase of high affinity epidermal growth factor (EGF) binding sites on TNF-α-treated fibroblasts probably contributes to the growth stimulatory effect of TNF (Palombella et al., 1987).

TNF-α is also identical to the differentiation inducing factor (DIF), which induces human myelogenous leukemic cells to differentiate into the monocyte-macrophage pathway. DIF is produced by T cells of human leukemia cell lines with macrophage characteristics (Takeda et al., 1986). Both human rec TNF-α and natural TNF-β induce monocytic differentiation of human myeloid cell lines (Trinchieri et al., 1986).

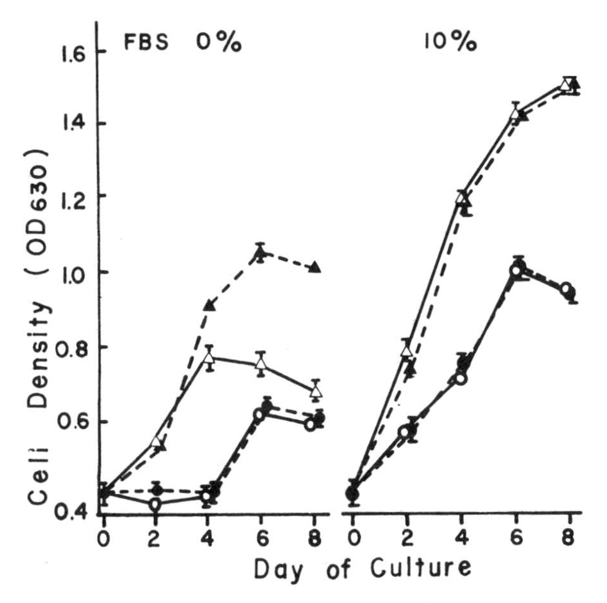

Figure 13.1 The stimulating effect of TNF-α on the proliferation of human FS-4 cells is amplified by antibodies to Hu IFN-β. The amplifying effect of the antibodies is greatly influenced by the fetal bovine serum (FBS) concentration of the culture medium and is most pronounced when no serum is used (left panel). When 10% serum is present in the culture medium (right panel), antibodies to Hu IFN-β no longer stimulate TNF-induced cell proliferation. This is almost certainly due to the presence of various growth factors in the medium, which already act as IFN antagonists, making the addition of antibodies superfluous (see, for example, Chapter 7, Fig. 7.3). △—△: plus TNF-α; ▲—▲: plus TNF-α and anti-IFN antibodies. ○—○: controls without TNF; ●—●: controls without TNF but with anti-IFN antibodies. (Adapted from Kohase et al., 1986.)

3. Properties of TNF-β

TNF-β, originally known as "lymphotoxin," was first characterized as a lymphokine having nonspecies specific cytotoxic activity for fibroblasts and for transformed cells (Granger and Kolb, 1968; Ruddle and Waksman, 1968). TNF-β can be produced by mitogen-activated T cells, as well as by cytotoxic and T-helper cells and by some lymphoblastoid cell lines (Aggarwal et al., 1985b; Schmid et al., 1986). Target cell destruction by TNF-β is characterized by the release from the destroyed cells of DNA digested into discretely sized fragments that are multiples of 200 bp, as opposed to the large molecular weight DNA fragments released from target cells lysed by complement and antibody (Russell et al., 1980, 1982; Schmid et al., 1986).

TNF-β and TNF-α can stimulate osteoclastic bone resorption, and in this respect their activity resembles that of parathyroid hormone and of IL-1. The mechanism by which TNF stimulates bone resorption is unknown, and there are indications that it occurs probably more by stimulating osteoblastic cells to produce a factor that causes bone resorption than by directly stimulating osteoclast activity (Bertolini et al., 1986; Thomson et al., 1987).

The significant production of TNF-β by human myeloma cells appears as a possible cause, or at the least as a contributing factor, for the extensive bone destruction and hypercalcemia that occur in patients with myeloma. This lymphoproliferative disorder is characterized by bone destruction accompanied by susceptibility to fracture and hypercalcemia. What activates the TNF-β gene in the tumor cells is not known (Garrett et al., 1987).

B. Similarities Between the Biological Activities of TNFs and IFNs

In addition to similarities in induction systems, for example, LPS in macrophages or Sendai virus in peripheral blood leukocytes (Berent et al., 1986), there are a number of interesting similarities in the biological effects of TNFs and IFNs.

1. Effects on Cell Proliferation

Like IFNs, TNFs inhibit cell proliferation and have cytotoxic and cytostatic activity, but they have been shown to stimulate cell growth exceptionally (Sugarman et al., 1985; Vilcek et al., 1986).

In general, the activity of TNF-α is much less species-specific than that of IFNs, although a certain degree of species specificity does exist in that Hu TNF-α is more cytotoxic for human than for mouse cells, whereas on mouse cells Mu TNF-α is more growth inhibitory and cytotoxic than Hu TNF-α. This is a reflection of different affinities for the TNF-α receptor, since the affinity is higher for the receptor of the corresponding species (Fransen et al., 1986a; Smith et al., 1986).

Both TNF-α and IFN-γ suppress colony formation of different lineages derived from progenitor cells in cultures of human bone-marrow cells (Lu et al., 1986).

2. Antitumor Effects

TNF-α has antitumor activity in mice on a number of transplantable tumors. Although some tumor cells are resistant to the direct cytotoxic effects of TNF-α in vitro, their development is, nevertheless, inhibited in vivo. This is strongly suggestive that part of the antitumor effect of TNF-α occurs via host-mediated mechanisms (Haranaka et al., 1984; Gresser et al., 1986). Likewise, IFN-α/β can exert antitumor effects in vivo in mice inoculated with cells that are resistant to the antiproliferative activity of IFN in vitro.

3. Antiviral Effects

TNF-α and TNF-β are able to induce antiviral activity against different RNA and DNA viruses in a wide variety of cells. This effect, however, is not as universal as that of IFNs, and many cell lines are resistant to the TNF-induced antiviral effect. Like IFNs, TNF-α induces 2-5A synthetase activity, and activation of the 2-5A pathway is an important point of similarity between these two classes of cytokines. Part of the antiviral effect of TNF-α takes place via the induction of IFN-β and can be abrogated by polyclonal anti-IFN-β antibodies. The IFN-β-mediated effect is clearly distinct from the direct antiviral effect of TNF-α, which is not accompanied by the transcription of IFN genes nor abrogated in the presence of anti-IFN-α or IFN-β antibodies (Kohase et al., 1986; Mestan et al., 1986; Wong and Goeddel, 1986; Van Damme et al., 1987b).

4. Endogenous Pyrogenic Activity

Like IL-1 and IFN-α/β, TNF-α acts as an endogenous pyrogen by stimulating the hypothalamic thermoregulatory center (Dinarello et al., 1986).

5. Effects on Macrophage Function

Like IFN-γ, TNF-α inhibits the intracellular replication of the tropomastigote forms of *Trypanosoma cruzi* in murine peritoneal macrophages (De Titto et al., 1986) and, like IFNs, TNF-α and TNF-β activate phagocytic activity and antibody-dependent cellular cytotoxicity of human polymorphonuclear neutrophils (Shalaby et al., 1985).

6. Expression of MHC Antigens

We have seen in Chapter 9 that activation of the expression of class I and class II MHC antigens is an important function of IFNs. Under certain conditions TNF-α also induces or augments the expression of MHC class I and class II antigens (Chang and Lee, 1986; Pober et al., 1986; Pfizenmaier et al., 1987). Furthermore, TNF-α and IFN-γ have similar activities on human vascular endothelial cell monolayers and cause elongation, rearrangement of actin filaments, and loss of stainable fibronectin matrix. These in vitro changes are correlates of immunologically mediated vascular responses in vivo, as found, for example, in delayed hypersensitivity responses. This supports the hypothesis that both TNF-α and IFN-γ are mediators of delayed hypersensitivity (Stolpen et al., 1986).

7. Effects on Adipocyte Metabolism

The ability of TNF-α to inhibit anabolic processes in adipocytes is also a property of all three IFN species; it results in a stimulated secretion of free fatty acids by adipocytes and an inhibition of anabolic lipid metabolism in these cells (Patton et al., 1986).

C. Interactions of TNFs and IFNs

IFNs and TNF interact at various levels.

1. The Production of Both TNF-α and TNF-β Can Be Stimulated by IFN-γ

Endotoxin-stimulated macrophages represent the major source of TNF-α in mice. Peritoneal macrophages of endotoxin-resistant C3H/HeJ mice having the *Lps*^{d} mutation at the *Lps* locus on chromosome 4 (see Chapter 10) produce little or no TNF-α when stimulated by endotoxin, whereas macrophages from mice lacking this mutation produce high amounts of TNF-α. This lack of production by *Lps*^{d} macrophages is the result of two events. At low LPS concentrations, the TNF-α gene is not transcribed; at higher LPS concentrations the gene is transcribed but the resulting mRNA is not translated (Beutler et al., 1986). Treatment of such endotoxin-resistant macrophages with Mu IFN-γ corrects the defect at both the transcriptional and posttranscriptional levels, since in these macrophages the TNF-α gene is transcribed upon induction by LPS and the resulting mRNA is translated. IFN-γ does not induce TNF-α synthesis by itself, nor does it markedly augment TNF-α synthesis by LPS-induced macrophages from mice with the normal allele at the *Lps* locus. Somehow IFN-γ succeeds in mitigating the effect of the *Lps*^{d} allele (Beutler et al., 1986b).

In human peripheral blood mononuclear cells, IL-2 or mitogens induce both TNF-α and TNF-β. IFN-γ by itself does not induce TNF-α or TNF-β, but when treatment of the cells with IL-2 or mitogens is combined with rec IFN-γ, a synergistic action ensues, and the synthesis of TNF-α and TNF-β is enhanced (Nedwin et al., 1985b; Svedersky et al., 1985). Similarly, the enhanced toxicity of IFN-γ-treated human monocytes for tumor cells appears to be due to stimulation of TNF-α synthesis (Philip and Epstein, 1986). The in vitro priming effect on TNF synthesis by IFN-γ is also observed in vivo in the mouse. Moreover, natural Mu IFN-α/β or rec Mu IFN-β also prime the production of TNF in mice (Satoh et al., 1986). Conversely, TNF-α can act as a costimulator of IL-2-dependent IFN-γ production (Scheurich et al., 1987).

2. TNF and IFNs Sometimes Act Synergistically

In view of the similarity of many biological effects of TNFs and IFNs, and since they bind to separate cell-surface receptors, it comes as no surprise that at times they act synergistically. Moreover, in murine and human cells

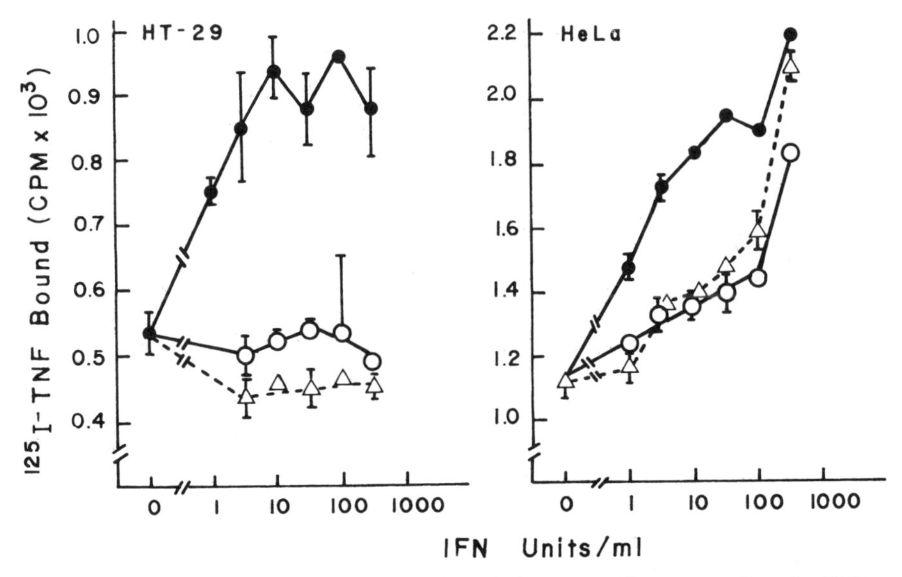

Figure 13.2 IFNs can enhance the expression of plasma membrane receptors for TNF-α. In human colon carcinoma HT-29 cells (left panel), only IFN-γ (●) increases the number of TNF receptors, whereas Hu leukocyte IFN (○) or Hu IFN-β (△) are without effect. In human HeLa cells (right panel), all three IFN species can augment the high affinity binding of TNF-α. (Adapted from Tsujimoto et al., 1986.)

the number of specific TNF receptors is increased by treatment with IFN-γ, and in some human cell lines TNF receptors are also increased after treatment with Hu IFN-β or highly purified natural Hu IFN-α (Fig. 13.2) (Aggarwal et al., 1985a; Tsujimoto et al., 1986; Aggarwal and Eessalu, 1987). This is probably the result of an IFN-stimulated enhanced transcription of TNF receptor mRNA, followed by an increased expression of receptor protein (Tsujimoto and Vilcek, 1986). Only the number of TNF receptors, but not the binding affinity, is influenced by the IFN treatment. Such an increase probably accounts at least for part of the synergism of IFNs and TNF, but other mechanisms are undoubtedly involved, since in some cells synergistic growth inhibition by combined IFN-γ TNF-α treatment can occur without a concomitant increase of TNF receptors (Ruggiero et al., 1986). Synergism of IFN and TNF has been observed for the following activities:

1. Monocytic differentiation of human myeloid cell lines (Trinchieri et al., 1986).
2. The antiproliferative effect of Hu IFN-γ on human neoplastic cells is synergistically enhanced by natural TNF-β (Lee et al., 1984; Stone-Wolff et al., 1984). Concentrations of rec IFN-γ and rec TNF-α that individually only slightly inhibit the replication of different human tumor cell lines act synergistically when combined and cause significant inhibition of cell proliferation (Ruggiero et al., 1986).

3. The antitumor effect in mice of Hu TNF-α is augmented in a synergistic way by rec Mu IFN-γ (Brouckaert et al., 1986). Similarly, growth inhibition of human tumor xenografts in nude mice is significantly enhanced by combining rec Hu TNF-α with Hu lymphoblastoid IFN or IFN-γ (Balkwill et al., 1986), and the cytotoxic action of TNF-α on many different human tumor cells in vitro is boosted by IFN-γ (Fransen et al., 1986b). Moreover, IFN-γ acts synergistically with either TNF-α or TNF-β to activate the killing of tumor cells or parasites by murine macrophages (Esparza et al., 1987).

   IFN-induced increased target cell sensitivity may be one of the contributing factors in this synergism, since in vitro, treatment of human HT-29 colon adenocarcinoma cells with IFN-γ increases their susceptibility to TNF cytotoxicity and to monocyte killing (Feinman et al., 1987).
4. The antiviral action of TNF-α and β on cells in culture is enhanced synergistically by IFN-γ (Wong and Goeddel, 1986).
5. Both TNF-α and Mu IFN-γ induce the expression of MHC class II antigens on cells of a murine myelomonocytic line, and a combination of the two cytokines results in a synergistic augmentation of the antigen expression (Fig. 13.3) (Chang and Lee, 1986). TNF-α also enhances the IFN-γ-induced expression of MHC class I and II HLA genes in various human tumor cells, which could be one way by which TNF and IFN act synergistically in vivo to favor tumor rejection (Pfizenmaier et al., 1987).

D. Interaction of TNFs and Other Cytokines Related to IFN Production and Action

1. Induction of IL-1

In contrast to its effects on some tumor cell lines, TNF-α is not toxic for cultured human vascular endothelial cells. It binds to high affinity receptors on these cells and induces the synthesis and release of IL-1 (Nawroth et al., 1986). TNF-α also induces the synthesis of membrane-associated IL-1 in human fibroblasts (Le et al., 1987). This represents another connection to the IFN system, since IL-1 is an inducer of IFN-β, and furthermore, IL-1 also stimulates IL-2 production, which upregulates IFN-γ production.

2. Induction of GM-CSF

TNF-α stimulates the production of granulocyte-monocyte colony stimulating factor (GM-CSF) in normal human lung fibroblasts and vascular endothelial cells, as well as in cells of malignant tissues (Munker et al., 1986; Koeffler et al., 1987). This growth factor stimulates proliferation and differentiation of granulocyte-monocyte and eosinophil stem cells and acts as an antagonist of the IFN-α/β-mediated suppression of granulocyte-mono-

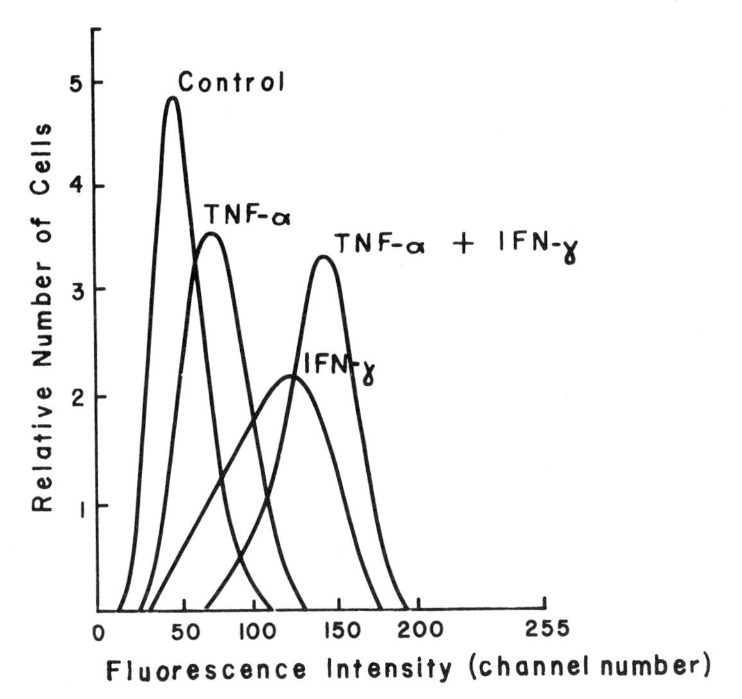

Figure 13.3 Hu TNF-α acts synergistically with Mu IFN-γ to augment the expression of MHC class II antigens on cells of a murine macrophage cell line (since TNF is not species specific, Hu TNF-α can be used for such an experiment). The figure represents a fluorescence histogram for class II antigen expression on control cells and cells stimulated with the two cytokines separately and combined. (Adapted from Chang and Lee, 1986.)

cyte colony formation (see Chapter 8). Thus, TNF-α induces cytokines such as IL-1 and IL-2, which stimulate IFN synthesis and growth factors like GM-CSF that can act as IFN antagonists.

Small-cell lung cancers are tumors that most likely originate from macrophages or their precursors, and the antigen Leu-M3 is a differentiation marker for mononuclear phagocytes. GM-CSF and also IFN-γ enhance the expression of Leu-M3 and HLA-DR antigens on cells from small-cell lung cancer lines and inhibit the replication of these cells. This shows that both IFN-γ, and indirectly TNF-α via induction of GM-CSF, can promote differentiation, resulting in growth inhibition, of cells of myeloid lineage (Ruff et al., 1986).

3. Induction of B-Cell Differentiation Factor (BCDF or BSF-2) Alias IFN-β2

TNF-α also induces the synthesis of a B-cell growth and differentiation factor, responsible for the final maturation of B cells into immunoglobulin secreting cells, which also acts as a growth factor for plasmacytomas/hybridomas. This factor, first isolated from mitogen-stimulated T cells, has been called a T-cell lymphokine (Hirano et al., 1985). This definition is too restricted, since the structure of BSF-2 is identical to that of Hu IFN-β2, also called 26K protein, which can be made by many cells, including diploid fibroblasts, and therefore is a cytokine (Billiau, 1986; Hirano et al., 1986; Zilberstein et al., 1986). BSF-2 is discussed in more detail in Chapter 11.

II. INTERLEUKIN-1

The name IL-1 covers a group of cytokines with similar biological activities derived by posttranslational processing from two distinct gene products, IL-1α and IL-1β. These cytokines are involved in a wide range of immune and inflammatory reactions. Their production is stimulated by IFNs α, β, and γ, and, conversely, IL-1 itself is an inducer of IFN-β in fibroblasts and a stimulator of IFN-γ synthesis by T cells via the induction of IL-2.

A. Structure and Genetics

The nucleotide sequence of IL-1α cDNA predicts a polypeptide of 271 amino acids in humans and of 270 amino acids in mice. For IL-1β the predicted sequence is 269 amino acids in both humans and mice. Human and murine IL-1α display 62% homology at the amino acid level, and human and murine IL-1β 67%. The α and β forms of IL-1 are more distantly related in each species, since they share about 30% homology in humans and 22% in mice. The region of strongest homology between IL-1α and IL-1β resides at the carboxy terminal half of the polypeptide, which is also the region of biological activity (Auron et al., 1984; Lomedico et al., 1984; Furutani et al., 1985; March et al., 1985; DeChiara et al., 1986; Gray et al., 1986; Rosenwasser et al., 1986).

Human and murine IL-1α and IL-1β are synthesized as precursors of almost identical length, with a molecular weight of about 33 kDa. In macrophages, these precursors are sometimes secreted as such, without further modification, but more frequently they are processed by enzymatic cleavage to smaller forms of distinct molecular weight and charge, resulting in microheterogeneity. Processing of the 33 kDa precursors gives mainly rise to 17 kDa molecules that result from the removal of about 115 amino acids, but smaller molecular weight forms are also formed. The IL-1α forms are acidic, with an isoelectric point of approximately 5, and the IL-1β forms are neutral, with an isoelectric point of 7. IL-1β is the most commonly found species in monocytes and macrophages (Lomedico et al., 1984; Giri et al., 1985; Wood et al., 1985; Gray et al., 1986). For recent biological studies rec IL-1α and IL-1β have been used, but in older work native IL-1, which is a mixture of IL-1α and IL-1β has been used, and hence is referred to as IL-1 without making a distinction between the two forms. This situation is comparable to the many studies that have been performed with a mixture of Mu IFN-α/β rather than with the separate molecular species.

Two distinct genes code for Hu IL-1α and IL-1β. The genomic organization of these two genes is quite similar with respect to the number and position of exon boundaries; each gene contains seven exons separated by six introns. The existence of two distantly related IL-1 genes coding for functionally similar proteins suggests a common evolutionary origin, and the possibility has been raised that the IL-1β gene is derived from the IL-1α gene, maybe via an RNA-mediated duplication-transposition event. An interspecies comparison of the Hu IL-1α and Mu IL1-β genes shows conservation of the coding sequences at the intron-exon boundaries, which suggests that these genes diverged from a common ancestor (Table 13.3) (Clark et al., 1986; Furutani et al., 1986; Telford et al., 1986).

Analysis of human–mouse somatic cell hybrids and in situ hybridization studies assign the Hu IL-1β gene to the long arm of human chromosome 2, at position 2q13-2q21, between two fragile sites (Webb et al., 1986).

B. The Receptor for IL-1

IL-1 molecules act on target cells via specific plasma membrane receptor proteins. IL-1α and IL-1β share a common high affinity receptor, which explains the similarity of their biological activities. High and low affinity receptors have been described, but only the high affinity receptors appear to be internalized. The general level of the IL-1 receptor expression at the cell surface is considerably lower than that of receptors for other cytokines such as IL-2 or the epidermal growth factor and varies from a few hundred to a few thousand, which is comparable to the relatively low number of IFN receptors that are found on most cells. Receptors for IL-1 are present on many cell types, such as T cells, B cells, monocytes, fibroblasts, and epithelial cells (Dower et al., 1985; Bird et al., 1986; Dower et al., 1986a,b; Matshushima et al., 1986; Chin et al., 1987).

The functional dichotomy between helper and cytotoxic T cells is reflected by the presence or absence of IL-1 receptors: IL-1 receptors are expressed on cells of the murine L3T4$^+$ T-cell subset, corresponding to T-helper cells (see Chapter 11), but they are absent on cells of the Lyt2$^+$ T-cell subset, corresponding to cytotoxic-suppressor T cells. T cells have both high and low affinity receptors for IL-1 (Lowenthal and MacDonald, 1986, 1987).

C. The Production of IL-1

1. Many Cells Produce IL-1

Although originally thought to be a monokine, exclusively made by activated macrophages, IL-1α and IL-1β can in fact be produced by many cell types, such as dendritic cells, neutrophils, T and B cells, endothelial cells, fibro-

TABLE 13.3 Intron-Exon Organization of Human and Murine IL-1 Genes: Alignment of Deduced Amino Acid Sequences Showing Exon Junctions (Arrows)

Human α	MAKVPDMFED	LKNCYSENEE	DSSSIDHLSL	NQK.SFYHVS	Y.GPLHEGCM	DQSVSLSISE
Murine α	MAKVPDLFED	LKNCYSENED	YSSAIDHLSL	NQK.SFYDAS	Y.GSLHETCT	DQFVSLRTSE
Human β	MAEVPKLASE	MMAYYSGNED	DLFFEADGPK	QMKCSFQDLD	L.CPL.....	DGGIQLRISD
Murine β	MATVPELNCE	MPPFDSD.EN	DLFFEVDGPQ	KMKGCFQTFD	LGCP......	DESIQLQISQ
Human α	TSKTSKLTFK	ESM.VVV.AT	..NGKVLKKR	RLSLSQSITD	DDLEAIANDS	.EEEII....
Murine α	TSKMSNFTFK	ESR.VTVSAT	SSNGKILKKR	RLSFSETFTE	DDLQSITHDL	..EETI....
Human β	HH.YSKG.FR	QAASVVV.AM	...DK.LRKM	LVPCPQTFQE	NDLSTFFPFI	FEEEPIFFDT
Murine β	QH.INKS.FR	QAVSLIV.AV	...EK.LWQL	PVSFPWTFQD	EDMSTFFSFI	FEEEPILCDS
Human α	.......KPR	SAPFSFLSNV	KYNFMRIIKY	EFI.LNDALN	QSIIRANDQ.	YLTAAALHN.
Murine α	.......QPR	SAPYTYQSDL	RYKLMKLVRQ	KF.VMNDSLN	QTIYQDVDKH	YLSTTWL.ND
Human β	W..DNEAYVH	DAP.VR.S.L	NCTLRDSQ.Q	KSLVMSG...	..PYELKALH	.LQG..Q.DM
Murine β	WDDDDNLLVC	DVP.IRQ..L	HYRLRDEQ.Q	KSLVLSD...	..PYELKALH	.LNG..Q.NI
Human α	LDEAVKFDMG	AYKSSK.DDA	KITVILRISK	TQLYVTAQ.D	EDQPVLLKEM	PEIPKTIT..
Murine α	LQQEVKFDMY	AY.SSGGDDS	KYPVTLKISD	SQLFVSAQ.G	EDQPVLLKEL	PETPKLIT..
Human β	EQQVV.FSMS	FVQGEESND.	KIPVALGLKE	KNLYLSCVLK	DDKPTLQLES	VH.PKNYPKK
Murine β	NQQVI.FSMS	FVQGEPSND.	KIPVALGLKG	KNLYLSCVMK	DGTPTLQLES	VD.PKQYPKK
Human α	GSETNLLFFW	WTHGTKNYFT	SVAHPNLFIA	TKQ..DYWVC	L..AGGPPSI	TDFQILENQA
Murine α	GSETDLIFFW	KSINSKNYFT	SAAYPELFIA	TK..EQSRVH	L..ARGLPSM	TDFQI....S
Human β	KMEKRFVFNK	IEINNKLEFE	SAQFPNWYIS	TSQAENMPVF	LGGTKGGQDI	TDFTMQFVSS
Murine β	KMEKRFVFNK	IEVKSKVEFE	SAEFPNWYIS	TSQAEHKPVF	L.GNNSGQDI	IDFTMESVSS

Reference: Telford et al., 1986.

The one-letter amino acid code is explained in Chapter 2, Table 2.5.

blasts, keratinocytes, astrocytes, and microglial cells. LPS or immune complexes, and also interaction with T cells, trigger IL-1 production in macrophages and monocytes, and furthermore, LPS induces IL-1 in many other cell types (Kupper et al., 1986; Oppenheim et al., 1986; Acres et al., 1987). The major producer cell, however, is the macrophage (Koide and Steinman, 1987).

2. *IFNs Enhance IL-1 Production*

Fresh cultures of human monocytes release IL-1 upon stimulation by LPS or polyrIrC, but progressively lose this ability upon maintenance in vitro. In the presence of Hu IFN-α2 or Hu IFN-β, and especially in the presence of Hu IFN-γ, IL-1 production is significantly boosted in aged cultures, and furthermore, the level of production is maintained for a longer time. But even in freshly isolated macrophages, LPS-induced IL-1 production is boosted when IFN-γ is added simultaneously (Arenzana-Seisdedos et al., 1985; Newton, 1985; Eden and Turino, 1986). Prior exposure of murine macrophages to Mu IFN-α/β or of human macrophages to Hu leukocyte IFN also boosts subsequent IL-1 induction by LPS or CSF-1 (Herman et al., 1984; Candler et al., 1985). Since LPS, polyrIrC, and CSF-1 also induce IFN-α/β in macrophages, the possibility is raised that under natural conditions IL-1 production is under the influence of endogenous IFN production. In fact, under certain experimental conditions, Hu IFN-γ and Hu IFN-α can directly induce IL-1 production in monocyte cultures (Fig. 13.4) (Rhodes et al., 1986). However, whether or not IFN-γ can directly induce IL-1 production is controversial. Indeed, under different experimental conditions, IL-1 is directly induced in human monocytes only by Hu IFN-α but not by Hu IFN-γ (Gerrard et al., 1987). Endotoxin is such a potent IL-1 inducer that the

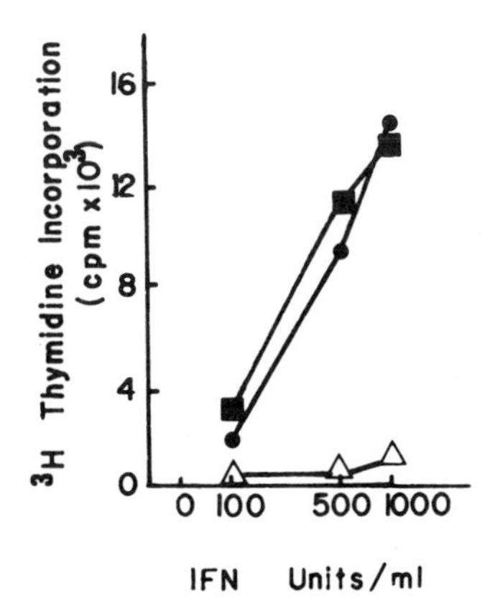

Figure 13.4 Stimulation of IL-1 production in human monocytes by Hu IFN-γ (■) or Hu IFN-αI (●). In this particular assay Hu IFN-α2 (△) was without effect. IL-1 production was followed with a functional assay, by measuring the stimulation of thymidine incorporation in murine thymocytes. (Adapted from Rhodes et al., 1986).

presence of trace amounts can complicate the interpretation of induction experiments.

The stimulating effect of IFN-γ on IL-1 production by monocytes is due to enhanced transcription of the IL-1 gene(s), as shown by nuclear run-off experiments (Collart et al., 1986). IFN-γ can also stimulate LPS-induced IL-1 production in other cells, such as human endothelial cells taken from human umbilical cord veins (Miossec and Ziff, 1986). IL-1 and IFN-γ are thus connected in a positive feedback loop, since IL-1 stimulates production of IL-2, which itself is an enhancer of IFN-γ production (see also Chapters 10 and 11).

D. The Pleiotropic Activities of IL-1

Contrary to IFNs, the IL-1 cytokines do not display species specificity, and IL-1 from one mammalian species can act on cells of another mammalian species. Species specificity is lacking to such an extent that IL-1 even crosses phylogenic barriers, and Hu IL-1 is active on fish lymphocytes (Hamby et al., 1986). Apparently, the functional domains of the IL-1 proteins have been highly conserved throughout evolution, and an invertebrate organism as primitive as the starfish produces an IL-1-like protein that stimulates murine lymphocytes and is inhibited by an antibody raised to human IL-1 (Beck and Habicht, 1986).

A wide variety of activities are attributed to IL-1. These range from T-cell activation and other immunomodulating activities to inflammatory interactions with the brain, bone, muscle, cartilage, and liver. It could be that different functional domains of the molecule are involved in the immunomodulating and in the inflammatory activities, respectively. One can indeed synthesize short peptide fragments corresponding to different regions of human or murine IL-1 and show that a fragment of nine residues, corresponding to amino acids 163–171, has T-cell activating ability, but does not induce prostaglandin synthesis and is not pyrogenic (Antoni et al., 1986).

1. *Immunomodulating Effects*

a. Early T-Cell Development

The thymus is the primary site of T-cell differentiation (see Chapter 11), and IL-1 plays a role in the maturation of T cells in this organ. High and low affinity receptors for IL-1 are present on thymocytes, and thymocyte proliferation is triggered by IL-1. Stimulated proliferation of thymocytes actually was the first assay to measure the biological activity of this cytokine. For T-cell maturation to take place, the expression of MHC class II antigens on accessory cells in the thymus is required, and thymic T cells will proliferate optimally in the presence of IL-1 and MHC class II expressing accessory cells. IL-1 thus causes the expansion of thymocytes with specificity for MHC class II self antigens and contributes to the generation of self-restricted T cells (Rock and Benacerraf, 1984). This expansion takes place via IL-2 induction, and immature murine thymocytes of the Lyt-2^- and L3T4^- phenotype can be induced by rec IL-1α or rec IL-1β to secrete IL-2 and express IL-2 receptors (Howe et al., 1986). Furthermore, IL-1 also stimulates the expression of MHC class II antigens, but probably in an indirect way via the production of IFN-γ (DeLuca and Mizel, 1986).

b. T-cell activation

IL-1 was originally described as a lymphocyte-activating factor with the acronym LAF (Gery et al., 1972). The interaction of T-helper cells with antigen-presenting accessory cells results in the production of IL-1, some of which remains associated with the macrophage membrane, probably to facilitate interaction with T cells in close contact. IL-1 interacts with the IL-1 receptor on T-helper cells and induces in these cells the production of both IL-2 and its receptor. Subsequently, IL-2 stimulates the proliferation of helper- as well as cytotoxic T-cell subsets. In addition to IL-2 and its receptor, the expression of the protooncogenes c-*myc* and c-*fos* is also stimulated in IL-1-treated T cells (Gery et al., 1972; Kurt-Jones et al., 1985; Williams et al., 1985; Kovacs et al., 1986; Lowenthal et al., 1986a; MacDonald and Nabholz, 1986; Unanue and Allen, 1986; Weaver and Unanue, 1986; Lowenthal and MacDonald, 1987).

There is, however, an opposing view of IL-1 activity on T cells, which maintains that IL-1 primarily influences T-cell function via accessory dendritic cells. It is true that the accessory function of dendritic cells increases substantially after exposure to IL-1 (Koide et al., 1987).

c. Stimulation of B-Cell Growth and Differentiation

Several growth-promoting and differentiation factors are involved in B-cell proliferation and antibody production (see Chapter 11). Of these, IL-1 enhances the proliferative response of murine B lymphocytes that are activated by anti-IgM antibodies or by dextran sulfate (Howard et al., 1983; Booth and Watson, 1984). Rec Mu IL-1α stimulates the growth and maturation of murine B cells into Ig-secreting cells (Pike and Nossal, 1985; Chiplunkar et al., 1986).

This B-cell differentiation activity of IL-1 may well be the result of the activity of BSF-2 or IFN-β2 induced by IL-1 in some cells.

2. Induction of IFN-β and IFN-β2

Hu IL-1β induces low levels of Hu IFN-β in some human cells, thereby causing the establishment of an antiviral state in these cells. As with the IFN inducer polyrIrC, the level of induction is significantly boosted in the presence of cycloheximide (Van Damme et al., 1983, 1985; Billiau et al., 1985; Van Damme et al., 1987a).

Furthermore, IL-1β also induces the formation of Hu IFN-β2 (Content et al., 1985; Van Damme et al., 1987a). The capacity to induce Hu IFN-β2 is shared by other cytokines and growth factors such as TNF or PDGF. Hu IFN-β2, also known as BSF-2 or BCDF, was described independently as an IFN, as a B-cell differentiation factor, and as a T-cell-derived hybridoma growth factor (Hirano et al., 1985; Billiau, 1986; Hirano et al., 1986; Zilberstein et al., 1986; Revel and Zilberstein, 1987; Kohase et al., 1987; Van Snick et al., 1987; Van Damme et al., 1987b). This substance is involved in the regulation of B-cell growth and differentiation and has little or no antiviral activity.

3. *IL-1 as an Inflammatory Mediator*

The inflammatory response is triggered by a variety of injurious or infectious agents and is designed to rid the organism of the aggression, a process often accompanied by tissue damage. Inflammation is one of the primary mechanisms to combat foreign intruders and to eliminate damaged tissue and is characterized by four cardinal signs: heat, redness, swelling, and pain. Inflammation mobilizes humoral, cellular, vascular, and neurological elements.

In the acute inflammatory response local vessel dilatation and increased vascular permeability is followed by exudation of plasma proteins and neutrophils, causing the swelling. The neutrophil is the predominant type in acute inflammation, but other cells also participate in the inflammatory reaction. The predominant types in chronic inflammation are T cells and macrophages. Other important cells involved are mast cells and basophils, secreting vasoactive mediators. In addition to vasoactive mediators, many other mediators participate in the inflammatory response, as either effector substances involved in tissue destruction and regeneration or mobilizing agents for participating cells.

IL-1 is an important local mediator of acute and chronic inflammation, and furthermore, is also involved in the systemic reaction such as the induction of fever and release of acute phase proteins that accompany inflammation.

a. Local Reactions. Release of histamine is instrumental in vasodilatation, which facilitates the access of mediators and leukocytes to the site of injury. Basophils and mast cells are the two major sources of histamine, and IL-1 induces histamine release by these cells, via a pathway that is different from the IgE receptor pathway (Subramanian and Bray, 1987). It is interesting that this activity is shared with IFNs, which can also stimulate histamine release by basophils and mast cells (see Chapter 12).

Vascular endothelium normally acts to inhibit coagulation and thrombosis, but during inflammation endothelial cells may actively promote coagulation. On human endothelial cells IL-1 as well as TNF induce surface changes, referred to as procoagulant activity, that favor local microcoagulation and increase the adhesiveness for neutrophils, monocytes, and lymphocytes. Thus, IL-1, by augmenting the adhesive properties of endothelium, stimulates transendothelial passage and infiltration of leukocytes into the site of inflammation (Bevilacqua et al., 1984; Cavender et al., 1986; Schleimer and Rutledge, 1986).

Prostaglandins are also important inflammatory mediators, but unlike histamine, prostaglandins are not stored, rather, they are continuously synthesized in membranes from precursors that have been cleaved from membrane phospholipids by phospholipases. During inflammation membrane phospholipids are cleaved by phospholipases to make arachidonic acid, which is a precursor of the vasoactive agents prostaglandin PGE_2 and leukotrienes. In human synovial cells or rabbit articular chondrocytes, phospholipase activity is stimulated by IL-1, and also by TNF, thereby providing the necessary metabolites for the synthesis of these two important meditors (Chang et al., 1986; Godfrey et al., 1987).

This mechanism probably is also instrumental during arthritis, an inflammation of joints. Rheumatoid arthritis is characterized by the progressive destruction of joint structure, accompanied by the typical signs of inflammation. The degradation of joint structure in inflammatory synovitis such as rheumatoid arthritis is a complex process in which many cell types and mediators are involved. One of its characteristics is degradation of collagen as a result of collagenase activity. The latter is stimulated by IL-1, which significantly boosts procollagenase synthesis by rheumatoid synovial fibroblasts (Dayer et al., 1981; McCroskery et al., 1985).

Joint fluids of patients with rheumatoid arthritis contain elevated levels of IL-1. Macrophages of rats with collagen-induced arthritis, an experimental model obtained by immunizing rats with a mixture of type II collagen and incomplete Freund's adjuvant, also release increased amounts of IL-1 (Phadke et al., 1986). These elevated levels of IL-1 are probably directly involved in the erosive process of arthritis, since inoculation of IL-1 directly into the knee joints of rabbits induces the accumulation of polymorphonuclear and mononuclear leukocytes into the joint space, and causes the loss of proteoglycan from the articular cartilage (Pettipher et al., 1986). The latter is due the activity of proteoglycanase, one of several metalloproteinases that are stimulated by IL-1 and that can also be engaged in normal bone metabolism (Bunning et al., 1986; Schnyder et al., 1987).

Several pathological conditions, accompanied by chronic inflammation such as rheumatoid arthritis and periodontal disease are accompanied by increased bone resorption. Rec Mu IL-1α and native Hu IL-1β have been shown to be potent stimulators of bone resorption in vitro. This occurs via the activation of osteoclasts, monocyte-derived cells engaged in bone resorption. The IL-1-mediated activation of osteoclasts is inhibited by IFN-γ (Gowen et al., 1983; Dewhirst et al., 1985; Gowen and Mundy, 1986). Direct inhibition of Il-1-mediated osteoclast activity may possibly contribute to the clinical improvement of rheumatoid arthritis patients as a result of systemic administration of rec Hu IFN-γ (Obert et al., 1986).

These observations obviously do not preclude a function for IL-1 in normal turnover and metabolism of bone tissue, resulting in a dynamic equilibrium between synthesis and destruction.

During chronic inflammation, fibrosis of the surrounding tissue is a common occurrence, resulting in granulomatous tissue. IL-1 appears to be involved also in this reaction, since under certain conditions it can act as a growth factor for fibroblasts (Schmidt et al., 1982). In this aspect it is an IFN antagonist, since IFNs usually inhibit growth and replication of fibroblasts. The induction of IFN-β by IL-1, therefore, can be considered a part of the negative feedback loop also observed with other cell-proliferation-inducing agents.

Tuberculosis is a chronic disease that is characterized by the formation of granulomatous lesions. Stimulation of monocytes by mycobacterial protein antigens results in the release of IL-1, which suggests that IL-1 is involved in the formation of these fibrotic lesions (Wallis et al., 1986).

b. Systemic Effects *Induction of fever:* The febrile response to infection is a fundamental component of host defense. IL-1 is important in this respect, since it is an endogenous pyrogen, which like TNF and IFN-α, directly stimulates hypothalamic prostaglandin PGE_2 synthesis (Dinarello et al., 1985).

Induction of Acute Phase Proteins: The acute phase response is a characteristic pattern of changes in the concentration of several plasma proteins as a result of infection, inflammation, or tissue damage. These proteins are mainly produced in the liver, but also to some extent in extrahepatic tissues. They consist to a large extent of proteinase inhibitors, coagulation proteins, transport proteins, and complement proteins. Within hours after an acute phase stimulus, the plasma concentrations of C-reactive protein and serum amyloid A are increased more than a 100-fold, but the concentration of most other acute phase proteins, including several complement proteins, is increased up to threefold. The plasma concentration of some other proteins, for example, high- and low-density lipoproteins and albumin, is decreased during the acute phase response (Pepys and Baltz, 1983). These changes in plasma protein concentrations result from the activity of regulatory mediators, among which IL-1 appears to occupy the most important place, since it can cause the release of many, if not all, acute phase proteins (Sztein et al., 1981; Ramadori et al., 1985). One of the acute phase proteins is an MHC class III protein, complement factor B, a 93 kDa serine protease that shows many functional similarities to the complement protein C2. The latter, although structurally very similar and encoded by a gene that is closely linked to the one coding for factor B, is not enhanced by IL-1 during the acute phase (Perlmutter et al., 1986). However, whether or not IL-1 influences complement gene expression, including C2, appears to be greatly influenced by genotype and tissue, at least judging from results obtained in mice (Falus et al., 1987).

III. INTERLEUKIN-2

This lymphokine, also called T-cell growth factor because it was discovered as an inducer of T-cell clonal expansion, is vital for the T-cell function in general and for IFN-γ production in particular. Furthermore, the role of IL-2 is not limited to stimulation of T-cell clonal expansion and function, since, like IFNs, it also affects B cell, NK cell, and monocyte function.

A. Structure and Genetics

IL-2 is an inducible 15.5 kDa glycoprotein, made by T cells. It is not produced by resting T cells, but is rapidly synthesized and secreted following activation by mitogens or antigens in the presence of IL-1, which, as discussed in the preceding section, is the product of activated macrophages (Morgan et al., 1976; Gillis and Smith, 1977; Larsson et al., 1980; Smith et al., 1980; Robb et al., 1983).

The nucleotide sequence of IL-2 cDNA predicts a polypeptide of 153 aa in humans and of 169 aa in the mouse, including a 20 aa signal peptide, which is cleaved off during secretion of the molecule (Table 13.4) (Devos et al., 1983; Taniguchi et al., 1983; Kashima et al., 1985; Yokota et al., 1985). The single copy human IL-2 gene is on chromosome 4, bands q26–28 (Seigel et al., 1984). The coding regions for human and murine IL-2 share about 76% homology at the nucleotide level and the gene products 63% at the amino acid level. The murine coding sequence contains an unusual stretch of CAG repeats, encoding 12 glutamines (see Table 13.4). The coding regions are separated by three introns, of comparable position in humans and mice

TABLE 13.4 Alignment of the Murine (Mu) and Human (Hu) IL-2-Deduced Amino Acid Sequences

	10	20	30	40	50
Mu	MYSMQLASCV	TLTLVLLVNS	APTSSSTSSS	TAEAQQQQQ	QQQQQQHLEQ
Hu	MYRMQLLSCI	ALSLALVTNS	APTSSSTKKT	QLQ____	____LEH
	10	20	30	33	36
	60	70	80	90	100
Mu	LLMALQELLS	RMENYRNLKL	PRMLTFKFYL	PKQATELKDL	QCLELELGPL
Hu	LLLDLQMILN	GINNYKNPKL	TRMLTFKFYM	PKKATELKHL	QCLEEELKPL
	46	56	66	76	86
	110	120	130	140	150
Mu	RHVLDLTQSK	SFQLEDAENF	ISNIRVTVVK	LKGSDNTFEC	QFDDESATVV
Hu	EEVLNLAQSK	NFHLRPRDL_	ISNINVIVLE	LKGSETTFMC	EYADETATIV
	96	105	115	125	135
	160	169			
Mu	DFLRRWIAFC	QSIISTSPQ			
Hu	EFLNRWITFC	QSIISTLT			
	145	153			

References. Devos et al., 1983; Taniguchi et al., 1983; Fuse et al., 1984.

The putative signal sequence ends at the serine residue in position 20.

The one-letter amino acid code is explained in Chapter 2, Table 2.5.

(Fujita et al., 1983; Fuse et al., 1984; Holbrook et al., 1984). The IL-2 structural gene contains in its 5′ upstream region nucleotide sequences that are homologous to sequences found in the 5′ flanking regions of other T-cell specific genes such as the IFN-γ and the IL-2 receptor gene. This sequence, which spans about 200 bp, can function in an orientation-independent manner and seems to control induced T-cell specific gene expression. Interestingly, this sequence also has homology to long terminal repeat sequences of the AIDS virus HIV-1, which replicates preferentially in T-helper cells (Fujita et al., 1986). The chromatin pattern of the 5′ upstream region of the IL-2 gene, furthermore, is different in T cells from what it is in cells of nonhematopoietic origin, since in T cells the 5′ region of the IL-2 gene contains three regions of hypersensitivity to DNase I. Also, an additional DNase hypersensitive site develops as a result of induction, whereas non-IL-2 inducible cells display a different pattern of sensitivity to DNase in this region (Siebenlist et al., 1986). Apparently, during T-cell differentiation the chromatin pattern around the promoter region of the gene undergoes some changes, resulting in accessibility and inducibility of the gene. Indeed, the 5′-flanking region contains a 275-bp sequence with the characteristics of an inducible transcriptional enhancer, which can in either orientation enhance transcription from a heterologous promoter. This region is likely to be a target for signal transduction molecules in T cells (Durand et al., 1987).

The 3′ flanking regions of the IL-2 genes also display some interesting features. The human, murine, and bovine IL-2 genes share as much sequence homology in this region as they do in their coding regions. This evolutionary conserved similarity in structural features of a noncoding region of the gene in three different species is highly suggestive of an involvement of the 3′ flanking region in the regulation of gene expression. In addition, a tandemly repeated sequence, (TATT)n, is also found in the 3′ flanking regions of several other cytokine genes such as the genes for IL-1, IFN-α, and IFN-β and IFN-γ (Reeves et al., 1986). The nonrandom occurrence of repeated sequences in the 3′ flanking regions of these genes is highly suggestive of some evolutionary conserved regulatory function, common to the genes for these immunoregulatory proteins.

B. The Receptor for IL-2

1. Structure and Genetics

IL-2 binds to high affinity receptors (Kd = 10^{-11} M) at the same concentration that promotes T-cell cycle progression in vitro (Robb et al., 1981). These high affinity receptors appear to consist of a heterodimer formed by two noncovalently linked cell-surface molecules, the 55 kDa, Tac-antigen expressing chain, and the 75 kDa chain (Sharon et al., 1986; Smith, 1987).

The 55 kDa component of the IL-2 receptor contains a cell-surface epitope reacting with a murine monoclonal antibody, known as "anti-Tac," because, before its function was known, the epitope was referred to as "T-cell activation" antigen (Leonard et al., 1982; Robb and Greene, 1983). This receptor component consists of a 33 kDa peptide precursor that is post-translationally glycosylated into the mature form having an apparent MW of 55 kDa. The amino acid sequence deduced from the cDNA corresponds to a 272-aa polypeptide including a 21 aa signal peptide (Cosman et al., 1984; Leonard et al., 1984; Nikaido et al., 1984). The intracytoplasmic domain of the 55 kDa IL-2-binding protein is only 13 aa long and has no tyrosine kinase activity. Another protein complexed with the 55 kDa protein appears, therefore, to be required for high affinity binding and signal transduction (Greene and Leonard, 1986; Kondo et al., 1986). Analysis of IL-2-binding proteins expressed by human T-cell lines does indeed reveal the existence of two different receptor molecules, one having the Tac epitope, with a MW of 55 kDa, and a second, larger one, with a MW of 75 kDa. The presence of both proteins is required for high affinity binding, indicating that the IL-2 receptor does consist of a complex of two distinct binding proteins (Sharon et al., 1986; Teshigawara et al., 1987).

Cells that express predominantly or even exclusively the 75 kDa subunit of the receptor are able to mediate endocytosis of surface-bound IL-2 as rapidly as cells having the heterodimeric receptor (Robb and Greene, 1987).

The single copy gene encoding the human IL-2 55 kDa receptor chain contains eight exons, spanning more than 25 kbs, and is situated on chromosome 10, bands p14–15 (Ishida et al., 1985; Leonard et al., 1985a,b). The corresponding murine IL-2 55 kDa receptor chain sequence shares 72% nucleotide and 61% amino acid homology with the human receptor. A comparison of both species reveals several conserved regions in the transmembrane and intracytoplasmic domains, suggesting that these regions are important for receptor function and regulation (Miller et al., 1985; Shimizu et al., 1985).

The expression of IL-2-binding sites is not limited to T cells. Cortical and medullar thymocytes, B cells, and cells of the monocyte-macrophage lineage can also be induced to express IL-2-binding sites. This is strongly in favor of a role for IL-2 in the function of these cells (Waldmann et al., 1984; Ceredig et al., 1985; Hermann et al., 1985).

2. Expression

The specific high affinity membrane receptors for IL-2 are absent on resting T cells but are rapidly synthesized following the interaction of mitogens or antigens with the T-cell antigen receptor-CD3 complex (Cantrell and Smith, 1983; Meuer et al., 1984) (see Chapter 11). This is in contrast to many ligand-receptor systems, in which the receptor does not have to be induced together with its ligand but is already present on the plasma membrane. Thus, production of both IL-2 and its receptor are prerequisites for the T-cell immune response.

Mitogen or antigen induction of IL-2 receptor expression is a transcriptional event that is accompanied by an increase in intracellular Ca^{++} and involves the translocation of protein kinase C from the cytoplasm to the plasma membrane. After mitogen addition, transcription of the IL-2 receptor and IFN-γ genes precedes transcription of the IL-2 structural gene (Imboden et al., 1985; Krönke et al., 1985).

T-cell DNA synthesis as well as mitosis, dependent on activation of the T-cell antigen receptor complex, is critically influenced by signals that are generated from the interaction of IL-2 and its receptor. Three factors are critical for T-cell cycle progression: IL-2 concentration, IL-2 receptor density, and the duration of the ligand-receptor interaction (Cantrell and Smith, 1984). After binding, the ligand-receptor complex is internalized by receptor-mediated endocytosis through the invagination of coated pits, and the internalized IL-2 is rapidly degraded. As in other cytokine systems, including IFNs, it remains unclear whether endocytosis has a role in signal transduction, and the mechanism by which IL-2 transmits its growth-promoting signal is unknown. In vitro treatment of IL-2-dependent T-cell lines with IL-2 results in a rapid stimulation of inositol phospholipid metabolism. This suggests that inositol phospholipid-derived metabolites such as diacylglycerol and inositol triphosphate are part of the mechanism by which certain IL-2 signals are transduced (Fujii et al., 1986; Lowenthal et al., 1986b; Weissman et al., 1986; Bonvini et al., 1987; Evans et al., 1987).

The effect of IL-2 on its high affinity binding is down-regulation (Smith and Cantrell, 1985). The 55 kDa receptor chain, however, is up-regulated. This up-regulation requires new RNA and protein synthesis, and transcription assays performed with isolated nuclei show increased receptor mRNA accumulation. When the synthesis of IL-2 or its binding to the receptor is inhibited, the number of IL-2 receptors decreases significantly (Reem and Yeh, 1984; Depper et al., 1985).

Stimulation of antigen-specific murine T-cell clones by the appropriate antigen results in the enhanced expression of IL-2 receptor, probably via IL-2 production, and furthermore, in the production of a soluble form of the receptor that is released into the culture supernatant and can still bind IL-2 efficiently. This raises the possibility that the IL-2 receptor molecule has other, unknown, immunoregulatory functions (Rubin et al., 1986; Wagner et al., 1986). The shedding of cell-surface molecules is not a very rare event, and, among others, it also occurs with MHC antigens.

The IL-2 receptor expression can escape normal regulatory controls in malignancy. For example, in human T-cell lymphotropic virus I (HTLV-1) associated adult T-cell leukemia, leukemic cells constitutively express large numbers of IL-2 receptors (Waldmann, 1986). This IL-2 receptor expression is due to an HTLV-1 protein, $p40^x$, encoded by the pX sequence of the virus, which acts as a transactivator of the long terminal repeat region. Transfection of some cells, like the Jurkat T-cell line, but not of other human T- or B-cell lines, is sufficient to induce the expression of the IL-2 receptor Tac receptor molecule (Inoue et al., 1986).

C. IL-2 Regulates IFN-γ Synthesis

Resting peripheral blood T cells can be activated to proliferate by stimulation with polyclonal mitogens such as PHA or phorbol myristate acetate. Such mitogen or antigen stimulation also induces the synthesis of specific immunoregulatory proteins, via the increased expression of specific mRNAs.

There is some contradiction in the literature, and although the bulk of evidence seems to suggest that IL-2 by itself is not an IFN-γ inducer but is a necessary requirement for subsequent mitogen or antigen-induced IFN-γ production, there are indications that IL-2 sometimes directly induces IFN-γ production, especially in NK cells.

In any case, IL-2 plays an essential role in the production of IFN-γ, and T cells cannot produce IFN-γ unless they express the IL-2 receptor (Farrar et al., 1981; Kasahara et al., 1983; Reem and Yeh, 1984, 1985; Vilcek et al., 1985).

After mitogen stimulation of T cells, IL-2 mRNA accumulates very rapidly, within 60 minutes, together with an early peak of IFN-γ mRNA, followed by a second later peak of IFN-γ mRNA. The IL-2 55 kDa subunit receptor mRNA synthesis starts later, but continues to accumulate throughout the duration of mitogen stimulation. Dexamethasone, a synthetic glucocorticoid, inhibits the formation of both IL-2 and IFN-γ mRNA, which may in part explain some of the immunosuppressive and anti-inflammatory effects of glucocorticoids (Arya et al., 1984; Grabstein et al., 1986).

When spleen cell and thymocyte cultures, or purified human peripheral mononuclear cells, are stimulated with mitogens or with LPS, the addition of IL-2 enhances IFN-γ production by at least one order of magnitude (Fig. 13.5) (Blanchard et al., 1986). There can be, however, considerable variability in the magnitude of IFN-γ response and its enhancement by IL-2 treatment in cells from different donors (Yamamoto et al., 1982; Pearlstein et al., 1983). This points to the existence of genetic factors influencing the sensitivity to IL-2 and subsequent IFN-γ production. It is also possible that some of the IL-2 effect is mediated through stabilization of IFN-γ mRNA.

The cells that produce IFN-γ upon stimulation by IL-2 have been identified as T cells and NK cells (Kasahara et al., 1983; Weigent et al., 1983). Can IFN-γ production by T cells proceed in the absence of IL-2? It appears that both antigen-specific and mitogen-induced IFN-γ production occur mainly as a result of IL-2 activity, since IFN production is largely blunted by the addition of a specific antibody to the human IL-2 receptor Tac antigen (Reem and Yeh, 1985; Vilcek et al., 1985). Once IFN-γ synthesis has started, IL-2

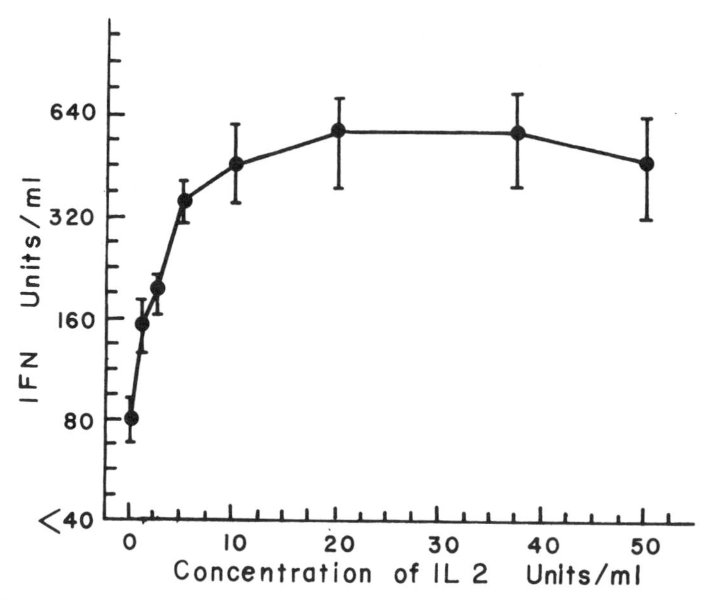

Figure 13.5 IL-2 stimulates the production of Mu IFN-γ in mouse splenocyte cultures induced with LPS. (Adapted from Blanchard et al., 1986.)

receptors can be up-regulated by IFN-γ itself, thus providing a positive feedback loop (Herrmann et al., 1985).

Histamine influences a variety of T-cell, B-cell, and NK-cell activities, and it can suppress various activities of immune cells. Histamine suppresses the production of IL-2 and IFN-γ by human mononuclear cells, and this inhibition can be reversed by the addition of exogenous IL-2, which indicates that histamine primarily acts at the level of IL-2 synthesis (Dohlsten et al., 1986).

D. IL-2 Also Stimulates IFN-α/β Synthesis by Bone-Marrow Cells

Mouse bone-marrow cells stimulated by alloantigens produce cytotoxic effector T cells, as well as IL-2 and Mu IFN-α/β; this is one of the few examples of antigen-induced IFN-α/β synthesis. The alloantigen-induced IFN synthesis apparently requires IL-2 activity, since it can be inhibited with antisera to IL-2. Moreover, addition of recombinant or pure natural IL-2 to mouse bone-marrow cultures—in the absence of alloantigen-stimulation—results in the production of high levels of IFN-α/β activity. This production is inhibited by monoclonal antibodies to the 55-kDa receptor subunit, which is a clear indication for the prime role of IL-2 in this reaction (Reyes et al., 1986). This antigen-induced mechanism of IFN induction provides an explanation for the presence of circulating IFN-α/β during graft-versus-host disease. Its function is unknown.

E. IL-2 Stimulates Cytolytic Effector Cell Function and Induces IFN-γ Production

IL-2 stimulates the generation of cytolytic effector cells, which can be T cells, NK cells, monocytes, and probably other, less well-defined subsets of killer cells. The combined effects of IL-2 on these different cytolytic effector cells undoubtedly contribute to the antitumor activity of IL-2 in mice and humans (Grimm et al., 1982; Rosenberg et al., 1984; Koo and Manyak, 1986; Shiloni et al., 1986; Thompson et al., 1986).

1. NK Cells

Highly purified populations of large granular lymphocytes, consisting mainly of NK cells, can be maintained in culture in the presence of IL-2. In such cultures, IL-2 stimulates cellular proliferation, induces the formation of IFN-γ, and enhances cytotoxic activity. The boosting of NK activity is dependent on the production of IFN-γ and can be prevented by monoclonal anti-IFN-γ antibodies. Moreover, in the absence of any proliferative signal, IL-2 can function as an immunoenhancing agent and can stimulate NK activity via IFN-γ production. In some NK-target cell systems, however, the kinetics of enhancement of cytotoxicity after IL-2 treatment are faster than those of IFN-γ production, and monoclonal anti-IFN-γ antibodies do not suppress this effect. This shows that IL-2 can enhance NK-cell activity through an IFN-γ-independent mechanism, as observed, for example, in AIDS patients. The fact that activation can take place with or without IFN-γ is not too surprising, since it can be assumed that there are different pathways leading to NK-cell activation (Weigent et al., 1983; Ortaldo et al., 1984; Trinchieri et al., 1984; Rook et al., 1985).

Furthermore, IL-2 stimulates NK-cell activity against virus-infected target cells. Guinea-pig HSV-2 genital infection represents the animal model that most closely resembles human HSV-2 genital infection. In this model, rec Hu IL-2, which does not have the species-specific characteristic of IFNs, exerts protective activity against HSV-2 infection. This antiviral activity is mainly due to stimulation of NK cells by the administered IL-2, probably via induction of IFN-γ (Weinberg et al., 1986).

2. T Cells and Monocytes

In addition to NK cells, other types of cytotoxic effector cells are stimulated by IL-2. The generation of cytolytic T cells as a result of IL-2 treatment has been correlated with the induction of IFN-γ in cytolytic precursor cells, and human monocytes display a substantially increased cytotoxic activity as a response to rec Hu IL-2. IFN-γ induces the appearance of IL-2 receptors on human peripheral blood monocytes, as measured by anti-Tac antibody and the binding of IL-2 (Holter et al., 1986; Simon et al., 1986; Malkovsky et al., 1987).

F. Activation of B Cells by IL-2 and IFN-γ

IL-2 is not only a T-cell growth factor, but it is also a B-cell growth factor, and both activities occur at very similar concentrations. Appropriately activated B cells—for example, by LPS and anti-Ig treatment—express IL-2 receptors and certain B cells respond to sequential stimulation by IL-2 and IFN-γ with terminal differentiation into Ig-secreting cells. This can be shown in different ways; for example, an Epstein-Barr virus transformed B-cell line is somewhat stimulated into Ig secretion by treatment with IL-2 alone, whereas IFN-γ has no effect. If the two agents are combined, however, with IL-2 given initially, followed by IFN-γ, the number of Ig-secreting cells increases about fivefold. A comparable synergism in stimulating Ig secretion is observed when highly purified human peripheral B cells, after stimulation with *Staphylococcus aureus* Cowan I, are treated with IL-2 and IFN-γ. Rec IL-2 alone, without help from IFN-γ, can also stimulate anti-IgM-activated B-cell growth, but this effect is purely proliferative and does not engender terminal differentiation (Zubler et al., 1984; Suzuki and Cooper, 1985; Defrance et al., 1986; Jelinek et al., 1986; Nakagawa et al., 1986, Waldmann, 1986).

REFERENCES

Acres, R. B., Larsen, A., and Conlon, P. J. IL1 expression in a clone of human T cells. *J. Immunol.* 138: 2132–1236 (1987).

Aggarwal, B. B. and Eessalu, T. E. Induction of receptors for tumor necrosis factor-α by interferons is not a major mechanism for their synergistic cytotoxic response. *J. Biol. Chem.* 262: 10000–10007 (1987).

Aggarwal, B. B., Moffat, B., and Harkins, R. N. Human lymphotoxin. Production by a lymphoblastoid cell line, purification and initial characterization. *J. Biol. Chem.* 259: 689–691 (1984).

Aggarwal, B. B., Eessalu, T. E., and Hass, P. E. Characterization of receptors for human tumor necrosis factor and their regulation by γ-interferon. *Nature* 318: 665–667 (1985a).

Aggarwal, B. B., Henzel, W. J., Moffat, B., Kohr, W. J., and Harkins, R. N. Primary structure of human lymphotoxin derived from 1788 lymphoblastoid cell line. *J. Biol. Chem.* 260: 2334–2344 (1985b).

Antoni, G., Presentini, R., Perin, F., Tagliabue, A., Ghiara, P., Censini, S., Volpini, G., Villa, L., and Boraschi, D. A short synthetic peptide fragment of human interleukin 1 with immunostimulatory but not inflammatory activity. *J. Immunol.* 137: 3201–3204 (1986).

Arenzana-Seisdedos, F., Virelizier, J. L., and Fiers, W. Interferons as macrophage-activation factors. III. Preferential effects of interferon-γ on the interleukin 1 secretory potential of fresh or aged human monocytes. *J. Immunol.* 134: 2444–2448 (1985).

Arya, S. K., Wong-Staal, F., and Gallo, R. C. Dexamethasone-mediated inhibition of human T cell growth factor and γ-interferon messenger RNA. *J. Immunol.* 133: 273–276 (1984).

Auron, P. E., Webb, A. C., Rosenwasser, L. J., Mucci, S. F., Rich, A., Wolff, S. M., and Dinarello, C. A. Nucleotide sequence of human monocyte interleukin 1 precursor cDNA. *Proc. Natl. Acad. Sci.* (USA) 81: 7907–7911 (1984).

Balkwill, F. R., Lee, A., Aldam, G., Moodie, E., Thomas, J. A., Tavernier, J., and Fiers, W. Human tumor xenografts treated with recombinant human tumor necrosis factor alone or in combination with interferons. *Cancer Res.* 46: 3990–3993 (1986).

Beck, G. and Habicht, G. S. Isolation and characterization of a primitive interleukin-1-like protein from an invertebrate, Asterias forbesi. *Proc. Natl. Acad. Sci.* (USA) 83: 7429–7433 (1986).

Berent, S. L., Torczynski, R. M., and Bollon, A. P. Sendai virus induces high levels of tumor necrosis factor mRNA in human peripheral blood leukocytes. *Nucl. Acids Res.* 14: 8997–9015 (1986).

Bertolini, D. R., Nedwin, G. E., Bringman, T. S., Smith, D. D., and Mundy, G. R. Stimulation of bone resorption and inhibition of bone formation in vitro by human tumour necrosis factors. *Nature* 319: 516–518 (1986).

Beutler, B. and Cerami, A. Cachectin and tumor necrosis factor as two sides of the same biological coin. *Nature* 320: 584–588 (1986).

Beutler, B. and Cerami, A. Cachectin: More than a tumor necrosis Factor. *N. Eng. J. Med.* 316: 379–385 (1987).

Beutler, B., Krochin, N., Milsark, I. W., Luedke, C., and Cerami, A. Control of cachectin (tumor necrosis factor) synthesis: Mechanisms of endotoxin resistance. *Science* 232: 977–980 (1986a).

Beutler, B., Tkacenko, V., Milsark, I., Krochin, N., and Cerami, A. Effect of γ-interferon on cachectin expression by mononuclear phagocytes. Reversal of the lps^d (endotoxin resistance) phenotype. *J. Exp. Med.* 164: 1791–1796 (1986b).

Bevilacqua, M. P., Pober, J. S., Majeau, G. R., Cotran, R. S., and Gimbrone, M. A. Interleukin 1 (IL-1) induces biosynthesis and cell surface expression of procoagulant activity in human vascular endothelial cells. *J. Exp. Med.* 160: 618–623 (1984).

Billiau, A. BSF-2 is not just a differentiation factor. Letter to the Editor. *Nature* 324: 415 (1986).

Billiau, A., Opdenakker, G., Van Damme, J., De Ley, M., Volckaert, G., and Van Beeumen, J. Interleukin 1: Amino acid sequencing reveals microheterogeneity and relationship with an interferon-inducing monokine. *Immunol. Today* 6: 235–236 (1985).

Bird, T. A. and Saklatvala, J. Identification of a common class of high affinity receptors for both types of porcine interleukin-1 on connective tissue cells. *Nature* 324: 263–266 (1986).

Blanchard, D. K., Djeu, J. Y., Klein, T. W., Friedman, H., and Stewart II, W. E. Interferon-γ induction by lipopolysaccharide: Dependence on interleukin 2 and macrophages. *J. Immunol.* 136: 963–970 (1986).

Bonvini, E., Ruscetti, F. W., Ponzoni, M., Hoffman, T., and Farrar, W. L. Interleukin 2 rapidly stimulates synthesis and breakdown of polyphosphoinositides in interleukin 2-dependent, murine T-cell lines. *J. Biol. Chem.* 262: 4160–4164 (1987).

Booth, R. J. and Watson, J. D. Interleukin 1 induces proliferation in two distinct B cell subpopulations responsive to two different murine B cell growth factors. *J. Immunol.* 133: 1346–1349 (1984).

Brouckaert, P. G. G., Leroux-Roels, G. G., Guisez, Y., Tavernier, J., and Fiers, W. In vivo antitumour activity of recombinant human and murine TNF, alone and in combination with murine IFN-γ, on a syngeneic murine melanoma. *Int. J. Cancer* 38: 763–769 (1986).

Bunning, R. A. D., Van Damme, J., Richardson, H. J., Hughes, D. E., Opdenakker, G., Billiau, A., and Russell, R. G. G. Homogeneous interferon-β-inducing 22K factor (IL-1β) has connective tissue cell stimulating activities. *Biochem. Biophys. Res. Comm.* 139: 1150–1157 (1986).

Candler, R. V., Rouse, B. T., and Moore, R. N. Regulation of interleukin 1 production by α and β interferons: Evidence for both direct and indirect enhancement. *J. Interferon Res.* 5: 179–189 (1985).

Cantrell, D. A. and Smith, K. A. Transient expression of interleukin 2 receptors. Consequences for T cell growth. *J. Exp. Med.* 158: 1895–1911 (1983).

Cantrell, D. A. and Smith, K. A. The interleukin-2 T-cell system: A new cell growth model. *Science* 224: 1312–1316 (1984).

Caput, D., Beutler, B., Hartog, K., Thayer, R., Brown-Shimer, S., and Cerami, A. Identification of a common nucleotide sequence in the 3′-untranslated region of mRNA molecules specifying inflammatory mediators. *Proc. Natl. Acad. Sci.* (USA) 83: 1670–1674 (1986).

Carswell, E. A., Old, L. J., Kassel, R. L., Green, S., Fiore, N., and Williamson, B. An endotoxin-induced serum factor that causes necrosis of tumors. *Proc. Natl. Acad. Sci.* (USA) 72: 3666–3670 (1975).

Cavender, D. E., Haskard, D. O., Joseph, B., and Ziff, M. Interleukin 1 increases the binding of human B and T lymphocytes to endothelial cell monolayers. *J. Immunol.* 136: 203–207 (1986).

Ceredig, R., Lowenthal, J. W., Nabholz, M., and MacDonald, H. R. Expression of interleukin-2 receptors as a differentiation marker on intrathymic stem cells. *Nature* 314: 98–100 (1985).

Chang, J., Gilman, S. C., and Lewis, A. J. Interleukin 1 activates phospholipase AP_2 in rabbit chondrocytes: a possible signal for IL 1 action. *J. Immunol.* 136: 1283–1287 (1986).

Chang, R. J. and Lee, S. H. Effects of interferon-γ and tumor necrosis factor-α on the expression of an Ia antigen on a murine macrophage cell line. *J. Immunol.* 137: 2853–2856 (1986).

Chin, J., Cameron, P. M., Rupp, E., and Schmidt, J. A. Identification of a high-affinity receptor for native human interleukin 1β and interleukin 1α on normal human lung fibroblasts. *J. Exp. Med.* 165: 70–86 (1987).

Chiplunkar, S., Langhorne, J., and Kaufmann, S. H. E. Stimulation of B cell growth and differentiation by murine recombinant interleukin 1. *J. Immunol.* 137: 3748–3752 (1986).

Clark, B. D., Collins, K. L., Gandy, M. S., Webb, A. C., and Auron, P. E. Genomic sequence for human prointerleukin 1β: Possible evolution from a reverse transcribed prointerleukin 1α gene. *Nucl. Acids Res.* 14: 7897–7914 (1986).

Collart, M. A., Belin, D., Vassalli, J. D., De Kossodo, S., and Vassalli, P. γ-interferon enhances macrophage transcription of the tumor necrosis factor/cachectin, interleukin 1, and urokinase genes, which are controlled by short-lived repressors. *J. Exp. Med.* 164: 2113–2118 (1986).

Content, J., De Wit, L., Poupart, P., Opdenakker, G., Van Damme, J., and Billiau, A. Induction of a 26-kDa-protein mRNA in human cells treated with an interleukin-1-related, leukocyte-derived factor. *Eur. J. Biochem.* 152: 253–257 (1985).

Cosman, D., Cerretti, D. P., Larsen, A., Park, L., March, C., Dower, S., Gillis, S., and Urdal, D. Cloning, sequence and expression of human interleukin-2 receptor. *Nature* 312: 768–771 (1984).

Cuturi, M. C., Murphy, M., Costa-Giomi, M. P., Weinman, R., Perussia, B., and Trinchieri, G. Independent regulation of tumor necrosis factor and lymphotoxin production by human peripheral blood lymphocytes. *J. Exp. Med.* 165: 1581–1594 (1987).

Dayer, J. M., Stephenson, M. L., Schmidt, E., Karge, W., and Krane, S. M. Purification of a factor from human blood monocyte-macrophages which stimulates the production of collagenase and prostaglandin E_2 by cells cultured from rheumatoid synovial tissues. *FEBS Lett.* 124: 253–256 (1981).

DeChiara, T. M., Young, D., Semionow, R., Stern, A. S., Batula-Bernardo, C., Fiedler-Nagy, C., Kaffka, K. L., Kilian, P. L., Yamazaki, S., Mizel, S. B., and Lomedico, P. T. Structure-function analysis of murine interleukin 1: Biologically active polypeptides are at least 127 amino acids long and are derived from the carboxyl terminus of a 270-amino acid precursor. *Proc. Natl. Acad. Sci.* (USA) 83: 8303–8307 (1986).

Defrance, T., Aubry, J. P., Vanbervliet, B., and Banchereau, J. Human interferon-γ acts as a B cell growth factor in the anti-IgM antibody co-stimulatory assay but has no direct B cell differentiation activity. *J. Immunol.* 137: 3861–3867 (1986).

Degliantoni, G., Murphy, M., Kobayashi, M., Francis, M. K., Perussia, B., and Trinchieri, G. Natural killer (NK) cell-derived hematopoietic colony-inhibiting activity and NK cytotoxic factor. Relationship with tumor necrosis factor and synergism with immune interferon. *J. Exp. Med.* 162: 1512–1530 (1985).

DeLuca, D. and Mizel, S. B. I-A positive nonlymphoid cells and T cell development in murine fetal thymus organ cultures: Interleukin 1 circumvents the block in T cell differentiation induced by monoclonal anti-I-A antibodies. *J. Immunol.* 137: 1435–1441 (1986).

Depper, J. M., Leonard, W. J., Drogula, C., Kronke, M., Waldmann, T. A., and Greene, W. C. Interleukin 2 (IL-2) augments transcription of the IL-2 receptor gene. *Proc. Natl. Acad. Sci.* (USA) 82: 4230–4234 (1985).

De Titto, E. H., Catterall, J. R., and Remington, J. S. Activity of recombinant tumor necrosis factor on toxoplasma gondii and trypanosoma cruzi. *J. Immunol.* 137: 1342–1345 (1986).

Devos, R., Plaetinck, G., Cheroutre, H., Simons, G., Degrave, W., Tavernier, J., Remaut, E., and Fiers, W. Molecular cloning of human interleukin 2 cDNA and its expression in *E. coli*. *Nucl. Acids Res.* 11: 4307–4323 (1983).

Dewhirst, F. E., Stashenko, P. P., Mole, J. E., and Tsuramachi, T. Purification and partial sequence of human osteoclast-activating factor: Identity with interleukin 1β. *J. Immunol.* 135: 2562–2568 (1985).

Dinarello, C. A., Bernheim, H. A., Cannon, J. G., Lopreste, G., Warner, S. J. C., Webb, A. C., and Auron, P. E. Purified, ^{35}S-met, ^{3}H-leu-labelled human monocyte interleukin-1 (IL-1) with endogenous pyrogen activity. *Br. J. Rheum.* 24(suppl. 1): 59–64 (1985).

Dinarello, C. A., Cannon, J. G., Wolff, S. M., Bernheim, H. A., Beutler, B., Cerami, A., Figari, I. S., Palladino, M. A., Jr., and O'Connor, J. V. Tumor necrosis factor (cachectin) is an endogenous pyrogen and induces production of interleukin 1. *J. Exp. Med.* 163: 1433–1450 (1986).

Dohlsten, M., Sjogren, H. O., and Carlsson, R. Histamine inhibits interferon-γ production via suppression of interleukin 2 synthesis. *Cell. Immunol.* 101: 493–501 (1986).

Dower, S. K., Call, S. M., Gillis, S., and Urdal, D. L. Similarity between the interleukin 1 receptors on a murine T-lymphoma cell line and on a murine fibroblast cell line. *Proc. Natl. Acad. Sci.* USA 83: 1060–1064 (1986a).

Dower, S. K., Kronheim, S. R., Hopp, T. P., Cantrell, M., Deeley, M., Gillis, S., Henney, C. S., and Urdal, D. L. The cell surface receptors for interleukin-1α and interleukin-1β are identical. *Nature* 324: 266–268 (1986b).

Dower, S. K., Kronheim, S. R., March, C. J., Conlon, P. J., Hopp, T. P., Gillis, S., and Urdal, D. L. Detection and characterization of high affinity plasma membrane receptors for human interleukin 1. *J. Exp. Med.* 162: 501–515 (1985).

Durand, D. B., Bush, M. R., Morgan, J. G., Weiss, A., and Crabtree, G. R. A 275 basepair fragment at the 5′ end of the interleukin 2 gene enhances expression from a heterologous promoter in response to signals from the T cell antigen receptor. *J. Exp. Med.* 165: 395–407 (1987).

Eden, E. and Turino, G. M. Interleukin-1 secretion by human alveolar macrophages stimulated with endotoxin is augmented by recombinant immune (γ) interferon. *Am. Rev. Respir. Dis.* 133: 455–460 (1986).

Esparza, I., Männel, D., Ruppel, A., Falk, W., and Krammer, P. H. Interferon-α and lymphotoxin or tumor necrosis factor act synergistically to induce macrophage killing of tumor cells and schistosomula of *Schistosoma mansoni*. *J. Exp. Med.* 166: 589–594 (1987).

Evans, S. W., Beckner, S. K., and Farrar, W. L. Stimulation of specific GTP binding and hydrolysis activities in lymphocyte membrane by interleukin-2. *Nature* 325: 166–168 (1987).

Falus, A., Beuscher, H. U., Auerbach, H. S., and Colten, H. R. Constitutive and IL1-regulated murine complement gene expression is strain and tissue specific. *J. Immunol.* 138: 856–860 (1987).

Farrar, W. L., Johnson, H. M., and Farrar, J. J. Regulation of the production of immune interferon and cytotoxic T lymphocytes by interleukin 2. *J. Immunol.* 126: 1120–1125 (1981).

Feinman, R., Henriksen-DeStefano, D., Tsujimoto, M., and Vilcek, J. Tumor necrosis factor is an important mediator of tumor cell killing by human monocytes. *J. Immunol.* 138: 635–640 (1987).

Fransen, L., Muller, R., Marmenout, A., Tavernier, J., Van Der Heyden, J., Kawashima, E., Chollet, A., Tizard, R., van Heuverswyn, H., van Vliet, A., Ruysschaert, M. R., and Fiers, W. Molecular cloning of mouse tumor necrosis factor cDNA and its eukaryotic expression. *Nucl. Acids Res.* 13: 4417–4429 (1985).

Fransen, L., Ruysschaert, M. R., Van Der Heyden, J., and Fiers, W. Recombinant tumor necrosis factor: Species specificity for a variety of human and murine transformed cell lines. *Cell. Immunol.* 100: 260–267 (1986a).

Fransen, L., Van der Heyden, J., Ruysschaert, R., and Fiers, W. Recombinant tumor necrosis factor: Its effect and its synergism with interferon-γ on a variety of normal and transformed human cell lines. *Eur. J. Cancer Clin. Oncol.* 22: 419–426 (1986b).

Fujii, M., Sugamura, K., Sano, K., Nakai, M., Sugita, K., and Hinuma, Y. High-affinity receptor-mediated internalization and degradation of interleukin 2 in human T cells. *J. Exp. Med.* 163: 550–562 (1986).

Fujita, T., Shibuya, H., Ohashi, T., Yamanishi, K., and Taniguchi, T. Regulation of human interleukin-2 gene: Functional DNA sequences in the 5′ flanking region for the gene expression in activated T lymphocytes. *Cell* 46: 401–407 (1986).

Fujita, T., Takaoka, C., Matsui, H., and Taniguchi, T. Structure of the human interleukin 2 gene. *Proc. Natl. Acad. Sci.* (USA) 80: 7437–7441 (1983).

Furutani, Y., Notake, M., Fukui, T., Ohue, M., Nomura, H., Yamada, M., and Nakamura, S. Complete nucleotide sequence of the gene for human interleukin 1α. *Nucl. Acids Res.* 14: 3167–3179 (1986).

Furutani, Y., Notake, M., Yamayoshi, M., Yamagishi, J. I., Nomura, H., Ohue, M., Furuta, R., Fukui, T., Yamada, M., and Nakamura, S. Cloning and characterization of the cDNAs for human and rabbit interleukin-1 precursor. *Nucl. Acids Res.* 13: 5869–5882 (1985).

Fuse, A., Fujita, T., Yasumitsu, H., Kashima, N., Hasegawa, K., and Taniguchi, T. Organization and structure of the mouse interleukin-2 gene. *Nucl. Acids Res.* 12: 9323–9331 (1984).

Garrett, I. R., Durie, B. G. M., Nedwin, G. E., Gillespie, A., Bringman, T., Sabatini, M., Bertolini, D. R., and Mundy, G. R. Production of lymphotoxin, a bone-resorbing cytokine, by cultured human myeloma cells. *N. Engl. J. Med.* 317: 526–532 (1987).

Gerrard, T. L., Siegel, J. P., Dyer, D. R., and Zoon, K. C. Differential effects of interferon-α and interferon-γ on interleukin 1 secretion by monocytes. *J. Immunol.* 138: 2535–2540 (1987).

Gery, I., Gershon, R. K., and Waksman, B. H. Potentiation of the T-lymphocyte response to mitogens. I. The responding cell. *J. Exp. Med.* 136: 128–142 (1972).

Gillis, S. and Smith, K. A. Long term culture of tumour-specific cytotoxic T cells. *Nature* 268: 154–156 (1977).

Giri, J. G., Lomedico, P. T., and Mizel, S. B. Studies on the synthesis and secretion of interleukin 1. I. A 33,000 molecular weight precursor for interleukin 1. *J. Immunol.* 134: 343–349 (1985).

Godfrey, P. W., Johnson, W. J., and Hoffstein, S. T. Recombinant tumor necrosis factor and interleukin-1 both stimulate human synovial cell arachidonic acid release and phospholipid metabolism. *Biochem. Biophys, Res. Comm.* 142: 235–241 (1987).

Gowen, M. and Mundy, G. R. Actions of recombinant interleukin 1, interleukin 2, and interferon-γ on bone resorption in vitro. *J. Immunol.* 136: 2478–2482 (1986).

Gowen, M., Wood, D. D., Ihrie, E. J., McGuire, M. K. B., and Russell, R. G. G. An interleukin 1 like factor stimulates bone resorption in vitro. *Nature* 306: 378–380 (1983).

Grabstein, K., Dower, S., Gillis, S., Urdal, D., and Larsen, A. Expression of interleukin 2, interferon-γ, and the IL2 receptor by human peripheral blood lymphocytes. *J. Immunol.* 136: 4503–4508 (1986).

Granger, G. A. and Kolb, W. P. Lymphocyte in vitro cytotoxicity: Mechanisms of immune and non-immune small lymphocyte mediated target L cell destruction. *J. Immunol.* 101: 111–120 (1968).

Grau, G. E., Fajardo, L. F., Piguet, P. F., Allet, B., Lambert, P. H., and Vassalli, P. Tumor necrosis factor (cachectin) as an essential mediator in murine cerebral malaria. *Science* 237: 1210-1212 (1987).

Gray, P. W., Aggarwal, B. B., Benton, C. V., Bringman, T. S., Henzel, W. J., Jarrett, J. A., Leung, D. W., Moffat, B., Ng P., Svedersky, L. P., Palladino, M. A., and Nedwin, G. E. Cloning and expression of cDNA for human lymphotoxin, a lymphokine with tumor necrosis activity. *Nature* 312: 721–724 (1984).

Gray, P. W., Chen, E., Li, C. B., Tang, W. L., and Ruddle, N. The murine tumor necrosis factor-β (lymphotoxin) gene sequence. *Nucl. Acids Res.* 15: 3937 (1987).

Gray, P. W., Glaister, D., Chen, E., Goeddel, D. V., and Pennica, D. Two interleukin 1 genes in the mouse: Cloning and expression of the cDNA for murine interleukin 1. *J. Immunol.* 137: 3644–3648 (1986).

Greene, W. C. and Leonard, W. J. The human interleukin-2 receptor. *Ann. Rev. Immunol.* 4: 69–95 (1986).

Gresser, I., Belardelli, F., Tavernier, J., Fiers, W., Podo, F., Federico, M., Carpinelli, G., Duvillard, P., Prade, M., Maury, C., Bandu, M. T., and Maunoury, M. T. Anti-tumor effects of interferon in mice injected with interferon-sensitive and interferon-resistant Friend leukemia cells. V. Comparisons with the action of tumor necrosis factor. *Int. J. Cancer* 38: 771–778 (1986).

Grimm, E. A., Mazumder, A., and Rosenberg, S. A. In vitro growth of cytotoxic human lymphocytes. V. Generation of allospecific cytotoxic lymphocytes to nonimmunogenic antigen by supplementation of in vitro sensitization with partially purified T-cell growth factor. *Cell. Immunol.* 70: 248–259 (1982).

Hamby, B. A., Huggins, E. M., Jr., Lachman, L. B., Dinarello, C. A., and Sigel, M. M. Fish lymphocytes respond to human IL-1. *Lymphokine Res.* 5: 157–162 (1986).

Haranaka, K., Satomi, N., and Sakurai, A. Antitumor activity of murine tumor necrosis factor (TNF) against transplanted murine tumors and heterotransplanted human tumors in nude mice. *Int. J. Cancer* 34: 263–267 (1984).

Herman, J., Kew, M. C., and Rabson, A. R. Defective interleukin-1 production by monocytes from patients with malignant disease. Interferon increases IL-1 production. *Cancer Immunol. Immunother.* 16: 182–185 (1984).

Herrmann, F., Cannistra, S. A., Levine, H., and Griffin, J. D. Expression of interleukin 2 receptors and binding of interleukin 2 by γ interferon-induced human leukemic and normal monocytic cells. *J. Exp. Med.* 162: 1111–1116 (1985).

Hirano, T., Taga, T., Nakano, N., Yasukawa, K., Kashiwamura, S., Shimizu, K., Nakajima, K., Pyun, K. H., and Kishimoto, T. Purification to homogeneity and characterization of human B-cell differentiation factor (BCDF or BSFp-2). *Proc. Natl. Acad. Sci.* (USA) 82: 5490–5494 (1985).

Hirano, T., Yasukawa, K., Harada, H., Taga, T., Watanabe, Y., Matsuda, T., Kashiwamura, S. I., Nakajima, K., Koyama, K., Iwamatsu, A., Tsunasawa, S., Sakiyama, F., Matsui, H., Takahara, Y., Taniguchi, T., and Kishimoto, T. Complementary DNA for a novel human interleukin (BSF-2) that induces B lymphocytes to produce immunoglobulin. *Nature* 324: 73–76 (1986).

Holbrook, N. J., Smith, K. A., Fornace, A. J., Jr., Comeau, C. M., Wiskocil, R. L., and Crabtree, G. R. T-cell growth factor: Complete nucleotide sequence and organization of the gene in normal and malignant cells. *Proc. Natl. Acad. Sci.* (USA) 81: 1634–1638 (1984).

Holter, W., Grunow, R., Stockinger, H., and Knapp, W. Recombinant interferon-γ induces interleukin 2 receptors on human peripheral blood monocytes. *J. Immunol.* 136: 2171–2175 (1986).

Howard, M., Mizel, S. B., Lachman, L., Ansel, J., Johnson, B., and Paul, W. E. Role of interleukin 1 in anti-immunoglobulin-induced B cell proliferation. *J. Exp. Med.* 157: 1529–1543 (1983).

Howe, R. C., Lowenthal, J. W., and MacDonald, H. R. Role of interleukin 1 in early T cell development: LYT-2-L3T4-thymocytes bind and respond in vitro to recombinant IL1. *J. Immunol.* 137: 3195–3200 (1986).

Imboden, J. B., Weiss, A., and Stobo, J. D. The antigen receptor on a human T cell line initiates activation by increasing cytoplasmic free calcium. *J. Immunol.* 134: 663–665 (1985).

Inoue, J. -I., Seiki, M., Taniguchi, T., Tsuru, S., and Yoshida, M. Induction of interleukin 2 receptor gene expression by $p40^x$ encoded by human T-cell leukemia virus type 1. *EMBO J.* 5: 2883–2888 (1986).

Ishida, N., Kanamori, H., Noma, T., Nikaido, T., Sabe, H., Suzuki, N., Shimizu, A., and Honjo, T. Molecular cloning and structure of the human interleukin 2 receptor gene. *Nucl. Acids Res.* 13: 7579–7589 (1985).

Jadus, M. R., Schmunk, G., Djeu, J. Y., and Parkman, R. Morphology and lytic mechanisms of interleukin 3-dependent natural cytotoxic cells: Tumor necrosis factor as a possible mediator. *J. Immunol.* 137: 2774–2783 (1986).

Jelinek, D. F., Splawski, J. B., and Lipsky, P. E. The roles of interleukin 2 and interferon-γ in human B cell activation, growth and differentiation. *Eur. J. Immunol.* 16: 925–932 (1986).

Kasahara, T., Hooks, J. J., Dougherty, S. F., and Oppenheim, J. J. Interleukin 2 mediated immune interferon (IFN-γ) production by human T cells and T cell subsets. *J. Immunol.* 130: 1784–1789 (1983).

Kashima, N., Nishi-Takaoka, C., Fujita, T., Taki, S., Yamada, G., Hamuro, J., and Taniguchi, T. Unique structure of murine interleukin-2 as deduced from cloned cDNAs. *Nature* 313: 402–406 (1985).

Koeffler, H. P., Gasson, J., Ranyard, J., Souza, L., Shepard, M. and Munker, R. Recombinant human TNF-γ stimulates production of granulocyte colony-stimulating factor. *Blood* 70: 55–59 (1987).

Kohase, M., Henriksen-Destefano, D., May, L. T., Vilcek, J., and Sehgal, P. B. Induction of β2-interferon by tumor necrosis factor: A homeostatic mechanism in the control of cell proliferation. *Cell* 45: 659–666 (1986).

Kohase, M., May, L. T., Tamm, I., Vilcek, J., and Sehgal, P. B. A cytokine network in human diploid fibroblasts: Interactions of β-interferons, tumor necrosis factor, platelet-derived growth factor, and interleukin-1. *Mol. Cell. Biol.* 7: 273–280 (1987).

Koide, S. L., Inaba, K., and Steinman, R. M. Interleukin enhances *T*-dependent immune responses by amplifying the function of dendritic cells. *J. Exp. Med.* 165: 515–530 (1987).

Koide, S. and Steinman, R. M. Induction of murine interleukin 1: Stimuli and responsive primary cells. *Proc. Natl. Acad. Sci.* (USA) 84: 3802–3806 (1987).

Kondo, S., Shimizu, A., Saito, Y., Kinoshita, M., and Honjo, T. Molecular basis for two different affinity states of the interleukin 2 receptor: Affinity conversion model. *Proc. Natl. Acad. Sci.* (USA) 83: 9026–9029 (1986).

Koo, G. C. and Manyak, C. L. Generation of cytotoxic cells from murine bone marrow by human recombinant IL-2. *J. Immunol.* 137: 1751–1756 (1986).

Kovacs, E. J., Oppenheim, J. J., and Young, H. A. Induction of c-fos and c-myc expression in T lymphocytes after treatment with recombinant interleukin lα. *J. Immunol.* 137: 3649–3651 (1986).

Kronke, M., Leonard, W. J., Depper, J. M., and Greene, W. C. Sequential expression of genes involved in human T lymphocyte growth and differentiation. *J. Exp. Med.* 161: 1593–1598 (1985).

Kupper, T. S., Ballard, D. W., Chua, A. O., McGuire, J. S., Flood, P. M., Horowitz, M. C., Langdon, R., Lightfoot, L., and Gubler, U. Human keratinocytes contain mRNA indistinguishable from monocyte interleukin lα and β mRNA. Keratinocyte epidermal cell-derived thymocyte-activating factor is identical to interleukin l. *J. Exp. Med.* 164: 2095–2100 (1986).

Kurt-Jones, E. A., Beller, D. I., Mizel, S. B., and Unanue, E. R. Identification of a membrane-associated interleukin l in macrophages. *Proc. Natl. Acad. Sci.* (USA) 82: 1204–1208 (1985).

Larsson, E. L., Iscove, N. N., and Coutinho, A. Two distinct factors are required for induction of T-cell growth. *Nature* 283: 664–666 (1980).

Le, J., Weinstein, D., Gubler, U., and Vilcek, J. Induction of membrane-associated interleukin 1 by tumor necrosis factor in human fibroblasts. *J. Immunol.* 138: 2137–2142 (1987).

Lee, S. H., Aggarwal, B. B., Rinderknecht, E., Assisi, F., and Chiu, H. The synergistic anti-proliferative effect of γ-interferon and human lymphotoxin. *J. Immunol.* 133: 1083–1086 (1984).

Leonard, W. J., Depper, J. M., Crabtree, G. R., Rudikoff, S., Pumphrey, J., Robb, R. J., Kronke, M., Svetlik, P. B., Peffer, N. J., Waldmann, T. A., and Greene, W. C. Molecular cloning and expression of cDNAs for the human interleukin-2 receptor. *Nature* 311: 626–631 (1984).

Leonard, W. J., Depper, J. M., Kanehisa, M., Kronke, M., Peffer, N. J., Svetlik, P. B., Sullivan, M., and Greene, W. C. Structure of the human interleukin-2 receptor gene. *Science* 230: 633–639 (1985a).

Leonard, W. J., Depper, J. M., Uchiyama, T., Smith, K. A., Waldmann, T. A., and Greene, W. C. A monoclonal antibody that appears to recognize the receptor for human T-cell growth factor; partial characterization of the receptor. *Nature* 300: 267–269 (1982).

Leonard, W. J., Donlon, T. A., Lebo, R. V., and Greene, W. C. Localization of the gene encoding the human interleukin-2 receptor on chromosome 10. *Science* 228: 1547–1549 (1985b).

Lomedico, P. T., Gubler, U., Hellman, C. P., Dukovich, M., Giri, J. G., Pan, Y.-C. E., Collier, K., Semionow, R., Chua, A. O., and Mizel, S. B. Cloning and expression of murine interleukin-l cDNA in *E. coli*. *Nature* 312: 458–462 (1984).

Lowenthal, J. W., Cerottini, J. C., and MacDonald, H. R. Interleukin l-dependent induction of both interleukin 2 secretion and interleukin 2 receptor expression by thymoma cells. *J. Immunol.* 137: 1226–1231 (1986a).

Lowenthal, J. W. and MacDonald, H. R. Binding and internalization of interleukin l by T cells. Direct evidence for high- and low-affinity classes of interleukin l receptor. *J. Exp. Med.* 164: 1060–1074 (1986).

Lowenthal, J. W., MacDonald, H. R., and Iacopetta, B. J. Intracellular pathway of interleukin 2 following receptor-mediated endocytosis. *Eur. J. Immunol.* 16: 1461–1463 (1986b).

Lowenthal, J. W. and MacDonald, H. R. Expression of interleukin l receptors is restricted to the L3T4 subset of mature T lymphocytes. *J. Immunol.* 138: 1–3 (1987).

Lu, L., Welte, K., Gabrilove, J. L., Hangoc, G., Bruno, E., Hoffman, R., and Broxmeyer, H. E. Effects of recombinant human tumor necrosis factor α, recombinant human γ-interferon, and prostaglandin E on colony formation of human hematopoietic progenitor cells stimulated by natural human pluripotent colony-stimulating factor, pluripoietin α, and recombinant erythropoietin in serum-free cultures. *Cancer Res.* 46: 4357–4361 (1986).

MacDonald, H. R. and Nabholz, M. T-cell activation. *Ann. Rev. Cell Biol.* 2: 231–253 (1986).

Malkovsky, M., Loveland, B., North, M., Asherson, G. L., Gao, L., Ward, P., and Fiers, W. Recombinant interleukin-2 directly augments the cytotoxicity of human monocytes. *Nature* 325: 262–265 (1987).

March, C. J., Mosley, B., Larsen, A., Cerretti, D. P., Braedt, G., Price, V., Gillis, S., Henney, C. S., Kronheim, S. R., Grabstein, K., Conlon, P. J., Hopp, T. P., and Cosman, D. Cloning, sequence and expression of two distinct human interleukin-l complementary DNAs. *Nature* 315: 641–647 (1985).

Marmenout, A., Fransen, L., Tavernier, J., Van der Heyden, J., Tizard, R., Kawashima, E., Shaw, A., Semon, M. J., Muller, R., Ruysschaert, M. R., Van Vliet, A., and Fiers, W. Molecular cloning and expression of human tumor necrosis factor and comparison with mouse tumor necrosis factor. *Eur. J. Biochem.* 152: 515–522 (1985).

Matsushima, K., Akahoshi, T., Yamada, M., Furutani, Y., and Oppenheim, J. J. Properties of a specific interleukin l (ILl) receptor on human Epstein Barr virus-transformed B lymphocytes: Identity of the receptor for ILl-α and ILl-β. *J. Immunol.* 136: 4496–4502 (1986).

McCroskery, P. A., Arai, S., Amento, E. P., and Krane, S. M. Stimulation of procollagenase synthesis in human rheumatoid synovial fibroblasts by mononuclear cell factor/interleukin 1. *FEBS Lett.* 191: 7–12 (1985).

Mestan, J., Digel, W., Mittnacht, S., Hillen, H., Blohm, D., Möller, A., Jacobsen, H., and Kirchner, H. Antiviral effects of recombinant tumor necrosis factor in vitro. *Nature* 323: 816–819 (1986).

Meuer, S. C., Hussey, R. E., Cantrell, D. A., Hodgdon, J. C., Schlossman, S. F., Smith, K. A., and Reinherz, E. L. Triggering of the T3-Ti antigen-receptor complex results in clonal T-cell proliferation through an interleukin 2-dependent autocrine pathway. *Proc. Natl. Acad. Sci.* (USA) 81: 1509–1513 (1984).

Miller, J., Malek, T. R., Leonard, W. J., Greene, W. C., Shevach, E. M., and Germain, R. N. Nucleotide sequence and expression of a mouse interleukin-2 receptor cDNA. *J. Immunol.* 134: 4212–4217 (1985).

Miossec, P. and Ziff, M. Immune interferon enhances the production of interleukin 1 by human endothelial cells stimulated with lipopolysaccharide. *J. Immunol.* 137: 2848–2852 (1986).

Morgan, D. A., Ruscetti, F. W., and Gallo, R. Selective in vitro growth of T lymphocytes from normal human bone marrow. *Science* 193: 1007–1008 (1976).

Müller, U., Jongeneel, C. V., Nedospasov, S., Lindahl, F., and Steinmetz, M. Tumor necrosis factor and lymphotoxin genes map close to H-2D in the mouse major histocompatibility complex. *Nature* 325: 265–267 (1987).

Munker, R., Gasson, J., Ogawa, M., and Koeffler, H. P. Recombinant human TNF induces production of granulocyte-monocyte colony-stimulating factor. *Nature* 323: 79–82 (1986).

Nakagawa, T., Nakagawa, N., Volkman, D. J., and Fauci, A. S. Sequential synergistic effect of interleukin 2 and interferon-γ on the differentiation of a Tac-antigen positive B cell line. *J. Immunol.* 136: 164–168 (1986).

Nawroth, P. P., Bank, I., Handley, D., Cassimeris, J., Chess, L., and Stern, D. Tumor necrosis factor/cachectin interacts with endothelial cell receptors to induce release of interleukin 1. *J. Exp. Med.* 163: 1363–1375 (1986).

Nedospasov, S. A., Hirt, B., Shakhov, A. N., Dobrynin, V. N., Kawashima, E., Accolla, R. S., and Jongeneel, C. V. The genes for tumor necrosis factor (TNF-α) and lymphotoxin (TNF-β) are tandemly arranged on chromosome 17 of the mouse. *Nucl. Acids Res.* 14: 7713–7725 (1986).

Nedwin, G. E., Naylor, S. L., Sakaguchi, A. Y., Smith, D., Jarrett-Nedwin, J., Pennica, D., Goeddel, D. V., and Gray, P. W. Human lymphotoxin and tumor necrosis factor genes: Structure, homology and chromosomal localization. *Nucl. Acids Res.* 13: 6361–6373 (1985a).

Nedwin, G. E., Svedersky, L. P., Bringman, T. S., Palladino, M. A., Jr., and Goeddel, D. V. Effect of interleukin 2, interferon-γ and mitogens on the production of tumor necrosis factors α and β. *J. Immunol.* 135: 2492–24497 (1985b).

Newton, R. C. Effect of interferon on the induction of human monocyte secretion of interleukin-1 activity. *Immunology* 56: 441–449 (1985).

Nikaido, T., Shimizu, A., Ishida, N., Sabe, H., Teshigawara, K., Maeda, M., Uchiyama, T., Yodoi, J., and Honjo, T. Molecular cloning of cDNA encoding human interleukin-2 receptor. *Nature* 311: 631–635 (1984).

Obert, H. J. Treatment of rheumatoid arthritis with interferon-γ: Results of clinical trials in Germany. *J. Interferon Res.* (Abstracts) 6(suppl. l): 37 (1986).

Oppenheim, J. J., Kovacs, E. J., Matsushima, K., and Durum, S. K. There is more than one interleukin l. *Immunol. Today* 7: 45–56 (1986).

Ortaldo, J. R., Mason, A. T., Gerard, J. P., Henderson, L. E., Farrar, W., Hopkins R. D., III, Herberman, R. B., and Rabin, H. Effects of natural and recombinant IL 2 on regulation of IFN-γ production and natural killer activity: Lack of involvement of the TAC antigen for these immunoregulatory effects. *J. Immunol.* 133: 779–783 (1984).

Ortaldo, J. R., Mason, L. H., Mathieson, B. J., Liang, S. M., Flick, D. A., and Herberman, R. B. Mediation of mouse natural cytotoxic activity by tumour necrosis factor. *Nature* 321: 700–702 (1986).

Palombella, V. J., Yamashiro, D. J., Maxfield, F. R., Decker, S. J., and Vilcek, J. Tumor necrosis factor increases the number of epidermal growth factor receptors on human fibroblasts. *J. Biol. Chem.* 262: 1950–1954 (1987).

Patton, J. S., Shepard, H. M., Wilking, H., Lewis, G., Aggarwal, B. B., Eessalu, T. E., Gavin, L. A., and Grunfeld, C. Interferons and tumor necrosis factors have similar catabolic effect on 3T3 Ll cells. *Proc. Natl. Acad. Sci.* (USA) 83: 8313–8317 (1986).

Pearlstein, K. T., Palladino, M. A., Welte, K., and Vilcek, J. Purified human interleukin-2 enhances induction of immune interferon. *Cell. Immunol.* 80: 1–9 (1983).

Pennica, D., Nedwin, G. E., Hayflick, J. S., Seeburg, P. H., Derynck, R., Palladino, M. A., Kohr, W. J., Aggarwal, B. B., and Goeddel, D. V. Human tumour necrosis factor: Precursor structure, expression and homology to lymphotoxin. *Nature* 312: 724–729 (1984).

Pepys, M. B. and Baltz, M. L. Acute phase proteins with special reference to C-reactive protein and related proteins (pentaxins) and serum amyloid A protein. *Adv. Immunol.* 34: 141–212 (1983).

Perlmutter, D. H., Goldberger, G., Dinarello, C. A., Mizel, S. B., and Colten, H. R. Regulation of class III major histocompatibility complex gene products by interleukin-1. *Science* 232: 850–852 (1986).

Peters, P. M., Ortaldo, J. R., Shalaby, M. R., Svedersky, L. P., Nedwin, G. E., Bringman, T. S., Hass, P. E., Aggarwal, B. B., Herberman, R. B., Goeddel, D. V., and Palladino, M. A., Jr. Natural killer-sensitive targets stimulate production of TNF-α but not TNF-β (lymphotoxin) by highly purified human peripheral blood large granular lymphocytes. *J. Immunol.* 137: 2592–2598 (1986).

Pettipher, E. R., Higgs, G. A., and Henderson, B. Interleukin l induces leukocyte infiltration and cartilage proteoglycan degradation in the synovial joint. *Proc. Natl. Acad Sci.* (USA) 83: 8749–8753 (1986).

Pfizenmaier, K., Scheurich, P., Schlüter, C., and Krönke, M. Tumor necrosis factor enhances HLA-A,B,C and HLA-DR gene expression in human tumor cells. *J. Immunol.* 138: 975–980 (1987).

Phadke, K., Carlson, D. G., Gitter, B. D., and Butler, L. D. Role of interleukin 1 and interleukin 2 in rat and mouse arthritis models. *J. Immunol.* 136: 4085–4091 (1986).

Philip, R. and Epstein, L. B. Tumor necrosis factor as immunomodulator and mediator of monocyte cytotoxicity induced by itself, γ-interferon and interleukin-1. *Nature* 323: 86–89 (1986).

Pike, B. L. and Nossal, G. J. V. Interleukin 1 can act as a B-cell growth and differentiation factor. *Proc. Natl. Acad. Sci.* (USA) 8153–8157 (1985).

Pober, J. S., Gimbrone, M. A., Jr., Lapierre, L. A., Mendrick, D. L., Fiers, W., Rothlein, R., and Springer, T. A. Overlapping patterns of activation of human endothelial cells by interleukin 1, tumor necrosis factor, and immune interferon. *J. Immunol.* 137: 1893–1896 (1986).

Ramadori, G., Sipe, J. D., Dinarello, C. A., Mizel, S. N., and Colten, H. R. Pretranslational modulation of acute phase hepatic protein synthesis by murine recombinant interleukin 1 (IL-1) and purified human IL-1. *J. Exp. Med.* 162: 930–942 (1985).

Reem, G. H. and Yeh, N. H. Interleukin 2 regulates expression of its receptor and synthesis of γ interferon by human T lymphocytes. *Science* 225: 429–430 (1984).

Reem, G. H. and Yeh, N. H. Regulation by interleukin 2 of interleukin 2 receptors and γ-interferon synthesis by human thymocytes: Augmentation of interleukin 2 receptors by interleukin 2. *J. Immunol.* 134: 953–958 (1985).

Reeves, R., Spies, A. G., Nissen, M. S., Buck, C. D., Weinberg, A. D., Barr, P. J., Magnuson, N. S., and Magnuson, J. A. Molecular cloning of a functional bovine interleukin 2 cDNA. *Proc. Natl. Acad. Sci.* (USA) 83: 3228–3232 (1986).

Revel, M. and Zilberstein, A. Interferon-β2 living up to its name. *Nature* 325: 581–582 (1987).

Reyes, V. E., Ballas, Z. K., Singh, H., and Klimpel, G. R. Interleukin 2 induces interferon α/β production in mouse bone marrow cells. *Cell. Immunol.* 102: 374–385 (1986).

Rhodes, J., Yvanyi, J., and Cozens, P. Antigen presentation by human monocytes: Effects of modifying major histocompatibility complex class II antigen expression and interleukin 1 production by using recombinant interferons and corticosteroids. *Eur. J. Immunol.* 16: 370–375 (1986).

Robb, R. J. and Greene, W. C. Direct demonstration of the identity of T cell growth factor binding protein and the TAC antigen. *J. Exp. Med.* 158: 1332–1337 (1983).

Robb, R. J. and Greene, W. C. Internalization of interleukin 2 is mediated by the β chain of the high-affinity interleukin 2 receptor. *J. Exp. Med.* 165: 1201–1206 (1987).

Robb, R. J., Kutny, R. M., and Chowdhry, V. Purification and partial sequence analysis of human T-cell growth factor. *Proc. Natl. Acad. Sci.* (USA) 80: 5990–5994 (1983).

Robb, R. J., Munck, A., and Smith, K. A. T cell growth factor receptors. Quantitation, specificity, and biological relevance. *J. Exp. Med.* 154: 1455–1474 (1981).

Rock, K. L. and Benacerraf, B. The role of Ia molecules in the activation of T lymphocytes. IV. Basis of the thymocyte IL 1 response and its possible role in the generation of the T cell repertoire. *J. Immunol.* 132: 1654–1662 (1984).

Rook, A. H., Hooks, J. J., Quinnan, G. V., Lane, H. C., Manischewitz, J. F., Macher, A. M., Masur, H., Fauci, A. S., and Djeu, J. Y. IL 2 enhances the natural killer cell activity of acquired immunodeficiency syndrome patients through a γ-interferon-independent mechanism. *J. Immunol.* 134: 1503–1507 (1985).

Rosenberg, S. A., Grimm, E. A., McGrogan, M., Doyle, M., Kawasaki, E., Koths, K., and Mark, D. F. Biological activity of recombinant human interleukin-2 produced in the *E. coli*. *Science* 223: 1412–1415 (1984).

Rosenwasser, L. J., Webb, A. C., Clark, B. D., Irie, S., Chang, L., Dinarello, C. A., Gehrke, L., Wolff, S. M., Rich, A., and Auron, P. E. Expression of biologically active human interleukin 1 subpeptides by transfected simian COS cells. *Proc. Natl. Acad. Sci.* (USA) 83: 5243–5246 (1986).

Rubin, L. A., Jay, G., and Nelson, D. L. The released interleukin 2 receptor binds interleukin 2 efficiently. *J. Immunol.* 137: 3841–3844 (1986).

Ruddle, N. H. and Waksman, B. H. Cytotoxicity mediated by soluble antigen and lymphocytes in delayed hypersensitivity. III. Analysis of mechanism. *J. Exp. Med.* 128: 1267–1279 (1968).

Ruff, M. R., Farrar, W. L., and Pert, C. B. Interferon-γ and granulocyte/macrophage colony-stimulating factor inhibit growth and induce antigens characteristic of myeloid differentiation in small-cell lung cancer cell lines. *Proc. Natl. Acad. Sci.* (USA) 83: 6613–6617 (1986).

Ruggiero, V., Tavernier, J., Fiers, W., and Baglioni, C. Induction of the synthesis of tumor necrosis factor receptors by interferon-γ. *J. Immunol.* 136: 2445–2450 (1986).

Russell, J. H., Masakowski, V. R., and Dobos, C. B. Mechanisms of immune lysis. I. Physiological distinction between target cell death mediated by cytotoxic T lymphocytes and antibody plus complement. *J. Immunol.* 124: 1100–1105 (1980).

Russell, J. H., Masakowski, V., Rucinsky, T., and Philips, G. Mechanisms of immune lysis. III. Characterization of the nature and kinetics of the cytotoxic T lymphocyte-induced nuclear lesion in the target. *J. Immunol.* 128: 2087–2094 (1982).

Satoh, M., Shimada, Y., Inagawa, H., Minagawa, H., Kajikawa, T., Oshima, H., Abe, S., Yamazaki, M., and Mizuno, D. Priming effect of interferons and interleukin 2 on endogenous production of tumor necrosis factor in mice. *Jpn. J. Cancer Res.* 77: 342–344 (1986).

Scheurich, P., Thoma, B., Ucerr, U., and Pfizenmaier, K. Immunoregulatory activity of recombinant human tumor necrosis factor (TNF)-α: Induction of TNF receptors on human T cells and TNF-α-mediated enhancement of T cell responses. *J. Immunol.* 138: 1786–1790 (1987).

Schleimer, R. P. and Rutledge, B. K. Cultured human vascular endothelial cells acquire adhesiveness for neutrophils after stimulation with interleukin 1, endotoxin, and tumor-promoting phorbol diesters. *J. Immunol.* 136: 649–654 (1986).

Schmid, D. S., Tite, J. P., and Ruddle, N. H. DNA fragmentation: Manifestation of target cell destruction mediated by cytotoxic T-cell lines, lymphotoxin-secreting helper T-cell clones, and cell-free lymphotoxin-containing supernatant. *Proc. Natl. Acad. Sci.* (USA) 83: 1881–1885 (1986).

Schmidt, J. A., Mizel, S. B., Cohen, D., and Green, I. Interleukin 1, a potential regulator of fibroblast proliferation. *J. Immunol.* 128: 2177–2182 (1982).

Schnyder, J., Payne, T., and Dinarello, C. A. Human monocyte or recombinant interleukin 1's are specific for the secretion of a metalloproteinase from chondrocytes. *J. Immunol.* 138: 496–503 (1987).

Seigel, L. J., Harper, M. E., Wong-Staal, F., Gallo, R. C., Nash, W. G., and O'Brien, S. J. Gene for T-cell growth factor: Location on human chromosome 4q and feline chromosome Bl. *Science* 223: 175–178 (1984).

Shalaby, M. R., Aggarwal, B. B., Rinderknecht, E., Svedersky, L. P., Finkle, B. S., and Palladino, M. A., Jr. Activation of human polymorphonuclear neutrophil functions by interferon-γ and tumor necrosis factors. *J. Immunol.* 135: 2069–2073 (1985).

Sharon, M., Klausner, R. D., Cullen, B. R., Chizzonite, R., and Leonard, W. J. Novel interleukin-2 receptor subunit detected by cross-linking under high-affinity conditions. *Science* 234: 859–863 (1986).

Shaw, G. and Kamen, R. A conserved AU sequence from the 3′ untranslated region of GM-CSF mRNA mediates selective mRNA degradation. *Cell* 46: 659–667 (1986).

Shiloni, E., Lafreniere, R., Mule, J. J., Schwarz, S. L., and Rosenberg, S. A. Effect of immunotherapy with allogeneic lymphokine-activated killer cells and recombinant interleukin 2 on established pulmonary and hepatic metastases in mice. *Cancer Res.* 46: 5633–5640 (1986).

Shimizu, A., Kondo, S., Takeda, S. I., Yodoi, J., Ishida, N., Sabe, H., Osawa, H., Diamantstein, T., Nikaido, T., and Honjo, T. Nucleotide sequence of mouse IL-2 receptor cDNA and its comparison with the human IL-2 receptor sequence. *Nucl. Acids Res.* 13: 1505–1516 (1985).

Siebenlist, U., Durand, D. B., Bressler, P., Holebrook, N. J., Norris, C. A., Kamoun, M., Kant, J. A., and Crabtree, G. R. Promoter region of interleukin-2 gene undergoes chromatin structure changes and confers inducibility on chloramphenicol acetyltransferase gene during activation of T cells. *Mol. Cell. Biol.* 6: 3042–3049 (1986).

Simon, M. M., Hochgeschwender, U., Brugger, U., and Landolfo, S. Monoclonal antibodies to interferon-γ inhibit interleukin-2 dependent induction of growth and maturation in lectin/antigen-reactive cytolytic T lymphocyte precursors. *J. Immunol.* 136: 2755–2762 (1986).

Smith, K. A. The two-chain structure of high-affinity IL-2 receptors. *Immunol. Today* 8: 11–13 (1987).

Smith, K. A. and Cantrell, D. A. Interleukin 2 regulates its own receptors. *Proc. Natl. Acad. Sci.* (USA) 82: 864–868 (1985).

Smith, K. A., Lachman, L. B., Oppenheim, J. J., and Favata, M. F. The functional relationship of the interleukins. *J. Exp. Med.* 151: 1551–1556 (1980).

Smith, R. A., Kirstein, M., Fiers, W., and Baglioni, C. Species specificity of human and murine tumor necrosis factor. A comparative study of tumor necrosis factor receptors. *J. Biol. Chem.* 261: 14871–14874 (1986).

Spies, T., Morton, C. C., Nedospasov, S., Fiers, W., Pious, D., and Strominger, J. L. Genes for the tumor necrosis factors α and β are linked to the human major histocompatibility complex. *Proc. Natl. Acad. Sci.* (USA) 83: 8699–8702 (1986).

Stolpen, A. H., Guinan, E. C., Fiers, W., and Pober, J. S. Recombinant tumor necrosis factor and immune interferon act singly and in combination to reorganize human vascular endothelial cell monolayers. *Am. J. Pathol.* 123: 16–24 (1986).

Stone-Wolff, D. S., Yip, Y. K., Chroboczek Kelker, H., Le, J., Henriksen-Destefano, D., Rubin, B. Y., Rinderknecht, R., Aggarwal, B. B., and Vilcek, J. Interrelationships of human interferon-γ with lymphokin and monocyte cytotoxin. *J. Exp. Med.* 159: 828–843 (1984).

Stutman, O., Paige, C. J., and Figarella, E. F. Natural cytotoxic cells against solid tumors in mice. I. Strain and age distribution and target cell susceptibility. *J. Immunol.* 121: 1819–1826 (1978).

Subramanian, N. and Bray, M. A. Interleukin 1 releases histamine from human basophils and mast cells in vitro. *J. Immunol.* 138: 271–275 (1987).

Sugarman, B. J., Aggarwal, B. B., Hass, P. E., Figari, I. S., Palladino, M. A., Jr., and Shepard, H. M. Recombinant human tumor necrosis factor-α: Effects on proliferation of normal and transformed cells in vitro. *Science* 230: 943–945 (1985).

Suzuki, T. and Cooper, M. Comparison of the expression of IL 2 receptors by human T and B cells: Induction by the polyclonal mitogens, phorbol myristate acetate, and anti-μ antibody. *J. Immunol.* 134: 3111–3119 (1985).

Svedersky, L. P., Nedwin, G. E., Goeddel, D. V., and Palladino, M. A., Jr. Interferon-γ enhances induction of lymphotoxin in recombinant interleukin 2-stimulated peripheral blood mononuclear cells. *J. Immunol.* 134: 1604–1608 (1985).

Sztein, M. B., Vogel, S. N., Sipe, J. D., Murphy, P. A., Mizel, S. B., Oppenheim, J. J., and Rosenstreich, D. L. The role of macrophages in the acute-phase response: SAA inducer is closely related to lymphocyte activating factor and endogenous pyrogen. *Cell. Immunol.* 63: 164–176 (1981).

Takeda, K., Iwamoto, S., Sugimoto, H., Takuma, T., Kawatani, N., Noda, M., Masaki, A., Morise, H., Arimura, H., and Konno, K. Identity of differentation inducing factor and tumor necrosis factor. *Nature* 323: 338–340 (1986).

Taniguchi, T., Matsui, H., Fujita, T., Takaoka, C., Kashima, N., Yoshimoto, R., and Hamuro, J. Structure and expression of a cloned cDNA for human interleukin-2. *Nature* 302: 305–310 (1983).

Telford, J. L., Macchia, G., Massone, A., Carinci, V., Palla, E. and Melli, M. The murine interleukin 1β gene: Structure and evolution. *Nucl. Acids Res.* 14: 9955–9963 (1986).

Teshigawara, K., Wang, H.-M., Kato, K., and Smith, K. A. Interleukin 2 high-affinity receptor expression requires two distinct binding proteins. *J. Exp. Med.* 165: 223–238 (1987).

Thompson, B. M., Mundy, G. R., and Chambers, T. J. Tumor necrosis factor α and β induce osteoblastic cells to stimulate osteoclastic bone resorption. *J. Immunol.* 138: 775–779 (1987).

Thompson, J. A., Peace, D. J., Klarnet, J. P., Kern, D. E., Greenberg, P. D., and Cheever, M. A. Eradication of disseminated murine leukemia by treatment with high-dose interleukin 2. *J. Immunol.* 137: 3675–3680 (1986).

Tracey, K. J., Beutler, B., Lowry, S. F., Merryweather, J., Wolpe, S., Milsark, I. W., Hariri, R. J., Fahey, T. J., III, Zentella, A., Albert, J. D., Shires, G. T., and Cerami, A. Shock and tissue injury induced by recombinant human cachectin. *Science* 234: 470–474 (1986).

Trinchieri, G., Kobayashi, M., Rosen, M., Loudon, R., Murphy, M., and Perussia, B. Tumor necrosis factor and lymphotoxin induce differentiation of human myeloid cell lines in synergy with immune interferon. *J. Exp. Med.* 164: 1206–1225 (1986).

Trinchieri, G., Matsumoto-Kobayashi, M., Clark, S. C., Seehra, J., London, L., and Perussia, B. Response of resting human peripheral blood natural killer cells to interleukin 2. *J. Exp. Med.* 160: 1147–1169 (1984).

Tsujimoto, M. and Vilcek, J. Tumor necrosis factor receptors in HeLa cells and their regulation by interferon-γ. *J. Biol. Chem.* 261: 5384–5388 (1986).

Tsujimoto, M., Yip, Y. K., and Vilcek, J. Interferon-γ enhances expression of cellular receptors for tumor necrosis factor. *J. Immunol.* 136: 2441–2444 (1986).

Unanue, R. R. and Allen, P. M. Biochemistry and biology of antigen presentation by macrophages. *Cell. Immunol.* 99: 3–6 (1986).

Urban, J. L., Shepard, H. M., Rothstein, J. L., Sugarman, B. J., and Schreiber, H. Tumor necrosis factor: A potent effector molecule for tumor cell killing by activated macrophages. *Proc. Natl. Acad. Sci.* (USA) 83: 5233–5237 (1986).

Van Damme, J., Billiau, A., De Ley, M., and De Somer, P. An interferon-β-like or interferon-inducing protein released by mitogen-stimulated human leukocytes. *J. Gen. Virol.* 64: 1819–1822 (1983).

Van Damme, J., Cayphas, S., Opdenakker, G., Billiau, A., and Van Snick, J. Interleukin 1 and poly(rI).poly(rC) induce production of a hybridoma growth factor by human fibroblasts. *Eur. J. Immunol.* 17: 1–7 (1987a).

Van Damme, J., De Ley, M., Opdenakker, G., Billiau, A., De Somer, P., and Van Beumen, J. Homogeneous interferon-inducing 22K factor is related to endogenous pyrogen and interleukin 1. *Nature* 314: 266–268 (1985).

Van Damme, J., De Ley, M., Van Snick, J., Dinarello, C. A., and Billiau, A. The role of interferon-β1 and the 26-kDa protein (interferon-β2) as mediators of the antiviral effect of interleukin 1 and tumor necrosis factor. *J. Immunol.* 139: 1867–1872 (1987b).

Van Damme, J., Opdenakker, G., Simpson, R. J., Rubira, M. R., Cayphas, S., Vink, A., Billiau, A., and Van Snick, J. Identification of the human 26-kD protein, interferon β2 (IFN-β2) as a B cell hybridoma/plasmacytoma growth factor induced by interleukin 1 and tumor necrosis factor. *J. Exp. Med.* 165: 914–919 (1987c).

Van Snick, J., Vink, A., Cayphas, S., and Uyttenhove, C. Interleukin-HP1, a T cell-derived hybridoma growth factor that supports the in vitro growth of murine plasmacytomas. *J. Exp. Med.* 165: 641–649 (1987).

Vilcek, J., Henriksen-Destefano, D., Siegel, D., Klion, A., Robb, R. J., and Le, J. Regulation of IFN-γ induction in human peripheral blood cells by exogenous and endogenously produced interleukin 2. *J. Immunol.* 135: 1851–1856 (1985).

Vilcek, J., Palombella, V. J., Henriksen-Destefano, D., Swenson, C., Feinman, R., Hirai, M., and Tsujimoto, M. Fibroblast growth enhancing activity of TNF and its relationship to other polypeptide growth factors. *J. Exp. Med.* 163: 632–643 (1986).

Wagner, D. K., York-Jolley, J., Malek, T. R., Berzofsky, J. A., and Nelson, D. L. Antigen-specific murine T cell clones produce soluble interleukin 2 receptor on stimulation with specific antigens. *J. Immunol.* 137: 592–596 (1986).

Waldmann, T. A. The structure, function, and expression of interleukin-2 receptors on normal and malignant lymphocytes. *Science* 232: 727–732 (1986).

Waldmann, T. A., Goldman, C. K., Robb, R. J., Depper, J. M., Leonard, W. J.,

Sharrow, S. O., Bongiovanni, K. F., Korsmeyer, S. J., and Greene, W. C. Expression of interleukin 2 receptors on activated human B cells. *J. Exp. Med.* 160: 1450–1466 (1984).

Wallis, R. S., Fujiwara, H., and Ellner, J. J. Direct stimulation of monocyte release of interleukin 1 by mycobacterial protein antigens. *J. Immunol.* 136: 193–196 (1986).

Weaver, C. T. and Unanue, E. R. T-cell induction of membrane IL-1 on Macrophages. *J. Immunol.* 137: 3868–3873 (1986).

Webb, A. C., Collins, K. L., Auron, P. E., Eddy, R. L., Naki, H., Byers, M. G., Haley, L. L., Henry, W. M., and Shows, T. B. Interleukin-1 gene (IL1) assigned to long arm of human chromosome 2. *Lymphokine Res.* 5: 77–85 (1986).

Weigent, D. A., Stanton, G. J., and Johnson, H. M. Interleukin 2 enhances natural killer cell activity through induction of γ interferon. *Infect. Immun.* 41: 992–997 (1983).

Weinberg, A., Basham, T. Y., and Merigan, T. C. Regulation of guinea-pig immune functions by interleukin 2: Critical role of natural killer activity in acute HSV-2 genital infection. *J. Immunol.* 137: 3310–3317 (1986).

Weissman, A. M., Harford, J. B., Svetlik, P. B., Leonard, W. L., Depper, J. M., Waldmann, T. A., Greene, W. C., and Klausner, R. D. Only high-affinity receptors for interleukin 2 mediate internalization of ligand. *Proc. Natl. Acad. Sci.* (USA) 83: 1463–1466 (1986).

Wong, G. H. W. and Goeddel, D. V. Tumor necrosis factors α and β inhibit virus replication and synergize with interferons. *Nature* 323: 819–822 (1986).

Wood, D. D., Bayne, E. K., Goldring, M. B., Gowen, M., Hamerman, D., Humes, J. L., Ihrie, E. J., Lipsky, P. E., and Staruch, M. J. The four biochemically distinct species of human interleukin 1 all exhibit similar biologic activities. *J. Immunol.* 134: 895–903 (1985).

Yamamoto, J. K., Farrar, W. L., and Johnson, H. M. Interleukin 2 regulation of mitogen induction of immune interferon (IFN-γ) in spleen cells and thymocytes. *Cell. Immunol.* 66: 333–341 (1982).

Yokota, T., Arai, N., Lee, F., Rennick, D., Mosmann, T., and Arai, K. I. Use of a cDNA expression vector for isolation of mouse interleukin 2 cDNA clones: Expression of T-cell growth-factor activity after transfection of monkey cells. *Proc. Natl. Acad. Sci.* (USA) 82: 68–72 (1985).

Zilberstein, A., Ruggieri, R., Korn, J. H., and Revel, M. Structure and expression of cDNA and genes for human interferon-β2, a distinct species inducible by growth-stimulatory cytokines. *EMBO J.* 5: 2529–2537 (1986).

Zubler, R. H., Lowenthal, J. W., Erard, F., Hashimoto, N., Devos, R., and MacDonald, H. R. Activated B cells express receptors for, and proliferate in resonse to, pure interleukin 2. *J. Exp. Med.* 160: 1170–1183 (1984).

14 THE EFFECTS OF INTERFERONS ON TUMOR CELLS

One of the great challenges of IFN research is to understand the multiple ways by which IFNs influence the behavior of tumor cells. For this, we have to gain an insight into the complexities of the regulatory interactions of IFNs with many other cytokines, growth factors, and oncogenes, and with the various effector cells of the immune system. The effects of IFNs on tumor cells can indeed directly result from an action of IFN on the cell via the IFN receptors, or they can occur indirectly, via the activation of several effector mechanisms. An investigation of these problems may help to define whether there is a natural role for IFNs in tumor surveillance and will lead to a more rational and efficient use of IFNs as antitumor agents in the clinic.

Other chapters in this book describe properties of IFNs that have bearing upon their antitumor activity in vivo. These include the stimulation of MHC class I and class II antigen expression (Chapter 9), macrophage activation (Chapter 10), effects on T and B lymphocytes and on NK cells (Chapter 11), and interactions with other cytokines, for example, IL-2 and TNF (Chapter 13). The present chapter, therefore, is mainly devoted to the direct effects of IFNs on tumor cells and will only briefly mention the antitumor effects resulting from the activation of effector mechanisms.

I. DIRECT EFFECTS OF IFNs ON TUMOR CELLS

A. Antiproliferative Effect

Just as they do in normal cells, IFNs slow down the growth and proliferation of tumor cells by causing a prolongation of the cell cycle. There are, however, striking differences in sensitivity to this direct antiproliferative effect, and some tumor cells are extremely sensitive whereas others can be totally resistant (Tsuruo et al., 1982). In this respect, tumor cells are not different from non-malignant cells, which also display disparities in sensitivity that can reach several orders of magnitude. This aspect of IFN activity is treated in detail in the chapter on the antiproliferative effects of IFNs (Chapter 7). One important point, relevant to the antitumor activity, is the fact that IFN-resistant mutants can be isolated from IFN-sensitive cell lines (Gresser et al. 1974). Such resistant mutants sometimes retain their sensitivity to the antiviral effects, which means that resistance is not necessarily due to the absence of receptors for IFN (Lin et al., 1982).

For most in vitro studies on the antiproliferative action of IFNs, continuous lines of human or murine tumor cells are used. One of the most widely employed, the EBV-transformed human lymphoblastoid Daudi cell line provides a good example of cells that are extremely sensitive to the antiproliferative action of Hu IFN-α or Hu IFN-β (but not of Hu IFN-γ), whereas the Namalwa line, another EBV-transformed lymphoblastoid line, is virtually resistant. The sensitivity of many other tumor cell lines lies in between these two extremes. The study of established tumor cell lines is necessary for obtaining information concerning the mode of action and the induction of resistant mutants, but the study of fresh tumor specimens is more representative of the situation in the clinic. The theoretical possibility that many established tumor cell lines have become resistant to IFNs because of selection during in vitro or in vivo maintenance appears unlikely since freshly isolated tumor cells can also be resistant. When one examines the antiproliferative activity of different natural or cloned Hu IFNs on freshly isolated tumor cells, the range of sensitivity forms a continuum from complete inhibition to total resistance, with no obvious correlation between the tumor cell type and the degree of inhibition by IFNs. Also, within the same type of tumor, for example, human ovarian carcinoma, sensitivity to Hu leukocyte

IFN is not related to the histology or grade of the tumor, or to the stage of the disease. It is clear that one cannot predict on the basis of cell type alone what the response to IFN will be. Moreover, and somewhat paradoxically, the replication of cells derived from some fresh tumors is stimulated rather than inhibited by the presence of IFNs. Growth stimulation of these cells preferentially occurs at low dosages of IFN, corresponding to a few hundred units per milliliter, and is observed both with Hu IFN-β and Hu leukocyte IFN (Bradley and Ruscetti, 1981; Epstein and Marcus, 1981; Schlag et al., 1982; Ludwig et al., 1983).

The direct antiproliferative action of IFNs on tumor cells conceivably contributes to the antitumor effects in vivo, but in view of the concentrations of IFN that are required locally, in the tumor, it is evident that such a direct effect can only be responsible for a minor part of the antitumor activity, unless IFN is inoculated directly into the tumor site. Evidence for a direct cytostatic effect of IFN on the growth and development of various tumor cells in vivo has been provided by the inhibition of human tumor growth in nude mice treated with Hu lymphoblastoid IFN, Hu leukocyte IFN, or rec Hu IFN-α2. In some cases this treatment can lead to complete regression of the tumors, presumably due to the direct action of IFN on the cells since one would not expect Hu IFN to act via murine effector cells (Balkwill et al., 1983a,b; Masuda et al., 1983; Rivière and Hovanessian, 1983; Bauer et al., 1986). In the clinic direct inoculation of IFN into tumors can result in a decrease of the tumor mass (Sawada et al., 1981; Nakagawa et al., 1983).

B. Reversion of the Transformed Phenotype

The intriguing possibility that IFN treatment can progressively reverse the phenotype of some malignant cells, and redirect these cells toward "normality," is raised by several observations on the behavior of malignant cells cultured in the continuous presence of IFNs. When X-ray transformed, murine fibroblasts are cultivated and passaged in the continuous presence of Mu IFN-α/β, a stepwise, progressive reversion of the transformed phenotype to the nontransformed phenotype occurs. The cells are no longer tumorigenic in nude mice, and their morphology changes from fibroblastic to epitheloid. However, when the cells are no longer maintained under IFN, they revert back to the transformed phenotype and become tumorigenic again (Brouty-Boyé and Gresser, 1981). Prolonged maintenance of murine sarcoma virus-transformed mouse fibroblasts under Mu IFN-α/β has comparable effects: Such cells show a decreased saturation density, form fewer foci when plated on monolayers of normal cells, and are more anchorage-dependent as shown by decreased colony formation when grown in soft agar (Hicks et al., 1981). After being continuously maintained for many generations in the presence of Mu IFN-α/β, BALB/c embryonic fibroblasts transformed by Moloney sarcoma virus recover a normal phenotype, lose their capacity to form colonies in agar, and are unable to produce tumors in nude mice (Chany-Fournier, 1983; Sergiescu et al., 1986).

Reversion of transformed properties in vitro, however, is not always accompanied by reduced tumorigenicity and can, in fact, sometimes result in enhanced tumorigenicity. Mu IFN-α/β reverses the phenotype of murine Kirsten virus transformed cells, but potentiates the tumorigenic phenotype (Brouty-Boyé, 1986). The nature of the transforming agent seems to influence the potential reversion effect of IFNs, which is not observed in every system examined. Although the morphology of methylcholanthrene-transformed murine fibroblasts is reversed to the normal phenotype by prolonged IFN treatment, the tumorigenic potential of these cells in syngeneic mice is not diminished. The morphology and tumorigenic potential of SV40-transformed cells remains unchanged in spite of the prolonged presence of high doses of Mu IFN-α/β, and the tumorigenic potential is even stimulated at low IFN dosage! Since usually prolonged in vitro exposure to IFN is required, encompassing many cell generations, it is not easy to distinguish between the selection of certain cell types present in the population and a true phenomenon of phenotypic reversal, not based on genetic selection.

Whether or not IFN treatment of tumor-bearing animals or of patients induces reversion of the malignant phenotype of some tumor cells, and if so, whether such reversion is significant enough to be of therapeutic benefit, remains an open question. The phenomenon is undoubtedly of great theoretical interest, but in view of the duration of exposure to IFN and the IFN levels required to obtain it, one would not expect it to play a significant therapeutic role. Tumor cell lines derived from human osteosarcomas and from bladder, colon, and ovarian carcinomas have been serially passaged in the presence of Hu leukocyte or lymphoblastoid IFN. In no instance, though, was there a full reversal of tumor cell phenotype, resulting in suppression of tumorigenicity in nude mice. At best, a reduction of tumorigenicity has been obtained, and moreover, a line of bladder tumor cells appeared even more tumorigenic in the nude mice than the untreated cells (Brouty-Boyé, 1986). Manifestly, the change of phenotype is not consistently unidirectional toward decreased malignancy.

C. Effects of IFNs on Oncogene Expression

One possible mechanism of phenotype reversion by IFNs is their effect on the expression of viral or cellular oncogenes (Table 14.1). Oncogenes are viral or cellular genes that cause normal cells to acquire characteristics of malignantly transformed cells by altering the regulation of cellular functions such as replication and differentiation. The products of oncogenes are varied. They can be growth factors, either normal or altered. They can also be receptors for growth factors, signal transducers, such as tyrosine kinases, and nuclear effectors, corresponding to mitogenic signals (Bishop, 1985; Klein and Klein, 1985).

Normal cells harbor genes, called protooncogenes or c-*onc*, which share a high degree of sequence homology with the oncogenes of retroviruses or v-*onc*. The viral oncogenes thus appear to have arisen by the transduction of

TABLE 14.1 Effects of IFNs on Oncogene Expression

Oncogene	Cell	IFN	Effect on mRNA Levels	Reference
c-myc	Murine 3T3	MuIFN-α/β	Down	Einat et al., 1985a
	Murine 3T3	MuIFN-α/β	Slightly up	Tominaga and Lengyel, 1985
	Human leukemic cell lines:			
	Daudi	Hu leukocyte IFN	Down	Einat et al., 1985b
	HL-60	Hu leukocyte IFN	Up	Einat et al., 1985b
	U-937	Hu leukocyte IFN	Up	Einat et al., 1985b
	Daudi, sensit. to antiprol. effect	Hu lymphoblastoid IFN	Down	Dron et al., 1986
	Daudi, sensit. to antiprol. effect	Hu lymphoblastoid IFN	Down	McMahon et al., 1986
	Daudi resistant to antiprol. effect	Hu lymphoblastoid IFN	None	Dron et al., 1986
	Daudi resistant to antiprol. effect	Hu lymphoblastoid IFN	None	McMahon et al., 1986
	Daudi, back revertants from resistant to sensitive	Hu lymphoblastoid IFN	None	Dron et al., 1986
	Daudi	Hu IFN-β	Down	Jonak and Knight, 1984 Knight et al., 1985
		Rec Hu-IFNα2	Down	Dani et al., 1985
		Rec Hu-IFNβ	Down	
c-fos	Murine 3T3 stimulated by PDGF	Mu IFN-α/β	Down	Einat et al., 1985a
c-Ha-*ras*	Murine 3T3	Mu IFN-α/β	Down	Samid and Friedman, 1986
	Human bladder carcinoma	Hu IFN-β	Down (but no effect on c-Ki-*ras*!)	Soslau et al., 1984

the cellular protooncogenes from the genomes of vertebrate cells into retroviral genomes (Bishop, 1983). During this process the c-*onc* coding sequences have become altered, creating an abnormal gene, regulated by the viral control elements, independently of the normal cellular regulatory circuits. Thus, the v-*onc* gene product can be functionally different from the normal gene product and is no longer subject to the stringent regulation that is operative in normal cells. The cellular protooncogenes, c-*onc,* can also become activated and altered in several ways. Activation can take place by retroviral enhancer-promoter or enhancer insertion into the c-*onc* gene. Only a small part of the retroviral genome, containing the long, terminal repeat region (LTR) is necessary to fulfill this function. Activation can also result from DNA rearrangements, chromosomal translocations, or point mutations. Abnormal expression and/or abnormal function are the two key conditions leading to malignant transformation.

Over 40 oncogenes have been identified in humans and mice, but we will only discuss those that have been shown to be influenced by IFNs. We shall first discuss the *myc* and the *fos* oncogenes, because they belong to the growth factor-activated competence genes, and are directly relevant to effects of IFNs on the cell cycle.

1. The myc *Oncogene*

The c-*myc* protooncogene is the cellular homologue of the v-*myc* oncogene of avian myelocytomatosis virus, an acute avian leukemia virus. The c-*myc* gene product is a nuclear protein that can bind to dsDNA and is present in avian, murine, and human cells. Its actual function is unknown, but studies using isolated nuclei treated with antibodies to the c-*myc* protein suggest that it is involved in DNA synthesis (Studzinski et al., 1986). The gene has been highly conserved throughout evolution, and sequences closely related to avian c-*myc* are found from Drosophila to vertebrates. The gene is on chromosome 8 in humans and on chromosome 15 in the mouse (Sheiness and Bishop, 1979; Crews et al., 1982; Donner et al., 1982; Neel et al., 1982; Colby et al., 1983). The expression of c-*myc* is correlated with proliferative activity of the cell, since mitogenic stimulation of lymphocytes or of fibroblasts by growth factors leads to a sharp, up to 40-fold, increase in c-*myc* mRNA levels. The expression of c-*myc* drops as cells enter the resting stage, but a basic level of c-*myc* protein synthesis persists during all phases of the cell cycle in proliferating cells (Kelly et al., 1983; Hann et al., 1985; Reed et al., 1986).

Furthermore, the expression of c-*myc* plays a role in malignant transformation, because, while c-*myc* by itself is not transforming in rodent fibroblasts in vitro, it cooperates with the *ras* oncogene to transform primary rat embryo fibroblasts into focus-forming cells that give rise to tumors in nude mice (Land et al., 1983). Similarly, cooperation of the v-*myc* oncogene with the v-*mil* oncogene of the MH2 avian carcinoma virus elicits malignant transformation of chicken neuroretinal cells and of macrophages. This requirement for a second oncogene in chick cells, however, is not absolute but depends on culture conditions, and v-*myc* by itself can be sufficient to transform chick neuroretina cells (Béchade et al., 1985; Graf et al., 1986; Casalbore et al., 1987). Structural alterations, amplifications, and translocations of the c-*myc* gene are observed in a wide range of human tumors, such as myelocytomatosis, myeloid leukemia, pancreatic adenocarcinoma, colon carcinoma, breast carcinoma, and Burkitt's lymphoma as well as in murine plasmacytomas. In most Burkitt lymphomas and in murine plasmacytomas, the c-*myc* gene has been translocated into the immunoglobulin heavy chain locus, resulting in a t(12:15) translocation in the mouse and a t(8:14) translocation in humans. These translocations do not alter the *myc* coding region, but *myc* expression becomes constitutive, presumably because it is now regulated by the heavy chain locus (Mushinski et al., 1983; Saito et al., 1983; Buick and Pollak, 1984; Marx, 1984; Bishop, 1985; Eick et al., 1985; Erisman et al., 1985; Escot et al., 1986; Yamada et al., 1986).

Moreover, transgenic mice that carry the normal murine c-*myc* gene, but with the promoter regions replaced by the mouse mammary tumor virus promoter, show a strong tendency to develop malignant mammary tumors. Transgenic mice carrying the c-*myc* gene with its own promoter do not develop such cancers. Transgenic mice bearing the c-*myc* gene coupled to the lymphoid cell-specific IgH enhancer first have benign polyclonal overproliferation of B cells, almost invariably followed by B-cell malignancies. This is consistent with the view that the aberrant *myc* expression contributes to malignant transformation, maybe by increasing the proportion of cells that are less mature and have greater proliferative capacity (Stewart et al., 1984; Adams et al., 1985; Langdon et al., 1986). Thus, abnormal expression of c-*myc* is clearly involved in malignant transformation, which explains the interest in investigating the effect of IFNs on the expression of this gene.

Stimulation of quiescent mouse fibroblasts from G1 into the S phase by the platelet-derived growth factor (PDGF) activates a number of genes, known as competence genes because the stimulated cells become "competent" to engage in DNA synthesis. The expression of several genes belonging to the competence family, including c-*myc*, is inhibited when the cells are treated with Mu IFN-α/β at the time of growth factor stimulation. In addition to c-*myc*, the genes for β-actin and ornithine decarboxylase and another proto-oncogene, c-*fos* (see subsequent discussion), are also inhibited by IFN (Einat et al., 1985a). It is, therefore, not possible to decide whether the inhibition of the c-*myc* expression results from a direct IFN-induced effect on the c-*myc* gene, or occurs via a cascade effect resulting from the inhibition of the expression of other competence genes that are activated before c-*myc*. Yet, whatever the molecular mechanism and the prime cause of this inhibition, it does offer the beginning of an insight into one possible pathway of antiproliferative action of IFNs on normal cells, since the c-*myc* gene product is a stimulator of cell growth and replication, as shown in BALB/c 3T3 cells that are transfected with mammary tumor virus promoter-c-*myc* recombinant plasmids (Armelin et al., 1984).

Moreover, the effect of IFN depends on the timing of its addition. When quiescent BALB/c 3T3 cells are pretreated with Mu IFN-β for 48 hours before being stimulated by PDGF, rather than being exposed to the growth factor and IFN at the same time, the steady-state level of c-*myc* mRNA is not reduced, but somewhat enhanced, although the proliferative response of such cells to the growth factor is, nevertheless, inhibited (Tominaga and Lengyel, 1985).

IFNs can also down-regulate the c-*myc* expression in some malignant cells. In Burkitt's lymphoma Daudi cells the c-*myc* gene has been activated by a translocation from chromosome 8 to 14. In Daudi cells whose growth has been inhibited by Hu IFN-β, a more than 75% reduction in c-*myc* mRNA occurs (Fig. 14.1). This reduction in mRNA levels occurs as early as 3 hours after the addition of Hu IFN-β and precedes the IFN-induced inhibition of cell growth, suggesting a correlation between inhibition of c-*myc* and cessation of cell proliferation. As a result of the reduction of the levels of c-*myc* mRNA, the amount of c-*myc* protein is significantly reduced in Hu IFN-β-treated Daudi cells (Jonak et al., 1987). Whether the decreased levels of c-*myc* mRNA are the result of decreased transcription rates or of decreased stability is presently a matter of debate; the two mechanisms are not mutually exclusive (Jonak and Knight, 1984; Dani et al., 1985; Einat et al., 1985b; Jonak et al., 1985; Knight et al., 1985; McMahon et al., 1986).

It is clear that in IFN-treated cells the expression of c-*myc* undergoes variations, as measured by hybridization signals with cDNA probes, but the interpretation of these variations is not always simple. It is not easy to distinguish between the possibility that the decreased expression of c-*myc* is

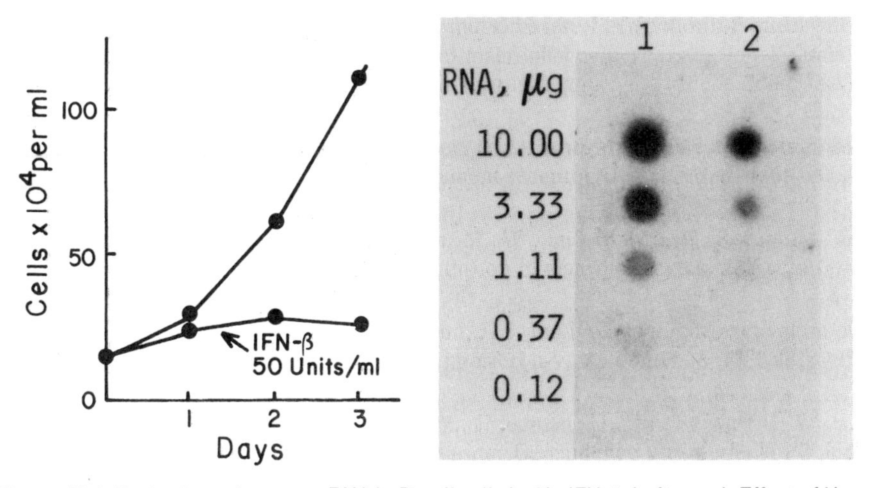

Figure 14.1 Reduction of c-*myc* mRNA in Daudi cells by Hu IFN-β. Left panel: Effect of Hu IFN-β on proliferation of the cells. Right panel: Dot blot analysis of c-*myc* expression in Daudi cells after 24 hours exposure to 50 IFN units/ml of IFN-β; indicated quantities of poly(A)$^+$ RNA from control (row 1) and IFN-β-treated (row 2) were hybridized with ^{32}P-labeled c-myc probe. (From Jonak and Knight, 1984, with permission.)

instrumental in arresting cell growth or that it is just a result of the latter, and one can only look for suggestive correlations. Moreover, such correlations are not consistently found, and in some Daudi cell mutants that reverted from an IFN-resistant phenotype back to sensitivity, cell proliferation can be inhibited by Hu lymphoblastoid IFN without a concomitant decrease in levels of c-*myc* mRNA measured 24 hours after the beginning of the IFN treatment (Dron et al., 1986). Of course, measuring the expression of a gene with a cDNA probe in mRNA blots does not readily detect a mutant gene that might still be expressed but would be less active than the parental gene. Moreover, both the coding and noncoding strands of the c-*myc* gene are transcribed, and the transcription of each strand is regulated independently (Kindy et al., 1987). As a consequence of this, if double-stranded cDNA is used as a probe in Northern blots, the presence of the antisense mRNA will also be revealed. This may therefore not provide a true reflection of the transcription of the c-*myc* gene into functional mRNA. Moreover, regulation of c-*myc* could conceivably also take place at the posttranscriptional level. There is indeed evidence for transcriptional and posttranscriptional control of c-*myc* expression, mainly by the stabilization of mRNA. It seems quite possible that at least part of the increase in c-*myc* mRNA levels after stimulation by growth factors is due to the stabilization of the normally highly unstable c-*myc* mRNA (Blanchard et al., 1985; Endo and Nadal-Girard, 1986). Therefore, in addition to measuring c-*myc* mRNA levels, it is important to determine at the same time the actual levels of c-*myc* protein in IFN-treated cells.

2. The fos Oncogene

The c-*fos* protooncogene is present in the genome of many vertebrates; in the mouse the gene is located on chromosome 12 and in humans on chromosome 14 (Barker et al., 1984, D'Eustachio, 1984). The corresponding v-*fos* oncogene is present in the FBJ murine osteosarcoma virus, first isolated by Finkel, Biskis, and Jinkins (Finkel et al., 1966; Curran et al., 1982). The normal c-*fos* gene does not transform cells, but the addition of a transcriptional enhancer and disruption of the 3′ end activate its transforming potential. Like c-*myc*, c-*fos* belongs to the competence genes, activated in quiescent mouse fibroblasts by growth factors. Induction of the c-*fos* gene precedes activation of the c-*myc* gene (Cochran et al., 1984; Greenberg and Ziff, 1984; Müller et al., 1984). The c-*fos* gene product has a molecular weight of approximately 55 kDa. The protein undergoes extensive post-translational modifications and is mainly present in the nucleus, but its function is not known. It is constitutively expressed at high concentrations in bone-marrow cells and differentiated macrophages, which suggests a functional role for c-*fos* in the differentiation of cells derived from the hematopoietic system (Gonda and Metcalf, 1984; Müller et al., 1984). Interestingly, in quiescent terminally differentiated macrophages, the expression of c-*fos* is induced by the macrophage-specific growth factor CSF-1, which is also an IFN-α/β inducer in these cells (Müller et al., 1985).

Since the growth factors PDGF and CSF-1 activate the c-*fos* gene prior to the IFN-α/β genes, the down-regulatory effect of IFN-α/β on the c-*fos* expression suggests a negative feedback loop. The increase in c-*fos* mRNA occurring after PDGF stimulation of quiescent 3T3 mouse fibroblasts is significantly reduced by concomitant exposure of the cells to Mu IFN-α/β. This is a very rapid IFN effect, since the increase in c-*fos* mRNA occurring within 20 minutes after the addition of PDGF can still be inhibited if the IFN is added together with the growth factor (Einat et al., 1985a).

3. The ras *Oncogenes*

The human and rodent cellular protooncogenes c-*ras* are the homologues of the oncogenes of the Harvey (v-Ha-*ras*) and Kirsten (v-Ki-*ras*) sarcoma viruses. There are four members of the human *ras* gene family, c-Ha-*ras,* c-Ki-*ras,* N-*ras,* and R-*ras*. N-*ras* was first isolated from a human neuroblastoma line and R-*ras* has been isolated from a human genomic library; these two protooncogenes have not been found associated with a retrovirus. C-Ha-*ras* is located on human chromosome 11, c-Ki-*ras* on chromosome 12, and R-*ras* on chromosome 19 (McBride et al., 1982; Sakaguchi et al., 1983; Lowe et al., 1987). In the mouse, the members of the c-*ras* gene family are also on different chromosomes: N-*ras* is on chromosome 3, Ki-*ras* on 6, and R-*ras* and Ha-*ras* are syntenic on 7 (Pravtcheva et al., 1983; George et al., 1984; Ryan et al., 1984). All members of the *ras* gene family encode antigenically related proteins with a molecular weight of about 21 kDa, which have been named p21s. These polypeptides diverge from one another principally in their carboxy terminal regions. They are membrane bound, largely to the plasma membrane, and they bind guanine nucleotides and display GTPase activity. Proteins of the *ras* gene family are highly conserved evolutionary, since they are present in Dictyostelium, yeast, and Drosophila (Shilo and Weinberg, 1981; McGrath et al., 1984; Shilo and Hoffman, 1984). The yeast *ras* protein functions as a regulator of adenyl cyclase. Its inactivation leads to defects in spore germination, that can be cured by the homologous mammalian *ras* gene (Weinberg, 1985). In mammalian cells, the function of the p21 c-*ras* proteins is not known, but there is enough evidence to indicate that their activity is required for cell division, which can be triggered by *ras* proteins and blocked by antibodies to these proteins. Ha-*ras* p21 proteins of human or murine origin, when microinjected into quiescent murine 3T3 cells, stimulate cellular growth and division. On the contrary, after microinjection of monoclonal antibodies to cellular *ras* protein, murine 3T3 cells are blocked from entering the S phase of the cell cycle (Stacey and Kung, 1984; Mulcahy et al., 1985).

Ras genes are activated to oncogenes by either point mutations or linkage to a long terminal repeat and occasionally by amplification of their copy numbers. Spontaneously arising point mutations have been observed in naturally occurring human neoplasias. The mutations that affect amino acid residues 12, 13, or 61 of the p21 protein confer to the latter the ability to transform cells at very low levels. Only high levels of normal p21, resulting from overexpression, lead to transformation (Chang et al., 1982; Cooper, 1982; Land et al., 1983; Seeburg et al., 1984; Weinberg, 1985). Activation of Ki-*ras* oncogenes has been detected in various human neoplasias and often represents single point mutations within codon 12 of the human Ki-*ras* locus (Cooper, 1982; Pulciani et al., 1982; Capon et al., 1983; Santos et al., 1984; Rodenhuis et al., 1987). Activated Ha-*ras* genes are present in malignant liver tumors in mice, and furthermore, are also found in benign liver tumors, indicating that activation is not always synonymous with malignancy (Reynolds et al., 1986). Like the expression of c-*myc,* the expression of *ras* oncogenes, either endogenous or transfected, can be influenced by IFNs.

When murine 3T3 cells, transformed with the human Ha-*ras*1 gene, are cultured in the continuous presence of Mu-IFN-α/β, revertant colonies arise after 10 to 20 cell generations. Such revertants no longer grow in soft agar, and contrary to the transformed line from which they are derived, they do not give rise to tumors in nude mice. In these revertant cells the transfected c-Ha-*ras* DNA is still present, unchanged in genomic distribution and quantity as compared to the parental tumor cells. There is, however, a significant reduction in c-Ha-*ras*-specific mRNA and of the c-Ha-*ras* p21 protein. This is not the result of a general depression of mRNA synthesis by IFN, since the levels of actin and collagen mRNA are not affected in the revertants (Samid et al., 1984). Nuclear run-off transcription assays show that transcription of the c-Ha-*ras* genes is inhibited in the IFN-treated revertants, and removal of IFN from the cultures is followed by a gradual elevation of *ras* transcription. After cessation of IFN-treatment, a small fraction of the cells regains the transformed phenotype, but the majority keeps the revertant phenotype during many cell generations, in spite of renewed high transcription of the c-Ha-*ras* gene and the resulting high levels of p21 *ras* protein. Manifestly, in these persistent revertant cells, overexpression of the c-Ha-*ras* gene does not suffice to provoke a return to the transformed state. This is borne out by the fact that the revertant cells have become resistant to malignant transformation by transfection with a c-Ha-*ras*1 gene encoding for a mutated p21. This important control shows that the lack of malignant behavior, in spite of the high expression of c-Ha-*ras* in the revertant cells, is probably not the result of a change in the Ha-*ras* gene. What are the IFN-induced changes that make high expression of the oncogenic c-Ha-*ras* no longer oncogenic? Since these changes persist for over 40 cell generations after withdrawal of IFN-treatment, they are manifestly the result of a selection of resistant cells, unless one accepts the provocative hypothesis that Mu IFN-α/β itself has caused a genetically or epigenetically stable change in the transformed cells (Samid et al., 1986; Samid and Friedman, 1986).

In addition to influencing the expression of transfected c-*ras* oncogenes, IFNs can also affect the expression of endogenous c-*ras* oncogenes in their normal chromosomal location. Hu IFN-β decreases the expression of c-Ha-*ras* in human bladder carcinoma cells, but not the expression of the c-Ki-*ras* gene. The reduced c-Ha-*ras* transcription occurs prior to the antiproliferative response of the carcinoma cells. Whether this is just a temporal correlation or implies a cause and effect relationship is not known (Soslau et al., 1984).

4. The mos Oncogene

The c-*mos* protooncogene is the cellular homologue of the v-*mos* oncogene of Moloney sarcoma virus. In mice, c-*mos* is on chromosome 4 and in humans on chromosome 8 (Canaani et al., 1979; Swan et al., 1982; Caubet et al., 1985). The *mos* gene codes for a cytoplasmic protein, displaying serine/threonine kinase activity. Its actual function is not known, but it is intriguing that there is a low degree of homology between the murine proepidermal growth factor and c-*mos* protein (Baldwin, 1985). In most mouse tissues the c-*mos* gene is either not expressed or is expressed at very low levels. Significant expression, however, occurs in mouse embryonic tissues as well as in the testes and ovaries, suggesting a transcription mediated by hormonal control (Propst and Vande Woude, 1985). The normal c-*mos* gene undergoes rearrangements in myeloma cells, and as a result, becomes transcriptionally active and acquires transforming potential for murine fibroblasts (Rechavi et al., 1982; Gattoni-Celli et al., 1983).

The long-term treatment of Moloney murine sarcoma virus-transformed murine fibroblasts with Mu IFN-α/β results in the selection of a cell population that has reverted to a nonmalignant state and that is stable even after many cell generations in the absence of IFN. In spite of this, v-*mos* gene expression is not inhibited in these cells, comparable to what is observed in the revertant c-*myc*- and Ha-*ras*-transformed cells, which also continue to express the transforming genes after IFN treatment has been interrupted (Sergiescu et al., 1986).

Reversion of phenotype from malignancy to nonmalignancy by IFN without the extinction of oncogene expression, thus, can occur with many different oncogenes. As discussed earlier, however, only measurements of the actual gene products, rather than mRNA levels, will answer the question whether in cells, derived from IFN-treated populations, oncogenes can be overexpressed without causing reversal to the malignant state.

5. General Comment

Evidence that the products of oncogenes are involved in malignant transformation is overwhelming. Moreover, in some human tumors the degree of protooncogene amplification appears directly related to the degree of malignancy (Slamon et al., 1987). The effects of IFNs on oncogene expression offer, therefore, an interesting avenue for exploring the antitumor actions of IFNs. This may at the same time contribute to a better understanding of oncogene function. For example, how can one explain the continuous high expression of some oncogenes in persistent revertant cells?

D. Elimination of Extrachromosomal Viral Genomes in Papillomavirus Transformed Cells

Papillomaviruses are responsible for a wide variety of usually benign tumors in humans, including juvenile laryngeal papilloma, condyloma acuminata, flat warts of the uterine cervix, and common warts. These viruses are also frequently associated with cervical carcinomas, and one suspects that they may be causally involved (Lehn et al., 1985; Beaudenon et al., 1986). Papillomatous lesions respond rather well to IFN treatment; Hu leukocyte IFN offers an efficient treatment of juvenile laryngeal papillomatosis (see Strander, 1986), and rec Hu IFN-α2, directly injected into genital warts, is an effective form of therapy for condyloma acuminata (Eron et al., 1986).

Bovine papillomavirus type 1 (BPV-1) causes fibropapillomas in cattle and can also transform cells of other species, including mice. BPV-1-transformed murine cells are anchorage-independent and form tumors in athymic or syngeneic mice. Such transformed cells contain multiple copies of nonintegrated supercoiled circular BPV-1 DNA molecules in their nuclei. Treatment of these cells with Mu IFN-α/β causes a decrease in the average number of plasmid viral genomes and results in the elimination of BPV-1 genomes from some transformed cells, which become revertants with the characteristics of normal cells (Turek et al., 1982).

This clear-cut example of cells cured from a transforming infection offers a possible clue to the efficacy of IFN treatment of papillomatous lesions.

E. Effects on Differentiation

1. Malignant Cells

One of the characteristics of malignant cells is deviation from the normal differentiation pattern toward a more undifferentiated state. Whatever redirects such cells toward differentiation is likely to decrease their malignant potential, since fully differentiated cells are more subject to the mechanisms that govern cell growth and replication in harmony with the other cells in the body. IFNs have the potential to cooperate with other agents to stimulate the differentiation of some malignant cells. The most extensively documented example of this capacity concerns the effects of Mu IFN-α/β on Friend erythroleukemic cells. In this particular instance, IFN treatment does not direclty result in differentiation, but renders the leukemic cells more responsive to a differentiation-inducing chemical. Friend leukemic cells are erythroid precursors that, as a result of Friend leukemia virus infection, have been blocked at the erythroblast stage. Treatment with certain chemicals, for example, dimethylsulfoxide (DMSO), can put these cells back on the track to differentiation and convert them into hemoglobin-synthesizing cells. When Friend erythroleukemia cells undergo DMSO treatment, they differentiate toward the normoblast stage, resulting in complete cessation of cell division. DMSO-induced differentiation is significantly augmented when the cells are treated with low doses of Mu IFN-α/β. This is a highly dose-dependent phenomenon, since high amounts of IFN have the opposite effect and block commitment to differentiation (Rossi, 1985). The definition of what constitutes a low- or a high-IFN amount is obviously relative and can only be determined in a particular experimental setting (see also Appendix).

Stimulation by IFNs of chemically induced differentiation is not limited to cells of erythroid lineage, and the effects of differentiation inducers like DMSO or transretinoic acid are also enhanced when murine or human myeloid leukemia cells are subjected to IFN-α or IFN-β treatment. As a result of the treatment, leukemia cell conversion to granulocytes or macrophages is facilitated (Tomida et al., 1980, 1982; Spooner et al., 1985).

IFNs by themselves, in the absence of any other differentiation-inducing agent, also have the potential to redirect tumor cells toward a more differentiated state. This is the case for the highly IFN sensitive Daudi cell line, a B-cell lymphoblastoid line that is transformed by EBV. Growth inhibition of Daudi cells by Hu lymphoblastoid or rec IFN-α2 is accompanied by plasmacytoid differentiation and refractoriness to growth factors and there are indications that the observed growth inhibition is a consequence of cell differentiation (Exley et al., 1987a,b). This may be a reflection of the property of IFN-α to act under certain conditions as B-cell differentiation factor (see Chapter 11).

The success of IFN-α treatment in hairy cell leukemia, an uncommon form of leukemia that is usually B cell derived, is possibly related to the B-cell differentiation-inducing potential of IFN-α or IFN-β. In many patients with this form of leukemia IFN-α treatment results in the disappearance of hairy cells from the peripheral blood, and sometimes also from the bone marrow, and restores the normal levels of platelets, granulocytes, monocytes, and hemoglobin. The disappearance of hairy cells has been ascribed to the capacity of Hu IFN-α or IFN-β to stimulate the lymphomyeloid stem cells of these patients toward the myelomonocytic lineage, which reduces the excessive formation of partially mature B cells with the phenotype of hairy cells. Hu IFN-γ, which has no therapeutic effect in hairy cell leukemia, does not exert this effect on lymphomyeloid stem cells (Quesada et al., 1984; Chebath et al., 1986; Ratain et al., 1987).

Cells from other leukemias of B-cell lineage can also be directed toward differentiation by IFN treatment. Hu lymphoblastoid IFN as well as Hu IFN-β and Hu IFN-γ can induce blast transformation and plasmacytoid differentiation in chronic B-lymphocytic leukemia cells, which is another indication that abrogation of maturation arrest contributes to the IFN-induced remissions obtained in some patients with B-cell malignancies (Fig. 14.2). It is important to stress that such effects are obtained with cells from some, but not from all, patients; what causes the difference in response is not known (Einhorn et al., 1985; Ostlünd et al., 1986).

Not only leukemic cells, but also solid tumors can be directed toward

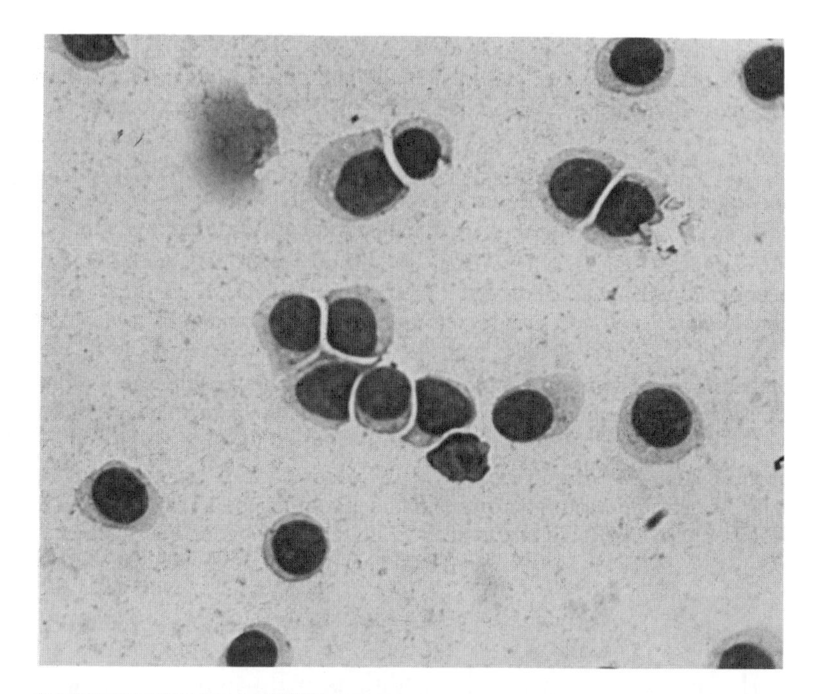

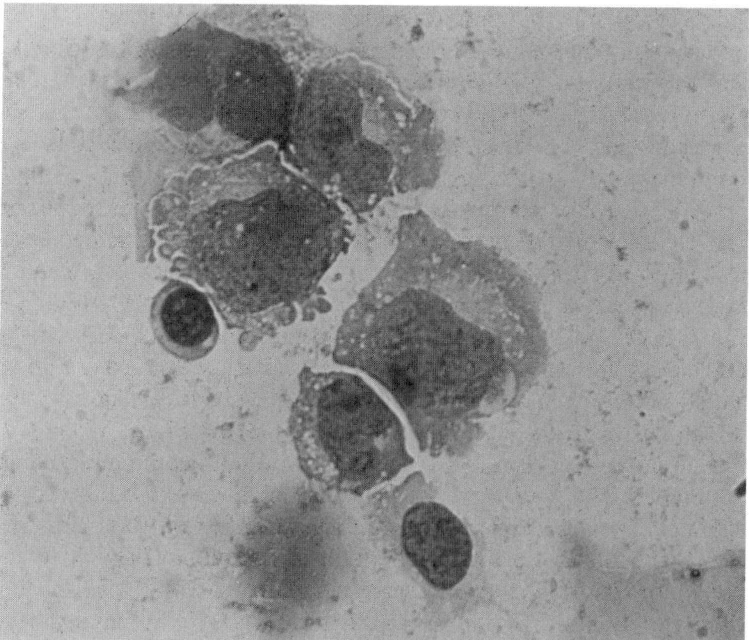

Figure 14.2 Differentiation of chronic B-lymphocytic leukemia cells into blasts and plasmacytoid cells after treatment with 5,000 units of Hu lymphoblastoid IFN. Upper panel: Without IFN, showing a homogenous picture of small regular lymphocytes. Lower panel: After IFN-treatment, showing varying degrees of morphological transformation into blasts and plasmacytoid cells. (From Ostlund et al., 1986, with permission.)

differentiation by IFN. Hu leukocyte IFN, administered to nude mice bearing xenografts of human osteosarcomas induces the growth arrest of the osteosarcoma cells. Inhibition of tumor growth is only maintained provided IFN is given daily, and in some cases the tumor tissue is replaced by normal bone and marrow tissues. The most likely explanation is that the blocking of cellular proliferation by IFN secondarily leads to tumor differentiation (Brosjo et al., 1987).

2. *Normal Cells*

Can IFNs also induce or accelerate differentiation in nonmalignant cells? Stimulation of the MHC class II antigen expression on cells of the immune system, as well as stimulation of T and B cell function, can be compared to differentiation-inducing effects, since they enable cells to perform their specialized activities. Yet, when such effects are only transient, as is often the case, they should not be considered as real induction of differentiation. However, the Hu IFN-α2-induced rearrangement in a cytotoxic T lymphocyte clone of the TCRA gene, which codes for the α chain of the human T-cell receptor, provides an indisputable example of IFN-induced T-cell differentiation (Chen et al., 1986).

IFN-γ as well has differentiation-inducing potential, and, for example, cells of human promyelocytic cell lines are induced to differentiate along the monocytic pathway by Hu IF-γ. Apparently, the differentiation induced by IFN-γ is not secondary to cell growth arrest, but rather results directly from induction of a proportion of the cells along a pathway of terminal differentiation, similar to that of normal monocytes (Dayton et al., 1985). Cells of the Epstein-Barr virus-transformed lymphoblastic CB-cell line, which secretes low levels of IgG, can be induced to differentiate into terminally differentiated IgG-producing cells by a combined treatment with IL-2 and IFN-γ, which act sequentially and synergistically (Nakagawa et al., 1986). The differentiation-inducing potential of IFN-γ is not restricted to cells of the immune system, since in cultures of human keratinocytes, rec Hu-IFN γ causes a shift from small, basal-type cells to large, mature-type keratinocytes. Mu IFN-α/β similarly stimulates keratin synthesis in cultured mouse epidermal cells (Sperry et al., 1985; Nickoloff et al., 1986).

One should not get the impression from the foregoing that IFNs always stimulate differentiation, as there are several examples of blocked differentiation pathways as a result of IFN treatment. Inhibition of lipid synthesis in insulin-dependent conversion of murine 3T3-Li cells to adipocytes can be obtained by treatment with Mu IFN-α/β. Also, as mentioned previously, high doses of Mu IFN-α/β inhibit rather than stimulate the induction of hemoglobin synthesis in mouse erythroleukemic cells (Keay and Grossberg, 1980; Rossi, 1985). Hu IFN-α or Hu-IFN-β can also inhibit maturation of monocytes to macrophages, as determined by changes in morphology, phagocytosis, and three different lysosomal activities (Lee and Epstein, 1980). A further striking example of inhibition of differentiation is provided by the blocking activity exerted by Mu IFN-γ on Mu IL-4-induced B-cell proliferation and Ig synthesis (Chapter 11).

The differentiation-stimulating activities of IFNs are theoretically relevant to their antitumor action, since inducing differentiation redirects cells toward normality. With a few exceptions, such as hairy cell leukemia, however, there are no direct indications that stimulation of differentiation plays an important role in the antitumor activity of IFNs in vivo.

Whether or not IFNs act as differentiation-inducing agents or cooperate with such agents during embryonic development is an interesting but unanswered question, which has received relatively little attention (Saxen, 1985; Clemens and McNurlan, 1985).

F. Enhanced Expression of Cell-Surface Antigens

IFNs increase the expression of MHC class I antigens on the plasma membrane of tumor cells, and studies in animal models leave little doubt that the enhanced expression of MHC class I antigens can augment tumor rejection. How important these effects are for clinical applications can, again, not be decided. The possibility that the enhanced expression of MHC class I and class II antigens on the plasma membrane of tumor cells facilitates elimination of these cells has been discussed at length in Chapter 9.

Stimulation of MHC antigen expression at the cell surface results from a direct effect of IFNs on the tumor cell, which, however, influences the potential survival of the tumor cell indirectly, via the different effector mechanisms implicated in tumor cell rejection.

II. INDIRECT ANTITUMOR EFFECTS VIA ACTIVATION OF EFFECTOR MECHANISMS

The growth and development of tumor cells that are resistant to the antiproliferative effects of Mu IFN-α/β in vitro can still be inhibited in IFN-treated mice. When such tumor cells are taken out of the IFN-treated animals for determination of their in vitro sensitivity, they still display resistance to the direct antiproliferative effect of IFN, which excludes the possibility that revertant mutants have been selected in vivo (Gresser, 1985). This is the best demonstration that at least some of the antitumor activity of IFNs takes place indirectly via host-dependent mechanisms. What are these mechanisms? We have only a very incomplete answer to this question, almost totally obtained from the study of mice receiving murine transplantable tumor cells (Gresser, 1982).

The elimination of tumor cells involves several effector systems, the most important of which consist of macrophages, cytotoxic T cells, and natural cytotoxic cells, including NK cells. The multiple effects of IFNs on these effector mechanisms, and the interactions of IFNs and the other cytokines involved, have been treated in detail in Chapters 9, 10, 11, and 13. Experimental models show that any one of these systems can be involved in transplantable tumor rejection by IFN-treated mice. Elimination of T cells by antimouse thymocyte globulin under certain conditions significantly impedes the therapeutic effect of IFN-α/β in BALB/c mice that have received transplantable tumor cells, and the Mu-IFN-α/β therapeutic effect is also much weaker in T-cell defective BALB/c nude mice than in immunologically competent animals (Kataoka et al., 1984). Yet, even in nude mice, Mu IFN-α/β still can decrease xenograft survival, probably via the stimulation of NK cells (Balkwill et al., 1983a). Yet, in some other murine host-tumor combinations, there is little evidence for the involvement of any cytotoxic cell in the elimination of the tumor cells as a result of treatment with Mu IFN-α/β. In the absence of any obvious host-cell infiltrate, tumor cells, insensitive to the direct antiproliferative effect of IFN, nevertheless, undergo necrosis after IFN treatment (Belardelli et al., 1982, 1983).

Obviously, there is no unifying hypothesis to account for the ensemble of observations dealing with the antitumor effects of IFNs in mice. It is evident that, similar to the antiviral activities of IFNs, the antitumor effects result from pleiotropic IFN activity, exerted either directly on the tumor cells or via the host. Some of these activities are manifestly still unknown.

III. RELEVANCE OF THE MOUSE MODEL STUDIES TO THE CLINICAL APPLICATIONS OF IFNs IN CANCER PATIENTS

Two important differences between mouse model studies and therapeutic trials complicate extrapolations from the bench or the animal room to the clinic.

The first difference lies in the IFN preparations employed. Virtually all work on tumors in mice has been performed with Mu IFN-α/β, which consists of a mixture of Mu IFN-β and various undefined Mu IFN-α species (see Appendix). Moreover, unless pure IFN is employed, the presence of other cytokines and substances cannot be excluded, and is, in fact, probable. The use of Hu leukocyte or lymphoblastoid IFN in humans poses the same problem, since the relative proportion of the different Hu IFN-α species is not known and is bound to vary between preparations. Other cytokines, such as TNF, can be present in more or less significant amounts, depending on the degree of purification (Wallach et al., 1986). The more recent clinical use of recombinant α and γ IFNs obviates this problem of interpretation, since only one molecularly well-characterized species is concerned. But how does one compare results obtained in murine models with an undetermined mixture of several murine IFN species with results obtained in patients with either an undefined mixture of Hu IFN-α species or a single IFN-α species? All IFN-α species are not quite identical, as they can differ in some biological activities, including the antiproliferative and immunomodulating properties (Chapter 2). This problem does not come up with IFN-γ, which represents only one molecular species in both mice and humans.

The second and far more important difference between mouse tumor model studies and the clinic lies in the nature of the tumors and the timing of the treatment. In mice, with very few exceptions, transplantable tumors are used, and the IFN treatment usually starts within hours after inoculation of the tumor cells. This ideal setting is absent from the clinic, in which treatment of primary tumors is often installed a long time after the tumor has arisen. When these conditions are mimicked in mice, for example, when the effects of Mu IFN-α/β on spontaneous mammary tumors are tested after the tumor has reached a palpable size, complete tumor regression is not obtained (De Clercq et al., 1982). IFN-α/β treatment of AKR mice, when started after the clinical diagnosis of lymphoma, results in a significant prolongation of survival time, but does not stem the disease (Gresser et al., 1976).

Yet, in spite of these problems, some extrapolation from the mouse to the clinic appears justified by the outcome of the many different clinical trials with Hu IFNs of various origins. These results can best be summarized as follows: Some tumors in some patients sometimes regress after IFN treatment. Conversely, this also means that many patients with many different tumors do not respond at all to IFN treatment (Borden, 1983; Quesada and Gutterman, 1983; Gresser, 1985; Strander, 1986). The great variability of results when the same type of tumor, for example, breast carcinoma, is treated in different patients is not understood. Genetically defined differences in the individual response to IFNs are probably involved, but it is unlikely that this is the main explanation. Tumor diversity is another possibility, as histologically similar tumors can still be very different in other aspects such as oncogene amplification and expression, degree of aneuploidy, and cell-surface antigens.

For some benign or malignant tumors, such as juvenile laryngeal papilloma, benign phase chronic myelogenous leukemia or hairy cell leukemia, IFN treatment has turned out to be of great value. When the conditions that are necessary for successful IFN therapy are understood, the clinical applications of IFNs and also of other cytokines will probably improve (Quesada et al., 1986; Smalley and Borden, 1986; Strander, 1986; Talpaz et al., 1986).

IV. IS THERE A NATURAL ROLE FOR IFNs IN TUMOR SURVEILLANCE?

Maybe we should start by asking whether the organism is at all endowed with a tumor surveillance system, and how efficient it is. Is there a finely tuned mechanism that recognizes and tries to eliminate transformed cells as soon as they arise, or can the organism only react when the malignant condition is more advanced? This is still a highly debated issue, with no final answer. There are certainly many effector systems that can be mobilized and stimulated against autochtonous tumors, but it is not clear at what time of tumor development these become activated. For instance, nude mice, with a defective T-cell system because the thymus is lacking, have no higher incidence of spontaneous or chemically induced tumors than normal mice (Outzen and Custer, 1975; Stutman, 1978). But nude mice still have natural cytoxic cells such as NK cells and also have macrophages, which are involved in natural resistance to tumor formation.

A natural role for IFNs in tumor surveillance is suggested by the observation that anti-IFN globulins can stimulate growth and proliferation of tumor cells in vivo. Baby hamster kidney cells or human HeLa cells form tumors when inoculated into athymic nude mice. When these cells, however, are persistently infected with RNA viruses like mumps or measles virus, they either fail to grow in nude mice or remain as stationary nodules at the site of inoculation. The resistance of the mice to these virus-infected tumor cells results from an active process of the host, probably involving mobilization of NK cells. The same tumors, however, grow easily and even establish metastases when the mice have received immunoglobulins directed against Mu IFN-α/β. Furthermore, such immunoglobulin treatment abolishes the resistance of the nude mice to primary human prostatic carcinomas, which normally are very difficult to establish as tumor lines in nude mice. This clearly shows that, in nude mice, Mu IFN-α or IFN-β, and maybe both, play a natural role in defense against xenogeneic tumors (Reid et al., 1981). Moreover, proliferation of some murine transplantable tumors, inoculated into immunocompetent mice, is also stimulated by the treatment of the recipient mice with anti-Mu IFN-α/β antibodies (Gresser et al., 1983). Such experiments demonstrate the existence of a role for endogenous IFN-α/β in the resistance to many different transplantable tumors and suggest, but do not prove, a possible role for IFNs in the resistance to naturally occurring tumors as well.

REFERENCES

Adams, J. M., Harris, A. W., Pinkert, C. A., Corcoran, L. M., Alexander, W. S., Cory, S., Palmiter, R. D., and Brinster, R. L. The c-*myc* oncogene driven by immunoglobulin enhancers induces lymphoid malignancy in transgenic mice. *Nature* 318: 533–538 (1985).

Armelin, I. A., Armelin, M. C. S., Kelly, K., Stewart, T., Leder, P., Cochran, B. H., and Stiles, C. D. Functional role for c-*myc* in mitogenic response to platelet-derived growth factor. *Nature* 310: 655–660 (1984).

Baldwin, G. S. Epidermal growth factor precursor is related to the translation product of the Moloney sarcoma virus oncogene *mos*. *Proc. Natl. Acad. Sci.* (USA) 82: 1921–1925 (1985).

Balkwill, F. R., Moodie, E. M., Freedman, V., Lane, E. B., and Fantes, K. H. An animal model system for investigating the antitumor effect of human interferon. *J. Interferon Res.* 3: 319–326 (1983a).

Balkwill, F. R., Moodie, E. M., Mowshowitz, S., and Fantes, K. H. An animal model system for investigating the anti-tumor effect of human IFNs on human tumours. In: *The Biology of the Interferon System 1983,* E. De Maeyer and H. Schellekens, Eds., Elsevier Science Publishers, Amsterdam, pp. 443–448 (1983b).

Barker, P. E., Rabin, M., Watson, M., Breg, W. R., Ruddle, F. H., and Verma, I. M. Human c-*fos* oncogene mapped within chromosomal region 14q21→q31. *Proc. Natl. Acad. Sci.* (USA) 81: 5826–5830 (1984).

Bauer, H. C. F., Nilson, O. S., Brosjo, O., Brostrom, L. A., Strander, H., and Tribukait, B. IFN-α induced growth inhibition of eleven osteosarcoma xenografts in nude mice. In: *The Biology of the Interferon System 1985,* W. E. Stewart II and H. Schellekens, Eds., Elsevier Science Publishers, Amsterdam, pp. 397–400 (1986).

Beaudenon, S., Kremsdorf, D., Croissant, O., Jablonska, S., Wain-Hobson, S., and Orth, G. A novel type of human papillomavirus associated with genital neoplasias. *Nature* 321: 246–249 (1986).

Bechade, C., Calothy, G., Pessac, B., Martin, P., Cool, J., Denhez, F., Saule, S., Ghysdael, J., and Stehelin, D. Induction of proliferation or transformation of neuroretina cells by the *mil* and *myc* viral oncogenes. *Nature* 316: 559–562 (1985).

Belardelli, F., Gresser, I., Maury, C., Duvillard, P., Prade, M. and Maunoury, M. T. Antitumor effects of interferon in mice injected with interferon-sensitive and interferon-resistant Friend leukemia. III. Inhibition of growth and necrosis of tumors implanted subcutaneously. *Int. J. Cancer* 31: 649–653 (1983).

Belardelli, F., Gresser, I., Maury, C., and Maunoury, M. T. Anti-tumor effects of interferon in mice injected with interferon-sensitive and interferon-resistant Friend leukemia cells. II. Role of host mechanisms. *Int. J. Cancer* 30: 821–825 (1982).

Bishop, J. M. Cellular oncogenes and retroviruses. *Ann. Rev. Biochem* 52: 301–354 (1983).

Bishop, J. M. Viral oncogenes. *Cell* 42: 23–38 (1985).

Blanchard, J. M., Piechaczyk, M., Dani, C., Chambard, J. C., Franchi, A., Pouyssegur, J., and Jeanteur, P. c-*myc* gene is transcribed at high rate in Go-arrested fibroblasts and is posttranscriptionally regulated in response to growth factors. *Nature* 317: 443–445 (1985).

Borden, E. C. Interferons and cancer: How the promise is being kept. In: *Interferon 5,* I. Gresser, Ed., Academic Press, London, pp. 43–83 (1983).

Bradley, E. C. and Ruscetti, F. W. Effect of fibroblast, lymphoid, and myeloid interferons on human tumor colony formation in vitro. *Cancer Res.* 41: 244–249 (1981).

Brosjo, O., Bauer, H. C. F., Brostrom, L. A., Nilsson, O. S., Reinholt, F. P., and Tribukait, B. Growth inhibition of human osteosarcomas in nude mice by human interferon-α: Significance of dose and tumor differentiation. *Cancer Res.* 47: 258–262 (1987).

Brouty-Boye, D. Interferon and the tumour cell phenotype. In: *Interferon 7,* I. Gresser, Ed., Academic Press, London, pp. 145–165 (1986).

Brouty-Boye, D. and Gresser, I. Reversibility of the transformed and neoplastic phenotype. I. Progressive reversion of the phenotype of X-ray-transformed C3H/ 10T1/2 cells under prolonged treatment with interferon. *Int. J. Cancer* 28: 165–173 (1981).

Buick, R. N. and Pollak, M. N. Perspectives on clonogenic tumor cells, stem cells, and oncogenes. *Cancer Res.* 44: 4909–4918 (1984).

Canaani, E., Robbins, K. C., and Aaronson, S. A. The transforming gene of Moloney murine sarcoma virus. *Nature* 282: 378–383 (1979).

Capon, D. J., Seeburg, P. H., McGrath, J. P., Hayflick, J. S., Edman, U., Levinson, A. D., and Goeddel, D. V. Activation of Ki-*ras*2 gene in human colon and lung carcinomas by two different point mutations. *Nature* 304: 507–513 (1983).

Casalbore, P., Agostini, E., Alema, S., Falcone, G., and Tato, F. The v-*myc* oncogene is sufficient to induce growth transformation of chick neuroretina cells. *Nature* 326: 188–190 (1987).

Caubet, J. F., Mathieu-Mahul, D., Bernheim, A., Larsen, C. J., and Berger, R. Human proto-oncogene c-*mos* maps to 8q11. *EMBO J.* 4: 2245–2248 (1985).

Chang., E. H., Furth, M. E., Scolnick, E. M., and Lowy, D. R. Tumorigenic transformation of mammalian cells induced by a normal human gene homologous to the oncogene of Harvey murine sarcoma virus. *Nature* 297: 479–485 (1982).

Chany-Fournier, F. Rôle de l'interféron dans la réversion phénotypique des cellules transformées: Perte du caractère de malignité. *Path. Biol.* 31: 199–213 (1983).

Chebath, J., Benech, P., Mory, Y., Mallucci, L., Michalevicz, R., and Revel, M. IFN and (2′-5′) oligo A synthetase in cell growth and in differentiation of hematopoietic cells. In: *Interferons as Cell Growth Inhibitors and Antitumor Factors,* UCLA Symposia on Molecular and Cellular Biology New Series, R. M. Friedman, T. Merigan, and T. Sreevalsan, Eds., Alan R. Liss, New York, Vol. 50, pp. 351–363 (1986).

Chen, L. K., Mathieu-Mahul, D., Bach, F. H., Dausset, J., Bensussan, A., and Sasportes, M. Recombinant interferon-α can induce rearrangement of T-cell antigen receptor α-chain genes and maturation to cytotoxicity in T-lymphocyte clones in vitro. *Proc. Natl. Acad. Sci.* (USA) 83: 4887–4889 (1986).

Clemens, M. J. and McNurlan, M. A. Regulation of cell proliferation and differentiation by interferons. *Biochem. J.* 226: 345–360 (1985).

Cochran, B. H., Zullo, J., Verma, I. M., and Stiles, C. D. Expression of the c-*fos* gene and of a *fos*-related gene is stimulated by platelet-derived growth factor. *Science* 226: 1080–1082 (1984).

Colby, W. W., Chen, E. Y., Smith, D. H., and Levinson, A. D. Identification and nucleotide sequence of a human locus homologous to the v-*myc* oncogene of avian myelocytomatosis virus MC29. *Nature* 301: 722–725 (1983).

Cooper, G. M. Cellular transforming genes. *Science* 218: 801–806 (1982).

Crews, S., Barth, R., Hood, L., Prehn, J., and Calame, K. Mouse c-*myc* oncogene is located on chromosome 15 and translocated to chromosome 12 in plasmacytomas. *Science* 218: 1319–1321 (1982).

Curran, T., Peters, G., Van Beveren, C., Teich, N. M., and Verma, I. M. FBJ murine osteoarcoma virus: Identification and molecular cloning of biologically active proviral DNA. *J. Virol.* 44: 674–682 (1982).

Dani, C., Mechti, N., Piechaczy, M., Lebleu, B., Jeanteur, P., and Blanchard, J. M.

Increased rate of degradation of *c-myc* in interferon-treated Daudi cells. *Proc. Natl. Acad. Sci.* (USA) 82: 4896–4899 (1985).

Dayton, E. T., Matsumoto-Kobayashi, M., Perussia, B., and Trinchieri, G. Role of immune interferon in the monocytic differentiation of human promyelocytic cell lines induced by leukocyte conditioned medium. *Blood* 66: 583–594 (1985).

De Clercq, E., Zhang, Z. X., Huygen, K., and Leyten, R. Inhibitory effect of interferon on the growth of spontaneous mammary tumors in mice. *J. Nat. Cancer Inst.* 69: 653–657 (1982).

D'Eustachio, P. A genetic map of mouse chromosome 12 composed of polymorphic DNA fragments. *J. Exp. Med.* 160: 827–838 (1984).

Donner, P., Greiser-Wilke, I., and Moelling, K. Nuclear localization and DNA binding of the transforming gene product of avian myelocytomatosis virus. *Nature* 296: 262–266 (1982).

Dron, M., Modjtahedi, N., Brison, O., and Tovey, M. G. Interferon modulation of c-*myc* expression in cloned Daudi cells: Relationship to the phenotype of interferon resistance. *Mol. Cell. Biol.* 6: 1374–1378 (1986).

Eick, D., Piechaczyk, M., Henglein, B., Blanchard, J. M., Traub, B., Kofler, E., Wiest, S., Lenoir, G. M., and Bornkamm, G. W. Aberrant c-*myc* RNAs of Burkitt's lymphoma cells have longer half-lives. *EMBO J.* 4: 3717–3725 (1985).

Einat, M., Resnitzky, D., and Kimchi, A. Inhibitory effects of interferon on the expression of genes regulated by platelet-derived growth factor. *Proc. Natl. Acad. Sci.* (USA) 82: 7608–7612 (1985a).

Einat, M., Resnitzky, D., and Kimchi, A. Close link between reduction of c-*myc* expression by interferon and G_0/G_1 arrest. *Nature* 313: 597–600 (1985b).

Einhorn, S., Robert, K.-H., Ostlund, L., Juliusson, G., and Biberfeld, P. Interferon induces proliferation and differentiation in primary chronic lymphocytic leukemia cells. In: *The Biology of the Interferon System 1984,* H. Kirchner and H. Schellekens, Eds., Elsevier Science Publishers, Amsterdam, pp. 293–297 (1985).

Endo, T. and Nadal-Girard, B. Transcriptional and posttranscriptional control of c-*myc* during myogenesis: Its mRNA remains inducible in differentiated cells and does not suppress the differentiated phenotype. *Mol. Cell. Biol.* 6: 1412–1421 (1986).

Epstein, L. B. and Marcus, S. G. Review of experience with interferon and drug sensitivity testing of ovarian carcinoma in semisolid solid agar culture. *Cancer Chemother. Pharmacol.* 6: 273–277 (1981).

Erisman, M. D., Rothberg, P. G., Diehl, R. E., Morse, C. C., Spandorfer, J. M., and Astrin, S. M. Deregulation of c-*myb* gene expression in human colon carcinoma is not accompanied by amplification or rearrangement of the gene. *Mol. Cell. Biol.* 5: 1969–1976 (1985).

Eron, L. J., Judson, F., Tucker, S., Prawer, S., Mills, J., Murphy, K., Hickey, M., Rogers, M., Flannigan, S., Hien, N., Katz, H. I., Goldman, S., Gottlieb, A., Adams, K., Burton, P., Tanner, D., Taylor, E., and Peets, E. Interferon therapy for condylomata acuminata. *N. Engl. J. Med.* 315: 1059–1064 (1986).

Escot, C., Theillet, C., Lidereau, R., Spyratos, F., Champeme, M. H., Gest, J., and Callahan, R. Genetic alteration of the c-*myc* protooncogenes (MYC) in human primary breast carcinomas. *Proc. Natl. Acad. Sci.* (USA) 83: 4834–4838 (1986).

Exley, R., Gordon, J., and Clemens, M. Induction of B-cell differentiation antigens in interferon-α or phorbol ester-treated Daudi cells is impaired by inhibitors of ADP-ribosyltransferase. *Proc. Natl. Acad. Sci.* (USA) 84: 6467–6470 (1987a).

Exley, R., Nathan, P., Walker, L., Gordon, J., and Clemens, M. J. Anti-proliferative effects of interferons on Daudi Burkitt-lymphoma cells: Induction of cell differentiation and loss of response to autocrine growth factors. *Int. J. Cancer* 40: 53–57 (1987b).

Finkel, M. P., Biskis, B. O., and Jinkins, P. B. Virus induction of osteosarcomas in mice. *Science* 151: 698–701 (1966).

Gattoni-Celli, S., Hsiao, W. L. W., and Weinstein, I. B. Rearranged c-*mos* locus in a MOPC 21 murine myeloma cell line and its persistence in hybridomas. *Nature* 306: 795–796 (1983).

George, D. L., Scott, F., De Martinville, B., and Francke, U. Amplified DNA in Y1 mouse adrenal tumor cells: Isolation of cDNAs complementary to an amplified c-Ki-*ras* gene and localization of homologous sequences to mouse chromosome 6. *Nucl. Acids Res.* 12: 2731–2743 (1984).

Gonda, T. J. and Metcalf, D. Expression of *myb, myc* and *fos* proto-oncogenes during the differentiation of a murine myeloid leukaemia. *Nature* 310: 249–251 (1984).

Graf, T., Weizsaecker, F. V., Grieser, S., Coll, J., Stehelin, D., Patschinsky, T., Bister, K., Bechade, C., Calothy, G., and Leutz, A. v-*mil* induces autocrine growth and enhanced tumorigenicity in v-*myc* transformed avian macrophages. *Cell* 45: 357–364 (1986).

Greenberg, M. K. and Ziff, E. B. Stimulation of 3T3 cells induces transcription of the c-*fos* proto-oncogene. *Nature* 311: 433–438 (1984).

Gresser, I. How does interferon inhibit tumor growth? *Phil. Trans. R. Soc. (Lond.) B299: 69–76 (1982).*

Gresser, I. How does interferon inhibit tumor growth? In: *Interferon* 6, I. Gresser, Ed., Academic Press, London, pp. 93–126 (1985).

Gresser, I., Bandu, M. T., and Brouty-Boye, D. Interferon and cell division. IX. Interferon-resistant L1210 cells: Characteristics and origin. *J. Nat. Cancer. Inst.* 52: 553–559 (1974).

Gresser, I., Belardelli, F., Maury, C., Maunoury, M. T., and Tovey, M. G. Injection of mice with antibody to interferon enhances the growth of transplantable murine tumors. *J. Exp. Med.* 158: 2095–2107 (1983).

Gresser, I., Maury, C., and Tovey, M. Interferon and murine leukemia. VII. Therapeutic effect of interferon preparations after diagnosis of lymphoma in AKR mice. *Int. J. Cancer* 17: 647–651 (1976).

Hann, S. R., Thompson, C. B., and Eisenman, R. N. c-*myc* oncogene protein synthesis is independent of the cell cycle in human and avian cells. *Nature* 314: 366–369 (1985).

Hicks, N. J., Morris, A. G., and Burke, D. C. Partial reversion of the transformed phenotype of murine sarcoma virus-transformed cells in the presence of interferon. *J. Cell. Sci.* 49: 225–236 (1981).

Jonak, G. J., Anton, E. D., Fahey, D., Friedland, B. K., Cheng, Y. S. E., and

Knight, E., Jr. Interferon-mediated regulation of the c-*myc* gene in Burkitt lymphoma cells. In: *The Biology of the Interferon System 1984,* H. Kirchner and H. Schellekens, Eds., Elsevier, Amsterdam, pp. 199–204 (1985).

Jonak, G. J., Friedland, B. K., Anton, E. D., and Knight E., Jr. Regulation of c-*myc* RNA and its proteins in Daudi cells by interferon-β. *J. Interferon Res.* 7: 41–52 (1987).

Jonak, G. J. and Knight, E. Selective reduction of c-*myc* mRNA in Daudi cells by human β-interferon. *Proc. Natl. Acad. Sci.* (USA) 81: 1747–1750 (1984).

Kataoka, T., Oh-Hashi, F., Sakurai, Y., Usuki, K., and Ida, N. Relative contribution of antiproliferative and host immunity-associated activity of mouse interferon in murine tumor therapy. *Cancer Res.* 44: 5661–5665 (1984).

Keay, S. and Grossberg, S. E. Interferon inhibits the conversion of 3T3-L1 mouse fibroblasts into adipocytes. *Proc. Natl. Acad. Sci.* (USA) 77: 4099–4103 (1980).

Kelly, K., Cochran, B. H., Stiles, C. D., and Leder, P. Cell-specific regulation of the c-*myc* gene by lymphocyte mitogens and platelet-derived growth factor. *Cell.* 35: 603–610 (1983).

Kindy, M. S., McCormack, J. E., Buckler, A. J., Levine, R. A., and Sonenshein, G. E. Independent regulation of transcription of the two strands of the c-*myc* gene. *Mol. Cell. Biol.* 7: 2857–2862 (1987).

Klein, G. and Klein, E. Evolution of tumours and the impact of molecular oncology. *Nature* 315: 190–195 (1985).

Knight, E., Jr., Anton, E. D., Fahey, D., Friedland, B. K., and Jonak, G. J. Interferon regulates c-*myc* gene expression in Daudi cells at the post-transcriptional level. *Proc. Natl. Acad. Sci.* (USA) 82: 1151–1154 (1985).

Land, H., Parada, L. F., and Weinberg, R. A. Tumorigenic conversion of primary embryo fibroblasts requires at least two cooperating oncogenes. *Nature* 304: 596–602 (1983).

Langdon, W. Y., Harris, A. W., Cory, S., and Adams, J. M. The c-*myb* oncogene perturbs B lymphocyte development in Eu-*myc* transgenic mice. *Cell* 47: 11–18 (1986).

Lee, S. H. S. and Epstein, L. B. Reversible inhibition by interferon of the maturation of human peripheral blood monocytes to macrophages. *Cell. Immunol.* 50: 177–190 (1980).

Lehn, H., Krieg, P., and Sauer, G. Papillomavirus genomes in human cervical tumors: Analysis of their transcriptional activity. *Proc. Natl. Acad. Sci.* (USA) 82: 5540–5544 (1985).

Lin, S. L., Greene, J. J., Ts'O, P. O. P., and Carter, W. A. Sensitivity and resistance of human tumour cells to interferon and rIn.rCn. *Nature* 297: 417–419 (1982).

Lowe, D. G., Capon, D. J., Delwart, E., Sakaguchi, A. Y., Naylor, S. L., and Goeddel, D. V. Structure of the human and murine R-*ras* genes, novel genes closely related to *ras* proto-oncogenes. *Cell* 48: 137–146 (1987).

Ludwig, C. U., Durie, B. G. M., Salmon, S. E., and Moon, T. E. Tumor growth stimulation in vitro by interferon. *Eur. J. Cancer* 19: 1625–1632 (1983).

Marx, J. L. Tumor-prone mice and *myc. Science* 226: 823 (1984).

Masuda, S., Fukuma, H., and Beppu, Y. Antitumor effect of human leukocyte interferon on human osteosarcoma transplanted into nude mice. *Eur. J. Cancer Clin. Oncol.* 19: 1521–1528 (1983).

McBride, O. W., Swan, D. C., Santos, E., Barbacid, M., Tronick, S. R., and Aaronson, S. A. Localization of the normal allele of T24 human bladder carcinoma oncogene to chromosome 11. *Nature* 300: 773–774 (1982).

McGrath, J. P., Capon, D. J., Goeddel, D. V., and Levinson, A. D. Comparative biochemical properties of normal and activated human ras p21 protein. *Nature* 310: 644–649 (1984).

McMahon, M., Stark, G. R., and Kerr, I. Interferon-induced gene expression in wild-type and interferon-resistant human lymphoblastoid (Daudi) cells. *J. Virol.* 57: 362–366 (1986).

Mulcahy, L. S., Smith, M. R., and Stacey, D. W. Requirement for *ras* proto-oncogene function during serum-stimulated growth of NIH 3T3 cells. *Nature* 313: 241–243 (1985).

Müller, R., Bravo, R., Burckhardt, J., and Curran, T. Induction of c-*fos* gene and protein by growth factors precedes activation of c-*myc*. *Nature* 312: 716–720 (1984).

Müller, R., Curran, T., Muller, D., and Guilbert, L. Induction of c-*fos* during myelomonocytic differentiation and macrophage proliferation. *Nature* 314: 546–548 (1985).

Mushinski, J. F., Bauer, S. R., Potter, M., and Reddy, E. P. Increased expression of *myc*-related oncogene mRNA characterizes most BALB/c plasmacytomas induced by pristane or Abelson murine leukemia virus. *Proc. Natl. Acad. Sci.* (USA) 80: 1073–1077 (1983).

Nakagawa, Y., Hirakawa, K., Ueda, S., Suzuki, K., Fukuma, S., Kishida, T., Imanishi, J., and Amagai, T. Local administration of interferon for malignant brain tumors. *Cancer Treat. Rep.* 67: 833–835 (1983).

Nakagawa, T., Nakagawa, N., Volkman, D. J., and Fauci, A. S. Sequential synergistic effect of interleukin 2 and interferon-γ on the differentiation of a TAC-antigen-positive B cell line. *J. Immunol.* 136: 164–168 (1986).

Neel, B. G., Jhanwar, S. C., Chaganti, R. S. K., and Hayward, W. S. Two human c-*onc* genes are located on the long arm of chromosome 8. *Proc. Natl. Acad. Sci.* (USA) 79: 7842–7846 (1982).

Nickoloff, B. J., Mahrle, G., and Morhenn, V. Ultrastructural effects of recombinant γ-interferon on cultured human keratinocytes. *Ultrastruct. Pathol.* 10: 17–21 (1986).

Ostlund, L., Einhorn, S., Robert K. H., Juliusson, G., and Biberfeld, P. Chronic B-lymphocytic leukemia cells proliferate and differentiate following exposure to interferon in vitro. *Blood* 67: 152–159 (1986).

Outzen, H. C. and Custer, R. P. Growth of human normal and neoplastic mammary tissues in the cleared mammary fat pad of the nude mouse. *J. Natl. Cancer Inst.* 55: 1461–1466 (1975).

Pravtcheva, D. D., Ruddle, F. H., Ellis, R. W., and Scolnick, E. M. Assignment of murine cellular Harvey *ras* gene to chromosome 7. *Somat. Cell Genet.* 9: 681–686 (1983).

Propst, F. and Vande Woude, G. F. Expression of c-*mos* protooncogene transcripts in mouse tissues. *Nature* 315: 516–518 (1985).

Pulciani, S., Santos, E., Lauver, V., Long, L. K., Aaronson, S. A., and Barbacid, M. Oncogenes in solid human tumors. *Nature* 300: 539–542 (1982).

Quesada, J. R. and Gutterman, J. U. Clinical study of recombinant DNA-produced leukocyte interferon (clone A) in an intermittent schedule in cancer patients. *J. Natl. Cancer Inst.* 70: 1041–1046 (1983).

Quesada, J. R., Gutterman, J. U., and Hersh, E. M. Treatment of hairy cell leukemia with α interferons. *Cancer* 57: 1678–1680 (1986).

Quesada, J. R., Reuben, J., Manning, J. T., Hersh, E. M., and Gutterman, J. U. α-interferon for induction of remission in hairy cell leukemia. *N. Engl. J. Med.* 310: 15–18 (1984).

Ratain, M. J., Golomb, H. M., Bardawil, R. G., Vardiman, J. W., Westbrook, C. A., Kaminer, L. S., Lembersky, B. C., Bitter, M. A., and Daly, K. Durability of responses to interferon α-2b in advanced hairy cell leukemia. *Blood* 69: 872–877 (1987).

Rechavi, G., Givol, D., and Canaani, E. Activation of a cellular oncogene by DNA rearrangement: Possible involvement of an IS-like element. *Nature* 300: 607–611 (1982).

Reed, J. C., Alpers, J. D., Nowell, P. C., and Hoover, R. G. Sequential expression of protooncogenes during lectin-stimulated mitogenesis of normal human lymphocytes. *Proc. Natl. Acad. Sci.* (USA) 83: 3982–3986 (1986).

Reid, L. M., Minato, N., Gresser, I., Holland, J., Kadish, A., and Bloom, B. R. Influence of anti-mouse interferon serum on the growth and metastasis of tumor cells persistently infected with virus and of human prostatic tumors in athymic nude mice. *Proc. Natl. Acad. Sci.* (USA) 78: 1171–1175 (1981).

Reynolds, S. H., Stowers, S. J., Maronpot, R. R., Anderson, M. W., and Aaronson, S. A. Detection and identification of activated oncogenes in spontaneously occurring benign and malignant hepatocellular tumors of the B6C3F1 mouse. *Proc. Natl. Acad. Sci.* (USA) 83: 33–37 (1986).

Rivière, Y. and Hovanessian, A. G. Direct action of interferon and inducers of interferon on tumors cells in athymic nude mice. *Cancer Res.* 43: 4596–4599 (1983).

Rodenhuis, S., Van de Wetering, M. L., Mooi, W. J., Evers, S. G., Van Zandwijk, N., and Bos, J. L. Mutational activation of the K-ras oncogene: A possible pathogenic factor in adenocarcinoma of the lung. *N. Engl. J. Med.* 317: 929–935 (1987).

Rossi, G. B. Interferons and cell differentiation. In: *Interferon* 6, I. Gresser, Ed., Academic Press, London, pp. 31–68 (1985).

Ryan, J., Hart, C. P., and Ruddle, F. H. Molecular cloning and chromosome assignment of murine N-*ras*. *Nucl. Acids Res.* 12: 6063–6072 (1984).

Saito, H., Hayday, A. C., Wiman, K., Hayward, W. S., and Tonegawa, S. Activation of the c-*myc* gene by translocation: A model for translational control. *Proc. Natl. Acad. Sci.* (USA) 80: 7476–7480 (1983).

Sakaguchi, A. Y., Naylor, S. L., Shows, T. B., Toole, J. J., McCoy, M., and

Weinberg, R. A. Human c-Ki-*ras*2 proto-oncogene on chromosome 12. *Science* 219: 1081–1083 (1983).

Samid, D., Chang, E. H., and Friedman, R. M. Biochemical correlates of phenotypic reversion in interferon-treated mouse cells transformed by a human oncogene. *Biochem. Biophys. Res. Comm.* 119: 21–28 (1984).

Samid, D., Flessate, D. M., Greene, J. J., Chang, E. H., and Friedman, R. M. Persisting revertants after interferon treatment of oncogene-transformed cells. In: *The Biology of the Interferon System 1985,* W. E. Stewart II and H. Schellekens, Eds., Elsevier Science Publishers, Amsterdam, pp. 327–332 (1986).

Samid, D. and Friedman, R. M. Transcriptional regulation of *ras* by interferon. In: *Interferon as Cell Growth Inhibitors and Antitumor Factors,* R. M. Friedman, T. Merigan, and T. Sreevalsan, Eds., UCLA Symposia on Molecular and Cellular Biology, New Series, Vol. 50, Alan Liss, New York, pp. 413–422 (1986).

Santos, E., Martin-Zanca, D., Reddy, E. P., Pierotti, M. A., Della Porta, G., and Barbacid, M. Malignant activation of a K-*ras* oncogene in lung carcinoma but not in normal tissue of the same patient. *Science* 223: 661–664 (1984).

Sawada, T., Takamatsu, T., Tanaka, T., Mino, M., Fujita, K., Kusunoki, T., Arizono, N., Fukuda, M., and Kishida, T. Effects of intralesional interferon on neuroblastoma: Changes in histology and DNA content distribution of tumor masses. *Cancer* 48: 2143–2146 (1981).

Saxen, L. Effect of interferon tested in a model system for organogenesis. *J. Interferon Res.* 5: 355–359 (1985).

Schlag, P., Wolfrum, J., Schreml, W., and Herfarth, C. Effect of fibroblast interferon on colony tumor growth of miscellaneous solid human tumors. *Klin. Wochenschr.* 60: 1455–1459 (1982).

Seeburg, P. H., Colby, W. W., Capon, D. J., Goeddel, D. V., and Levinson, A. D. Biological properties of human c-Ha-*ras*1 genes mutated at codon 12. *Nature* 312: 71–75 (1984).

Sergiescu, D., Gerfaux, J., Joret, A. M., and Chany, C. Persistent expression of v-*mos* oncogene in transformed cells that revert to nonmalignancy after prolonged treatment with interferon. *Proc. Natl. Acad. Sci.* (USA) 83: 5764–5768 (1986).

Sheiness, D. and Bishop, J. M. DNA and RNA from uninfected vertebrate cells contain nucleotide sequences related to the putative transforming gene of avian myelocytomatosis virus. *J. Virol.* 31: 514–521 (1979).

Shilo, B. Z. and Hoffmann, F. M. *Drosophila melanogaster* cellular oncogenes. *Cancer Surveys* 3: 299–320 (1984).

Shilo, B. Z. and Weinberg, R. A. DNA sequences homologous to vertebrate oncogenes are conserved in *Drosophila melanogaster. Proc. Natl. Acad. Sci.* (USA) 78: 6789–6792 (1981).

Slamon, D. J., Clark, G. M., Wong, S. G., Levin, W. J., Ullrich, A., and McGuire, W. L. Human breast cancer: Correlation of relapse and survival with amplification of the HER-2/neu oncogene. *Science* 235: 177–182 (1987).

Smalley, R. V. and Borden, E. C. Interferons: Current status and future directions of this prototypic biological. *Springer Semin. Immunopathol.* 9: 73–83 (1986).

Soslau, G., Bogucki, A. R., Gillespie, D., and Hubbell, H. R. Phosphoproteins altered by antiproliferative doses of human interferon-β in a human bladder carcinoma cell line. *Biochem. Biophys. Res. Comm.* 119: 941–948 (1984).

Sperry, P. J., Vigil, J. M., and Juarez, R. A. β-interferon modulates proliferation and differentiation of cultured mouse epidermal cells. *Clin. Res.* 33: 683A (1985).

Spooner, P. J. R., Grant, S., Pestka, S., Weinstein, I. B., and Fisher, P. B. Recombinant human interferon sensitizes resistant myeloid leukemic cells to induction of terminal differentiation. *J. Cell Biochem.* 9(suppl. 10): 127 (1985).

Stacey, D. W. and Kung, H. F. Transformation of NIH 3T3 cells by microinjection of Ha-*ras* p21 protein. *Nature* 310: 508–511 (1984).

Stewart, T. A., Pattengale, P. K., and Leder, P. Spontaneous mammary adenocarcinomas in transgenic mice that carry and express MTV/*myc* fusion genes. *Cell* 38: 627–637 (1984).

Strander, H. Interferon treatment of human neoplasia. In: *Advances in Cancer Research,* 46, Academic Press Orlando, G. Klein and S. Weinhouse, Eds., pp. 1–265 (1986).

Studzinski, G. P., Brelvi, Z. S., Feldman, S. C., and Watt, R. A. Participation of c-*myc* protein in DNA synthesis of human cells. *Science* 234: 467–470 (1986).

Stutman, O. In: *The Nude Mouse in Experimental and Clinical Research,* J. Fogh and B. Giovanella, Eds., Academic Press, New York, pp. 11–435 (1978).

Swan, D., Oskarsson, M., Keithley, D., Ruddle, F. H., D'Eustacchio, P., and Van de Woude, G. F. Chromosomal localization of the Moloney sarcoma virus mouse cellular (c-*mos*) sequence. *J. Virol.* 44: 752–754 (1982).

Talpaz, M., Kantarjian, H., McCredie, K., Trujillo, J. M., Keating, M. J., and Gutterman, J. U. Hematologic remission and cytogenetic improvement induced by recombinant human interferon α A in chronic myelogenous leukemia. *N. Engl. J. Med.* 314: 1065–1069 (1986).

Tomida, M., Yamamoto, Y., and Hozumi, M. Stimulation by interferon of induction of differentiation of mouse myeloid leukemic cells. *Cancer Res.* 40: 2919–2924 (1980).

Tomida, M., Yamamoto, Y., and Hozumi, M. Stimulation by interferon of induction of differentiation of human promyelocytic leukemia cells. *Biochem. Biophys. Res. Comm.* 104: 30–37 (1982).

Tominaga, S. I. and Lengyel, P. β-interferon alters the pattern of proteins secreted from quiescent and platelet-derived growth factor-treated BALB/c 3T3 cells. *J. Biol. Chem.* 260: 1975–1978 (1985).

Tsuruo, T., Iida, H., Tsukagoshi, S., Oku, T., and Kishida, T. Different susceptibilities of cultured mouse cell lines to mouse interferon. *Gann* 73: 42–47 (1982).

Turek, L. P., Byrne, J. C., Lowy, D. R., Dvoretzky, I., Friedman, R. M., and Howley, P. M. Interferon induces morphologic reversion with elimination of extrachromosomal viral genomes in bovine papillomavirus-transformed mouse cells. *Proc. Natl. Acad. Sci.* (USA) 79: 7914–7918 (1982).

Wallach, D., Cantell, K., Hirvonen, S., Toker, L., Aderka, D., and Holtmann, H. Presence of tumor necrosis factor and lymphotoxin in clinical interferon preparations derived from leukocytes. *J. Interferon Res.* 6(suppl. 1): 57 (1986).

Weinberg, R. A. The action of oncogenes in the cytoplasm and nucleus. *Science* 230: 770–776 (1985).

Willumsen, B. M., Papageorge, A. G., Hubbert, N., Bekesi, E., Kung, H. F., and Lowy, D. R. Transforming p21 *ras* protein: Flexibility in the major variable region linking the catalytic and membrane-anchoring domains. *EMBO J.* 4: 2893–2986 (1985).

Yamada, H., Yoshida, T., Sakamoto, H., Terada, M., and Sugimura, T. Establishment of a human pancreatic adenocarcinoma cell line (PSN-1) with amplifications of both c-*myc* and activated c-Ki-*ras* by a point mutation. *Biochem. Biophys. Res. Comm.* 140: 167–173 (1986).

15 THE GENETICS OF INTERFERON PRODUCTION AND ACTION

In addition to the structural genes coding for the different IFNs, a number of other genes, not linked to the structural genes, influence IFN production. These genes have been identified and studied in the mouse, but there is evidence that they may also be present in humans. Moreover, not only the production but also the action of IFNs is affected by host genotype, which can significantly influence the antiviral, immunomodulating, and antiproliferative activities of IFNs.

I. GENES INFLUENCING IFN PRODUCTION

A. Quantitative Control of IFN-α/β Production in the Mouse by *If* Loci

1. If-*Loci Are Inducer-Specific*

Several autosomal loci, designated as *If* loci, influence the levels of IFN-α/β production in mice after intravenous inoculation of viruses, such as Newcastle disease virus (NDV), Sendai virus, herpes simplex virus 1 (HSV-1), mouse mammary tumor virus, Friend leukemia virus, and murine cytomegalovirus (MCMV). *If* loci have high and low producer alleles, and the difference in IFN production between high and low responders can vary from 3- to almost a 100-fold, depending on which loci are concerned. Furthermore, *If* loci are inducer-specific in that they only influence IFN production when the corresponding inducer is involved. This means that mice of a given genotype can be at the same time high IFN responders for one virus and low or intermediate responders for another, depending on the alleles that are present at the corresponding loci. This possibility for divergence in the quantitative response to IFN induction, depending on the inducer at play, results from the fact that the inducer-specific *If* loci are not closely linked, but rather, segregate independently. For example, although BALB/c mice are low responders and C57BL/6 mice are high responders when induced with the NDV, HSV-1, and Sendai virus, recombinant inbred CXBH mice, derived from a cross between these two strains, are high responders for NDV, low responders for HSV-1, and intermediate responders for the Sendai virus. Another example of independent segregation of alleles at *If* loci is provided by mice of the C3H/He strain, which are low IFN responders for NDV but high responders for murine cytomegalovirus (MCMV).

Table 15.1 represents the loci influencing Mu IFN-α/β production with ten different inducers. *If-1,* influencing IFN induction by NDV, has been assigned to chromosome 3 and is located about 6cM distal to the minor histocompatibility locus H-28 (De Maeyer et al., 1975a; Mobraaten et al., 1984). The chromosomal assignment of the other *If* loci is unknown, except for one X-linked locus (see subsequent discussion).

The many different *If* loci are characterized by their quantitative effects on IFN synthesis, but since their mode of action or gene products are not known, it cannot be excluded that, upon further characterization, *If* loci will turn out to consist of a relatively heterogenous group of genes, influencing IFN production in multiple ways. Their effects are not limited to viral inducers; For example, polyrIrC-induced IFN production is also influenced, in the mouse and in the rat, with differences in IFN titers that can reach up to a 100-fold (Pugliese et al., 1980; Davis et al., 1984).

By influencing the levels of IFN production, murine *If* loci play a role in the resistance of mice to infection with viruses and other IFN-inducing agents, for example, parasites, by either directly inhibiting virus replication

TABLE 15.1 Influence of Mouse Genotype on Levels of IFN-α/β and IFN-γ Production

Inducer	Maximal ratio of high to low responders			*If* Loci Involved	References
		Strain of Highest Resp.	Strain of Lowest Resp.		
			Mu IFN-α/β		
NDV	30	C57BL/6	BALB/c	If-1 and If-x	De Maeyer and De Maeyer-Guignard, 1979
mMTV	3	C57BL/6	BALB/c	If-2	De Maeyer and De Maeyer-Guignard, 1979
Sendai virus	10	C57BL/6	BALB/c	If-3, If-4	De Maeyer and De Maeyer-Guignard, 1979
HSV-1	10	C57BL/6	BALB/c	At least three loci different from the preceding, one of which is *If*-x	Zawatzky et al., 1982
Friend leukemia	10	C57BL/6	BALB/c	Not determined	De Maeyer, unpublished
MCMV	8	C3H/He	BALB/c	Not determined	Grundy-Chalmer et al., 1982
LCMV	10	C3H	BALB/c	Not determined	Rivière et al., 1980
polyrI.rC	20	C57BL/6 C57BL/10	B10.RIII A	Not determined	Pugliese et al., 1980
Trypanosoma brucei rhodesiense	10	C57BL/10	C3H/HeB	Not determined	De Gee et al., 1985
Toxoplasma	15	C57BL/6	BALB/c	Not determined	Shirahata et al., 1986
			Mu IFN-γ		
PPD	10	C57BL/6	BALB/c		Huygen and Palfliet, 1983
Pokeweed mitogen	15	C57BL/6 AKR CBA	BALB/c A C3H/He		Virelizier, 1982
Toxoplasma	3	C57BL/6	BALB/c		Shirahata et al., 1986
Leishmania	>10	C57BL/6	BALB/c		Sadick et al., 1986

NDV: Newcastle disease virus; mMTV: murine mammary tumor virus; HSV-1: herpes simplex virus type I; MCMV: murine cytomegalovirus; LCMV: lymphocytic choriomeningitis virus; PPD: purified protein derivative.

or stimulating cell-mediated immunity via IFN production (Engler et al., 1982; Grundy-Chalmer et al., 1982; Brandt-Pedersen et al., 1983; De Maeyer and De Maeyer-Guignard, 1983; Allan and Shellam, 1985; De Gee et al., 1985; Shirahata et al, 1986).

2. If *Loci Are Expressed in Cells Derived from the Hemopoietic System*

This has been studied in detail only for *If-1,* but there is enough evidence to suggest that other *If* loci are also expressed in bone-marrow-derived cells. This does not mean that these loci are exclusively expressed on such cells, since under certain conditions *If* alleles can also be expressed in mouse embryo fibroblast cultures.

a. Transfer of Phenotype by Grafting Bone-Marrow Cells. Mice bearing the low responder allele at *If-1* ($If\text{-}1^l$) can be converted into high responders if their hemopoietic stem cells are replaced with stem cells derived from high responders of $If\text{-}1^h$ genotype. This is achieved by destroying the bone marrow by total body irradiation, followed by repopulation through grafting of MHC compatible bone-marrow cells of $If\text{-}1^h$ genotype. The ensuing restoration of the stem-cell function results in the replacement of the host's $If\text{-}1^l$ lymphocytes and macrophages by the donor's $If\text{-}1^h$ cells (De Maeyer et al., 1975b). Likewise, when the bone-marrow stem cells of B10.BR mice, which are low IFN-α/β responders to murine cytomegalovirus (MCM virus), are destroyed by irradiation and replaced by MHC compatible bone-marrow cells from high responder CBA mice, the low responder B10.BR mice become high IFN responders to MCM virus as a result of repopulation with CBA leukocytes (Allan and Shellam, 1985). This shows that the alleles of *If* loci are expressed in cells derived from hemopoietic stem cells, and that such cells contribute to IFN-α/β synthesis during viremia.

b. Macrophages Express High and Low Producer Alleles In Vitro. Macrophages produce IFN-α/β upon infection with many viruses (see Chapter 10), and they are important contributors to IFN production during viremia. In line with the observation that *If* loci are expressed on cells derived from the hemopoietic system, resident peritoneal macrophages of $If\text{-}1^l$ and $If\text{-}1^h$ mice, when induced with NDV in vitro, express the corresponding allele, and $If\text{-}1^l$ macrophages produce about ten times less Mu IFN-β than $If\text{-}1^h$ macrophages do (De Maeyer et al., 1979).

Does a high IFN producer have ten times more IFN-producing cells, or does it make ten times more IFN per cell? Using a Mu IFN-β probe, in situ hybridization of peritoneal macrophages from $If\text{-}1^h$ and from $If\text{-}1^l$ mice after in vivo or in vitro induction with NDV reveals an equal number of IFN-mRNA-producing cells in high and low responders. This suggests that *If* loci exert their effect by influencing the amount of IFN produced per cell, rather than the number of IFN-producing cells. This conclusion still leaves room for uncertainty, however, since it is based on the assumption that the presence of IFN mRNA in a cell, as detected by in situ hybridization, implies IFN secretion by that cell (De Maeyer and De Maeyer-Guignard, 1986).

Furthermore, macrophages express a locus with a quantitative effect on IFN-α/β production, which is not related to *If* loci, but influence the response to endotoxin in general. Splenic macrophages from Lps^d C3H/HeJ mice fail to produce Mu IFN-α/β upon induction with LPS, contrary to splenic macrophages from nondefective, Lps^n mice (Ascher et al., 1981). The *Lps* locus, on chromosome 4, not only influences IFN-α/β production, but it also affects the synthesis of other LPS-induced cytokines, such as TNF-α and IL-1 (see Chapter 13).

3. *An X-Linked* If *Locus Influences IFN-α/β* Production

Mendelian analysis of NDV- or HSV-1-induced circulating IFN production in BALB/c and C57BL/6 mice and their F1 and F2 progeny shows that, in addition to autosomal *If* loci, an X-linked *If* locus exerts a quantitative effect on the induction of IFN-α/β by these viruses. This X-linked difference in levels of early IFN production influences the resistance of mice to HSV-1 infection, as clearly shown by the different IFN titers and the resistance of male F1 progeny from reciprocal crosses between BALB/c and C57BL mice: The males produce less IFN (Fig 15.1) and are more suceptible to the infection when they receive their X chromosome from a BALB/c mother than when they receive it from a C57BL mother. Like *If-1,* the X-linked *If* locus (*If*-x) is expressed in peritoneal macrophages in vitro (Zawatzky et al., 1982; Brandt Pedersen et al., 1983; De Maeyer-Guignard et al., 1983; Ellermann-Eriksen et al., 1986a,b).

In view of the highly conserved nature of the X chromosome, a similar locus most probably also exists in humans, and because of the particular

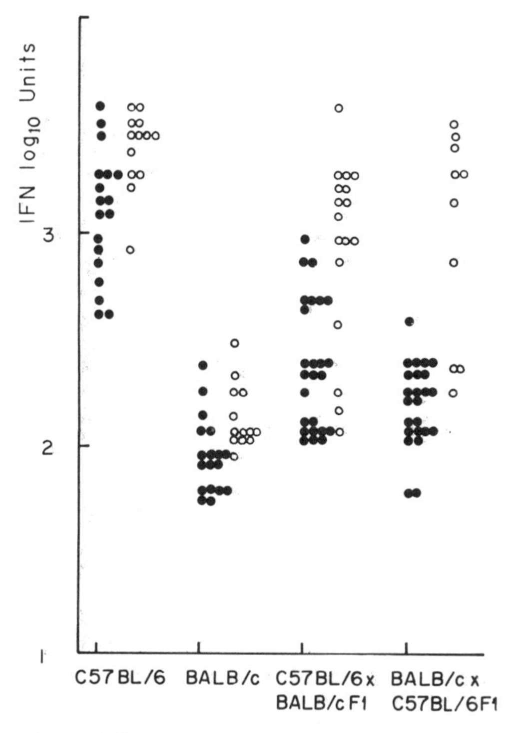

Figure 15.1 Evidence that an X-linked locus influences Mu IFN-α/β levels induced in mice by herpes simplex virus type 1. In the F1 progeny of the reciprocal crosses, the IFN titers of the females are significantly higher than those of the males and the male F1 progeny (black dots) produce more IFN when they receive their X chromosome from a C57BL/6 mother (C57BL/6 × BALB/c F1) than when they receive it from a BALB/c mother (BALB/c × C57BL/6 F1). Each point represents the IFN titer of one animal, measured between 2 and 3 hours after intravenous inoculation of virus (●: males; ○: females). (Adapted from Zawatzky et al., 1982.)

mode of inheritance of the X chromosome, one would expect a low or deficient X-linked influence on IFN production to result in more severe infections in males. There are indications from the clinic that infections with viruses belonging to the herpesviridae can be more severe in male than in female patients. Some X-linked genes in humans influence susceptibility to Epstein-Barr virus, which belongs to the herpes virus group. Males affected with the X-linked lymphoproliferative syndrome are uncommonly susceptible to infectious mononucleosis, which often has a fatal course in such patients (Purtilo, 1980). Furthermore, nasopharyngeal carcinoma, in which Epstein-Barr virus has been implicated as a causative agent, occurs mainly in genetically predisposed males (Purtilo, 1981). Moreover, patients with the Wiskott-Aldrich syndrome, which is inherited as an X-linked recessive trait and is characterized by congenital thrombocytopenia, are unusually susceptible to herpes simplex virus infections (Aldrich et al., 1954; St Geme et al., 1965). It is, therefore, likely that also in humans, herpes virus-induced IFN production has to some extent an X-linked component, although this remains to be established since the effect on IFN production of these human X-linked genes is not known.

4. The IFN Structural Genes Do Not Contribute to the Quantitative Regulation of IFN-α/β Production by *If* Loci

Although the different *If* loci are not closely linked, it is intriguing that mice of the BALB/c strain are low producers in every instance (Table 15.1). Could it be that some features of the BALB/c IFN-α and IFN-β structural genes contribute to the low IFN production of this strain? For example, one can envisage that less efficient inducible promoter or enhancer regions of these genes would result in lower inducibility, and thus contribute to low IFN production as compared to mice of some other genotypes.

Studies in C57BL/6 mice congenic with BALB/c at the Mu IFN-α gene cluster (*Ifa* locus) on chromosome 4 shows that this is not the case, since, even though these congenic mice carry the BALB/c structural genes, they still display the typical C57BL/6 high IFN production when probed with several viruses. Thus, the low IFN production of BALB/c mice is not directly due to some features of the structural genes, but instead results mainly from the presence of low producer alleles at *If* loci (De Maeyer-Guignard et al., 1986).

B. Mouse Genotype Also Affects the Levels of Mu IFN-γ Production

The in vitro induction of Mu IFN-γ by exposure of mouse spleen cells to mitogens reveals significant differences in levels of production, depending on the genotype of the cell donor. Pokeweed mitogen-stimulated spleen cells of BALB/c/J, A/J, or C3H/HeJ genotype produce less than ten times the amount of IFN-γ made by C57BL/6 or CBA spleen cells (Virelizier, 1982). In vivo, BCG-sensitized BALB/c mice, rechallenged with PPD, make less than ten times the amount of IFN-γ produced by sensitized C57BL/6 mice, and a comparable difference is observed when spleen cells from these two mouse strains are induced with PPD in vitro. An autosomal locus appears to exert a major effect on the difference between the two mouse strains, but its relation or linkage to other *If* loci is not known (Huygen and Palfliet, 1983, 1984). It is intriguing that also for IFN-γ induction, the BALB/c genotype confers low production as compared to other strains. The possibility has been raised that in this particular instance inhibitors of IFN-γ are coinduced in BALB/c mice (Neta, 1981).

C. Genes with Quantitative Effect on IFN Production Are Probably Also Present in Humans

We have seen that *If* loci are expressed in cells derived from the hemopoietic system, and, for example, can be determined in murine peritoneal macrophage cultures in vitro. There is also a good correlation between the magnitude of the in vivo IFN response after intravenous inoculation of a given inducer into mice and the levels of in vitro IFN production with the same inducer in peripheral blood suspensions, indicating that such suspensions can be used to determine high or low responder phenotypes (De Maeyer and De Maeyer-Guignard, 1970). This is relevant for studies in humans, in which circulating IFN production cannot be readily measured, but instead, peripheral blood leukocyte or monocyte cultures are used.

When virus-induced IFN-α/β production is measured in human leukocyte suspensions or in cultures derived from mononuclear cells, significant individual variations in the levels of IFN production are observed. For example, NDV-induced IFN-α production in mononuclear cell cultures from many different individuals shows a 100-fold difference between the highest and the lowest producers (Fig. 15.2). Individual differences of comparable magnitude are observed when the cultures are induced by phytohemagglutinin to make Hu IFN-γ (Bacon et al., 1983). Higher than tenfold individual differences in levels of IFN-α production are also observed when peripheral blood leukocyte suspensions are induced with influenza virus, and in this particular instance low responsiveness appears to be associated with the presence of the HLA-DR 2 antigen (Abb et al., 1983). Poor production of Hu IFN-α of peripheral blood mononuclear cells induced by either measles, rubella, herpes simplex, or mumps virus also seems to be associated with the HLA-DW2/DR2 antigen (Salonen et al., 1982). The individual responses of peripheral leukocyte suspensions derived from different individuals to IFN induction by either NDV, measles virus, polyrI.rC, or concanavalin A can also vary up to a 100-fold (Neighbour and Grayzel, 1981). Up to tenfold differences in Hu IFN-γ production have been observed in human spleen cell cultures derived from different individuals (Weck et al., 1983). However, a formal genetic analysis of the quantitative control of IFN-α/β or IFN-γ production, consisting of detailed pedigree studies, has never been performed in humans.

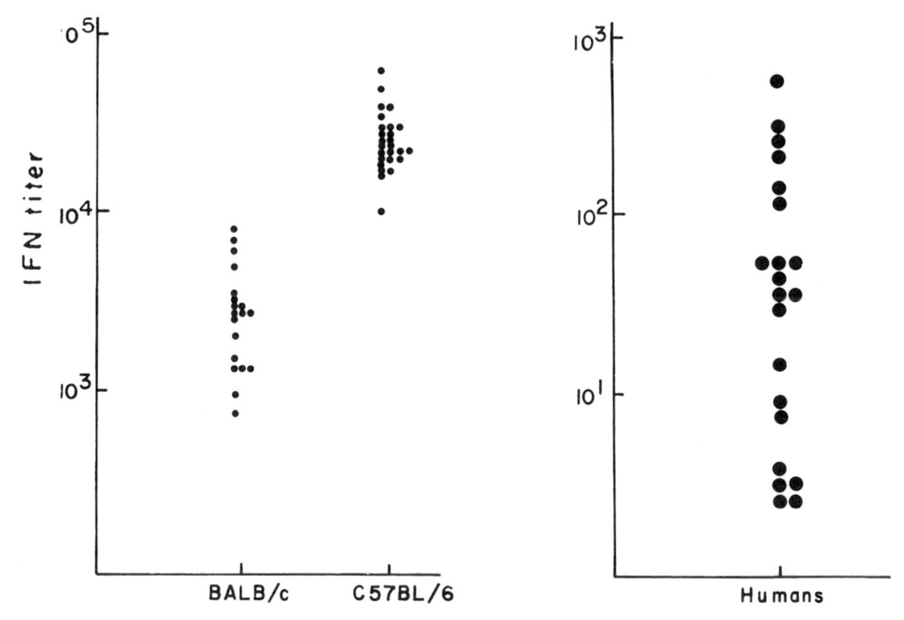

Figure 15.2 A comparison of the extent of individual Newcastle disease virus-induced IFN titers in high and low IFN-producing mice (left panel), and in leukocyte suspensions of different human individuals (right panel). In both instances the range spans about two orders of magnitude. Each point corresponds to the titer of one individual. (Adapted from De Maeyer and De Maeyer-Guignard, 1979, and from Bacon et al., 1983.)

Furthermore, there is some evidence that a low or deficient capacity for IFN production can be associated with enhanced susceptibility to infections. A small percentage of children with recurrent respiratory infections appears to have a deficient lymphocyte IFN production, since lymphocyte suspensions isolated from their peripheral blood consistently fail to produce IFN-α upon stimulation with NDV. Lymphocytes from some siblings of these patients also have deficient IFN production, but lymphocytes of the parents produce normal amounts (Isaacs et al., 1981). Clear evidence for defects in the IFN response of peripheral blood leukocytes from children with unduly frequent infections, however, is not always found (Bondestam et al., 1984). In mice, *If* loci are inducer-specific, meaning that an individual can be a low producer for one virus but a high producer for others. This raises the possibility that in humans the concept of high or low IFN production also has no meaning unless the corresponding inducer is indicated. Deficient or low IFN-α/β production may well have to be determined by employing as an inducer the virus that is responsible for the infection under study.

A selective defect in Hu IFN-γ production has been associated with a persistent Epstein-Barr virus infection in a child with hypergammaglobulinemia and immunoblastic proliferation. In this case, though, the primary genetic defect was probably not a deficiency in IFN-γ production (Virelizier et al., 1978).

II. GENES INFLUENCING IFN ACTION

The existence of considerable individual variation in response to IFNs is evident from the results of many clinical trials in which the antiviral and antitumor effects of IFNs α, β, and γ have been tested. Studies in mice show that there are, indeed, clear-cut effects of animal genotype on the various biological activities of IFNs. The genes involved have, with a few exceptions, however, not been identified or characterized. It is quite possible that some effects of genotype on IFN action are due to allelic variations of genes that have already been identified as important for IFN action, such as the 2-5A synthetase genes, or the IFN receptor genes. The significance and importance of allelism at these genes has not received much attention to date.

A. Effect of Genotype on Immunomodulation by IFNs

In addition to having different alleles at genes influencing the production of IFN-α/β and IFN-γ, BALB/c and C57BL/6 mice have different alleles also at loci influencing the immunomodulatory effects of IFNs. For example, the immunosuppressive effect of IFN-α/β on the afferent pathway of delayed hypersensitivity (see Chapter 12) is more pronounced in BALB/c than in C57BL/6 mice. BALB/c mice are also more sensitive to the inhibitory effect of IFN-α/β on the expression of delayed hypersensitivity, and about 25 times the amount of IFN effective in BALB/c mice is needed to obtain a comparable inhibition of expression in C57BL/6 mice (De Maeyer and De Maeyer-Guignard, 1980).

Stimulation by Mu IFN-α/β of phagocytosis by murine peritoneal macrophages is similarly affected by the genotype of the cell donor: BALB/c macrophages are much more readily stimulated than are macrophages of C57BL/6 origin (Degré et al., 1983). Furthermore, murine host genes also influence the efficacy of the stimulation of macrophage antitumor activity by Mu IFN-γ (Boraschi et al., 1984). Such effects of genotype possibly contribute to the considerable individual variation in the efficacy of monocyte stimulation, observed in patients after treatment with Hu lymphoblastoid IFN (Maluish et al., 1983).

Mitogen responsiveness of T and B lymphocytes, derived from Mu IFN-α/β-treated mice, is enhanced when the cells are derived from BALB/c mice, whereas a similar treatment has no enhancing effect in mice of the C57BL/6 genotype (Huygen and Palfliet, 1982).

There are also genes that control the degree of inducibility of MHC class II antigens by IFN-γ, thereby possibly influencing autoimmunity. This is suggested by the observation that rats or mice which belong to inbred strains that are highly susceptible to experimental autoimmune encephalitis display a genetically controlled IFN-γ-mediated hyperinducibility of class II antigens on astrocytes. Genes that cause an abnormally high expression of class II molecules on specific tissues could be instrumental in determining genetic susceptibility to autoimmune disease (Massa et al., 1987).

B. Effect of Genotype on the Antiproliferative Activity of IFNs

Mouse genotype affects the sensitivity to the antiproliferative effects of Mu IFN-α/β. The proliferation of erythropoietin-dependent committed erythroid precursors (CFU-E), originating from either bone marrow or fetal livers, and the proliferation of bone-marrow-derived macrophages, is more readily inhibited if the cells are of BALB/c or Swiss genotype than if they are derived from C57BL/6 mice (see Chapter 8, Fig 8.2) (Gallien-Lartigue et al., 1980; Dandoy et al., 1981). The sensitivity of a more primitive class of erythroid precursors, BFU/E, however, is identical for cells of BALB/c and C57BL/6 genotype, which suggests that the genes modulating the sensitivity of these cells to IFN action are only expressed after a certain stage of differentiation has been reached (Gallien-Lartigue et al., 1985).

C. Genes Influencing the Antiviral Action of IFNs

1. *Effect of Mouse Genotype on the Efficacy of Mu IFN-α/β Treatment of EMC and Herpes Simplex Virus Infection*

The complexity of the quantitative genetics of IFN action is demonstrated by a comparison of the antiviral activity of Mu IFN-α/β in BALB/c and C57BL/6 mice in vivo and in fibroblasts, derived from these strains, in vitro.

In vivo, EMC virus-infected BALB/c mice are more sensitive to the protective action of treatment with exogenous Mu IFN-α/β than are C57BL/6 mice (Fig. 15.3). F1 hybrids derived from these two strains are as sensitive to

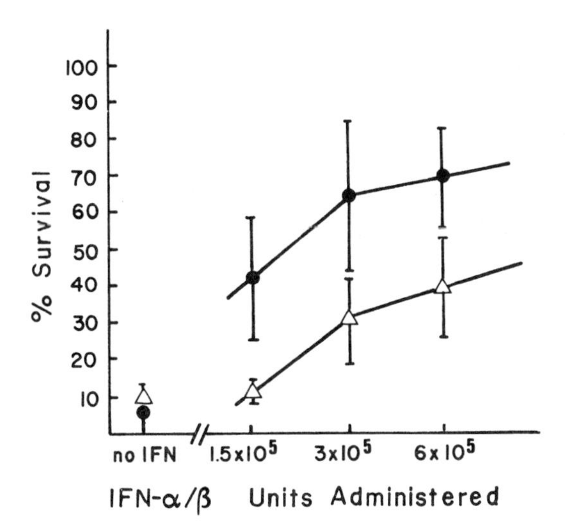

Figure 15.3 The effect of mouse genotype on the in vivo antiviral action of exogenous Mu IFN-α/β is shown by the dose-response curves of the protective effect of IFN treatment in encephalomyocarditis virus-infected BALB/c (●—●) and C57BL/6 (△—△) mice. The percentage of survival of different IFN-treatment groups on day 7 after infection are shown. For the same dose of IFN, the survival of BALB/c mice each time is superior to that of C57BL/6 mice. (Adapted from Dandoy et al., 1982.)

protection by IFN as are the BALB/c parents, indicating dominance of the greater sensitivity to IFN action. Multiple genes are involved in this effect, since none of the seven recombinant inbred strains derived from BALB/c and C57BL/mice has a parental pattern of response, but all display an intermediate protective effect of IFN. If only one locus were responsible, the segregation taking place in the recombinant inbred strains would have resulted in mice with either BALB/c- or C57BL/6-type behavior (Dandoy et al., 1982).

In vitro, primary cultures of embryo fibroblasts derived from these two strains display an opposite sensitivity, in that C57BL/6 fibroblasts are more readily protected by Mu IFN-α/β against HSV-2 and EMC virus infection than are BALB/c fibroblasts (Ellermann-Eriksen et al., 1986a; De Maeyer et al., unpublished). Aside from the fundamental problem posed by these apparently contradictory observations in vivo and in vitro, a practical consideration applies: It is not possible to predict the in vivo efficacy of IFN treatment from simple in vitro tests, even when measuring a relatively clear-cut phenomenon such as the inhibition of virus replication.

2. The Mx Locus: A Mu IFN-α/β Inducible Gene Determining Resistance to Influenza Virus Infection

The *Mx* locus provides an interesting example of genetic control of the sensitivity to IFN-α/β action that is specific only for members of a single virus family: the orthomyxoviruses.

When inoculated with the mouse-adapted neurotropic, pneumotropic, or hepatotropic variants of the influenza A virus, mice of most inbred strains succumb to infection. Mice of the A2G strain, however, are resistant, although they are as susceptible as mice of other strains to infection with the vesicular stomatitis virus, EMC virus, or HSV-1. Resistance is governed by a single gene, *Mx,* located on mouse chromosome 16 and characterized by two alleles: Mx^+, determining resistance, and Mx^-, determining susceptibility. Whereas homozygous Mx^+/Mx^+ mice develop resistance within a few days of birth, heterozygous Mx^+/Mx^- mice only develop resistance a few weeks after birth (Lindenmann, 1981; Staeheli et al., 1986c). Contrary to homozygous Mx^-/Mx^- animals, however, such heterozygous newborn mice can be easily protected from influenza virus infection by IFN-α/β treatment.

In vitro, cells isolated from resistant mice are as permissive for influenza virus as cells from genetically susceptible animals, but, like newborn mice, they are very easily protected by pretreatment with low amounts of IFN-α/β (but not with IFN-γ), whereas cells lacking the Mx^+ allele are only marginally protected. The difference in antiviral efficacy of IFN is only observed when the IFN-treated cells are challenged with the influenza virus but not with other viruses, against which *Mx* positive and negative cells are equally protected by IFN treatment (Haller et al., 1980; Horisberger et al., 1980; Haller, 1981; Haller et al., 1981; Arnheiter and Haller, 1983)

It is clear from the foregoing that the Mx^+ allele determines an IFN-α/β

inducible antiviral state, specifically directed against influenza virus replication, and specifically induced by Mu IFN-α/β but not by Mu IFN-γ. The natural resistance to influenza virus infection of Mx^+ mice is due to the induction by influenza virus of endogenous IFN-α/β, which then activates the *Mx* gene. Indeed, when Mx^+ mice are treated with anti-IFN-α/β serum and then infected with influenza virus, they lose their natural resistance and become fully susceptible to the infection (Haller et al., 1979; Staeheli et al., 1984).

The Mx^+ allele codes for a 75 kDa polypeptide, designated "*Mx* protein." The *Mx* protein is not constitutively expressed, but is induced by Mu IFN-α or IFN-β (Horisberger and Hochkeppel, 1985; Horisberger et al., 1983; Staeheli et al., 1985). Nuclear run-off experiments show that IFN regulates the *Mx* gene expression at the transcriptional level (Staeheli et al., 1986b). The protein accumulates in the nuclei of IFN-treated cells and inhibits influenza virus mRNA synthesis, probably by interfering with the transcription of the viral genome, which normally occurs in the nuclei of infected cells (Dreiding et al., 1985; Krug et al., 1985). Southern blot analysis of genomic DNA from different inbred mouse strains, using Mx^+ cDNA as probe, reveals that the Mx^- alleles result from a deletion in the *Mx* gene (Staeheli et al., 1986a). Obviously, the wild-type allele is Mx^+, and Mx^- is derived from it by deletion. The preponderance of the latter in most laboratory strains is probably due to a founder effect, since, as discussed in Chapter 2, most of the classical inbred mouse strains went through a genetic bottleneck when they were established. One cannot exclude, however, that, in the absence of influenza virus infection, the Mx^- allele confers an unknown selective advantage.

An 80 kD protein, antigenically related to the murine *Mx* protein, appears in human cells after treatment with Hu IFN-α but not with Hu IFN-γ. Unlike the murine *Mx* protein, which appears in the nucleus, the *Mx*-related protein in humans appears in the cytoplasm. The role played by this protein in the inhibition of influenza virus is not clear, since Hu IFN-γ, which does not induce it, is as efficient as IFN-α in conferring protection to influenza virus infection (Staeheli and Haller, 1985).

REFERENCES

Abb, J., Zander, H., Abb, H., Albert, E., and Deinhardt, F. Association of human leucocyte low responsiveness to inducers of interferon α with HLA-DR2. *Immunology* 49: 239–244 (1983).

Aldrich, R. A., Steinberg, A. G., and Campbell, D. C. Pedigree demonstrating sex-linked recessive condition characterized by draining ears, eczematoid dermatitis and bloody diarrhea. *Pediatrics* 13: 133–138 (1954).

Allan, J. E. and Shellam, G. R. Characterization of interferon induction in mice of resistant and susceptible strains during murine cytomegalovirus infection. *J. Gen. Virol.* 66: 1105–1112 (1985).

Arnheiter, H. and Haller, O. Mx gene control of interferon action: Different kinetics of the antiviral state against influenza virus and vesicular stomatitis virus. *J. Virol.* 47: 626–630 (1983).

Ascher, O., Apte, R. N., and Pluznik, D. H. Generation of lipopolysaccharide-induced interferon in spleen cell cultures. I. Genetic analysis and cellular requirements. *Immunogenetics* 12: 117–127 (1981).

Bacon, T. H., Devere-Tyndall, A., Tyrrell, D. A. J., Denman, A. M. and Ansell, B. M. Interferon system in patients with systemic juvenile chronic arthritis: In vivo and in vitro studies. *Clin. Exp. Immunol.* 54: 23–30 (1983).

Bondestam, M., Alm, G. V., and Foucard, T. Interferon production in children with undue susceptibility to infections. *Acta Poediatr. Scand.* 73: 197–202 (1984).

Boraschi, D., Censini, S., and Tagliabue, A. Macrophage antitumor activity: Impaired responsiveness to interferon-γ of macrophages from genetically defective mice. *Eur. J. Immunol.* 14: 1061–1063 (1984).

Brandt Pedersen, E., Haahr, S., and Mogensen, S. C. X-linked resistance of mice to high doses of herpes simplex virus type 2 correlates with early interferon production. *Infect. Immun.* 42: 740–746 (1983).

Dandoy, F., De Maeyer-Guignard, J., Bailey, D., and De Maeyer, E. Mouse genes influence antiviral action of interferon in vivo. *Infect. Immun.* 38: 89–93 (1982).

Dandoy, F., De Maeyer, E., and De Maeyer-Guignard, J. Antiproliferative action of interferon on murine bone-marrow derived macrophages is influenced by the genotype of the marrow donor. *J. Interferon Res.* 1: 263–270 (1981).

Davis, C. T., Blankenhorn, E. P., and Murasko, D. M. Genetic variation in the ability of several strains of rats to produce interferon in response to polyriboinosinic-polyribocytidylic acid. *Infect. Immun.* 43: 580–583 (1984).

De Gee, A. L. W., Sonnenfeld, G., and Mansfield, J. M. Genetics of resistance to the African trypanosomes. V. Qualitative and quantitative differences in interferon producing among susceptible and resistance mouse strains. *J. Immunol.* 134: 2723–2726 (1985).

Degré, M., Belsnes, K., Rollag, H., Beck, S., and Sonnenfeld, G. Influence of the genotype of mice on the effect of interferon on phagocytic activity of macrophages. *Proc. Soc. Exp. Biol. Med.* 173: 27–31 (1983).

De Maeyer, E. and De Maeyer-Guignard, J. A gene with quantitative effect on circulating interferon production: Further studies. *Ann. N.Y. Acad. Sci.* 173: 228–238 (1970).

De Maeyer, E. and De Maeyer-Guignard, J. Considerations on mouse genes influencing interferon production and action. In: *Interferon 1,* I. Gresser, Ed., Academic Press, London, pp. 75–100 (1979).

De Maeyer, E. and De Maeyer-Guignard, J. Host genotype influences immunomodulation by interferon. *Nature* 284: 173–175 (1980).

De Maeyer, E. and De Maeyer-Guignard, J. Delayed hypersensitivity to Newcastle disease virus in high and low interferon-producing mice. *J. Immunol.* 130: 2392–2396 (1983).

De Maeyer, E. and De Maeyer-Guignard, J. Interferon structural and regulatory genes in the mouse. *In: Interferons as Cell Growth Inhibitors and Antitumor Factors,* UCLA Symposia on Molecular and Cell Biology, New Series, Alan R. Liss, New York, Vol. 50, pp. 435–445 (1986).

De Maeyer, E., De Maeyer-Guignard, J., and Bailey, D. W. Effect of mouse genotype on interferon production. Lines congenic at the *If*-1 locus. *Immunogenetics* 1: 438–445 (1975a).

De Maeyer, E., Hoyez, M. C., De Maeyer-Guignard, J., and Bailey, D. W. Effect of mouse genotype on interferon production. *Immunogenetics* 8: 257–263 (1979).

De Maeyer, E., Jullien, P., De Maeyer-Guignard, J., and Demant, P. Effect of mouse genotype on interferon production. Distribution of *If*-1 alleles among inbred strains and transfer of phenotype by grafting bone marrow cells. *Immunogenetics* 2: 151–160 (1975b).

De Maeyer-Guignard, J., Dandoy, F., Bailey, D. W., and De Maeyer, E. Interferon structural genes do not participate in quantitative regulation of interferon production by *If* loci as shown in C57BL/6 mice that are congenic with BALB/c mice at the α interferon gene cluster. *J. Virol.* 58: 743–747 (1986).

De Maeyer-Guignard, J., Zawatzky, R., Dandoy, F., and De Maeyer, E. An X-linked locus influences early serum interferon levels in the mouse. *J. Interferon Res.* 3: 241–252 (1983).

Dreiding, P., Staeheli, P., and Haller, O. Interferon-induced protein Mx accumulates in nuclei of mouse cells expressing resistance to influenza virus. *Virology* 140: 192–196 (1985).

Ellermann-Eriksen, S., Justesen, J., and Mogensen, S. C. Genetically determined difference in the antiviral action of interferon in cells from mice resistant or susceptible to herpes simplex virus type 2. *J. Gen. Virol.* 67: 1859–1866 (1986a).

Ellermann-Eriksen, S., Liberto, M. C., Iannello, D., and Mogensen, S. C. X-linkage of the early in vitro α/β interferon response of mouse peritoneal macrophages to herpes simplex virus type 2. *J. Gen. Virol.* 67: 1025–1033 (1986b).

Engler, H., Zawatzky, R., Kirchner, H., and Armerding, D. Experimental infection of inbred mice with herpes simplex virus. IV. Comparison of interferon production and natural killer cell activity in susceptible and resistant adult mice. *Arch. Virol.* 74: 239–247 (1982).

Gallien-Lartigue, O., Carrez, D., De Maeyer, E., and De Maeyer-Guignard, J. Strain dependence of the antiproliferative action of interferon on murine erythroid precursors. *Science* 209: 292–293 (1980).

Gallien-Lartigue, O., De Maeyer-Guignard, J., Carrez, D., and De Maeyer, E. The antiproliferative effect of murine interferon α/β on early bone marrow-derived erythroid precursors (BFU/e). *J. Interferon Res.* 5: 347–354 (1985).

Grundy (Chalmer), J. E., Trapman, J., Allan, J. E., Shellam, G. R., and Melief, C. J. M. Evidence for a protective role of interferon in resistance to murine cytomegalovirus and its control by non-H-2-linked genes. *Infect. Immun.* 37: 143–150 (1982).

Haller, O. Antiviral activities of interferons: Evidence for control by host genes. *Behring Inst. Mitt.* 69: 1–9 (1981).

Haller, O., Arnheiter, H., Gresser, I., and Lindenmann, J. Genetically determined, interferon-dependent resistance to influenza virus in mice. *J. Exp. Med.* 149: 601–612 (1979).

Haller, O., Arnheiter, H., Gresser, I., and Lindenmann, J. Virus-specific interferon action. Protection of newborn Mx carriers against lethal infection with influenza virus. *J. Exp. Med.* 154: 199–203 (1981).

Haller, O., Arnheiter, H., Lindenmann, J., and Gresser, I. Host gene influences sensitivity to interferon action selectively for influenza virus. *Nature* 283: 660–662 (1980).

Horisberger, M. A. and Hochkeppel, H. K. An interferon-induced mouse protein involved in the mechanism of resistance to influenza viruses. Its purification to homogeneity and characterization by polyclonal antibodies. *J. Biol. Chem.* 260: 1730–1733 (1985).

Horisberger, M. A., Haller, O., and Arnheiter, H. Interferon-dependent genetic resistance to influenza virus in mice: Virus replication in macrophages is inhibited at an early step. *J. Gen. Virol.* 50: 205–210 (1980).

Horisberger, M. A., Staeheli, P., and Haller, O. Interferon induces a unique protein in mouse cells bearing a gene for resistance to influenza virus. *Proc. Natl. Acad. Sci.* (USA) 80: 1910–1914 (1983).

Huygen, K. and Palfliet, K. Influence of in vivo interferon treatment on mitogen responsiveness of T- and B-cells is dependent on the mouse genotype. *Immunol. Lett.* 5: 175–180 (1982).

Huygen, K. and Palfliet, K. In vitro production of γ interferon is dependent on the mouse genotype. *J. Interferon Res.* 3: 129–137 (1983).

Huygen, K. and Palfliet, K. Strain variation in interferon γ production of BCG-sensitized mice challenged with PPD. II. Importance of one major autosomal locus and additional sexual influences. *Cell. Immunol.* 85: 75–81 (1984).

Isaacs, D., Clarke, J. R., Tyrrell, D. A. J., Webster, A. D. B., and Valdman, H. B. Deficient production of leucocyte interferon (interferon-α) in vitro and in vivo in children with recurrent respiratory tract infections. *Lancet* ii: 950–952 (1981).

Krug, R. M., Shaw, M., Broni, B., Shapiro, G., and Haller, O. Inhibition of influenza viral mRNA synthesis in cells expressing the interferon-induced Mx gene product. *J. Virol.* 56: 201–206 (1985).

Lindenmann, J. The role of interferon in natural resistance. In: *Interferon 3,* I. Gresser, Ed., Academic Press, London, pp. 1–12 (1981).

Maluish, A. E., Ortaldo, J. R., Sherwin, S. A., Oldham, R. K., and Herberman, R. B. Changes in immune function in patients receiving natural leukocyte interferon. *J. Biol. Resp. Modif.* 2: 418–427 (1983).

Massa, P. T., Ter Meulen, V., and Vontana, A. Hyperinducibility of Ia antigen on astrocytes correlates with strain-specific susceptibility to experimental autoimmune encephalomyelitis. *Proc. Natl. Acad. Sci.* (USA) 84: 4219–4223 (1987).

Mobraaten, L. E., Bunker, H. P., De Maeyer-Guignard, J., De Maeyer, E., and Bailey, D. W. Location of histocompatibility and interferon loci on chromosome 3 of the mouse. *J. Hered.* 75: 233–234 (1984).

Neighbour, P. A. and Grayzel, A. I. Interferon production in vitro in leucocytes from patients with systemic lupus erythematosus and rheumatoid arthritis. *Clin. Exp. Immunol.* 45: 576–582 (1981).

Neta, R. Mechanisms in the in vivo release of lymphokines. II. Regulation of in vivo release of type II interferon (IFN-γ). *Cell. Immunol.* 60: 100–108 (1981).

Pugliese, A., Cortese, D., and Forni, G. Polygenic control of interferon production induced by poly I:C in vivo. *Arch. Virol.* 65: 83–87 (1980).

Purtilo, D. Epstein-Barr-virus induced oncogenesis in immune-deficient individuals. *Lancet* i: 300–303 (1980).

Purtilo, D. Immune deficiency predisposing to Epstein-Barr-virus-induced lymphoproliferative diseases: The X-linked lymphoproliferative syndrome as a model. *Adv. Cancer Res.* 34: 279–312 (1981).

Rivière, Y., Gresser, I., Guillon, J. C., Bandu, M. T., Ronco, P., Morel-Maroger, L., and Verroust, P. Severity of lymphocytic choriomeningitis virus disease in different strains of suckling mice correlates with increasing amounts of endogenous interferon. *J. Exp. Med.* 152: 633–640 (1980).

Sadick, M. D., Locksley, R. M., Tubbs C., and Raff, H. V. Murine cutaneous Leishmaniasis: Resistance correlates with the capacity to generate interferon-γ in response to Leishmania antigens in vitro. *J. Immunol.* 136: 655–661 (1986).

Salonen, R., Ilonen, J., Reunanen, M., and Salmi, A. Defective production of interferon-α associated with HLA-DW2 antigen in stable multiple sclerosis. *J. Neurol. Sci.* 55: 197–206 (1982).

Shirahata, T., Mori, A., Ishikawa, H., and Goto, H. Strain differences of interferon-generating capacity and resistance in toxoplasma-infected mice. *Microbiol. Immunol.* 30: 1307–1316 (1986).

Staeheli, P., Danielson, P., Haller, O., and Sutcliffe, J. G. Transcriptional activation of the mouse Mx gene by type I interferon. *Mol. Cell. Biol.* 6: 4770–4774 (1986b).

Staeheli, P., Dreiding, O., Haller, O., and Lindenmann, J. Polyclonal and monoclonal antibodies to the interferon-inducible protein Mx of influenza virus-resistant mice. *J. Biol. Chem.* 260: 1821–1825 (1985).

Staeheli, P. and Haller, O. Interferon-induced human protein with homology to protein Mx of influenza virus-resistant mice. *Mol. Cell. Biol.* 5: 2150–2153 (1985).

Staeheli, P., Haller, O., Boll, W., Lindenmann, J., and Weissmann, C. Mx protein: Constitutive expression in 3T3 cells transformed with cloned Mx cDNA confers selective resistance to influenza virus. *Cell* 44: 147–158 (1986a).

Staeheli, P., Horisberger, M. A., and Haller, O. Mx-dependent resistance to influenza viruses is induced by mouse interferons α and β but not γ. *Virology* 132: 456–461 (1984).

Staeheli, P., Pravtcheva, D., Lundin, L. G., Acklin, M., Ruddle, F., Lindenmann, J., and Haller, O. Interferon-regulated influenza virus resistance gene Mx is localized on mouse chromosome 16. *J. Virol.* 58: 967–969 (1986c).

St Geme, J. W., Jr., Prince, J. T., Burke, B. A., Good, R. A., Krivit, W. Impaired cellular resistance to herpes simplex virus in Wiskott-Aldrich syndrome. *N. Engl. J. Med.* 273: 229–232 (1965).

Virelizier, J. L. Murine genotype influences the in vitro production of γ (immune) interferon. *Eur. J. Immunol.* 12: 988–990 (1982).

Virelizier, J. L., Lenoir, G., and Griscelli, C. Persistent Epstein-Barr virus infection in a child with hypergammaglobulinaemia and immunoblastic proliferation associated with a selective defect in immune interferon secretion. *Lancet* ii: 231–234 (1978).

Weck, P. K., May, L., and Weck, C. J. γ interferon production by different populations of human splenic lymphocytes. *J. Interferon Res.* 3: 121–128 (1983).

Zawatzky, R., Kirchner, H., De Maeyer-Guignard, J., and De Maeyer, E. An X-linked locus influences the amount of circulating interferon induced in the mouse by herpes simplex virus type 1. *J. Gen. Virol.* 63: 325–332 (1982).

16 THE PRESENCE AND POSSIBLE PATHOGENIC ROLE OF INTERFERONS IN DISEASE

In the course of several diseases, some of no apparent viral origin, IFNs can be present in the circulation. This is the case for disease states accompanied by immune dysfunction, which are frequently characterized by the presence of IFN-γ and/or IFN-α. The presence of IFN-γ is indicative of T-cell recognition by antigens, and therefore is not unexpected in diseases with an autoimmune component. The presence of IFN-α is more ambiguous, since this IFN can be made by either antigen-stimulated lymphocytes or virus-infected cells. One of the problems here is to discern between the presence of IFNs as a symptom of disease or as active contributors to pathogenesis.

Although there is not the slightest doubt that the role of IFN-α/β in the great majority of viral infections is beneficial, in some diseases of known viral origin, IFN, produced as a result of the infection, paradoxically contributes to the disease. In this chapter we discuss some examples of virus infections in which IFNs become the enemy within and exacerbate the disease.

I. AUTOIMMUNE DISEASE

In the previous chapters we have seen that IFNs belong to a complex network of regulatory mediators, which are important for the activation of macrophages and other accessory cells, and for T- and B-cell functions. Because of the multiple interactions of IFNs and the immune system, or, more precisely, because all IFNs are part of the immune system, it is not surprising that autoimmune disease can be accompanied by anomalies of the IFN system. One of the major problems in this respect is to find out what is cause and effect and determine the pathogenic role of the anomalies of IFN production that are associated with autoimmune disease.

Autoimmune disease represents a pathological state resulting from the action of antibodies and T cells against self-antigens. This breakdown in tolerance to self can arise when either abnormal T-helper cells are induced or certain T-suppressor cells, responsible for self-tolerance, are inhibited. Contrary to the general opinion held a few years ago that auto-antibodies are incompatible with health, it has now become evident that the presence of low levels of auto-antibodies is quite common under normal conditions.

Newborn mice of many different inbred strains display a high frequency of natural antibodies to a variety of antigens, such as erythrocytes, thymocytes, actin, tubulin, and myosin (Cunningham, 1974; Martin and Martin, 1975; Dresser, 1978; Steele and Cunningham, 1978; Dighiero et al., 1985). Naturally occurring auto-antibodies of different specificities are also present in normal human serum (Dighiero et al., 1982; Guilbert et al., 1982).

Lymphocytes that are capable of making monoclonal auto-antibodies reacting with antigens from many different organs are apparently a common feature of the normal B-cell repertoire (Prabhakar et al., 1984). Manifestly, clones of self-recognizing lymphocytes are not inevitably deleted during neonatal life, and this lack of deletion contradicts that part of the clonal selection theory that predicts the destruction of "forbidden" autoreactive clones in early life (Burnet, 1959). Since most of the time the presence of auto-antibodies does not result in autoimmune disorder, there must be regulatory mechanisms that keep such antiself clones in check and prevent them from expanding and causing disease. Furthermore, it is not certain whether these naturally occurring antibodies are the same as those that are responsible for autoimmune disease.

B cells with the potential for making auto-antibodies are apparently not the exception but the rule. Autoimmunity as represented by self-reacting B cells, however, is quite different from autoimmune disease, in which the proliferation of self-reacting clones, for example, antireceptor antibodies, can become a source of pathology. These self-reacting clones not only consist of B cells, but also consist of cytotoxic T cells with the potential for destruction of specific organs such as the thyroid gland. Very often one finds at the same time evidence for T- and B-cell autoimmunity, for example, in juvenile-onset diabetes mellitus, in which both T and B cells are directed against antigens of the β cells in the islets of Langerhans.

Studies in thymectomized mice suggest that loss of suppressor cells can cause clones of antiself lymphocytes to expand, but the primary cause of inactivation of specific suppressor cells is unknown. It is certainly intriguing that the onset of autoimmune disease is often accompanied by a viral infection. Diabetes can be induced in mice by infection with a number of viruses, such as reovirus, encephalomyocarditis virus, and coxsackievirus B4 (Onodera et al., 1981, 1982). In mice of the SJL/J strain, infection with reovirus type 1 stimulates the formation of auto-antibodies that react with endocrine tissues and with nuclei (Haspel et al., 1983). By infecting and transforming normal human B cells, Epstein-Barr virus can trigger polyclonal expansion characterized by the production of multiple organ-reacting auto-antibodies (Garzelli et al., 1984). Since virus infection and activation of the IFN system are intimately related, the possibility of an involvement of the IFN system in certain autoimmune states merits serious consideration, especially in view of the immunomodulating effects of IFNs. Many questions about autoimmunity in general and its connection to the action of IFNs in particular are waiting to be answered.

A. Systemic Lupus Erythematosus

Systemic lupus erythematosus (SLE) is a complex, multisystem autoimmune disease. The name is derived from the bright facial rash that often accompanies the disease, and which bears some resemblance to the rash developing as a result of tuberculosis of the skin, known as lupus vulgaris. The great majority of cases occur in young adult females.

A characteristic feature of SLE is the deposit of immune complexes, frequently involving nucleic acids, in many tissues. The antinucleic acid antibodies can be directed against RNA as well as against single- or double-stranded DNA. In addition, lupus patients usually have auto-antibodies that are directed against many different tissues. The disease is characterized by exacerbations and remissions of unknown origin, with arthritis and nephritis frequently as dominant features. The clinical, serological, and pathological findings are extremely variable in different patients and at different times in the same patient. The disease, therefore, has often been characterized as multifactorial, resulting from a combination of different pathogenetic factors, but its primary cause is not known.

1. IFN-α Is Present in the Circulation of Many SLE Patients

A special type of IFN, also found in several other disease states, is often present in the serum of SLE patients. It is antigenically related to Hu IFN-α, but unlike all the Hu IFN-α species that have been structurally characterized so far, it is acid-labile. In many patients the more conventional acid-stable form of IFN-α is also found in the circulation, and the acid-stable and acid-labile forms can be present simultaneously (Skurkovich and Eremkina, 1975; Hooks et al., 1982; Preble et al., 1982; Ytterberg and Schnitzer, 1982). Hu IFN-α can also be present in the cerebrospinal fluid of SLE patients with neurological complications (Lebon et al., 1983).

The acid-labile IFN-α present in lupus patients and in autoimmune disease in general has not been purified or further characterized. Except for the fact that it is neutralized by anti-IFN-α sera and is inactivated at pH 2, nothing is known about its physicochemical properties, size, structure, or cellular origin. SLE patients with active disease have a higher frequency (about 70 to 80%) of interferonemia than patients in remission (about 10 to 40%), depending on the study and probably also on the sensitivity of the method used for detection. In some SLE patients, circulating IFN-α cannot be demonstrated, but increased 2-5A synthetase levels are found in their lymphocytes, which is symptomatic of a low IFN production, not accompanied by overspill into the circulation (Preble et al., 1983). The presence of increased 2-5A synthetase levels indicates that the circulating IFN-α is biologically active in these patients, since activation of the 2-5A pathway is one of the hallmarks of IFN-receptor interaction (see Chapter 5). The presence of circulating IFN provides an explanation for the observation that herpes zoster usually does not occur in lupus patients with active disease, but rather, occurs during remissions (Moutsopoulos et al., 1978). During active disease the circulating IFN-α apparently provides some protection against this infection.

Is the presence of IFN in any way involved in the pathogenesis of SLE, or is it just another symptom of immune dysfunction? There is no definite answer to this question, and we can only engage in theoretical considerations and evoke a few possibilities. Some observations indicate that the level of T-suppressor cells is abnormally low in SLE patients, as it is in mice of the NZB strain, which have an autoimmune disease resembling SLE in humans (Klassen et al., 1977; Cantor et al., 1978; Sakane et al., 1978). Low amounts of Mu IFN-α/β selectively block the T-suppressor circuit of delayed hypersensitivity in mice, without affecting T-effector cells, and preferential inhibition of suppressor cells by IFNs has also been observed in other systems (see Chapter 12). The presence of IFN in the circulation of lupus patients, therefore, could theoretically be either responsible for or, at the least, contribute to a decreased suppressor-cell activity. The amount of IFN produced by SLE patients appears sufficient to disturb immune function, if one can go by the figures obtained from patients who have been monitored for several months: Significant levels of acid-labile IFN are constantly present in their serum during clinically active disease (Hooks et al., 1980). Since IFN released into the circulation often results from an "overflow" of local production, concentrations could be considerably higher at the sites of production. A serum concentration of 128 units/ml is a value reported for many patients; this corresponds to a total of about 3.8×10^5 units being present at a given time if 3,000 ml is taken as the average plasma volume of the human adult. The normal clearance rate of IFN, when injected directly into the circulation, has been variously estimated to correspond to a half-life ranging from about 15 minutes to 1 hour at the most (Emödi et al., 1975; Bino et al., 1982; Borneman et al., 1985). After intramuscular injection of several million units of rec Hu IFN-αA, however, IFN is released slowly into the circulation and the serum half-life is much longer and ranges from 6 to 8 hours (Gutterman et al., 1982a). Assuming that the IFN-α clearance rates in SLE patients correspond to the slowest values observed after inoculation, one can estimate a production of at least 1.4×10^5 units per 8 hours, going on for several months. This represents a considerable output of IFN-α—much higher than the production of a relatively short duration that takes place during acute viral infections. This calculation admittedly is hypothetical because the actual clearance rate of circulating IFN in SLE patients is not known, but even a significantly lower than normal value still implies an IFN-α production unlike anything observed under normal conditions. The amount produced is certainly enough to exert a disturbing effect on immune functions and to cause some of the clinical symptoms such as fever, general malaise, and muscle pain, which are the side effects of IFN-α administration to patients.

The hypothesis that the acid-labile IFN-α could inhibit suppressor-cell activity and exert other immunomodulating effects implies that it has the same biological activity as other IFNs, a likely assumption, but without extensive experimental support.

A substantial number of SLE patients have decreased NK-cell activity (Hoffman, 1980; Strannegard et al., 1982; Tsokos et al., 1982). Since repeated administration of Hu IFN-α to patients, or of Mu IFN-α/β to mice, results in down-regulation of NK-cell activity (see Chapter 11), the impairment of NK function in SLE patients most likely results from the presence of IFN-α in the circulation. Indeed, elevated levels of circulating IFN-α in SLE patients seem to be correlated with the degree of unresponsiveness to stimulation by IFN of their NK cells in vitro (Sibbitt et al., 1985).

2. *IFN-Induced Tubuloreticular Inclusions*

The endoplasmic reticulum of reticuloendothelial cells and peripheral blood lymphocytes from patients with SLE frequently contains inclusions of unknown origin, called tubuloreticular or lupus inclusions, consisting of abnormal microtubular structures that can be viruslike in appearance. Such structures are not limited to SLE, but can be found in other autoimmune diseases and in immunodeficiency disorders. Human leukocyte IFN and Hu IFN-β can induce similar inclusions in normal lymphocytes in vitro, which raises the possibility that the circulating IFN-α of lupus patients is responsible for these inclusions (Rich, 1981). This eventuality receives additional support from the appearance of similar inclusions in lymphocytes and monocytes of individuals receiving IFN-α treatment (Preble and Friedman, 1983; Rich et al., 1983). Moreover, similar tubular aggregates, associated with the endoplasmic reticulum of hepatocytes, are induced in newborn mice that were treated with high amounts of IFN-α/β (Fig. 16.1). Infection with lymphocytic choriomeningitis virus results in the appearance of similar lesions, at the time of high interferonaemia (Moss et al., 1982). Tubular structures also exist in some mouse lymphoma cells in culture; upon treatment of these cells with Mu IFN-α/β, the structures undergo a profound alteration and clusters of viruslike particles appear (Hochman et al., 1985). In spite of the suggestive morphology, however, there is no convincing evidence otherwise that these structures represent viruses, and the clinical and biological significance of IFN-induced tubuloreticular lesions remains to be determined.

The continuous production and release into the circulation of acid-labile IFN-α in lupus patients does not imply that the capacity for IFN production in general is abnormally high in such patients. Quite to the contrary, and paradoxically, peripheral blood leukocytes from SLE patients produce less IFN-α/β than do leukocytes from healthy individuals when induced in vitro with the Newcastle disease virus, measles virus, or polyrIrC, and they also produce less IFN-γ after mitogen induction. In fact, as compared to normal responders, an important fraction of SLE patients score as nonresponders in such in vitro production assays (Neighbour and Grayzel, 1981; Strannegard

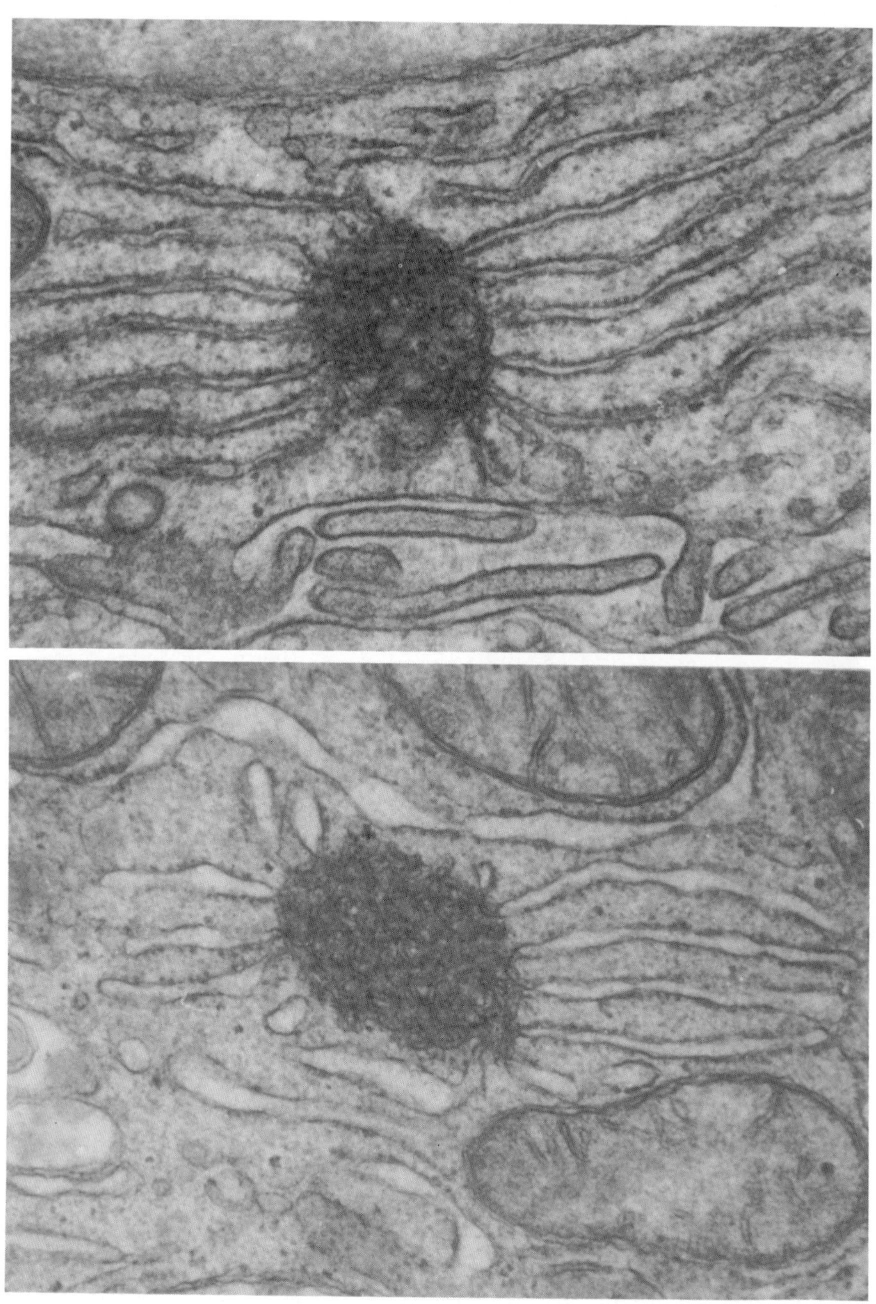

Figure 16.1 The presence of tubular aggregates in livers from newborn mice treated with Mu IFN-α/β or infected with lymphocytic choriomeningitis virus. Upper panel: Cytoplasm of an hepatocyte from a 4-day-old Swiss mouse after four daily injections of IFN. Note the apparent continuity of the membranes of the tubular aggregate with those of the granular endoplasmic reticulum. Uranyl acetate and lead citrate; magnification 46,000. Lower panel: Cytoplasm of an hepatocyte from a 4-day-old C3H mouse after neonatal injection of lymphocytic choriomeningitis virus. Note again the apparent continuity of the tubular aggregate with the granular endoplasmic reticulum. Uranyl acetate and lead citrate. Magnification 46,000. (From Moss et al., 1982, with permission.)

et al., 1982; Tsokos et al., 1982; Preble et al., 1983). The reason for this hyporesponsiveness is obscure. In cell cultures, when low levels of spontaneous IFN production are present, exactly the reverse is usually observed: Virus-induced IFN production is enhanced as a result of the priming effect, whereby pretreatment with low doses of IFN boosts subsequent IFN production. When relatively high amounts of circulating IFN are induced in animals, however, a decreased production upon subsequent induction is frequently observed. This may be the case for SLE patients as a result of prolonged exposure to endogenous IFN-α. The effect of acid-labile IFN-α on IFN induction in general could be one of hyporesponsiveness, but it is also possible that the impaired IFN production in vitro by leukocytes of SLE patients is due to other factors.

Lupus patients make auto-antibodies to a wide variety of antigens, and the presence of anti-IFN-α antibodies of the IgG type has been detected in a few patients, out of many examined (Panem et al., 1982). The presence of these antibodies is most likely responsible for the IFN-α-containing immune deposits in the kidneys of lupus patients (Panem et al., 1983; Panem, 1984).

What could be the cause of the continuous presence of Hu IFN-α—acid-labile or other—in lupus patients, and in patients with other autoimmune diseases as well? Normally, IFN of any species is not present in the circulation of healthy individuals; the presence of IFN-α or IFN-β usually denotes acute viral infection, and the presence of IFN-γ is a sign of antigen recognition by T cells. In the latter case, the levels produced are usually so low that one would not expect to find IFN in the serum, except in some animal model studies in which conditions for massive production have been worked out. Acid-stable Mu IFN-α/β, however, is sometimes produced in mice as the result of antigen recognition, which means that the presence of acid-stable IFN-α is not invariably a sign of viral infection, but can also be an indicator of immune activity (Nakane et al., 1982). Whereas mitogenic lectins usually induce IFN-γ, nonmitogenic lectins can induce Mu IFN-β in mouse spleen cells, again demonstrating that the presence of an acid-stable IFN is not automatically a sign of virus infection (Ito et al., 1984). Acid-labile IFN-α has been found in human lymphocyte cultures stimulated by *Corynebacterium parvum,* in human leukocytes induced by Sendai virus, and in influenza virus-induced lymphocytes from individuals previously vaccinated against this virus (Vilcek et al., 1980; Balkwill et al., 1983; Matsuoka et al., 1985). This shows that this particular IFN species can be made as a result of specific and nonspecific immune stimulation by viral and nonviral antigens, but does not exclude the possibility of other ways of production. For example, infection of a thymic lymphosarcoma line with a feline retrovirus results in the production of an acid-labile IFN-α; therefore, the presence of acid-labile IFN-α could also imply retroviral activity (Yamamoto et al., 1986). Is the circulating acid-labile IFN-α in SLE patients then the result of the continuous stimulation of their lymphocytes by autoantigens or maybe by a retrovirus? We certainly do not know what serves as the stimulus for IFN production in lupus patients, but antigenic stimulation of lymphocytes sensitized to self-antigens appears as one serious possibility.

3. Different Ways in Which IFNs Could Contribute to the Pathogenesis of SLE

One can envisage several ways by which the IFN that is produced in this manner then contributes to the disease in a positive feedback loop:

1. By forming immune complexes with the spontaneous auto-antibodies directed against Mu IFN-α that are present in New Zealand mice, and apparently also sometimes in humans.
2. By interfering with T-cell function, for example, suppressor cells, as discussed earlier.
3. By stimulating immunoglobulin production by B cells. B-cell hyperactivity is one of the dominant features of SLE, both in humans and in mice, and B-cell hyperactivity associated with SLE may be the result of abnormal responses to, as well as hyperproduction of, B-cell activating factors (Theofilopoulos et al., 1983). As discussed in Chapter 11, IFNs α, β, and γ can function as B-cell differentiation factors and the continuous presence of significant levels of IFN in the circulation could conceivably contribute to enhanced immunoglobulin synthesis. The accelerated appearance of anti-dsDNA antibodies in Mu IFN-α/β-treated mice that are prone to autoimmune disease shows that this possibility is more than theoretical (Sergiescu et al., 1981).
4. By enhancing expression of class I and by enhancing or inducing the expression of class II MHC cell-surface antigens, and thus stimulating auto-antigen specific T and B cells. Initiation of many immune responses depends on the presentation of antigen to T cells by antigen-presenting cells. The expression of MHC class II antigens is a condition *sine qua non* for a cell to function as an antigen presenting cell, and, contrary to MHC class I antigens that are normally expressed on all nucleated cells of the organism, the expression of class II antigens is usually limited to cells belonging to the immune system. In humans, macrophages constitutively express class II antigens, but many other potential antigen-presenting cells do not. Class II expression has to be induced or enhanced to favor interaction between antigen presenting cells and T cells.

A stimulating involvement of class II antigens in autoimmunity is revealed by the effect of treatment with antibodies specifically directed against MHC class II molecules. Such antibodies, when administered to mice with autoimmune disease, decrease the number of autoreactive B-cell clones (Klinman et al., 1986). Experimental allergic encephalitis is an inflammatory disease of the central nervous system, resulting in paralysis, which is caused by T cells that are autoreactive to myelin basic protein. Treatment of mice with ongoing disease with anti-MHC class II antibodies results in decreased morbidity and mortality, demonstrating that class II antigens are a stimulating factor in the pathogenesis of this autoimmune disease (Steinman et al., 1986). More specifically to the point of the present discussion, treatment of NZB/NZW F1 mice with antibodies directed against MHC class II antigens significantly increases the survival of animals that already exhibit signs of renal disease, one of the complications of the spontaneous autoimmune syndrome with strong resemblance to LE that develops in NZB/NZW F1 mice (see subsequent discussion) (Adelman et al., 1983).

As discussed in Chapter 9, IFNs, including IFN-α, are excellent inducers of MHC class II antigens on B cells, T cells, and accessory cells, and the continuous presence of IFN-α is bound to stimulate autoimmune reactions by up-regulating the expression of class II antigens.

4. What Can Be Learned from the Study of IFN in Animal Models for SLE?

a. New Zealand Mice. New Zealand black (NZB) mice and the F1 hybrids of NZB and New Zealand white (NZW) mice spontaneously develop an autoimmune disease that is comparable to lupus erythematosus in humans. The disease is characterized by antinuclear antibodies, the appearance of typical lupus erythematosus cells (LE cells), glomerulonephritis, and premature death due to renal failure (Lambert and Dixon, 1968; Huston and Steinberg, 1979).

Because of the great similarity of the animal disease to SLE in humans, an obvious question is: Does a special, acid-labile IFN-α, or any other IFN for that matter, appear in the circulation of these animals once the disease has developed? This has apparently—and surprisingly—not been investigated in great detail, and there is only one report in the literature that mentions in one sentence that there are no "significant" levels of circulating IFN in NZB/NZW F1 mice (Engleman et al., 1981). In section b we will discuss another mouse model, provided by the MRL/*lpr* strain, in which circulating IFN-γ is present when the disease develops. NZB mice, like mice of many other inbred strains, have low levels of spontaneous anti-Mu IFN-α and Mu IFN-β auto-antibodies, and these antibodies could conceivably mask the presence of low levels of circulating IFN-α (De Maeyer-Guignard and De Maeyer, 1986).

The NZB mouse models have been used to investigate the effects of IFNs on the spontaneous occurrence of the disease. Treatment of NZB and of NZB/NZW F1 mice with either Mu IFN-α/β or with Mu IFN-γ accelerates the onset of autoimmune disease and causes death at an earlier age than the untreated controls (Fig. 16.2) (Heremans et al., 1978; Adam et al., 1980; Engleman et al., 1981). Moreover, treatment of NZB/NZW F1 mice with anti-IFN-γ antibodies results in a significant remission of the autoimmune

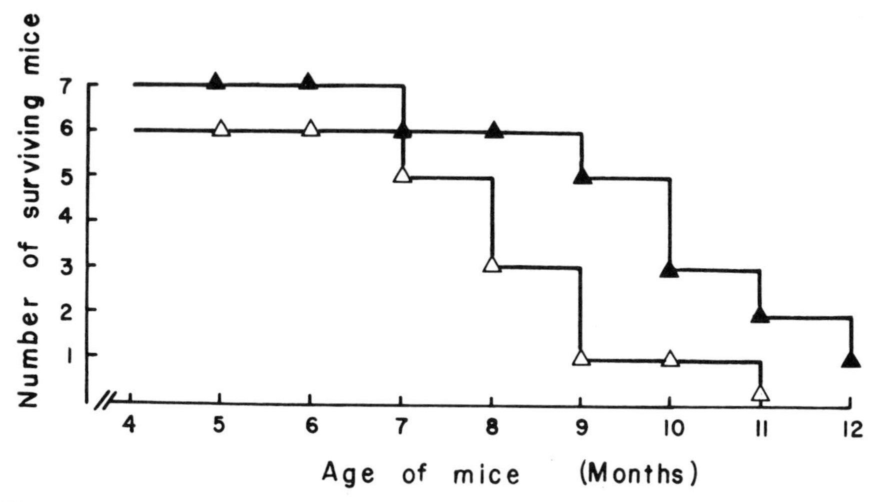

Figure 16.2 Mu IFN-γ accelerates the onset of autoimmune disease in NZB/NZW F1 mice and causes death at an earlier age than in untreated controls. Survival of NZB/NZW F1 mice with (△—△) and without (▲—▲) IFN-γ treatment. (Adapted from Engleman et al., 1981.)

disease and a prolonged survival of the animals, thereby showing that endogenous IFN-γ is instrumental in the pathogenesis of the disease (Jacob et al., 1987). In Mu IFN-α/β-treated NZB/NZW F1 mice, the level of anti-dsDNA antibodies begins to increase at 4 to 6 months of age, whereas untreated mice do not display similar levels until the age of 12 months (Sergiescu et al., 1981). This highlights the possibility that the presence of IFN in lupus patients does play a role in the pathogenesis of the disease, by either inhibiting suppressor-cell function, enhancing MHC class II expression, or stimulating B-cell differentiation as discussed earlier, or by some other unknown mechanism.

b. Other Lupus-Prone Mouse Strains. Mice of the MRL-*lpr*/*lpr* and of the BXSB strains also develop a lupuslike spontaneous autoimmune disease. In both strains multiple genes determine susceptibility to the disease, but a gene with major accelerating influence on the onset of disease is located on the Y chromosome in BXSB mice, and the recessive *lpr* (*lpr* stands for "lymphoproliferation") allele has a similar accelerating effect on the disease in MRL mice when homozygous. B cells of young BXSB males, but not B cells of MRL-*lpr*/*lpr* mice, are more responsive to B-cell growth and differentiation factors than normal B cells, and this hyperreactivity could explain polyclonal B-cell activation and high output of auto-antibodies in BXSB mice. B cells of the MRL-*lpr*/*lpr* strain appear to have a normal degree of responsiveness to accessory signals, but the *lpr* gene causes a massive proliferation of a T-cell subset that spontaneously secretes a B-cell differentiation factor (Theofilopoulos et al., 1983).

Contrary to the decreased IFN-γ production in cell suspensions derived from SLE patients, mitogen-induced IFN-γ production in MRL-*lpr/lpr* mice is comparable to that of control animals. The production of IL-2, however, is greatly decreased. In addition, four-month-old mice of the MRL-*lpr/lpr* strain, at an age when they develop clinical disease, have low levels of Mu IFN-γ in their circulation, in spite of the decreased production of IL-2. The continuous presence of IFN-γ could explain the increased number of MHC class II antigen-expressing macrophages, typical for autoimmune MRL-*lpr/lpr* mice (Dauphinée et al., 1981; Santoro et al., 1983; Steinberg et al., 1983; Rosenberg et al., 1984).

B. Behçet's Disease

Circulating IFN can be present in the serum of patients with several other diseases, in addition to SLE.

Behçet's disease is a disorder of possible autoimmune origin, characterized by typical eye lesions with retinal vasculitis, uveitis, and retinochoroiditis, as well as oral and genital inflammatory lesions. It occurs most frequently among Japanese and Mediterranean populations. The cause of this disease is unknown (Lehner, 1979). Patients having this disease have significant levels of circulating acid-labile IFN, consisting of a mixture of IFN-γ and acid-labile IFN-α (Hooks et al., 1982; Ohno et al., 1982). Cultures of peripheral blood mononuclear leukocytes obtained from patients with this disease produce spontaneous IFN-γ. Like all autoimmune diseases, Behçet's disease is characterized by spontaneous remissions and exacerbations. The spontaneous IFN-γ production in leukocyte cultures from convalescent patients is much higher than that in leukocyte cultures from patients in an exacerbation stage (Fujii et al., 1983). This correlation is exactly the opposite of the one observed in lupus patients, in which the incidence of circulating acid-labile IFN-α is higher during exacerbation than during remission. We are totally in the dark as to what this means in terms of pathogenesis, exacerbations, remissions, or any other aspect of the disease.

C. Insulin-Dependent Diabetes Mellitus

Diabetes mellitus is a disease in which glucose can no longer be assimilated, resulting in hyperglycemia and glycosuria. Breakdown of tissue proteins, lipids, and fatty acids is accelerated, leading to the production of excessive amounts of ketones. The accumulation of these acids in the blood, a condition called ketoacidosis, can bring the pH below the critical level of 7.4, with ensuing coma and death.

Insulin-dependent diabetes mellitus (type I diabetes) is characterized by a high incidence of antibodies against pancreatic islet cells. It is also called juvenile onset diabetes, but since its onset is not restricted to childhood or adolescence, the designation of insulin-dependent diabetes mellitus (IDDM) is to be preferred. Noninsulin-dependent diabetes (type II diabetes), also called maturity onset diabetes although it can affect juveniles, represents a different and heterogeneous group of disorders, and is strongly associated with obesity. This is by far the most common type of diabetes, since fewer than 25% of diabetics have insulin-dependent disease. We are only concerned with the latter form, which has an autoimmune origin and can be an aftermath of a virus infection.

1. Evidence for an Autoimmune Component in the Pathogenesis of IDDM

The polypeptide hormone insulin, which plays a key role in glucose transport and metabolism, is made exclusively in the β cells of the pancreatic islets of Langerhans. In IDDM or type I diabetes, the β cells disappear from the islets and insulin can no longer by synthesized, resulting in hyperglycemia and ketoacidosis.

There are good indications that the destruction of β cells is very often the result of an autoimmune process in which both auto-antibodies and cytotoxic T cells are involved. Auto-antibodies against surface or cytoplasmic structural features of β cells are present in many patients, often prior to the onset of overt diabetes. Some of these anti-islet cell antibodies have the capacity to mediate complement-dependent as well as antibody-dependent cellular cytotoxicity, two immune effector mechanisms that can be influenced by IFNs (Papadopoulos and Lernmark, 1983). Furthermore, anti-islet cell antibodies are also present in rats of the BB/W strain, a strain that offers an interesting animal model because of its propensity for spontaneously developing IDDM. This strain was accidentally discovered as a mutation in a commercial breeding colony in Canada (Biobreeding), and one of the main sites of breeding is now Worcester, Mass., hence BB/W. The presence of anti-islet auto-antibodies in BB rats strongly suggests a pathogenic role (Marliss et al., 1982).

In addition to anti-islet cell antibodies, spontaneous antibodies against insulin can also be present in insulin-dependent diabetics. These antibodies are not a result of immunization through insulin treatment, since they are present before treatment is initiated (Palmer et al., 1983).

Cytotoxic T cells, directed against β islets, appear involved in islet destruction, and the peripheral blood of many patients with IDDM contains cytotoxic cells that react in vitro with pancreatic antigens (Volpé, 1981). Moreover, leukocytic infiltration of damaged islets of Langerhans (insulitis) is a common histopathological feature associated with certain types of IDDM in humans.

As in lupus patients and in some other autoimmune diseases, significant amounts of IFN-α can be present in the circulation of many diabetics, especially when the serum is examined less than 3 months after onset of the disease. Also, about 30% of children with newly diagnosed disease have low levels of circulating IFN-γ (Tovo et al., 1984).

In the aforementioned rats of the BB strain, in which IDDM usually develops during or shortly after puberty, the islets of Langerhans are infiltrated by lymphocytes and macrophages, and diabetes can be prevented by neonatal thymectomy or by immunosuppressive treatment such as antilymphocyte antibodies or cyclosporin A. Furthermore, the disease can be transmitted to nondiabetic BB rats with concanavalin A activated spleen cells derived from diabetic animals. Moreover, autoreactive anti-islet specific T cells are clearly present in these diabetic rats since several T-cell lines that can be stimulated with rat islet cell antigens have been isolated from the spleen and pancreas of newly diabetic BB rats (Koevary et al., 1983; Prud'homme et al., 1985). Contrary to several murine models, so far there is no evidence that viruses are involved in the etiology of diabetes in BB rats.

2. Insulin-Dependent Diabetes Occurs Frequently After Virus Infection

A link between IDDM and some virus infections has been suspected for many years, since pancreatitis is a frequent complication of mumps, and there is good evidence for a temporal association between mumps and the onset of diabetes. In addition to mumps, other viruses have been suspected of an etiological involvement with IDDM, the most important of these being influenza, measles, rubella, cytomegalo, and hepatitis viruses, as well as several coxsackie B viruses. In humans, evidence for a causal involvement of virus infection and IDDM remains largely circumstantial and is mainly based on epidemiological studies and on the observation that there is a correlation between the onset of diabetes and a previous viral illness or a rise of antibodies to certain viruses, attesting of recent infection. Such studies suggest, for example, a strong link between congenital rubella and IDDM, since children with congenital rubella have over ten times the normal incidence of IDDM in the population (Maugh, 1975; Rayfield and Seto, 1978). Direct evidence that a virus infection can trigger diabetes comes from a few instances in which viruses have been identified in or isolated from the pancreas of patients with acute onset diabetes resulting in terminal ketoacidosis. The viruses that have been thus identified are coxsackie B4, cytomegalovirus, and rubella (Gladish et al., 1976; Yoon et al., 1979; Jenson et al., 1980; Patterson et al., 1981). A link between the clinic and animal model studies, furthermore, is provided by the observation that coxsackie B4 virus can become diabetogenic in mice after repeated passage in murine β cell cultures (Yoon et al., 1978). Moreover, two other coxsackie viruses, B3 and B5, freshly isolated from patients with IDDM, have been shown to induce diabetes in mice. The study of virally induced diabetes in mice, therefore, provides information that is of obvious relevance to the pathogenesis of diabetes in humans, and furthermore, makes it possible to investigate the involvement of autoimmunity and IFNs.

3. Virus-Induced IDDM in Mice

In addition to the aforementioned Coxsackie viruses, several other viruses have diabetogenic potential in mice, and IFNs can influence the pathogenesis of the disease.

a. EMC Virus-Induced Diabetes. The EMC virus (encephalomyocarditis virus), like coxsackie viruses, belongs to the picornavirus family. Although the EMC virus usually infects mice, it can also occasionally infect humans. In mice of certain inbred strains such as SWR and DBA/1, the M (myocardiotropic) variant of EMC virus infects and destroys β islet cells, resulting in diabetes. Furthermore, the M strain is also cardiotropic and neurotropic, and infected animals can die from myocarditis and encephalitis. Mice of other strains, for example, C57BL/6, AKR, and A, are resistant to the development of diabetes during EMC infection. The difference between resistant and susceptible strains is quantitative rather than qualitative, since even in resistant mouse strains some β islet cells can be infected, but not in sufficient quantity to cause clinical disease. Susceptibility to EMC-induced diabetes, but not to EMC infection, is controlled by one major, non-MHC-linked locus together with several other loci exerting minor effects (Craighead and McLane, 1968; Ross et al., 1976; Onodera et al., 1978).

Host genes manifestly influence the induction of diabetes by EMC virus, but viral genes are also important, since only the D (for diabetes) variant of the EMC M strain is capable of causing diabetes, whereas another variant, the B (nondiabetogenic) strain, fails to produce diabetes in infected mice. Both virus strains are antigenically similar, and the genomic difference seems limited to only one or a few nucleotides (Ray et al., 1983). These two EMC variants elicit indistinguishable antibody responses in infected mice, but they differ in their ability to induce IFN-α/β in vitro and in vivo. The nondiabetogenic B virus induces high titers of early appearing circulating Mu IFN-α/β, whereas infection with the D variant elicits lower titers of IFN, which, furthermore, appear later (Fig. 16.3). Also, in vitro, the IFN-inducing potential of the B variant is significantly higher than that of the D variant, and this difference between the two virus strains suggests that IFN is one of the factors limiting the number of β islet cells that become infected with EMC virus. In fact, when mice are infected with the high IFN-inducing, nondiabetogenic B variant and at the same time or somewhat later with the low IFN-inducing diabetogenic D variant, the triggering of diabetes by the latter is inhibited. Apparently, IFN induced by one virus protects against β islet destruction by the other (Yoon et al., 1980; Cohen et al., 1983). The protective role of endogenous IFN-α/β production in limiting the onset of diabetes can be directly demonstrated by the use of anti-IFN-α/β antibodies. When mice infected with the nondiabetogenic, high-IFN-inducing B variant are treated with anti-IFN antibodies during the infection, a significant percentage develops diabetes, whereas none of these mice develop diabetes in the absence of anti-IFN antibodies. Administration of Mu IFN-α/β to mice prior to or at the same time as infection with the low-IFN-inducing EMC D variant reduces the development of diabetes. This effect in all likelihood results from the general property of IFNs to limit viral replication (Yoon et al., 1983).

In addition to genotype, the onset of EMC-induced diabetes in some

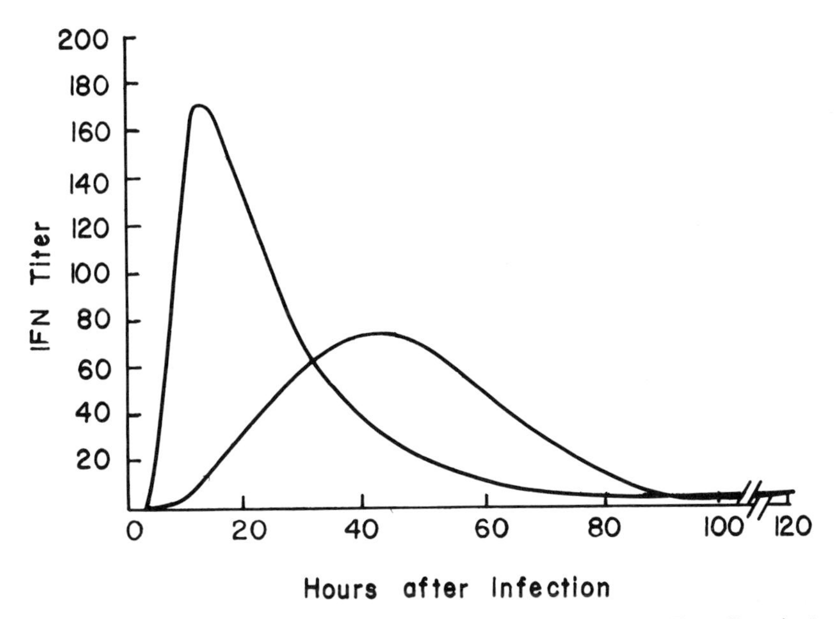

Figure 16.3 IFN levels in the circulation of mice infected with the B or D variants of encephalomyocarditis virus. The nondiabetogenic B virus induces high titers of early appearing IFN-α/β, peaking at 180 units, whereas the diabetogenic D variant elicits lower titers, peaking at 60 units, which furthermore appear later. (Adapted from Yoon et al., 1980).

inbred mouse strains is influenced by the sex of the animal. For example, in the DBA/2 strain, female individuals are more resistant than males (Boucher et al., 1975). This reflects the greater sensitivity of males to EMC infection in general, resulting from two different causes. The first of these is hormonal, since castrated males become less susceptible to EMC infection, and also develop less diabetes. The effect can be reversed with testosterone. Moreover, female mice, resistant to the diabetogenic effects of the D variant, develop diabetes to the same extent as males if they are pretreated with testosterone (Friedman et al., 1972; Morrow et al., 1980; Giron and Patterson, 1982). The second reason for the greater resistance to EMC infection in adult female mice are the higher IFN levels due to an X-linked locus that influences early circulating IFN-α/β production (see Chapter 15) (Zawatzky et al., 1982; Pozzetto and Gresser, 1985). It is intriguing, however, that susceptibility to induction of diabetes by streptozotocin, a methyl nitrosurea compound with specific β islet cell toxicity, is also greater in male mice and is related to testosterone levels (MacLaren et al., 1980; Kromann et al., 1982). This indicates that there is more to the higher incidence of EMC-induced diabetes in male mice than just an enhanced permissiveness for virus replication, and that β islet cell damage, by either viruses or chemicals, occurs more readily in males. Moreover, induction of diabetes in male mice can be the cumulative result of β cell damage caused by low doses of streptozotocin and an infection with subdiabetogenic doses of either EMC or Coxsackie virus B3 or B5 (Toniolo et al., 1980). It may be relevant to mention here that in humans, IDDM is also more prevalent in males than in females.

The diabetogenic potential of the D variant of EMC virus is most likely the direct result of β islet destruction caused by virus infection, and the evidence suggesting the involvement of an autoimmune mechanism in this particular virus-host system is limited. Contrary to other models, for example, the BB/W rat, immuno-suppression does not affect the establishment of diabetes in EMC-infected DBA/2 mice (Vialettes et al., 1983). Destruction of islet cells, however, is not an absolute requirement for virus-induced diabetes, which can also occur in the absence of cell lysis or inflammatory infiltrates, as shown in BALB/c mice that are chronically infected with lymphocytic choriomeningitis virus. In such mice, persistent infection of islet cells, with virions budding from β cells, is associated with chemical evidence for diabetes such as hyperglycemia and abnormal glucose tolerance (Oldstone et al, 1984). Thus, a noncytopathic virus can persist in β cells without killing them and without attracting immune cells, and nevertheless, cause a situation that is similar to early stages of diabetes.

In some virus–host animal models, however, an autoimmune component is clearly involved; one of these is the reovirus infected mouse.

b. Reovirus-Induced Autoimmune Diabetes Mellitus. Reoviruses are ubiquitous and can be found in vertebrates, invertebrates, and plants. The reovirus virion consists of a segmented, double-stranded RNA genome, surrounded by a protein shell that itself is enclosed in a protein capsid. Under natural conditions the viruses are believed to enter the host primarily through the gastrointestinal tract, and the first round of replication takes place in lymphoid tissue. The three mammalian reovirus serotypes produce distinct patterns of disease (Sharpe and Fields, 1985).

Newborn mice of the NIH Swiss or SJL/J strain, infected ip with reovirus type 1, develop transient diabetes accompanied by a runting syndrome. As a result of infection, inflammatory cells, viral antigens, and virus particles appear in the islets of Langerhans and in the anterior pituitary. Furthermore, the infection triggers a polyendocrine disease and induces auto-antibodies that react with insulin and growth hormone as well as with antigens in the islets of Langerhans, the anterior pituitary, and the gastric mucosa. Contrary to EMC virus-induced diabetes, reovirus-induced diabetes has an obvious autoimmune pathogenic component, since its development can be prevented by treatment of infected mice with antilymphocyte serum, antithymocyte serum, or cyclophosphamide. Reovirus-induced diabetes is probably the consequence of cumulative injury, resulting from the combination of the direct lytic effect of the virus and virus-induced autoimmune pathology (Onodera et al., 1981, 1982). Because of their dsRNA genome, reoviruses are very efficient IFN inducers, and the possibility exists that IFN production, in addition to its beneficial role in limiting virus infection, may be an aggravating factor in the pathogenesis of diabetes.

This has not been directly demonstrated, but is suggested by experimental results in EMC virus-infected mice and by the existence of a direct effect of IFN-α/β on insulin synthesis and of IFN-γ on MHC antigen expression by pancreatic beta cells.

We have seen that in the previously mentioned EMC-mouse model, early IFN-α/β production or early IFN administration have a beneficial effect because they limit virus replication. However, if administered four days after EMC virus infection, when the acute stage of infection is over but prior to the onset of insulitis, IFN-α/β or IFN-γ have an exacerbating effect on the pathogeneis of virus-induced diabetes (Gould et al., 1985). The reasons for this exacerbating effect could be manifold. In addition to the general stimulating effects of IFNs on autoimmunity, Mu IFN-α and IFN-β can inhibit glucose-stimulated proinsulin synthesis in isolated mouse pancreatic islet cells. Also, Hu lymphoblastoid IFN exerts a similar inhibitory effect on proinsulin synthesis in human islets of Langerhans in vitro (Rhodes and Taylor, 1984, 1985). IFN-γ has been shown to induce an expression of MHC class I antigens on human pancreatic β cells, which could favor enhanced targeting of autoreactive T cells to the β cells. This is more than a theoretical possibility, since in humans, enhanced expression of MHC class I antigens by pancreatic islet cells has been found in juvenile IDDM (Bottazo et al., 1985; Campbell et al., 1986).

II. AUTOIMMUNITY TO IFNs

A. In Humans

Natural antibodies to IFN-α, β, or γ do not seem to occur very frequently in humans. Spontaneous antibodies to IFN-α have been observed in a few individuals who never received IFN therapy, but the great majority of patients, screened before undergoing treatment, do not have antibodies (Mogensen et al., 1981; Gutterman et al., 1982b; Trown et al., 1983). A recent overview of the literature shows that of 280 patients with various autoimmune disorders, 3 did have anti-IFN-α antibodies, as did 2 out of 76 patients with various tumors (Panem, 1984). These figures show that the occurrence of auto-antibodies against IFN is a relatively rare event; it is, therefore, intriguing that the frequency of natural anti-IFN-α antibodies is significantly higher in human cord blood, where an incidence of 6% (3 out of 50 samples examined) has been reported (Trown et al., 1983).

As a result of treatment with Hu IFN-α, either the natural mixture produced by leukocytes or single species recombinant IFN, anti-IFN-α antibodies can appear. The frequency of appearance of these antibodies in cancer patients has been estimated at 25%. There does not seem to be a difference in therapeutic response of cancer patients to the IFN-α2 treatment, since the frequency of partial or complete remissions is about equal in antibody-negative and antibody-positive patients (Gutterman et al., 1982b; Hennes et al.,

1987; Itri et al., 1987). Administration of natural Hu IFN-β can also result in the appearance of specific antibodies. These antibodies are true auto-antibodies, since in addition to neutralizing a commercial preparation of Hu IFN-β made by a continuous line of human foreskin fibroblasts, they also neutralize the isogeneic IFN-β produced by fibroblasts derived from the individual displaying the antibodies (Vallbracht et al., 1981, 1982).

The presence of "spontaneous" anti-IFN antibodies in a few individuals and the fact that anti-IFN-α or anti-IFN-β antibodies can appear as a result of IFN treatment, shows that in humans, just as in mice, there are B-cell clones with the potential of making anti-IFN auto-antibodies.

With the exception of the anti-IFN antibodies in lupus patients, there is no obvious pattern of disease-relationship and anti-IFN antibodies. In lupus patients, the anti-IFN antibodies can cause immune-complex deposits in the kidney, because they combine with the acid-labile IFN-α that is present in the circulation of some of these patients. The occurrence of anti-IFN antibodies in lupus patients has probably no special significance but is part of the general propensity of such patients to make auto-antibodies against a wide variety of antigens.

B. In Animals

Natural antibodies of the IgM and IgG class to Mu IFN-α and Mu IFN-β are a common feature of many inbred mouse strains. Such antibodies neutralize the antiviral activity of C-243 cell-derived Mu IFN-α, which consists of a mixture of different, as yet undefined, IFN α subspecies. Murine IFN-β, of the same origin, is also neutralized, but it is not known whether the same antibodies neutralize both IFN species, or whether there are specific anti-α and anti-β antibodies. C-243 cells are derived from Swiss mice, and the IFN made by these cells is, therefore, allogeneic for all inbred mouse strains that are not of Swiss genotype. We can assume that the major antigenic components of all Mu IFN-αs are similar, but we do not have enough sequence data of IFNs from different mouse strains to make an assessment of antigenic variation caused by allelic differences. An assay, measuring neutralization of IFN of one genotype by serum of another genotype, therefore, is not necessarily a true measure of the titer of auto-antibodies. For this, isogeneic IFNs have to be used as an antigen, and when this is done, one still observes neutralization. Mice of the BALB/c and of the C57BL/6 strain have neutralizing antibodies to isogeneic IFN-α and IFN-β, prepared from fibroblasts of the same strains, from which it can be concluded that these antibodies are true auto-antibodies. Different neutralizing titers, however, are obtained when the same serum is assayed against isogeneic and various allogenic IFNs of the α as well as the β type. This suggests the existence of an antigenic variation between different allelic forms of Mu IFN-α and of Mu IFN-β.

Natural anti-IFN antibodies have been found in 14 different inbred mouse strains, out of 14 examined, and therefore are probably present in most, if not all, inbred strains of mice (De Maeyer-Guignard and De Maeyer, 1986). This is in striking contrast with the reported lack of natural anti-IFN antibodies in most human sera. Maybe mice are different from humans in this respect, or more sensitive assays will reveal a higher incidence of these antibodies in humans. The fact that in human cord blood a 6% incidence of anti-IFN antibodies has been reported suggests that in humans natural anti-IFN antibodies occur more frequently than is currently believed (Trown et al., 1983).

Do these spontaneous anti-IFN antibodies in mice have any function, and could they interfere with the antiviral role of IFN during infection with viruses? One can estimate that a one-month-old BALB/cBy mouse has enough antibodies in circulation to neutralize about 800 units of IFN-α and 3,200 units of IFN-β. From the IFN levels appearing in the circulation during acute viral infections, for example, with herpes simplex or encephalomyocarditis virus, one can estimate that at least 100 times more IFN is produced. The amount of anti-IFN antibody present can only neutralize a small percentage of this. We also know that the administration of high-titered anti-IFN serum, corresponding to several orders of magnitude more than what is present in the mouse, is required to counteract the protective effect of endogenous IFN production in several virus infections (Gresser et al., 1976). The level of natural anti-IFN antibodies, therefore, appears insufficient to impede the antiviral role of IFN during acute infection, but it is sufficient to neutralize lower amounts that are sometimes released as a result of more chronic infections. From an evolutionary point of view, one would not expect it to be sufficient for neutralizing large amounts of IFN produced during acute infection, since such mice would stand a good chance of having been eliminated by virus infection.

If the level of auto-antibodies against IFN is too modest to influence the outcome of acute virus infection, it is sufficient to neutralize small amounts of IFN-α/β, which would otherwise cause an interferonemia ranging from a few units to a few hundred units. Whether there is any physiological advantage to this is hard to say. In view of the different immunomodulatory properties of IFN-α/β, one can imagine that it may sometimes be useful to have local IFN production, without generalized distribution. The presence of natural anti-IFN antibodies explains why there can be evidence for continuous IFN activity in some mouse tissues, such as macrophages, without concomitant detection of IFN in tissues or serum.

Anti-IFN auto-antibodies are also present in rats; the serum of Lou/c rats contains antibodies of the IgG class that neutralize rat IFN-α and IFN-β (De Maeyer-Guignard et al., 1983, 1984). Rats belonging to other strains have not been examined for the presence of spontaneous anti-IFN antibodies.

What could be the function of these physiological auto-antibodies directed against IFN-α and IFN-β? This question should probably be considered in the broader context of the prevalence of autoantibodies in general, the function of which is not clear. It has been proposed that natural antiself antibodies have a positive role, either for removal of tissue degradation products or as part of a regulatory idiotypic-anti-idiotypic network (Guilbert et al., 1982). The presence of some regulatory mechanism, preventing autoimmune disease, is in any case implied by the low titers of auto-antibodies. In the case of Hu IFN-α, it has been possible to raise an anti-idiotype antibody, directed against an anti-IFN-α antibody, displaying IFN-like activity (Osheroff et al., 1985). Similar anti-idiotype antibodies have not yet been found under natural conditions, but the fact that they can be raised by immunization suggests the theoretical possibility of anti-idiotype antibodies that could regulate the level of anti-IFN auto-antibodies.

III. MULTIPLE SCLEROSIS

Clinically, multiple sclerosis is a highly variable disease, usually beginning between the second and fifth life decade. The major symptoms are sensory, visual, and motor dysfunction, and the disease, initially characterized by exacerbations and remissions, becomes progressively worse over the years. The primary pathology consists of macroscopic lesions, scattered throughout the white matter of the nervous system, which have given rise to the name for the disease. The lesions, characterized by demyelination of white matter, contain mononuclear cell infiltrates consisting predominantly of T cells and macrophages.

The etiology of multiple sclerosis is unknown. An autoimmune basis appears as one possibility, and a decrease in T-suppressor cells seems to be a characteristic feature (Huddlestone and Oldstone, 1979; Morimoto et al., 1987). A laboratory model of the disease consists of immunizing animals with central nervous system tissue, resulting in a disease known as experimental allergic encephalomyelitis. This experimental disease is most probably the result of cell-mediated autoimmunity, since it can be transmitted to syngeneic animals by lymphocytes. In view of our present ignorance, it cannot be excluded that there is more than one pathogenic mechanism for the clinical entity known as multiple sclerosis (McFarlin and McFarland, 1982a,b).

A. Defective IFN Production of Patients with Multiple Sclerosis

Mitogen-stimulated Hu IFN-γ production is impaired in peripheral blood leukocytes derived from patients with multiple sclerosis (MS), with many patients being nonresponders. The percentage of individuals whose leukocytes are nonresponders increases with an increasing degree of severity of disease (Neighbour et al., 1981; Vervliet et al., 1983, 1985b). Whether or not leukocytes from MS patients are found to be defective in IFN-γ production, however, depends to some extent on the mitogen used, and patients who are defective for concanavalin A or PHA-induced IFN induction can have normal responses when pokeweed mitogen is used instead. Defectiveness of IFN-γ production after stimulation with conA or PHA is not a special feature of MS patients since it is also observed in peripheral blood leukocytes of patients with a variety of other neurological diseases, including Alzheimer's dementia and Huntington's chorea. Moreover, it also occurs in patients recovering from cerebrovascular accidents or from craniotomy for brain tumors, which suggests that any injury to the brain can cause decreased IFN-γ responsiveness (Vervliet et al., 1984, 1985a).

Besides impaired IFN-γ production, defective IFN-α production has also been observed after stimulation of peripheral blood leukocytes from MS patients. Whether or not IFN production is diminished seems to depend to a large extent on the nature and the amount of virus used as an inducer, and also on the group of patients under study (Santoli et al, 1981; Neighbour et al., 1982; Tovell et al., 1983; Kamin-Lewis et al., 1984; Haahr et al., 1986). Newcastle disease virus induces normal amounts of IFN in peripheral blood leukocytes of MS patients. Vesicular stomatitis virus-induced IFN production, however, is significantly lower than that of normal individuals (Vervliet et al., 1984). It is revealing that, both for IFN-γ and IFN-α, whether or not MS patients have lowered or normal production depends entirely on the inducers that are tested. This is another illustration of the impossibility to define normal or decreased IFN production without specifying the nature and also the amount of inducer employed to probe the IFN-producing capacity. The problem is further complicated by the difficulty to define a standard of "normal" production, since even in disease-free individuals there can be differences in production reaching several orders of magnitude (see Chapter 15).

B. The Effect of IFNs on Multiple Sclerosis

Although the cause and pathogenic mechanism of MS are not well understood, experimental allergic encephalitis is considered to be a model presenting some similarities to this disease. With respect to cell-mediated immunity, there is a direct correlation between occurrence of delayed hypersensitivity to central nervous tissue antigens and the development of experimental allergic encephalitis in immunized animals (Paterson, 1977). IFNs can suppress the expression of delayed hypersensitivity (see Chapter 12), which could explain why rat IFN-α/β, when administered either systemically or by the intracerebroventricular route, decreases the severity and clinical symptoms of experimental allergic encephalitis in rats (Abreu, 1982; Abreu et al., 1983; Hertz and Deghenghi, 1985). In vitro correlates for delayed hypersensitivy reactions indicate that cell-mediated immunity against nervous tissue also exists in MS patients. Whether or not this can explain the apparent therapeutic benefit from the intrathecal administration of Hu IFN-β to MS patients is an open question (Jacobs et al, 1985, 1986, 1987).

In contrast to the beneficial effect of Hu IFN-β, the intravenous administration of Hu IFN-γ results in an increase of exacerbations in MS patients (Panitch et al., 1987). This is reminiscent of the IFN-γ-mediated exacerbation of autoimmune disease in mice (see section I.A.4) and supports the possibility that an autoimmune process is involved in the pathogenesis of MS.

IV. IFN-INDUCED PATHOLOGY IN SUCKLING MICE

We have seen in the previous sections how IFNs, through their stimulating effects on immune cells and surface antigen expression, can contribute to various diseases that have an immune component. There are a few exceptional instances, however, in which IFNs not merely act as contributing factors to disease but appear as the prime disease-inducing agents.

Massive amounts of Mu IFN-α/β, administered to newborn mice of different inbred strains during the first week of life, induce a specific pathological picture, characterized by the development of liver necrosis and glomerulonephritis. Organs other than the liver and kidneys appear to be spared from this effect, indicating that is not due to some general "cytotoxicity." The pathogenesis of IFN-induced liver necrosis is not known; it is accompanied by the appearance of tubuloreticular aggregates, as discussed earlier in the section on lupus erythematosus. When IFN treatment is halted before extensive liver degeneration has occurred, the suckling mice apparently recover, but die from glomerulonephritis in the months following the treatment. The mechanism of IFN-induced renal lesions is obscure. Electron microscopic examination reveals that the renal lesions develop very early after IFN treatment, and become progressively worse; a marked thickening of the glomerular basement membrane precedes the deposition of immunoglobulin and complement (Gresser, 1982). IFN-induced renal pathology offers a striking example of an adverse effect, with lethal clinical outcome occurring much later than the initiating treatment. Many virus infections are accompanied by the production of significant amounts of IFN-α and IFN-β. This raises the possibility that adverse effects can occur at a time when the relation with the original infection is no longer obvious.

Circulating IFN is to some extent eliminated by the kidneys (Bocci et al., 1984), but in mice other than newborns, as well as in patients receiving IFN treatment, there is little evidence for IFN-induced renal pathology, except for some exceptional cases. One patient, receiving treatment with rec Hu IFN-αA, developed acute interstitial nephritis and nephrotic syndrome, quite different from the IFN-induced renal disease in newborn mice (Averbuch et al., 1984).

V. THE ROLE OF IFN-α/β IN ARENAVIRUS-INDUCED PATHOLOGY

The members of the arenavirus family are characterized by a genome consisting of a segmented single-stranded negative sense RNA and by an outer lipoprotein membrane. A typical feature of arenavirus virions are host-cell ribosomes that have been included at the time of virus assembly. These ribosomes appear like grains of sand in electron-microscopical sections of virus particles, which has given rise to the family name "arena," derived from the Latin word for sand. The replication of arenaviruses takes place in the cytoplasm, and virus assembly occurs by budding from the plasma membrane. Cellular protein synthesis is not inhibited in arenavirus-infected cells, and infection is usually noncytolytic. This is an important consideration for understanding some of the effects of anti-IFN serum that will be discussed subsequently, which can increase virus replication and yet decrease virus-induced, IFN-mediated disease. The natural hosts of most arenaviruses are rodents, and humans are only occasionally infected. We will discuss the pathogenic role of arenavirus-induced IFN production in two mouse models, the lymphocytic choriomeningitis (LCM) virus-infected and the Pinchinde virus-infected mouse, and discuss the possibility that Junin virus-induced IFN production in humans contributes to the pathogenesis of Argentine hemorrhagic fever.

A. The Pathogenic Role of IFN-α/β in LCM Virus Infection of Adult Mice

Mice are the natural hosts for LCM virus, and apparently healthy animals can be lifelong carriers of the virus. Chronically infected mice shed the virus through the urine, which is probably how humans can become infected. LCM virus infection of humans is mostly inapparent, but can occasionally give rise to meningitis or meningoencephalitis.

When inoculated intracerebrally, many strains of LCM virus cause a severe and frequently lethal choriomeningitis in mice of different inbred strains. One of the most striking characteristics of this disease is that its lethal outcome is the result of an immune response directed against virus-infected cells. The pathogenic role of the immune system in LCM virus infection is demonstrated by the protection conferred by different immunosuppressive treatments. Administration of antilymphocyte serum, sublethal total body irradiation, or neonatal thymectomy, all treatments that would exacerbate most viral infections, protect mice against the lethal effects of the LCM virus (Buchmeier et al., 1980). Spleen cells from infected mice can transfer the lethal disease to virus-carrier adult mice, protected against illness by immunosuppressive treatment, and cytotoxic T cells, specifically reacting against LCM virus-infected cells in vitro, can be isolated from infected mice and transfer the disease (Doherty et al., 1976). It is, therefore, clear that T cells are crucially involved in the pathogenesis resulting from LCM virus infection, and there is evidence that both cytotoxic T cells and T cells involved in delayed hypersensitivity are instrumental in causing the inflammatory lesions that are typical of LCM virus-induced choriomeningitis (Zinkernagel et al., 1985).

In addition to acute, lethal disease caused by sensitized T cells, chronic, immune complex disease can occur in LCM virus-infected mice. This is due to the presence of antiviral antibodies that bind LCM virus antigens. The resulting immune complexes deposit in tissues and cause glomerulonephritis, focal hepatic necrosis and widespread lymphocyte infiltrations.

Destruction of virus-infected cells by sensitized T cells is responsible for the lethal syndrome that can develop after intracerebral inoculation of the LCM virus. Soon after infection, LCM virus-induced IFN-α/β synthesis starts, and some of the IFN is released into the circulation of the infected animals. We have seen in Chapter 11 that the cytotoxic activity of T cells against target cells that express viral antigens is significantly boosted by IFN-α and IFN-β. This is the result of both a direct action of these IFNs on T cells and the IFN-induced enhanced expression of MHC class I antigens at the target cell surface. Antigen-specific T-cell clones, devoid of lytic activity, acquire cytolytic activity after exposure to IFN-α or IFN-γ (Chen et al., 1986a,b). Moreover, it has been shown specifically in the case of LCM virus infection that Mu IFN-α/β treatment of LCM virus-infected target cells confers on these cells increased sensitivity to destruction by virus-specific, MHC class I restricted, cytotoxic T cells (Bukowski and Welsh, 1986). Evidently, IFN-α/β, induced as a result of viral infection, can up-regulate cytotoxic T-cell activity, and thereby stimulate the pathogenic process that in this particular virus infection is due to activated T cells. This is one instance in which IFN exacerbates rather than alleviates disease.

The pathogenic role of IFN is clearly demonstrated by treating LCM virus-infected adult mice with anti-Mu IFN-α/β globulins. Such treatment, when originated just before infection and continued for some days, protects the mice against the lethal effects of infection, in spite of the fact that at the same time virus production is boosted by several orders of magnitude. As a secondary effect of increased virus replication in these anti-IFN globulin treated animals, late IFN-α/β synthesis is significantly augmented (Saron et al., 1982a,b; Pfau et al., 1983).

What is the explanation for the therapeutic effect of the anti-IFN globulin treatment? Down-regulation of T-cell activation against LCM virus-infected cells has been observed in anti-IFN globulin-treated C3H mice, which confirms the role of IFN in this activation. The effect of anti-IFN-globulin treatment of infected mice on target cell sensitivity is not known, but one would also expect it to be down-regulatory. Paradoxically, the enhanced late IFN production in these antibody-treated mice, occurring several days after initial infection, could conceivably also contribute to down-regulation of T-cell activity, since IFN-α/β can cause a temporary anergy and inhibit the expression of delayed hypersensitivity against viral and nonviral antigens (see Chapter 12). Delayed hypersensitivity is believed to play a role in T-cell-mediated pathology of LCM virus-infected mice (Zinkernagel, 1976).

LCM virus-induced choriomeningitis is not the only model in which a therapeutic effect of anti-Mu IFN-α/β antibodies has been demonstrated. Influenza virus-induced encephalopathy in mice, a lethal inflammatory condition developing after intracerebral inoculation of influenza A virus, can also be treated by the administration of anti-IFN antibodies (Wabuke-Bunoti et al., 1986).

B. The Pathogenic Role of IFN-α/β in LCM Virus Infection in Suckling Mice

LCM virus infection of newborn mice results in stunted growth, liver cell degeneration, and death, a syndrome identical to that induced by the administration of potent IFN-α/β preparations to newborn mice, discussed in section IV. Moreover, mice surviving the acute disease develop a glomerulonephritis, which is identical to the one induced by IFN. The incidence of mortality in LCM virus-infected newborn mice is influenced by the virus strain and by the genotype of the host. The similarity between LCM virus-induced and IFN-induced disease suggests that the induction of endogenous IFN-α/β by the LCM virus plays an important pathogenic role in these animals. Several arguments strongly support this hypothesis. Treatment with anti-IFN-globulin of LCM-infected newborns reduces the incidence of mortality and delays the subsequent appearance of glomerulonephritis in those mice that survive the acute syndrome. In addition, there is an inverse correlation between the relative resistance of suckling mice from three different inbred strains and the level of IFN-α/β production triggered by infection with the LCM virus, whereas there is no correlation with levels of virus replication, which are identical in these different mouse strains. The low IFN producers are also more resistant to the subsequent development of glomerulonephritis. Thus, there is a direct correlation between the magnitude of the IFN response to the LCM virus infection and the manifestation of the acute and chronic LCM virus-induced disease (Rivière et al., 1977, 1980; Gresser, 1982; Rivière and Oldstone, 1986).

The property of inducing liver pathology via IFN production is not limited to LCM virus, and infection of newborn mice with Pichinde virus, another member of the arenavirus family, also results in extensive liver necrosis that can be prevented by anti-IFN globulin (Clark et al., 1986).

C. Correlation Between Endogenous IFN-α Levels and the Clinical Evolution of Patients with Argentine Hemorrhagic Fever

Argentine hemorrhagic fever is a disease caused by Junin virus, a member of the arenavirus family. The normal host for Junin virus is a small rodent,

Calomys musculinus. The virus is occasionally transmitted to humans, and the disease is rarely contagious between humans. The infection is characterized by bleedings and petechiae, and the case fatality rate is about 25%. During the first week of illness, corresponding to the time of acute disease, circulating IFN-α levels can reach extraordinarily high values, unlike anything seen during other virus infections, and serum IFN titers of up to 64,000 units/ml can be reached. For comparison, circulating IFN titers observed during some other virus infections, such as influenza or measles, at the most reach levels of a few hundred units per milliliter. The interferonemia lasts for several days and decreases during the late stages of the disease, around 10 days after the onset of illness. There is no correlation between IFN levels and viremia, and comparable titers of Junin virus are present in the serum of high and low IFN producers. There is however, a correlation between IFN levels and the clinical course of the disease, in that prognosis is worse in cases with high IFN levels (Fig. 16.4) (Levis et al., 1985). The main symptoms during the acute phase of illness are fever, fatigue, malaise, leukopenia, thrombocytopenia, and neurological and renal involvement, very much like the symptoms manifested by patients receiving high doses of IFN-α (Levis et al., 1984). The clearance rate of IFN-α in the serum of these patients is not known, but a serum titer of 64,000 units/ml at any given time corresponds to a total of about 2×10^8 units in circulation, which is at least

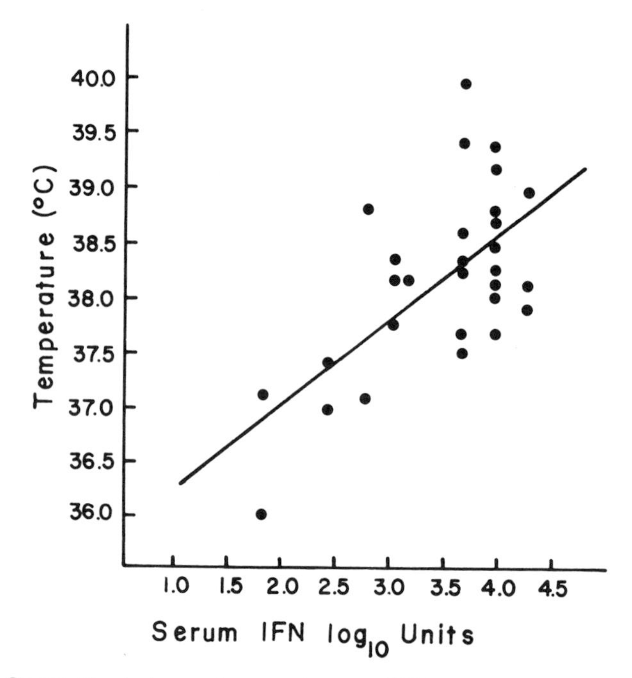

Figure 16.4 Correlation between fever and serum IFN levels in patients with Argentine hemorrhagic fever. (Adapted from Levis et al., 1985.)

ten times the amount of IFN-α that can be given to patients in a single dose without causing serious side effects (Scott, 1983; Mirro et al., 1985).

It is, therefore, possible that we are dealing here with a human counterpart of LCM virus-induced disease in mice, and that the extremely high IFN-α levels in some patients with Junin virus contribute to the pathogenesis and the fatal outcome of the disease.

VI. AIDS

The AIDS syndrome is caused by different strains of the distantly related HIV-1 (LAV-1/HTLV-III) and HIV-2 retroviruses, classified as retrolentiviruses, which mature by budding from the plasma membrane (Barré-Sinoussi et al., 1983; Popovic et al., 1984; Clavel et al., 1987; Guyader et al., 1987). HIV (standing for "human immunodeficiency virus") belongs to the subfamily of lentiviruses, which cause slowly developing diseases. HIV is transmitted by sexual contact, by blood products, and from mother to infants and preferentially infects T cells that express the CD4 cell-surface antigen (also called T4 antigen). The latter is a cell-surface glycoprotein of 62 kDa relative molecular mass, which is characteristic of a subset of T cells displaying helper and inducer activity (Dalgleish et al., 1984; Klatzmann et al., 1984). The replication of HIV-1, however, is not strictly limited to $CD4^+$ T cells, but can also take place in other cells like B cells, macrophages, dendritic cells, and microglial brain cells (Klatzmann and Gluckman, 1986). Due to the infection and eventual destruction of $CD4^+$ lymphocytes, the disease is characterized by a severe, selective depletion of T helper cells and a consequent reversal of the ratio of T helper to T suppressor cells (Kornfeld et al., 1982). As a result of the general immunodeficiency, one of the principal clinical expressions of AIDS is the development of Kaposi's sarcoma and of frequently fatal opportunistic infections that can be caused by many different agents, including viruses.

A. An Acid-Labile IFN-α Is Present in the Serum of Many AIDS Patients

Some of the immunological abnormalities in AIDS patients are similar to those present in patients with autoimmune diseases, such as lupus erythematosus, and the serum of AIDS patients frequently contains an acid-labile IFN-α, which is also found in the serum of many lupus patients. Since in neither AIDS nor lupus patients these circulating IFNs have been characterized molecularly, there is no way of knowing whether we are dealing with the same IFN in both diseases. In a minority of AIDS patients the circulating IFN-α is not acid labile, an observation that has also been made in lupus patients (DeStefano et al., 1982).

Furthermore, acid-labile IFN-α is also present in the serum of some individuals without the clinical disease, but infected with the AIDS virus as indicated by the presence of antibodies in their serum. The appearance of IFN can precede the onset of clinical disease by as much as 10 months, as observed in hemophiliacs who received the antihemophilic factor (factor VIII concentrate) contaminated with HIV-1 (Eyster et al, 1983). These patients probably have a mixture of normal IFN-α and of the acid-labile form, since the IFN activity in their serum is only partially inactivated at pH 2, but again, in the absence of complete molecular characterization, it is not possible to assess the relative proportion of each IFN.

The cellular origin and the inducer of these circulating IFNs, as well as their possible role in the pathology of the disease, are not known. Although a theoretical possibility, there is no experimental support for the hypothesis that the IFN is induced by the AIDS virus itself. Other retroviruses have been shown to be IFN inducers; for example, type C Friend leukemia virus and type B mouse mammary tumor virus induce circulating Mu IFN-α/β in mice (De Maeyer and De Maeyer-Guignard, 1970; De Maeyer et al., 1974). In addition, Visna-maedi virus of sheep, which, like HIV-1, belongs to the subfamily of retroviruses called lentiviruses because they cause diseases with long incubation periods, induces a substance with antiviral activity as a result of interaction of lymphocytes and infected macrophages. This substance, however, is acid stable, and is probably not a classical IFN (Narayan et al., 1985; Haase, 1986).

The presence of acid-labile IFN-α in the circulation of otherwise healthy individuals or in hemophiliacs seems to have predictive value for the subsequent development of AIDS (Buimovici-Klein et al., 1983; Eyster et al., 1983). Moreover, the presence of acid-labile IFN in the serum of AIDS patients also has predictive value for their survival if subsequently they develop Kaposi's sarcoma: The survival rates are much better for those without circulating IFN (Vadhan-Raj et al., 1986). This suggests the possibility that the circulating acid-labile IFN is one of the mediators of pathogenesis of the disease. There is, indeed, a striking similarity between some of the common symptoms in AIDS patients, such as fever, malaise, fatigue, and neutropenia, and the side effects of IFN-α treatment. This has led to the suggestion that these symptoms could in part result from the continuous presence of circulating acid-labile IFN-α in these patients (Krown, 1986). Dysfunction of the central nervous system is a prominent feature of many AIDS patients, and using techniques of in situ hybridization, the presence of HIV has been demonstrated in capillary endothelium and mononuclear cells of the brain rather than in astrocytes and neurons (Wiley et al., 1986). HIV-1 has been isolated from peripheral blood monocytes, human alveolar macrophages, and mononuclear phagocytes cultured from brain and lung tissues. Furthermore, the virus can infect CD4 positive, transformed cells of monocytic lineage in vitro (Gartner et al., 1986; Ho et al., 1986; Salahuddin et al., 1986; Asjo et al., 1987). Inflammatory macrophages are a potential source of

IFN, and IFN produced in this way could theoretically contribute to the neurological symptoms of these patients.

Since neurons and glial cells in human brain express the CD4 cell-surface molecule, they can serve as targets for HIV-1 (Maddon et al., 1986; Funke et al., 1987). It has been pointed out that other viruses, more specifically, measles virus and visna virus, also combine lymphotropism and neurotropism, and that, since the brain is an immunologically privileged site, it may provide a sanctuary for HIV-1 (Black, 1985).

In view of the multiple effects of IFNs on the immune system, it makes sense to think that the continuous presence of IFN-α in the circulation can have disturbing effects on many cell functions, but there is no proof that it actually contributes to the pathogenesis of the disease. Since circulating acid-labile IFN-α is a frequent component of autoimmune disease, the presence of such IFN may be indicative of an autoimmune component in AIDS, and auto-antibodies are indeed present in some AIDS patients (Wykoff et al., 1985). Among others, antilymphocyte auto-antibodies, directed against different subsets of T cells, have been found in the sera of many AIDS and pre-AIDS patients (Williams et al., 1984). The incidence of anti-IFN-α antibody formation is also higher in IFN-treated AIDS patients with Kaposi's sarcoma than in cancer patients in general (Itri et al., 1987).

It is clear that the continuous presence of acid-labile IFN in the circulation does not result in enhanced protection against opportunistic infections, including viral infections. The most commonly recurring viral infections in AIDS patients are due to cytomegalovirus, herpes simplex virus, varicella-herpes zoster virus, and Epstein-Barr virus, all viruses that are relatively resistant to IFN in any case.

B. Impaired IFN Production in Cells from AIDS Patients

The production of IFN-γ is greatly suppressed in suspensions of mitogen-stimulated mononuclear cells (a mixture of monocytes and lymphocytes) derived from AIDS patients (Murray et al., 1984). In HIV-1-infected individuals with persistent lymphadenopathy, but no other AIDS symptoms, the level of antigen-stimulated IFN-γ production is of prognostic value to determine the probability of development of acute disease. A deficient IFN-γ production is a prognostic sign of a definite and immediate risk for AIDS and opportunistic infection, and the probability of remaining free of AIDS-related infections is much higher if antigen-stimulated IFN-γ production is normal (Fig. 16.5) (Murray et al., 1985a).

The defect in IFN-γ production appears to be at the transcriptional level, since transcription of the IFN-γ gene is impaired in HIV-1-infected cells. Interestingly enough, this impairment does not reflect a general effect on transcription of lymphokine genes, since transcription of the IL-2 gene is not affected (Arya and Gallo, 1985). Nevertheless, like macrophages from healthy individuals, monocyte-derived as well as alveolar macrophages from

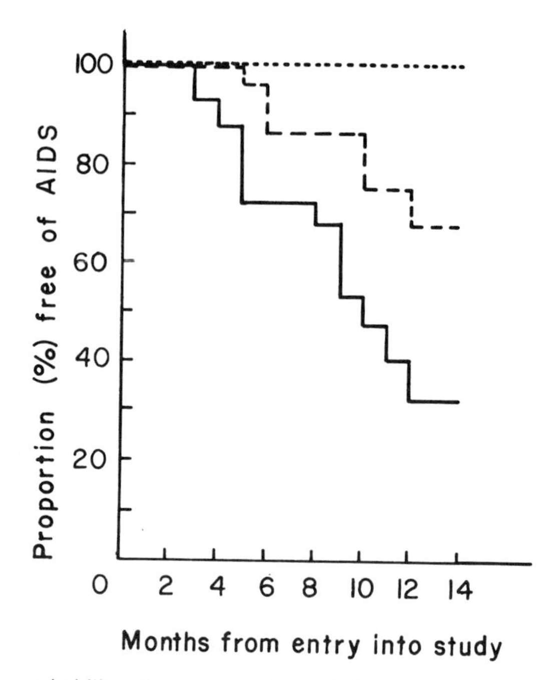

Figure 16.5 The probability of remaining free of AIDS-related opportunistic infections over time as a function of the capacity to produce IFN-γ in response to specific antigen in vitro. Dotted line: Normal or slightly impaired IFN-γ production. Dashed line: Severely impaired, but still detectable production. Solid line: Undetectable production. (Adapted from Murray et al., 1985a.)

AIDS patients can be activated in vitro by IFN-γ, which indicates that the production but not the action of IFN-γ is defective (Murray et al., 1985b; Murray et al., 1987). Impairment of IFN-γ synthesis is undoubtedly one of the contributing factors to the general dysfunction of the immune system in AIDS patients.

A defective Hu IFN-α production exists in macrophages derived from AIDS patients with opportunistic infections. When put in the presence of HSV-1-infected fibroblasts, mononuclear cells normally generate the production of IFN-α. When derived from AIDS patients with opportunistic infections, but not from AIDS patients without such infections, mononuclear cells make ten times less IFN-α than normal cells (Lopez et al., 1983).

Peripheral blood leukocytes from HIV-1-infected individuals—with or without AIDS—also make significantly (up to tenfold) less IFN-α upon stimulation by the influenza A virus (Abb et al., 1986).

It is intriguing that this situation bears some resemblance to what is observed in lupus—the continuous presence of an unusual form of IFN-α, combined with an impaired capacity to produce IFN when cells are isolated and studied in vitro. AIDS patients share several other symptoms with systemic lupus erythematosus patients—the already mentioned autoimmune phenomena, the imbalance of T-cell functions, B-cell hyperactivity, and interferon-related tuboreticular inclusions in circulating leukocytes (Grimley et al., 1984).

C. Sensitivity of HIV-1 to IFN-α and IFN-γ

In vitro, the replication of HIV-1 has been inhibited both by Hu IFN-α in peripheral blood mononuclear cells and by Hu IFN-γ in a line of human monocytes. Both these IFNs inhibit retrovirus replication in general, frequently by interfering with a late step in virion assembly (see Chapter 6) (Ho et al., 1985; Nakashima et al., 1986). In a human monocyte line persistently infected with HIV-1, Hu IFN-γ and GM-CSF (see Chapter 8) act synergistically and reduce reverse transcriptase activity almost completely (Hammer et al., 1986).

The sensitivity of HIV-1 to IFN is also illustrated by the fact that the isolation of one of the AIDS viruses, LAV-1, was facilitated by the addition of anti-IFN-α antibodies to the lymph node cultures from which it was recovered (Barré-Sinoussi et al., 1983). In general, the presence of antibodies to IFN-α/β increases in vitro retrovirus production by a factor of 10 to 50 (Barré-Sinoussi et al., 1979).

In vivo, however, HIV is quite resistant to IFN-treatment (Krown, 1986), which is not surprising, since the presence of circulating IFN-α in AIDS patients does not limit further development of the infection with HIV-1, but, as mentioned earlier, is a bad prognostic sign. Part of the resistance to the antiviral effect of IFN in vivo may be caused by an impairment of immune functions, since the antiviral action of IFNs in vivo not only results from a direct induction of the antiviral state in cells, but also results from activation of different compartments of the immune system. In view of the general deficiency in T-helper cells, this part of IFN action is obviously impaired in AIDS patients.

REFERENCES

Abb, J., Piechowiak, H., Zachoval, R., Zachoval, V., and Deinhardt, F. Infection with human T-lymphotropic virus type III and leukocyte interferon production in homosexual men. *Eur. J. Clin. Microbiol.* 5: 365–368 (1986).

Abreu, S. L. Suppression of experimental allergic encephalomyelitis by interferon. *Immunol. Comm.* 11: 1–7 (1982).

Abreu, S. L., Tondreau, J., Levine, S., and Sowinski, R. Inhibition of passive localized experimental allergic encephalomyelitis by interferon. *Int. Archs. Allergy Appl. Immun.* 2: 30–33 (1983).

Adam, C., Thoua, Y., Ronco, P., Verroust, P., Tovey, M., and Morel-Maroger, L. The effect of exogenous interferon: Acceleration of autoimmune and renal diseases in (NZB/W) F1 mice. *Clin. Exp. Immunol.* 40: 373–383 (1980).

Adelman, N. E., Watling, D. L., and McDevitt, H. O. Treatment of (NZBx NZW)F1 disease with anti-I-A monoclonal antibodies. *J. Exp. Med.* 158: 1350–1355 (1983).

Arya, S. K. and Gallo, R. C. Human T-cell growth factor (interleukin 2) and γ-interferon genes: Expression in human T-lymphotropic virus type III- and type I-infected cells. *Proc. Natl. Acad. Sci.* (USA) 82: 8691–8695 (1985).

Asjö, B., Ivhed, I., Gidlund, M., Fuerstenberg, S., Fenyo, E. V., Nilsson, K., and Wigzell, H. Susceptibility to infection by the human immunodeficiency virus (HIV) correlates with T4 expression in a parental monocytoid cell line and its subclones. *Virology* 157: 359–365 (1987).

Averbuch, S. D., Austin, H. A. III, Sherwin, S. A., Antonovych, T., Bunn, P. A., Jr. and Longo, D. L. Acute interstitial nephritis with the nephrotic syndrome following recombinant leukocyte A interferon therapy for mycosis fungoides. *N. Engl. J. Med.* 310: 32–35 (1984).

Balkwill, F. R., Griffin, D. B., Band, H. A., and Beverley, P. C. Immune human lymphocytes produce an acid-labile α-interferon. *J. Exp. Med.* 157: 1059–1063 (1983).

Barré-Sinoussi, F., Montagnier, L., Lidereau, R., Sisman, J., Wood, J., and Chermann, J. C. Enhancement of retrovirus production by anti-interferon serum. *Ann. Microbiol. Inst. Pasteur* 130B: 349–362 (1979).

Barré-Sinoussi, F., Chermann, J. C., Rey, F., Nugeyre, M. T., Chamaret, S., Gruest, J., Dauguet, C., Axler-Blin, C., Vezinet-Brun, F., Rouzioux, C., Rozenbaum, W., and Montagnier, L. Isolation of T-lymphotropic retrovirus from a patient at risk for acquired immune deficiency syndrome (AIDS). *Science* 220: 868–871 (1983).

Bino, T., Edery, H., Gertler, A., and Rosenberg, H. Involvement of the kidney in catabolism of human leukocyte interferon. *J. Gen. Virol.* 59: 39–45 (1982).

Black, P. H. HTLV-III, AIDS, and the brain. Editorials. *N. Engl. J. Med.* 313: 1538–1540 (1985).

Bocci, V., Maunsbach, A. B., and Mogensen, E. K. Autoradiographic demonstration of human ^{125}I-interferon α in lysosomes of rabbit proximal tubule cells. *J. Submicrosc. Cytol.* 16: 753–757 (1984).

Bornemann, L. D., Spiegel, H. E., Dziewanowska, Z. E., Krown, S. E., and Colburn, W. A. Intravenous and intramuscular pharmacokinetics of recombinant leukocyte A interferon. *Eur. J. Clin. Pharmacol.* 28: 469–471 (1985).

Bottazzo, G. F., Dean, B. M., McNally, J. M., Mackay, E. H., Swift, P. G. F., and Gamble, D. R. In situ characterization of autoimmune phenomena and expression of HLA molecules in the pancreas in diabetic insulitis. *N. Engl. J. Med.* 313: 353–360 (1985).

Boucher, D. W., Hayashi, K., Rosenthal, J., and Notkins, A. L. Virus-induced diabetes mellitus. III. Influence of the sex and strain of the host. *J. Infect. Dis.* 131: 462–466 (1975).

Buchmeier, M. J., Welsh, R. M., Dutko, F. J., and Oldstone, M. B. A. The virology and immunobiology of lymphocytic choriomeningitis virus infection. *Adv. Immunol.* 30: 275–331 (1980).

Buimovici-Klein, E., Lange, M., Klein, R. J., Cooper, L. Z., and Grieco, M. H. Is the presence of interferon predictive for AIDS? *Lancet,* 344 (1983).

Bukowski, J. F. and Welsh, R. M. Enhanced susceptibility to cytotoxic T lymphocytes of target cells isolated from virus-infected or interferon-treated mice. *J. Virol.* 59: 735–739 (1986).

Burnet, F. *The Clonal Selection Theory of Acquired Immunity,* Vanderbilt and Cambridge University Presses, Cambridge, England, (1959).

Campbell, I. L., Bizilj, K., Colman, P. G., Tuch, B. E., and Harrison, L. C. Interferon-γ induces the expression of HLA-A, B, C but not HLA-DR on human pancreatic β-cells. *J. Clin. Endocrinol. Metab.* 62: 1101–1109 (1986).

Cantor, H., McVay-Boudreau, L., Hugenberger, J., Naidorf, K., Shen, F. W., and Gershon, R. K. Immunoregulatory circuits among T-cell sets. II. Physiologic role of feedback inhibition in vivo: Absence in NZB mice. *J. Exp. Med.* 147: 1116–1125 (1978).

Chen, L. K., Mathieu-Mahul, D., Bach, F. H., Dausset, J., Bensussan, A., and Sasportes, M. Recombinant interferon-α can induce rearrangement of T-cell antigen receptor α-chain genes and maturation to cytotoxicity in T-lymphocyte clones in vitro. *Proc. Natl. Acad. Sci.* (USA) 83: 4887–4889 (1986a).

Chen, L. K., Tourvieille, B., Burns, G. F., Mathieu-Mahul, D., Sasportes, M., and Bensussan, A. Interferon: A cytotoxic T lymphocyte differentiation signal. *Eur. J. Immunol.* 16: 767–770 (1986b).

Clark, T., Gresser, I., Pfau, C., Moss, J., and Woodrow, D. Antibody to mouse α/β interferon abrogates Pichinde virus-induced liver lesions in suckling mice. *J. Virol.* 59: 728–730 (1986).

Clavel, F., Mansinho, K., Chamaret, S., Guetard, D., Favier, V., Nina, J., Santos-Ferreira, M. O., Champalimaud, J. L., and Montagnier, L. Human immunodeficiency virus type 2 infection associated with AIDS in west Africa. *N. Engl. J. Med.* 316: 1180–1185 (1987).

Cohen, S. H., Bolton, V., and Jordan, G. W. Relationship of interferon-inducing particle phenotype to encephalomyocarditis virus-induced diabetes mellitus. *Inf. Immun.* 42: 605–611 (1983).

Craighead, J. E. and McLane, M. F. Diabetes mellitus: Induction in mice by encephalomyocarditis virus. *Science* 162: 913–914 (1968).

Cunningham, A. J. Large numbers of cells in normal mice produce antibody components of isologous erythrocytes. *Nature* 252: 749–751 (1974).

Dalgleisch, A. G., Beverley, P. C. L., Clapham, P. R., Crawford, D. H., Greaves, M. F., and Weiss, R. A. The CD4 (T4) antigen is an essential component of the receptor for the AIDS retrovirus. *Nature* 312: 763–767 (1984).

Dauphinée, M. J., Kipper, S. B., Wofsy, D., and Talal, N. Interleukin 2 deficiency is a common feature of autoimmune mice. *J. Immunol.* 127: 2483–2487 (1981).

De Maeyer-Guignard, J., Cachard, A., and De Maeyer, E. Lou/c rats have natural anti-IFN antibodies. In: *The Biology of the Interferon System,* E. De Maeyer and H. Schellekens, Eds., Elsevier, Amsterdam, pp. 387–390 (1983).

De Maeyer-Guignard, J., Cachard-Thomas, A., and De Maeyer, E. Naturally occurring anti-interferon antibodies in Lou/c rats. *J. Immunol.* 133: 775–778 (1984).

De Maeyer, E. and De Maeyer-Guignard, J. A gene with quantitative effect on circulating interferon induction. Further studies. *Ann. N.Y. Acad. Sci.* 173: 228–238 (1970).

De Maeyer, E., De Maeyer-Guignard, J., Hall, W. T., and Bailey, D. W. A locus affecting circulating interferon levels induced by mouse mammary tumor virus. *J. Gen. Virol.* 23: 209–211 (1974).

De Maeyer-Guignard, J. and De Maeyer, E. Natural antibodies to interferon-α and interferon-β are a common feature of inbred mouse strains. *J. Immunol.* 136: 1708–1711 (1986).

De Stefano, E., Friedman, R. M., Friedman-Kien, A. E., Goedert, J. J., Henriksen, D., Preble, O. T., Sonnabend, J. A., and Vilcek, J. Acid-labile human leukocyte interferon in homosexual men with Kaposi's sarcoma and lymphadenopathy. *J. Infect. Dis.* 146: 451–455 (1982).

Dighiero, G., Guilbert, B., and Avrameas, S. Naturally occurring antibodies against nine common antigens in human sera. II. High incidence of monoclonal Ig exhibiting antibody activity against actin and tubulin and sharing antibody specificities with natural antibodies. *J. Immunol.* 128: 2788–2792 (1982).

Dighiero, G., Lymberi, P., Holmberg, D., Lundquist, I., Coutinho, A., and Avrameas, S. High frequency of natural autoantibodies in normal newborn mice. *J. Immunol.* 134: 765–771 (1985).

Doherty, P. C., Dunlop, M. B. C., Parish, C. R., and Zinkernagel, R. M. Inflammatory process in murine lymphocytic choriomeningitis virus is maximal in H-2K or H-2D compatible interactions. *J. Immunol.* 117: 187–190 (1976).

Dresser, D. W. Most IgM-producing cells in the mouse secrete auto-antibodies (rheumatoid factor). *Nature* 274: 480–482 (1978).

Emodi, G., Just, M., Hernandez, R., and Hirt, H. R. Circulating interferon in man after administration of exogenous human leukocyte interferon. *J. Nat. Cancer Inst.* 54: 1045–1049 (1975).

Engleman, E. G., Sonnenfeld, G., Dauphinee, M., Greenspan, J. S., Talal, N., McDevitt, H. O., and Merigan, T. C. Treatment of NZB/NZW F1 hybrid mice with mycobacterium bovis strain BCG or type II interferon preparations accelerates autoimmune disease. *Arthritis Rheum.* 24: 1396–1402 (1981).

Eyster, M. E., Goedert, J. J., Poon, M. C., and Preble, O. T. Acid-labile α interferon. A possible preclinical marker for the acquired immunodeficiency syndrome in hemophilia. *N. Engl. J. Med.* 309: 583–586 (1983).

Friedman, S. B., Grota, L. J., and Glasgow, L. A. Differential susceptibility of male and female mice to encephalomyocarditis virus: Effects of castration, adrenalectomy and the administration of sex hormones. *Infect. Immun.* 5: 637–644 (1972).

Fujii, N., Minagawa, T., Nakane, A., Kato, F., and Ohno, S. Spontaneous production of γ-interferon in cultures of T lymphocytes obtained from patients with Behçet's disease. *J. Immunol.* 130: 1683–1686 (1983).

Funke, I., Hahn, A., Rieber, E. P., Weiss, E., and Riethmuller, G. The cellular receptor (CD4) of the human immunodeficiency virus is expressed on neurons and glial cells in human brain. *J. Exp. Med.* 165: 1230–1235 (1987).

Gartner, S., Markovits, P., Markovitz, D. M., Kaplan, M. H., Gallo, R. C., and Popovic, M. The role of mononuclear phagocytes in HTLV-III/LAV infection. *Science* 233: 215–219 (1986).

Garzelli, C., Taub, F. E., Scharff, J. E., Prabhakar, B. S., Ginsberg-Fellner, F., and Notkins, A. L. Epstein-Barr virus-transformed lymphocytes produce monoclonal autoantibodies that react with antigens in multiple organs. *J. Virol.* 52: 722–725 (1984).

Giron, D. J. and Patterson, R. R. Effect of steroid hormones on virus-induced diabetes mellitus. *Infect. Immun.* 37: 820–822 (1982).

Gladish, R., Hofmann, W., and Waldherr, R. Myokarditis und insulitis nach Coxsackie-Virus-Infekt. *Z. Kardiol.* 65: 837–849 (1976).

Gould, C. L., McMannama, K. G., Bigley, N. J., and Giron, D. J. Exacerbation of the pathogenesis of the diabetogenic variant of encephalomyocarditis virus in mice by interferon. *J. Interferon Res.* 5: 33–37 (1985).

Gresser, I. Can interferon induce disease? In: *Interferon 3,* I. Gresser, Ed., Academic Press, London, pp. 95–127 (1982).

Gresser, I., Tovey, M. G., Maury, C., and Bandu, M. T. Role of interferon in the pathogenesis of virus diseases in mice as demonstrated by the use of anti-interferon serum. II. Studies with herpes simplex, Moloney sarcoma, vesicular stomatitis, Newcastle disease and influenza viruses. *J. Exp. Med.* 144: 1316–1323 (1976).

Grimley, P. M., Kang, Y. H., Frederick, W., Rook, A. H., Kostianowsky, M., Sonnabend, J. A., Macher, A. M., Quinnan, G. V., Friedman, R. M., and Masur, H. Interferon-related leukocyte inclusions in acquired immune deficiency syndrome: Localization in T cells. *Am. J. Clin. Pathol.* 81: 147–154 (1984).

Guilbert, B., Dighiero, G., and Avrameas, S. Naturally occurring antibodies against nine common antigens in human sera. Detection, isolation and characterization. *J. Immunol.* 128: 2779–2787 (1982).

Gutterman, J. U., Fine, S., Quesada, J., Horning, S., Levine, J. F., Alexanian, R., Bernhardt, L., Kramer, M., Spiegel, H., Colbrun, W., Trown, P., Merigan, T., and Dziewanowski, Z. Recombinant leukocyte A interferon: Pharmacokinetics, single-dose tolerance, and biological effects in cancer patients. *Ann. Intern. Med.* 96: 549–556 (1982a).

Gutterman, J. U., Quesada, J., and Fein, S. Clinical investigation of partially pure and recombinant DNA-derived leucocyte interferon in human cancer. In: *From Gene to Protein: Translation into Biotechnology,* F. Ahmad, Ed. Academic Press, New York, pp. 367–390 (1982b).

Guyader, M., Emerman, M., Sonigo, P., Clavel, F., Montagnicr, L., and Alizon, M. Genome organization and transactivation of the human immunodeficiency virus type 2. *Nature* 326: 662–669 (1987).

Haahr, S., Moller-Larsen, A., Justesen, J., and Pedersen, E. Interferon induction, 2′-5′ oligo A synthetase and lymphocyte subpopulations in out-patients with multiple sclerosis in a longitudinal study. *Acta Neurol. Scand.* 73: 345–351 (1986).

Haase, A. T. Pathogenesis of lentivirus infections. *Nature* 322: 130–136 (1986).

Hammer, S. M., Gillis, J. M., Groopman, J. E., and Rose, R. M. In vitro modification of human immunodeficiency virus infection by granulocyte-macrophage colony-stimulating factor and γ interferon. *Proc. Natl. Acad. Sci.* (USA) 83: 8734–8738 (1986).

Haspel, M. V., Onodera, T., Prabhakar, B. S., Horita, M., Suzuki, H., and Notkins, A. L. Virus-induced autoimmunity: Monoclonal antibodies that react with endocrine tissues. *Science* 220: 304–306 (1983).

Hennes, U., Jucker, W., Fischer, E. A., Krummenacher, Th., Palleroni, A. V., Trown, P. W., Linder-Ciccolunghi, S., and Rainisio, M. The detection of antibodies to recombinant interferon alpha-2a in human serum. *J. Biol. Standard.* 15: 231–244 (1987).

Heremans, H., Billiau, A., Colombatti, A., Hilgers, J., and De Somer, P. Interferon treatment of NZB mice: Accelerated progression of autoimmune disease. *Infect. Immun.* 21: 925–930 (1978).

Hertz, F. and Deghenghi, R. Effect of rat and β-human interferons on hyperacute experimental allergic encephalomyelitis in rats. *Agents Actions* 16: 397–403 (1985).

Ho, D. D., Rota, T. R., and Hirsch, M. Infection of monocytes/macrophages by human T lymphotropic virus type III. *J. Clin. Invest.* 77: 1712–1715 (1986).

Ho, D. D., Rota, T. R., Kaplan, J. C., Hartshorn, K. L., Andrews, C. A., Schooley, R. T., and Hirsch, M. *Lancet* i, 602–604 (1985).

Hochman, J., Mador, N., and Panet, A. Tubular structures in S49 mouse lymphoma are regulated through in vivo host-cell interaction and in vitro interferon treatment. *J. Cell. Biol.* 100: 1351–1356 (1985).

Hoffman, T. Natural killer function in systemic lupus erythematosus. *Arthritis Rheum.* 23: 30–35 (1980).

Hooks, J. J., Jordan, G. W., Cupps, T., Moutsopoulos, H. M., Fauci, A. S., and Notkins, A. L. Multiple interferons in the circulation of patients with systemic lupus erythematosus and vasculitis. *Arthritis Rheum.* 25: 396–400 (1982).

Hooks, J. J., Moutsopoulos, H. M., and Notkins, A. L. The role of interferon in immediate hypersensitivity and autoimmune diseases. *Ann. N.Y. Acad. Sci.* 350: 21–32 (1980).

Huddlestone, J. R. and Oldstone, M. B. A. T suppressor (T_G) lymphocytes fluctuate in parallel with changes in the clinical course of patients with multiple sclerosis. *J. Immunol.* 123: 1615–1618 (1979).

Huston, D. P. and Steinberg, A. D. Animal models of human systemic lupus erythematosus. *Yale J. Biol. Med.* 52: 289–305 (1979).

Ito, Y., Tsurudome, M., Yamada, A., and Hishiyama, M. Interferon induction in mouse spleen cells by mitogenic and nonmitogenic lectins. *J. Immunol.* 132: 2440–2444 (1984).

Itri, L. M., Campion, M., Dennin, R. A., Palleroni, A. V., Gutterman, J. U., Groopman, J. E., and Trown, P. W. Incidence and clinical significance of neutralizing antibodies in patients receiving recombinant interferon α-2a by intramuscular injection. *Cancer* 59: 668–674 (1987).

Jacob, C. O., van der Meide, P. H., and McDevitt, H. O. In vivo treatment of (NZB x NZW)F1 lupus-like nephritis with monoclonal antibody to γ-interferon. *J. Exp. Med.* 166: 798–803 (1987).

Jacobs, L., O'Malley, J. A., Freeman, A., Ekes, R., and Reese, P. A. Intrathecal interferon in the treatment of multiple sclerosis. Patient follow-up. *Arch. Neurol.* 42: 841–847 (1985).

Jacobs, L., O'Malley, J. A., Freeman, A., and Reese, P. Follow-up observations on multiple sclerosis patients who were treated with intrathecal interferon. In: *The Biology of the Interferon System 1985,* W. E. Stewart II and H. Schellekens, Eds., Elsevier Science Publishers, Amsterdam, pp. 497–504 (1986).

Jacobs, L., Salazar, A. M., Herdon, R., Reese, P. A., Freeman, A., Jozefowicz, R., Cuetter, A., Husain, F., Smith, W. A., Ekes, R., and O'Malley, J. A. Intrathecally administered natural human fibroblast interferon reduces exacerbations of multiple sclerosis. Results of a multicenter, double-blinded study. *Arch. Neurol.* 44: 589–595 (1987).

Jenson, A. B., Rosenberg, H. S., and Notkins, A. L. Virus-induced diabetes mellitus. XVII. Pancreatic islet cell damage in children with fatal viral infections. *Lancet* ii: 354–358 (1980).

Kamin-Lewis, R. M., Panitch, H. S., Merigan, T. C., and Johnson, K. P. Decreased interferon synthesis and responsiveness to interferon by leukocytes from multiple sclerosis patients given natural alpha interferon. *J. Interferon Res.* 4: 423–432 (1984).

Klassen, L. W., Krakauer, R. S., and Steinberg, A. D. Selective loss of suppressor cell function in New Zealand mice induced by NTA. *J. Immunol.* 119: 830–837 (1977).

Klatzmann, D., Champagne, E., Chamaret, S., Gruest, J., Guetard, D., Hercend, T., Gluckman, J. C., and Montagnier, L. T-lymphocyte T4 molecule behaves as the receptor for human retrovirus LAV. *Nature* 312: 767–768 (1984).

Klatzmann, D. and Gluckman, J. C. HIV infection: Facts and hypotheses. *Immunol. Today* 7: 291–296 (1986).

Klinman, D. M., Lefkowitz, M., Honda, M., Barrett, R., and Steinberg, A. D. Suppression of autoantibody production with anticlass II antibodies. In: *Immune intervention. 2. Anti-Ia Antibodies in the Treatment of Autoimmune Disease;* J. Brochier, J. Clot, and J. Sany, Eds., Academic Press, Orlando, pp. 87–100 (1986).

Koevary, S., Rossini, A., Stoller, W., Chick, W., and Williams, R. M. Passive transfer of diabetes in the BB/W rat. *Science* 220: 727–728 (1983).

Kornfeld, H., Stouwe, R. A. V., Lange, M., Reddy, M. M., and Grieco, M. H. T-lymphocyte subpopulations in homosexual men. *N. Engl. J. Med.* 307: 729–731 (1982).

Kromann, H., Christy, M., Lernmark, A., Nedergaard, M., and Nerup, J. The low dose streptozotocin murine model of type I (insulin dependent) diabetes mellitus: Studies in vivo and in vitro of the modulating effect of sex hormones. *Diabetologia* 22: 194–198 (1982).

Krown, S. E. AIDS and Kaposi's sarcoma: Interferons in pathogenesis and treatment. In: *Interferon 7,* I. Gresser, Ed., Academic Press, London, pp. 185–211 (1986).

Lambert, P. H. and Dixon, F. J. Pathogenesis of the glomerulonephritis of NZB/W mice. *J. Exp. Med.* 127: 507–521 (1968).

Lebon, P., Lenoir, G. R., Fischer, A., and Lagrue, A. Synthesis of intrathecal interferon in systemic lupus erythematosus with neurological complications. *Br. Med. J.* 287: 1165–1167 (1983).

Lehner, T. Immunopathology of Behçet syndrome. In: *Behçet Syndrome,* T. Lehner and C. G. Barnes, Eds., Academic Press, London, pp. 127–139 (1979).

Levis, S. C., Saavedra, M. C., Ceccoli, C., Falcoff, E., Feuillade, M. R., Enria, D. A. M., Maiztegui, J. I., and Falcoff, R. Endogenous interferon in Argentine hemorraghic fever. *J. Infect. Dis.* 149: 428–433 (1984).

Levis, S. C., Saavedra, M. C., Ceccoli, C., Feuillade, M. R., Enria, D. A., Maiztegui, J. I., and Falcoff, R. Correlation between endogenous interferon and the clinical evolution of patients with Argentine hemorraghic fever. *J. Interferon Res.* 5: 383–389 (1985).

Lopez, C., Fitzgerald, P. A., and Siegal, F. P. Severe acquired immune deficiency syndrome in male homosexuals: Diminished capacity to make interferon-α in vitro associated with severe opportunistic infections. *J. Infect. Dis.* 148: 962–966 (1983).

MacLaren, N. K., Neufeld, J., McLaughlin, J. V., and Taylor, G. Androgen sensitization of streptozotocin-induced diabetes in mice. *Diabetes* 29: 710–716 (1980).

Maddon, P. J., Dalgleish, A. G., McDougal, J. S., Clapham, P. R., Weiss, R. A., and Axel, R. The T4 gene encodes the AIDS virus receptor and is expressed in the immune system and the brain. *Cell* 47: 333–348 (1986).

Marliss, E. B., Nakhooda, A. F., Poussier, P., and Sima, A. A. F. The diabetic syndrome of the 'BB' wistar rat: Possible relevance to type 1 (insulin-dependent) diabetes in man. *Diabetologia* 22: 225–232 (1982).

Martin, W. J. and Martin, S. E. Thymus reactive IgM autoantibodies in normal mouse sera. *Nature* 254: 716–718 (1975).

Matsuoka, H., Kakui, Y., Tanaka, A., Cho, Y., and Imanishi, J. Acid-labile interferon produced in human peripheral leukocytes by induction with Sendai virus. *J. Gen. Virol.* 66: 2491–2494 (1985).

Maugh, T. H. Diabetes: Epidemiology suggest a viral connection. *Science* 188: 347–351 (1975).

McFarlin, D. E. and McFarland, H. F. Multiple sclerosis (first of two parts). *N. Engl. J. Med.* 307: 1183–1188 (1982a).

McFarlin, D. E. and McFarland, H. F. Multiple sclerosis (second of two parts). *N. Engl. J. Med.* 307: 1246–1251 (1982b).

Mirro, J., Jr., Kalwinsky, D., Whisnant, J., Weck, P., Chesney, C., and Murphy, S. Coagulopathy induced by continuous infusion of high doses of human lymphoblastoid interferon. *Cancer Treat. Rep.* 69: 315–317 (1985).

Mogensen, K. E., Daubas, P., Gresser, I., Sereni, D., and Varet, B. Patient with circulating antibodies to α-interferon. *Lancet* ii 1227–1228 (1981).

Morimoto, C., Hafler, D. A., Weiner, H. L., Letvin, N. L., Hagan, M., Daley, J., and Schlossman, S. F. Selective loss of the suppressor-inducer T-cell subset in progressive multiple sclerosis. Analysis with anti-2H4 monoclonal antibody. *N. Engl. J. Med.* 316: 67–72 (1987).

Morrow, P. L., Freedman, A., and Craighead, J. E. Testosterone effect on experimental diabetes mellitus in encephalomyocarditis (EMC) virus infected mice. *Diabetologia* 18: 247–249 (1980).

Moss, J., Woodrow, D. F., Sloper, J. C., Rivière, Y., Guillon, J. C., and Gresser, I. Interferon as a cause of endoplasmic reticulum abnormalities within hepatocytes in newborn mice. *Br. J. Exp. Pathol.* 63: 43–49 (1982).

Moutsopoulos, H. M., Gallagher, J. D., Decker, J. L., and Steinberg, A. D. Herpes zoster in patients with systemic lupus erythematosus. *Arthritis Rheum.* 21: 798–802 (1978).

Murray, H. W., Gellene, R. A., Libby, D. M., Rothermel, C. D., and Rubin, B. Y.

Activation of tissue macrophages from AIDS patients: in vitro response of AIDS alveolar macrophages to lymphokines and interferon-γ. *J. Immunol.* 135: 2374–2377 (1985b).

Murray, H. W., Hillman, J. K., Rubin, B. Y., Kelly, C. D., Jacobs, J. L., Tyler, L. W., Donelly, D. M., Carriero, S. M., Godbold, J. H. and Roberts, R. B. Patients at risk for AIDS-related opportunistic infections: Clinical manifestations and impaired γ interferon production. *N. Engl. J. Med.* 313: 1504–1510 (1985a).

Murray, H. W., Rubin, B. Y., Masur, H., and Roberts, R. B. Impaired production of lymphokines and immune (γ) interferon in the acquired immune deficiency syndrome. *N. Engl. J. Med.* 310: 883–889 (1984).

Murray, H. W., Scavuzzo, D., Jacobs, J. L., Kaplan, M. H., Libby, D. M., Schindler, J., and Roberts, R. B. In vitro and in vivo activation of human mononuclear phagocytes by interferon-γ. Studies with normal and AIDS monocytes. *J. Immunol.* 138: 2457–2462 (1987).

Nakane, A. and Minagawa, T. Induction of α and β interferons during the hyporeactive state of γ interferon by mycobacterium bovis BCG cell wall fraction in mycobacterium bovis BCG-sensitized mice. *Infect. Immun.* 36: 966–970 (1982).

Nakashima, H., Yoshida, T., Harada, S., and Yamamoto, N. Recombinant human interferon γ suppresses HTLV-III replication in vitro. *Int. J. Cancer* 38: 433–436 (1986).

Narayan, O., Sheffer, D., Clements, J. E., and Tennekoon, G. Restricted replication of lentiviruses. Visna viruses induce a unique interferon during interaction between lymphocytes and infected macrophages. *J. Exp. Med.* 162: 1954–1969 (1985).

Neighbour, P. A. and Grayzel, A. I. Interferon production in vitro by leucocytes from patients with systemic lupus erythematosus and rheumatoid arthritis. *Clin. Exp. Immunol.* 45: 576–582 (1981).

Neighbour, P. A., Grayzel, A. I., and Miller, A. E. Endogenous and interferon-augmented natural killer cell activity of human peripheral blood mononuclear cells in vitro. Studies of patients with multiple sclerosis, systemic lupus erythematosus or rheumatoid arthritis. *Clin. Exp. Immunol.* 49: 11–21 (1982).

Neighbour, P. A., Miller, A. E., and Bloom, B. R. Interferon responses of leukocytes in multiple sclerosis. *Neurology* 31: 561–566 (1981).

Ohno, S., Kato, F., Matsuda, H., Fujii, N., and Minagawa, T. Detection of γ interferon in the sera of patients with Behçet's disease. *Infect. Immun.* 36: 202–208 (1982).

Oldstone, M. B. A., Southern, P., Rodriguez, M., and Lampert, P. Virus persists in β cells of islets of Langerhans and is associated with chemical manifestations of diabetes. *Science* 224: 1440–1442 (1984).

Onodera, T., Ray, U. R., Melez, K. A., Suzuki, H., Toniolo, A., and Notkins, A. L. Virus-induced diabetes mellitus: Autoimmunity and polyendocrine disease prevented by immunosuppression. *Nature* 297: 66–68 (1982).

Onodera, T., Toniolo, A., Ray, U. R., Jenson, A. B., Knazek, R. A. and Notkins, A. L. Virus-induced diabetes mellitus. XX. Polyendocrinopathy and autoimmunity. *J. Exp. Med.* 153: 1457–1473 (1981).

Onodera, T., Yoon, J. W., Brown, K. S., and Notkins, A. L. Evidence for a single locus controlling susceptibility to virus-induced diabetes mellitus. *Nature* 274: 693–696 (1978.)

Osheroff, P. L., Chiang, T. R., and Manousos, D. Interferon-like activity in an anti-interferon anti-idiotypic hybridoma antibody. *J. Immunol.* 135: 306–313 (1985).

Palmer, J. P., Clemons, P., Lyen, K., Tatpati, O., Raghu, P. K., and Paquette, T. L. Insulin antibodies in insulin-dependent diabetics before insulin treatment. *Science* 222: 1337–1339 (1983).

Panem, S. Antibodies to interferon in man. In: *Interferons and the Immune System,* J. Vilcek and E. De Maeyer, Eds., Elsevier, Amsterdam, pp. 175–183 (1984).

Panem, S., Check, I. J., Henriksen, D., and Vilcek, J. Antibodies to α-interferon in a patient with systemic lupus erythematosus. *J. Immunol.* 129: 1–3 (1982).

Panem, S., Ordonez, N., and Vilcek, J. Renal deposition of α interferon in systemic lupus erythematosus. *Infect. Immun.* 42: 368–373 (1983).

Panitsch, H. S., Hirsch, R. L., Haley, A. S., and Johnson, K. P. Exacerbations of multiple sclerosis in patients treated with γ interferon. *Lancet* i: 893–894 (1987).

Papadopoulos, G. K. and Lernmark, A. The spectrum of islet cell antibodies. In: *Antoimmune Endocrine Disease,* T. F. Davies, Ed., Wiley, New York, pp. 167–180 (1983).

Paterson, P. Y. Autoimmune neurological disease: Experimental animal systems and implications for multiple sclerosis. In: *Autoimmunity: Genetic, Immunologic, Virologic and clinical aspects,* N. Talal, Ed., Academic Press, New York, pp. 643–692 (1977).

Patterson, K., Chandra, R. S., and Jenson, A. B. Congenital rubella insulitis, and diabetes mellitus in an infant. *Lancet* i: 1048–1049 (1981).

Pfau, C. J., Gresser, I., and Hunt, K. D. Lethal role of interferon in lymphocytic choriomeningitis virus-induced encephalitis. *J. Gen. Virol.* 64: 1827–1830 (1983).

Popovic, M., Sarngadharan, M. G., Read, E., and Gallo, R. C. Detection, isolation, and continuous production of cytopathic retroviruses (HTLV-III) from patients with AIDS and pre-AIDS. *Science* 224: 497–500 (1984).

Pozzetto, B. and Gresser, I. Role of sex and early interferon production in the susceptibility of mice to encephalomyocarditis virus. *J. Gen. Virol.* 66: 701–709 (1985).

Prabhakar, B. S., Saegusa, J., Onodera, T., and Notkins, A. L. Lymphocytes capable of making monoclonal autoantibodies that react with multiple organs are a common feature of the normal B cell repertoire. *J. Immunol.* 133: 2815–2817 (1984).

Preble, O. T., Black, R. J., Friedman, R. M., Klippel, J. H., and Vilcek, J. Systemic lupus erythematosus: Presence in human serum of an unusual acid-labile leukocyte interferon. *Science* 216: 429–431 (1982).

Preble, O. T. and Friedman, R. M. Biology of disease. Interferon-induced alterations in cells: Relevance to viral and nonviral diseases. *Lab. Invest.* 49: 4–18 (1983).

Preble, O. T., Rothko, K., Klippel, J. H., Friedman, R. M., and Johnston, M. I. Interferon-induced 2′-5′ adenylate synthetase in vivo and interferon production in vitro by lymphocytes from systemic lupus erythematosus patients with and without circulating interferon. *J. Exp. Med.* 157: 2140–2146 (1983).

Prud'homme, G. J., Colle, E., Fuks, A., Goldner-Sauve, A., and Guttman, R. D.

Cellular immune abnormalities and autoreactive T lymphocytes in insulin-dependent diabetes mellitus in rats. *Immunol. Today* 6: 160–162 (1985).

Ray, U. R., Aulakh, G. S., Schubert, M., McClintock, P. R., Yoon, J. W., and Notkins, A. L. Virus-induced diabetes mellitus. XXV. Difference in the RNA fingerprints of diabetogenic and nondiabetogenic variants of encephalomyocarditis virus. *J. Gen. Virol.* 64: 947–950 (1983).

Rayfield, E. J. and Seto, Y. Viruses and the pathogenesis of diabetes mellitus. *Diabetes* 27: 1126–1140 (1978).

Rhodes, C. J. and Taylor, K. W. Effect of human lymphoblastoid interferon on insulin synthesis and secretion in isolated human pancreatic islets. *Diabetologia* 27: 601–603 (1984).

Rhodes, C. J. and Taylor, K. W. Effect of interferon and double-stranded RNA on B-cell function in mouse islets of Langerhans. *Biochem. J.* 228: 87–94 (1985).

Rich, S. A. Human lupus inclusions and interferon. *Science* 213: 772–775 (1981).

Rich, S. A., Owens, T. R., Bartholomew, L. E., and Gutterman, J. U. Immune interferon does not stimulate formation of α and β interferon induced human lupus-type inclusions. *Lancet* i: 127–128 (1983).

Rivière, Y., Gresser, I., Guillon, J. C., Bandu, M. T., Ronco, P., Morel-Maroger, L., and Verroust, P. Severity of lymphocytic choriomeningitis virus disease in different strains of suckling mice correlates with increasing amounts of endogenous interferon. *J. Exp. Med.* 152: 633–640 (1980).

Rivière, Y., Gresser, I., Guillon, J. C., and Tovey, M. G. Inhibition by anti-interferon serum of lymphocytic choriomeningitis virus disease in suckling mice. *Proc. Natl. Acad. Sci. (USA)* 74: 2135–2139 (1977).

Rivière, Y. and Oldstone, M. B. A. Genetic reassortants of lymphocytic choriomeningitis virus: Unexpected disease and mechanism of pathogenesis. *J. Virol.* 59: 363–368 (1986).

Rosenberg, Y. J., Steinberg, A. D., and Santoro, T. J. The basis of autoimmunity in MRL-lpr/lpr mice: A role of self Ia-reactive T cells. *Immunol. Today* 5: 64–67 (1984).

Ross, M. E., Onodera, T., Brown, K. S., and Notkins, A. L. Virus-induced diabetes mellitus. IV. Genetic and environmental factors influencing the development of diabetes after infection with the M variant of encephalomyocarditis virus. *Diabetes* 25: 190–197 (1976).

Sakane, T., Steinberg, A. D., and Green, I. Failure of autologous mixed lymphocyte reactions between T and non T-cells in patients with systemic lupus erythematosus. *Proc. Natl. Acad. Sci.* (USA) 75: 3464–3468 (1978).

Salahuddin, S. Z., Rose, R. M., Groopman, J. E., Markham, P. D., and Gallo, R. C. Human T-lymphotropic virus type III (HTLV-III) infection of human alveolar macrophages. *Blood* 68: 281–284 (1986).

Santoli, D., Hall, W., Kastrukoff, L., Lisak, R. P., Perussia, B., Trinchieri, G., and Koprowski, H. Cytotoxic activity and interferon production by lymphocytes from patients with multiple sclerosis. *J. Immunol.* 126: 1274–1278 (1981).

Santoro, T. J., Benjamin, W. R., Oppenheim, J. J., and Steinberg, A. D. The cellular basis for immune interferon production in autoimmune MRL-*lpr*/*lpr* mice. *J. Immunol.* 131: 265–268 (1983).

Saron, M. R., Rivière, Y., Gresser, I., and Guillon, J. C. Prévention par les globulines anti-interféron de la mortalité des souris infectées par le virus de la chorioméningite lymphocytaire. I. Lésions histologiques, titres sériques du virus et de l'interféron. *Ann. Virol.* (Inst. Pasteur) 133E: 241–253 (1982a).

Saron, M. R., Rivière, Y., Gresser, I., and Guillon, J. C. Prévention par les globulines anti-interféron de la mortalité des souris infectées par le virus de la chorioméningite lymphocytaire. II. Etude de la réponse immunitaire. *Ann. Virol.* (Inst. Pasteur) 133E: 255–266 (1982b).

Scott, G. M. The toxic effects of interferon in man. In: *Interferon 5,* I. Gresser, Ed., Academic Press, London, Vol. 5, pp. 85–114 (1983).

Sergiescu, D., Cerutti, I., Kahan, A., Piatier, D., and Efthymiou, E. Isoprinosine delays the early appearance of autoimmunity in NZB/NZW F1 mice treated with interferon. *Clin. Exp. Immunol. 43: 36–45 (1981).*

Sharpe, A. H. and Fields, N. Pathogenesis of viral infections. Basic concepts derived from the reovirus model. *N. Engl. J. Med.* 312: 486–497 (1985).

Sibbitt, W. L., Jr., Gibbs, D. L., Kenny, C., Bankhurst, A. D., Searles, R. P., and Ley, K. D. Relationship between circulating interferon and anti-interferon antibodies and impaired natural killer cell activity in systemic lupus erythematosus. *Arthritis Rheum.* 28: 624–629 (1985).

Skurkovich, S. V. and Eremkina, E. I. The probable role of interferon in allergy. *Ann. Allergy* 35: 356–360 (1975).

Steele, E. J. and Cunningham, A. J. High proportion of Ig-producing cells making autoantibody in normal mice. *Nature* 274: 483–484 (1978).

Steinberg, E. B., Santoro, T. J., Chused, T. M., Smathers, P. A., and Steinberg, A. D. Studies of congenic MRL-lpr/lpr.xid mice. *J. Immunol.* 131: 2789–2795 (1983).

Steinman, L., Zamvil, S. S., O'Hearn, M., Schwartz, G., Sriram, S., Mitchell, D., and Waldor, M. D. Experience with anti-Ia therapy in the mouse and in the monkey and the applications to human protocols. In: *Immune Intervention. 2. Anti-Ia Antibodies in the Treatment of Autoimmune Disease,* J. Brochier, J. Clot, and J. Sany, Eds. Academic Press, London, pp. 109–128 (1986).

Strannegard, O., Hermodsson, S., and Westberg, G. Interferon and natural killer cells in systemic lupus erythematosus. *Clin. Exp. Immunol.* 50: 246–252 (1982).

Theofilopoulos, A. N., Prud'homme, G. J., Fieser, T. M., and Dixon, F. J. B-cell hypereactivity in murine lupus. II. Defects in response to and production of accessory signals in lupus-prone mice. *Immunol. Today* 4: 317–319 (1983).

Toniolo, A., Onodera, T., Yoon, J. W., and Notkins, A. L. Induction of diabetes by cumulative environmental insults from viruses and chemicals. *Nature* 288: 383–385 (1980).

Tovell, D. R., McRobbie, I. A., Warren, K. G., and Tyrrell, D. L. J. Interferon production by lymphocytes from multiple sclerosis and non-MS patients. *Neurology* 33: 640–643 (1983).

Tovo, P. A., Cerutti, F., Palomba, E., Salomone, C., and Pugliese, A. Evidence of circulating interferon-γ in newly diagnosed diabetic children. *Acta Poedriatr. Scand.* 73: 785–788 (1984).

Trown, P. W., Kramer, M. J., Dennin, R. A., Jr., Connell, E. V., Palleroni, A. V., Quesada, G., and Guttermann, J. U. Antibodies to human leucocyte interferons in cancer patients. *Lancet* i: 81–84 (1983).

Tsokos, G. C., Rook, A. H., Djeu, J. Y., and Balow, J. E. Natural killer cells and interferon responses in patients with systemic lupus erythematosus. *Clin. Exp. Immunol.* 50: 239–245 (1982).

Vadhan, Raj, S., Wong, G., Gnecco, C., Cunningham-Rundles, S., Krim, M., Real, F. X., Oettgen, H. F., and Krown, S. E. Immunological variables as predictors of prognosis in patients with Kaposi's sarcoma and the acquired immunodeficiency syndrome. *Cancer Res.* 46: 417–425 (1986).

Vallbracht, A., Treuner, J., Flehmig, B., Joester, K. E., and Niethammer, D. Interferon-neutralizing antibodies in a patient treated with human fibroblast interferon. *Nature* 289: 496–497 (1981).

Vallbracht, A., Treuner, J., Manncke, K. H., and Niethammer, D. Autoantibodies against human β interferon following treatment with interferon. *J. Interferon Res.* 2: 107–110 (1982).

Vervliet, G., Carton, H., and Billiau, A. Interferon-γ production by peripheral blood leucocytes from patients with multiple sclerosis and other neurological disease. *Clin. Exp. Immunol.* 59: 391–397 (1985a).

Vervliet, G., Carton, H., Meulepas, E., and Billiau, A. Interferon production by cultured peripheral leucocytes of MS patients. *Clin. Exp. Immunol.* 58: 116–126 (1984).

Vervliet, G., Claeys, H., Van Haver, H., Carton, H., Vermylen, C., Meulepas, E., and Billiau, A. Interferon production and natural killer (NK) activity in leukocyte cultures from multiple sclerosis patients. *J. Neurol. Sci.* 60: 137–150 (1983).

Vervliet, G., Deckmyn, H., Carton, H., and Billiau, A. Influence of prostaglandin E2 and indomethacin on interferon-γ production by cultured peripheral blood leukocytes of multiple sclerosis patients and healthy donors. *J. Clin. Immunol.:* 102–108 (1985b).

Vialettes, B., Baume, D., Charpin, C., De Maeyer-Guignard, J., and Vague, P. Assessment of viral and immune factors in EMC virus-induced diabetes: Effects of cyclosporin A and interferon. *J. Clin. Lab. Immunol.* 10: 35–40 (1983).

Vilcek, J., Sulea, I. T., Volvovitz, F., and Yip, Y. K. Characteristics of interferons produced in cultures of human lymphocytes by stimulation with corynebacterium parvum and phytohemagglutinin. In: *Biochemical Characterization of Lymphokines,* A. L. De Weck, F. Kristensen, and M. Landy, Eds. Academic Press, New York, pp. 323–329 (1980).

Volpé, R. Auto-immunity in diabetes mellitus. In: *Auto-Immunity in the Endocrine System,* Monographs on Endocrinology, F. Gross, M. Grumbach, A. Labhart, M. Lipsett, T. Mann, L. Samuels, and J. Zander, Eds., Springer-Verlag, Berlin, Vol. 20, pp. 112–145 (1981).

Wabuke-Bunoti, M. A. N., Bennink, J. R., and Plotkin, S. A. Influenza virus-induced encephalopathy in mice: Interferon production and natural killer cell activity during acute infection. *J. Virol.* 60: 1062–1067 (1986).

Wiley, C. A., Schrier, R. D., Nelson, J. A., Lampert, P. W., and Oldstone, M. B. A. Cellular localization of human immunodeficiency virus infection within the brains of AIDS patients. *Proc. Natl. Acad. Sci.* (USA) 83: 7089–7093 (1986).

Williams, R. C., Masur, H., and Spira, T. J. Lymphocyte-reactive antibodies in acquired immune deficiency syndrome. *J. Clin. Immunol.* 4: 118–123 (1984).

Wykoff, R. F., Pearl, E. R., and Saulsbury, F. T. Immunologic dysfunction in infants infected through transfusion with HTLV-III. *N. Engl. J. Med.* 312: 294–296 (1985).

Yamamoto, J. K., Ho, E., and Pedersen, N. C. A feline retrovirus induced T-lymphoblastoid cell-line that produces an atypical α type of interferon. *Vet. Immunol. Immunopathol.* 11: 1–19 (1986).

Yoon, J. W., Austin, M., Onodera, T., and Notkins, A. L. Virus-induced diabetes mellitus: Isolation of a virus from the pancreas of a child with diabetic ketoacidosis. *N. Engl. J. Med.* 300: 1173–1179 (1979).

Yoon, J. W., Cha, C. Y., and Jordan, G. W. The role of interferon in virus-induced diabetes. *J. Infect. Dis.* 147: 155–159 (1983).

Yoon, J. W., McClintock, P. R., Onodera, T., and Notkins, A. L. Virus-induced diabetes mellitus. XVIII. Inhibition by a nondiabetogenic variant of encephalmyocarditis virus. *J. Exp. Med.* 152: 878–892 (1980).

Yoon, J. W., Onodera, T., and Notkins, A. L. Virus-induced diabetes mellitus. XV. β cell damage and insulin-dependent hyperglycemia in mice infected with Coxsackie virus B4. *J. Exp. Med.* 148: 1068–1080 (1978).

Ytterberg, S. R. and Schnitzer, T. J. Serum interferon levels in patients with systemic lupus erythematosus. *Arthritis Rheum.* 25: 401–406 (1982).

Zawatzky, R., Kirchner, H., De Maeyer-Guignard, J., and De Maeyer, E. An X-linked locus influences the amount of circulating interferon induced in the mouse by herpes simplex virus type 1. *J. Gen. Virol.* 63: 325–332 (1982).

Zinkernagel, R. M. H-2 restriction of virus-specific T-cell-mediated effector functions in vivo. II. Adoptive transfer of delayed-type hypersensitivity to murine lymphocytic choriomeningitis virus is restricted by the K and D region of H-2. *J. Exp. Med.* 144: 776–787 (1976).

Zinkernagel, R. M., Leist, T., Hengartner, H., and Althage, A. Susceptibility to lymphocytic choriomeningitis virus isolates correlates directly with early and high cytotoxc T cell activity, as well as with footpad swelling reaction, and all three are regulated by H-2D. *J. Exp. Med.* 162: 2125–2141 (1985).

APPENDIX
INTERFERON UNITS AND NOMENCLATURE

Although this book is by no means a technical treatise but an overview of interferon biology, it is sometimes unavoidable to mention interferon units and to refer to "low" or "considerable" amounts of interferon. For the reader who is not familiar with interferon assays, a short definition will therefore help to understand the quantitative aspects of the multiple interferon activities.

A second point that may need some clarification is the designation of the different IFNs and IFN preparations, since the term IFN-α or IFN-β is not always correctly used in the literature.

WHAT IS AN INTERFERON UNIT?

Measuring inhibition of virus replication remains the most widely used method for interferon titration, and IFN amounts are usually expressed as antiviral units. Radioimmune assays may become the standard method for measuring interferon activity, but for the time being even the results of radioimmune assays are often converted into antiviral units. The number of interferon assays in use is about as numerous as there are different ways of measuring virus replication, but based on simultaneous titration of an international interferon reference preparation—when one is available—the results of each assay are converted into international reference units. The appropriate international reference preparations are designated by the World Health Organization.

The specific antiviral activity of most interferons lies between 10^8 and 10^9 antiviral units per milligram (mg) of protein. Therefore, one IFN unit is roughly equivalent to a concentration of 10^{-13} to 10^{-15} M, depending on which IFN is involved, and corresponds to an average of 1 to 10 picograms of protein. By "small" or "moderate" amounts of interferon, one usually means from 1 to 1,000 units, whereas "considerable" amounts can mean anything from 10^5 or 10^6 units and up.

A FEW WORDS ABOUT NOMENCLATURE

The standard abbreviation for interferon is IFN, the symbol used throughout the book. Based on molecular structure, antigenicity and mode of induction, one distinguishes three major IFN species, designated as IFN-α, IFN-β, and IFN-γ. Furthermore, each IFN is identified by a prefix according to its animal origin. For example, IFN of human origin is designated as "Hu IFN," and IFN of murine origin as "Mu IFN." When the IFN is of recombinant origin, the designation of the IFN species is preceded by "rec," for example, rec Hu IFN-α.

α and β IFNs that are usually acid stable were for a long time called type I IFNs. IFN-γ, which is acid-labile is sometimes referred to as immune IFN or type II IFN.

In both humans and mice, IFN-α consists of a family of molecules, and based on specific amino acid differences, each subtype has been classified as IFN-α1, α2, and so on. Depending on who did the original molecular cloning and characterization, however, some IFN-α subtypes have been labeled alphabetically as Hu IFN-αA, αB, and so on. As a result of this, the same Hu IFN-α subtype can sometimes be designated by either a number or a letter; for example, Hu IFN-α5 and αG are identical, and Hu IFN-α2 and αD differ by only one amino acid. We have listed the amino acid sequences of the different human and murine IFN-α subtypes and their numerical and/or alphabetical designation in Tables 2.1 and 2.4 of Chapter 2. The numerical rule also applies to IFN-β, in those species—for example, bovines—in which there are multiple IFN-β subtypes: These are then designated as Bo IFN-β1, β2, and so on.

When dealing with a recombinant IFN or with a purified and characterized natural IFN, the designation of species and subspecies should pose no problem. Many natural IFN preparations are not completely purified, however, and usually represent mixtures of different IFNs, consisting of several, undefined, IFN-α subtypes or of several IFN-α subtypes together with IFN-β. Two particular IFN preparations, Mu IFN-α/β and Hu leukocyte IFN, deserve a few words of explanation because they are widely used for experimentation and not always correctly referred to.

1. Mu IFN-α/β. Until very recently, most Mu IFN preparations consisted of a mixture of undefined Mu IFN-α subtypes together with Mu IFN-β, in varying proportions. We refer to these as "Mu IFN-α/β." There is, however, some confusion in the literature with regard to nomenclature of mouse IFNs, and in many publications Mu IFN-α/β is quite mistakenly referred to as Mu IFN-β. The reason for this confusion is the fact that most Mu IFN preparations are derived from mouse L or C243 cell cultures, and therefore, are sometimes called fibroblast IFNs. The term fibroblast IFN is then wrongly equated with IFN-β, because a very widely

used human IFN preparation, made in human diploid fibroblasts induced by polyrIrC, contains only Hu IFN-β. In general, however, and even in human fibroblasts when other inducers are used, fibroblasts make both IFN-α and IFN-β, and the equation of fibroblast IFN with IFN-β is not justified.

When the nature of the Mu IFN was obvious from its cellular origin and way of purification, we have used the correct designation while discussing the results of those publications in which the wrong nomenclature was used.

2. Hu Leukocyte IFN. A last but important point concerns the preparations that consist mainly of different, nondefined, Hu IFN-α subtypes derived from Sendai virus-induced buffy coats or from lymphoblastoid (usually Namalwa) cells. These are referred to in many publications as Hu IFN-α, and although such preparations are undoubtedly made up of various Hu IFN-α subtypes, since they are not totally purified they may contain other cytokines in addition to IFN-α. For example, the widely used human leukocyte IFN preparations, made according to the Cantell procedure, can contain TNF (tumor necrosis factor) and even sometimes small amounts of IFN-γ. These IFN preparations, therefore, should not be designated simply as Hu IFN-α, but as Hu leukocyte or lymphoblastoid IFN, with the implication that, unless totally purified, the presence of other cytokines cannot be excluded.

INDEX

A

B

C

E

J

K

L

M

R

S

T